Semiconductor Memories

Semiconductor Memories

Edited by
David A. Hodges
Associate Professor of Electrical Engineering
and Computer Sciences
University of California, Berkeley

A volume in the IEEE PRESS Selected Reprint Series, prepared under the sponsorship of the IEEE Computer Society.

IEEE PRESS

The Institute of Electrical and Electronics Engineers, Inc. New York

International Standard Book Numbers:

Clothbound: 0-87942-013-8
Paperbound: 0-87942-014-6

Library of Congress Card Number 72-86206

PRINTED IN THE UNITED STATES OF AMERICA

Contents

Part V: Reliability, Maintainability, and Testing

Part VI: System Examples

Part VII: Serial Memories and Content-Addressable Memories

Semiconductor Memories

Introduction

Interest in semiconductor memories has grown rapidly over the past three years. A major stimulus to increasing activity in this area was the mid-1971 delivery of the first IBM System 370 Model 145 computer, which employs a semiconductor main memory. Several recently announced small computers also use semiconductor main memories. It is now generally agreed that semiconductor memories will comprise a growing fraction of the computer main memory market.

Computer main storage is perhaps the most obvious area of application for semiconductor memories. However, many other technical opportunities are becoming apparent as low-cost semiconductor memories become available. Growing areas of application for these components are in terminals and peripheral controllers for computer systems, desk and pocket calculators, manufacturing process control equipment, and a wide variety of communication systems. Entirely new types of systems, never before economic, are being developed rapidly. The long-term total impact of semiconductor memories may be greater for these latter areas than for computer main memory.

The pace of development has been so rapid in semiconductor memory that many potential users have difficulty anticipating the direction and rate of future progress. System designers must be able to anticipate the characteristics of components one to three years in advance in order to arrive at competitive new designs. A major objective of this reprint book is to give the present or potential user some appreciation of the basic characteristics, limitations, and trends affecting semiconductor memories. No specific knowledge of digital memory technology is assumed.

This collection of readings is intended also to give the engineer involved in design or application of semiconductor memories specific information on various design approaches. Information on device, circuit, packaging, reliability, cost, and system interface considerations is included because all these matters are important to memory characteristics. Although parameters such as bit density, access time, power consumption, package characteristics, and failure rates may change drastically with time, the relative characteristics of alternative approaches are likely to change less rapidly. In this sense, certain principles of semiconductor memory design that are now evident are likely to remain as useful guides.

These readings are drawn primarily from technical literature and manufacturers' data published in the period 1969–1972. Material available in book form, such as transistor theory, basic circuit design, and standard integrated circuit process technology, is not covered here. Refer to the short bibliography at the end of Part I for references to this sort of background information. Manufacturers' data sheets are included in several instances, primarily because important specific design details are not available from other sources. Furthermore, the contrast of design concepts shown in manufactured products with those shown in the technical literature is sometimes highly instructive. When material to be included here was available from several sources, the article or data sheet providing the clearest, most complete information was selected. Space limitations did not permit inclusion of examples of all worthy design approaches. No endorsement of particular products is implied. Original material has been added in several places where no appropriate published material was available.

Publishers and manufacturers have been most generous in permitting reproduction of the copyrighted materials included herein.

Part I
Introduction to Semiconductor Memories

Editor's Introductory Comments and Definitions

System design engineers are usually concerned with the following parameters, evaluated at the system level, in considering alternative component technologies; the order of priority, of course, depends upon the application.

System performance.

Manufacturing cost.

Development schedule and cost.

Reliability and maintainability.

Component availability.

Many engineers now believe that semiconductor memories are likely to surpass all known technologies for electronic digital memory in these key characteristics within a very few years. This collection of reprints with added commentary is intended to help the reader identify the factors that determine the above parameters and the more promising directions for further development.

Basic memory terminology will be defined here as an aid to those without specific experience in this area. Semiconductor memory is also called monolithic memory, integrated circuit memory, large-scale integrated (LSI) memory, active memory, and transistor memory. These designations all refer to binary digital memories that employ an electronic circuit to store each bit. This circuit may be a bistable flip-flop, or some other circuit having two well-resolved states.

READ-WRITE (RW or R/W) memories permit data to be stored or retrieved at comparable intervals, usually in the range 10 ns–1 μs per operation for semiconductor memories. These memories are volatile; stored information is lost if power is lost. READ-ONLY memories and READ-MOSTLY memories (ROM and RMM) permit reading at these speeds, but the writing operation is restricted. READ-ONLY memories may be written only once and cannot be changed thereafter. READ-MOSTLY memories may be erased and written again, but the erase and write operations usually are much slower than the read operation. READ-ONLY and READ-MOSTLY memories are nonvolatile; stored information is retained when power supplies are removed.

Static memories are those that function correctly with no clock signals. Clocks may be used, but are required only to synchronize memory operation with other elements in a larger system. Dynamic memories require continuing clock signals for proper operation. Stored information will be lost from READ-WRITE dynamic memories if the required clock signals are interrupted. Dynamic memories are used despite this drawback because usually they are much cheaper and require less operating power than static memories.

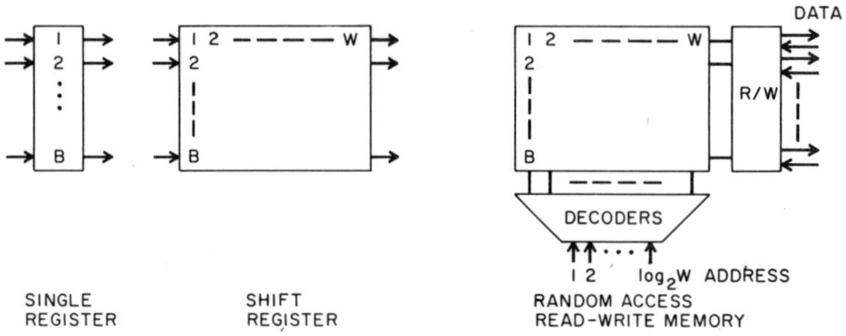

Fig. 1-1. Basic memory organization.

Destructive READ-OUT (DRO) memories sacrifice the stored information in the act of reading it; a restore (rewrite) cycle is required if the stored information is to be retained in the memory. In contrast, nondestructive READ-OUT (NDRO) memories retain the stored information despite any read operations. The time required for a restore operation is eliminated from the overall write cycle time. Most, but not all, semiconductor memories are NDRO.

Viewed from the system standpoint, digital memories are usually organized as W words, each of B bits, for a total storage capacity of $W \times B$ bits. Storage capacity may be stated in bytes, each commonly being of eight or nine bits. The B bits of one word are available in parallel (simultaneously in time) for reading or writing. Fig. 1-1 shows a comparison of three basic memory organizations. The single register is familiar to all digital designers. Its limitation when large storage capacity is desired is that two leads are required for each stored bit. A shift register provides more storage capacity with no increase in the number of leads by providing serial access to stored words. Data are available only in the same time sequence as they had when originally stored. The random-access READ-WRITE or READ-ONLY memory provides equal access time to each stored word, independent of the sequence in which words were stored. The main emphasis in the following pages is on random-access memories because of their leading significance.

Access time and cycle time are key concepts in memory technology. Access time is the time required to read out any arbitrarily chosen word from the memory. Cycle time is the reciprocal of the rate at which words may be addressed for reading or writing without danger of error. For semiconductor random-access memories, the access time is typically 0.5 to 1 times the cycle time.

An important advantage of semiconductor memory is that it is naturally modular. The minimum cost per bit is typically reached for modules of about 10^4 bits storage capacity, and the cost does not rise very rapidly for decreasing module sizes. In contrast, magnetic memories, such as ferrite core and plated wire, typically must be built in modules of 10^5 bits capacity or more to approach minimum cost per bit; cost rises very rapidly for decreasing module size. The reliability and maintainability of large memory systems is enhanced by building them up from small modules. The cost of small systems is reduced with use of low-cost small memory modules. Semiconductor memories are economical in small modules because the storage elements are much more compatible, electrically and physically, with other electronic circuits than are magnetic memory storage elements.

A short bibliography is offered below. In addition, most of the items included in this volume provide relevant references to the literature.

Bibliography

General Technology

[1] H. R. Camenzind, *Circuit Design for Integrated Electronics.* Reading, Mass.: Addison-Wesley, 1968.
[2] S. K. Ghandhi, *Theory and Practice of Microelectronics.* New York: Wiley, 1968.
[3] A. S. Grove, *Physics and Technology of Semiconductor Devices.* New York: Wiley, 1967.
[4] R. M. Warner and J. Fordemwalt, Ed., *Analysis and Design of Integrated Circuits.* New York: McGraw-Hill, 1965.

Semiconductor Memory

[1] J. Eimbinder, Ed., *Semiconductor Memories.* New York: Wiley-Interscience, 1971.
[2] D. A. Hodges, "Large-capacity semiconductor memory," *Proc. IEEE,* vol. 56, pp. 1148–1162, July 1968.

Semiconductor random-access memories

Random-access read–write semiconductor memories typify the revolution in computer memories—storage elements that promise to replace ferrite cores and plated wires in the seventies. Prices have dropped as yields have improved

L. L. Vadasz, H. T. Chua, A. S. Grove *Intel Corporation*

In the last few years, semiconductor random-access memory (RAM) components have been introduced into high-speed scratchpad applications and small buffer memory systems, where the performance or cost advantage of semiconductor components is greatest. More recently, developments in the technology of large-scale integrated (LSI) circuits have resulted in cost–performance characteristics of semiconductor RAMs that are competitive in computer main-frame memories as well. This article reviews the basic circuit concepts used in these components, representative products that are presently available, and some systems considerations involved in their use.

Semiconductor memories have been a part of high-performance computers for the last four to five years. Because of their high cost, these elements were used only in applications where no other memory component could provide the desired performance. Consequently, those applications were all related to high-performance bipolar semiconductor memory elements.

The first large semiconductor memory subsystem, the Cache memory, was reported by IBM in 1969.[1] Its major characteristics were a system access time of 60 ns and a minimum memory size of 112 kilobytes. The system used 64-bit bipolar memory chips.

Whereas early bipolar memory elements gave rise to the development of a new class of memory systems, the emerging metal oxide semiconductor (MOS) technology provided the first competitive semiconductor product in a market that was served by magnetic-core elements. The most vulnerable target was the small buffer memory of a few thousand bits per system, with access times of 1 to 5 μs. In such applications, the cost of using cores can be quite formidable, because of the high cost of overhead electronics needed for the core memory systems. Even early MOS memories were able to compete economically with such memory approaches. For instance, in one typical application of these MOS buffer memories, two small 16-pin dual-inline-package 256-bit static MOS memory components replaced a whole board of cores and associated electronics.

Another memory system also evolved in the past few years. This system is based not on RAMs, but rather on serial shift-register memory elements. Such memory systems find typical use in computer terminals where they can be used as refresh memories for CRT displays.

Examples of the emerging application of LSI main-frame memories are the use of high-speed bipolar memory circuits in the IBM System 370/145, and in the ILLIAC IV process element memory (Fig. 1), one quarter of which uses sixty-four 130-kilobit systems with a cycle time of less than 200 ns. The bulk of these systems use bipolar chips containing 256 bits of storage.[2] Other, less performance- and more cost-oriented applications are the Data General Supernova minicomputer, and the Four Phase IV/70, which use up to 1-μs MOS memory chips, typically containing 1024 bits of storage. Most of these systems were introduced commercially in the latter part of 1970 or first part of 1971. They are based on semiconductor memory components that are typical of the state of the art as of 1971.

The bistable flip-flop as a storage cell

Although there are many devices that may be used as storage elements, the bistable flip-flop, made of two cross-coupled inverters, has been the most widely adopted basic storage cell for both bipolar and static MOS circuits. The cell is simple to design, is inherently very high speed, and is generally insensitive to process parameter variations; these properties provide high-yield, low-cost components. In Fig. 2 the inverter is made of a transistor and a collector load resistor R_C. For an MOS circuit, R_C is made of another MOS transistor, which provides a small-area nonlinear resistor when its gate is tied to its drain.

In a flip-flop, one transistor is normally on, which keeps the other transistor off. When the "off" transistor is forced into the "on" state by an external signal, the "on" transistor turns off. The flip-flop can, therefore, have two stable states, and will remain in either of these states until an external signal is used to change its state. The two stable states may be used to store information by being interpreted as logical "1" and "0."

Bipolar memories. A gating arrangement gets information in and out of the circuit—writing or reading of the memory, respectively. Figure 3 shows some of the tech-

Reprinted from *IEEE Spectrum*, vol. 8, pp. 40–48, May 1971.

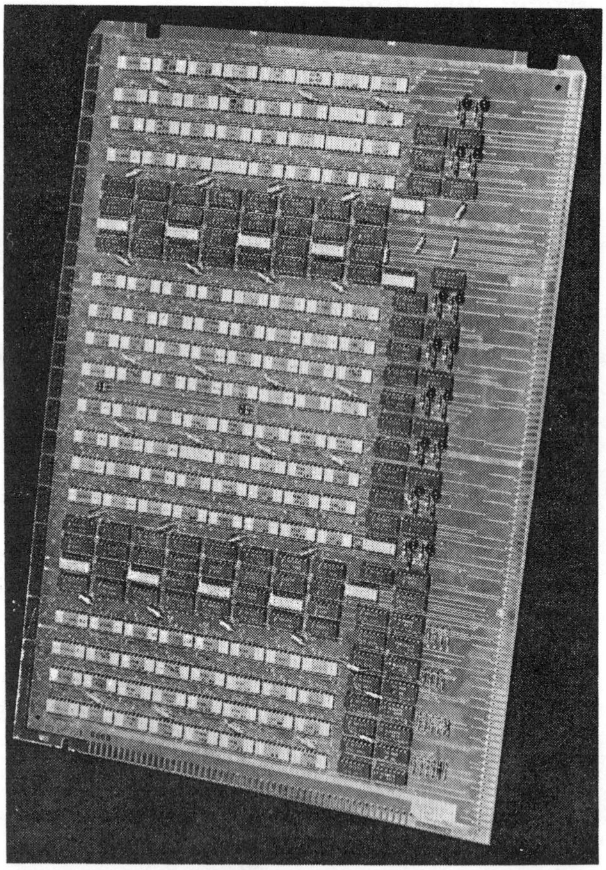

FIGURE 1. Memory module of the ILLIAC IV system. The 30.5- by 43-cm two-layer board contains 2048 words of 16 bits of memory, including data registers. The board contains 128 μL4100 256-bit bipolar memory devices. The memory system in which this board is used operates at a 200-ns read or write cycle time and a 188-ns maximum read access time. (Courtesy Fairchild Semiconductor)

FIGURE 2. Basic flip-flop memory cell that uses (A) bipolar elements and (B) MOS elements.

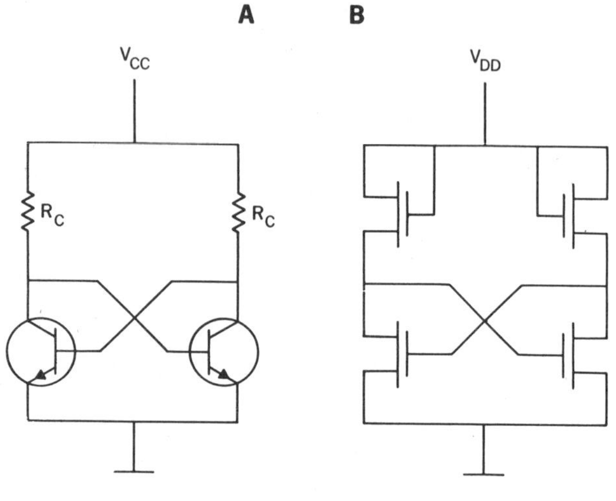

niques used for this purpose. Figure 3(A) illustrates a basic flip-flop made with two dual-emitter transistors.[3] One of the emitters from each transistor is tied to one of the bit lines. The remaining emitter of each transistor is tied to a common word line. Many flip-flops may be interconnected to form a large memory array where each flip-flop has a unique location, or address. A particular memory cell may be selected (addressed) for either writing or reading by applying proper signals to the word and bit lines. The word-line voltage is raised to read the content of a cell. The flip-flop current, which normally flows through the word line, transfers to one of the bit lines. A current-sensing amplifier detects the signal current. A cell is similarly selected for writing. Unbalancing the voltage at the two bit lines forces the flip-flop into the desired state. When the cell is not selected, the word-line voltage is low, and cell current flows through the word line. Under this condition, there is no signal current at the bit line, and the content of the cell is not sensed. Similarly, raising or lowering the voltage on the bit line will not affect the state of the flip-flop.

This circuit is very simple and consumes very little silicon area. It has been used successfully in many of the larger bipolar memory circuits, such as 64-bit random-access memory components by Intel, Intersil, Computer Microtechnology, and Raytheon. It is also used in Fairchild's 256-bit RAM, on which the ILLIAC IV memory module is based. The circuit is applicable to either the standard gold-doped TTL process or to the more novel Schottky process. The speed of accessing of the memory depends largely on the amount of current available from the cell for sensing. To maintain consistent and fast switching speed, the collector load resistors (R_C) of the flip-flop must not have wide variations in resistance.

The circuit shown in Fig. 3(B) operates in a different mode. This circuit is particularly suitable for the Schottky process because the two added gating Schottky diodes are made within the collector region of the flip-flop transistor, which keeps the increase in the size of the memory cell very small. A cell is selected for reading by lowering the voltage of its word line. Signals may be detected on the bit line through the Schottky diode. For writing, a cell is first selected. A large current feeds into one of the bit lines through the Schottky diode, which simultaneously turns on the "off" transistor and increases the collector load current of the previously "on" transistor, forcing it to turn off quickly. The access speed of the memory is not significantly affected by the size of the collector resistor R_C; therefore, R_C may be made large to reduce power dissipation. Low-power operation is further achieved in this circuit as a result of low voltage across the unselected cell. This circuit is used in Intel's 256-bit memory, the 3102. A collector bulk resistor, which provides a higher resistance for a given silicon area, is being used as R_C of the memory cell.

A third class of bipolar memory cells is shown in Fig. 3(C). This circuit is popularly used in emitter-coupled logic (ECL) circuits. A commercially available ECL memory product that uses this basic cell is Advanced Memory System's 64-bit memory. This circuit employs two layers of metal interconnections. Successful fabrication of an ECL memory circuit containing 128 bits using single-layer metal interconnection has also been reported.[4]

The memory cell operates as follows: A particular bit is selected by raising its word-line voltage. Writing or read-

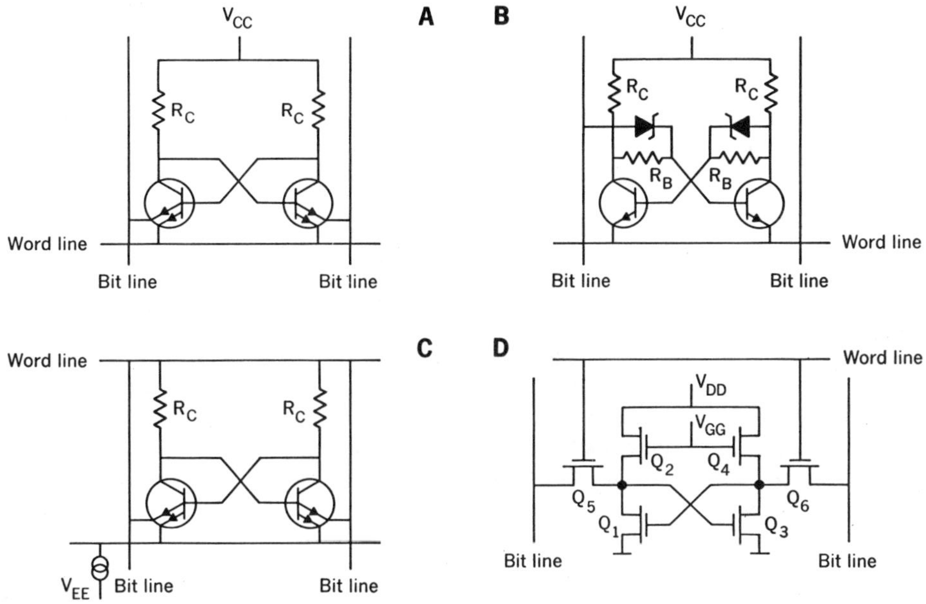

FIGURE 3. Various flip-flop memory cells. A—Basic flip-flop with two dual-emitter transistors. B—Flip-flop with two added gating Schottky diodes within the collector region. C—A basic cell for ECL circuits. D—An MOS flip-flop cell.

ing of the memory cell is very similar to that of the multi-emitter cell of Fig. 3(A), except that the voltage across the selected cell is higher than that across the unselected cell; therefore, a large sense current is available from the flip-flop when it is selected. When the flip-flop is not selected, the voltage across its supply terminal is quite low, and low standby power dissipation is thereby maintained.

MOS memories. The MOS flip-flop memory shown in Fig. 3(D) works as follows: The flip-flop formed by Q_1, Q_2, Q_3, and Q_4 is gated by transistors Q_5 and Q_6. To read the cell, the word line turns on Q_5 and Q_6, transferring the data of the flip-flop to the bit lines. To write into a cell, the word line again turns on Q_5 and Q_6 and now forces the cell into its proper logic state by establishing the proper voltage on the bit lines.[5]

The voltages V_{DD} and V_{GG} are the negative supply voltages for the cell for the case of *p*-channel MOS transistors. These voltages can be identical or different (V_{GG} more negative than V_{DD}), depending on a particular design. Having a common V_{DD} and V_{GG} line reduces the cell size. Having separate V_{DD} and V_{GG} lines, however, allows the user to switch to low-power standby mode in the time period when access is not made to the memory. This mode of operation will be further discussed in later sections. An MOS flip-flop storage cell uses six transistors per bit. This circuit is used in all of the commercially available 256-bit decoded MOS memory

FIGURE 4. Cell size of bipolar memory components as a function of time. Static and dynamic MOS cells are included for comparison.

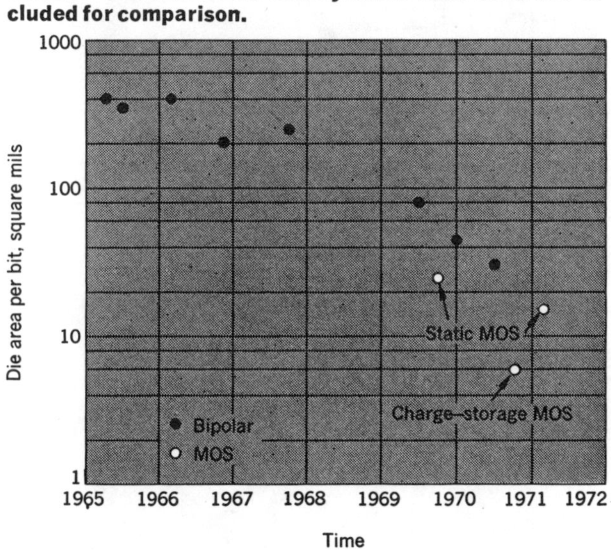

FIGURE 5. Power dissipation of bipolar memory component as a function of time. Static and dynamic MOS cells are included for comparison.

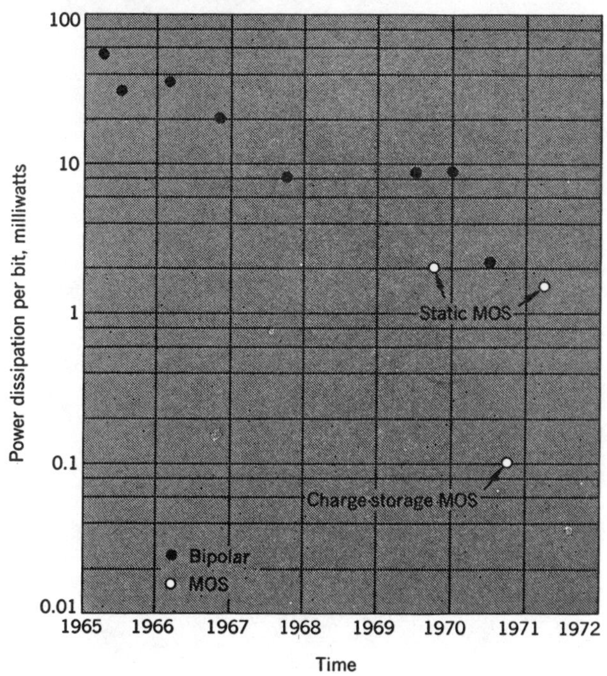

arrays—for example, Intel, Unisem, Intersil, and Computer Microtechnology.

Progress in bipolar semiconductor memory. Semiconductor memory became commercially available in its primitive form in early 1965. The memory chips made in those days typically contained four to eight bits per chip and were organized as linear select type, with several bits per word. Because of the high ratio of the number of sense amplifiers to the number of bits per chip, the memory cells had to be made complex enough to deliver level-compatible signals to the outside world without elaborate buffering and sensing. As the number of bits per chip increased, the memory cells became less complex and dissipated less power, while more sophistication was added to the drive and sense circuits.

Figures 4 and 5 show the cell size and power dissipation, respectively, of some of the typical memory chips made during the years 1965 to 1970. In 1965, the power dissipation was typically 30–40 mW, and the cell occupied approximately 400 square mils (0.258 mm^2) of silicon area. By 1970, commercially available cells consumed only 30 square mils of silicon area and dissipated 2 mW. Figure 6 shows a photomicrograph of a partially decoded 256-bit bipolar memory chip with such characteristics. Cell size of 20 square mils is practical using today's technology.

Hybrid memory systems

Even as bipolar memory cells decreased in size and in power consumption, a different approach to semiconductor memories was suggested, which used MOS flip-flops for storage.[6] Since the MOS process contains fewer steps, and since for a given set of tolerances it is capable of higher component density than the bipolar process, this approach promised to yield a lower-cost semiconductor storage. Figures 4 and 5 include the static MOS memory cell that became commercially available in 1969 for comparison. Both the smaller cell size and lower power consumption of MOS flip-flops are evident.

MOS devices are basically high-impedance devices, and their drive capability is limited. When their load is another small-geometry MOS device on the same chip, even the limited drive is sufficient to result in a generally satisfactory switching speed (100 ns or less). When, however, MOS devices must drive devices in another package, the greater loading results in a significant performance degradation. Thus, a desirable hybrid approach to semiconductor memories could be based on the use of MOS devices for storage and bipolar devices for the support circuitry: driving, sensing, and decoding.

The MOS storage cells, which make up the bulk of the system, are inherently less expensive to produce. The power consumed by the MOS cells is lower than the power that would be consumed by bipolar cells. The high-speed bipolar drive and sense circuits enable high-speed operations that MOS circuits alone cannot achieve. In addition, bipolar input and output circuits can be easily made level-compatible with any form of logic circuits now in existence.

Early hybrid memory systems were made by interconnecting MOS memory arrays and bipolar drive and sense circuits on a ceramic substrate by beam leads or solder bumps. The MOS chip—storage arrays of MOS flip-flops with minimum drive and sense circuitry—contained the highest number of bits per chip producible with reasonable yield in the given time period. The bipolar

FIGURE 6. Photomicrograph of a partially decoded 256-bit bipolar memory chip.

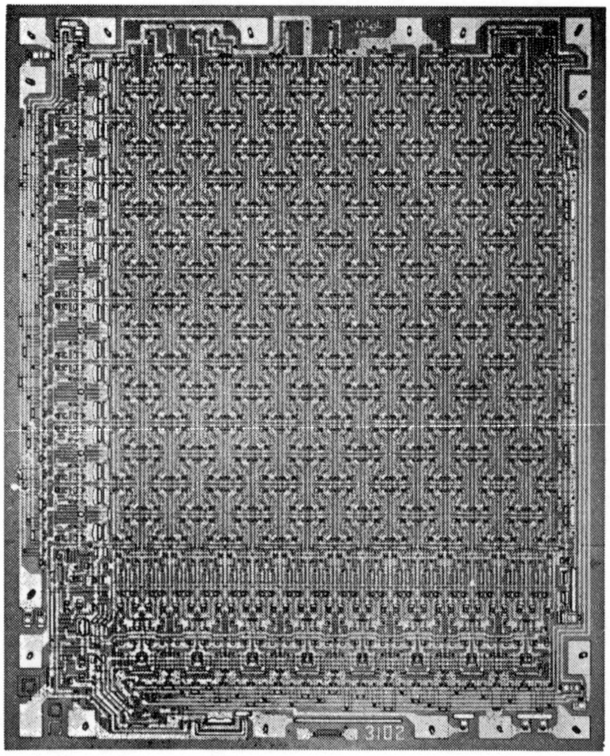

FIGURE 7. Shift in partitioning of semiconductor memories that use both MOS and bipolar devices.

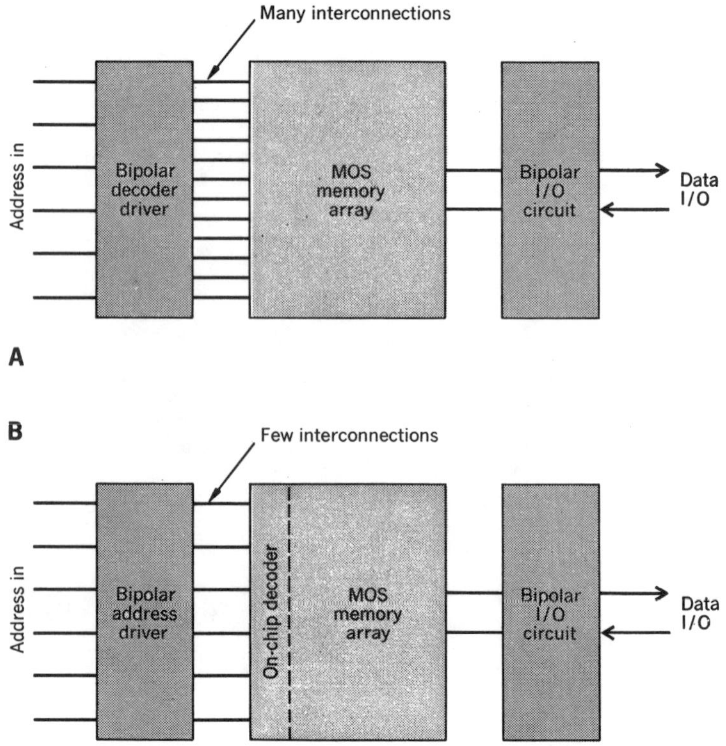

circuits performed the necessary decoding, level shifting, buffering, and sensing functions.

Some semiconductor companies see the hybrid system described above as the ultimate approach to utilizing fully the potential capability of both the bipolar and MOS technologies. Fairchild Semiconductor was the first to develop such a system.[7] Fairchild's system, called the Semiconductor Active Memory, is a 16-kilobit memory stack made up of 64-bit MOS memory arrays and bipolar drive and sense circuits. Motorola Semiconductor subsequently developed an 8-kilobit memory module, which claimed access times of less than 150 ns. Its basic storage array was a 256-bit MOS chip. The only commercially available multichip hybrid memory product is Computer Microtechnology's device, a 4-kilobit hybrid memory module built from 256-bit MOS chips. Its cycle time is rated at 400 ns.

These hybrid approaches were all partitioned in such a way that bipolar circuits were used for decoding. As a result, the storage chips had to have a large number of interconnections made to them—up to 40 for a 256-bit array. Because such a large number of interconnections is very cumbersome and expensive in the case of a multichip module, all of these approaches were based on some face-down bonding technique, such as flip-chip or beam-lead bonding.

These techniques have been in development many years. Although all of them were eventually shown to be technically feasible, they were not as effective in reducing assembly and package costs as had been hoped. As a result of high costs and of the development of MOS storage circuits other than the conventional flip-flop, the partitioning of MOS/bipolar memory has changed to permit decoding to be done on the MOS storage chips, with only the driving and sensing functions performed by bipolar circuits. This greatly reduces the number of required interconnections and permits the use of conventional packages and assembly techniques without excessive costs. Figure 7 illustrates this shift in partitioning, which took place as more area-efficient storage techniques developed.

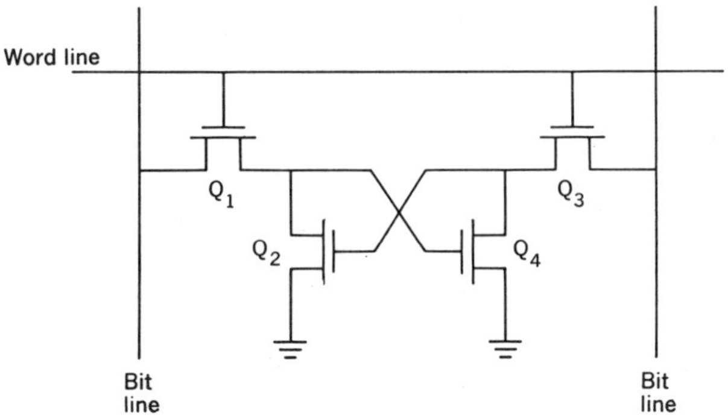

FIGURE 8. Four-transistor-per-bit charge-storage MOS memory cell.

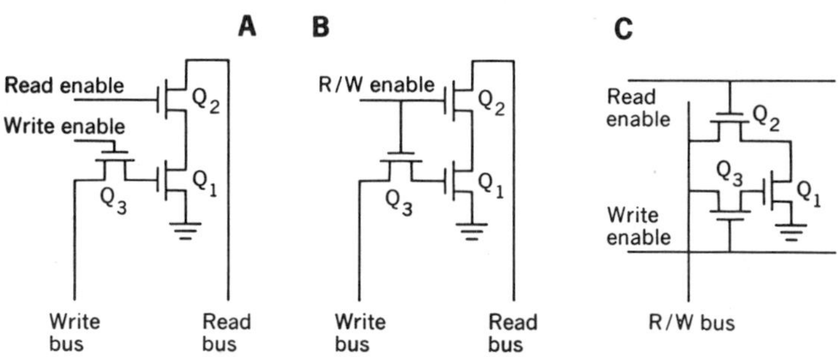

FIGURE 9. Various charge-storage MOS memory cell configurations that use three transistors per bit.

FIGURE 10. Organization of a fully decoded 1024-bit charge-storage MOS memory chip.

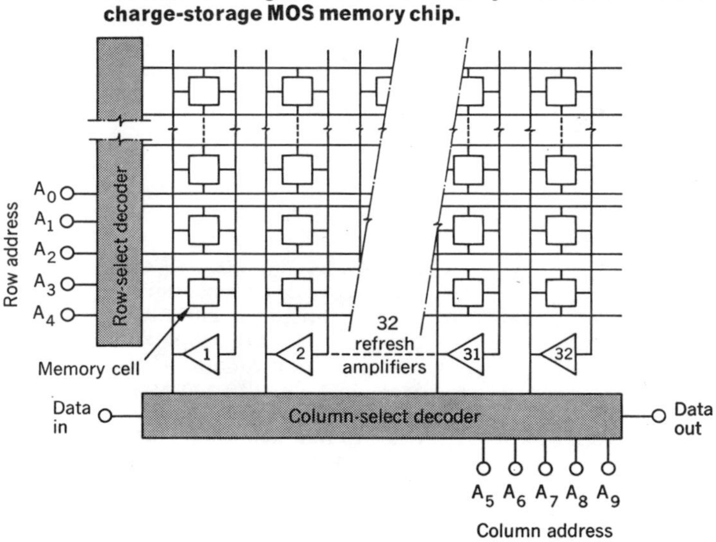

Charge-storage memory elements

The major disadvantages of bistable flip-flop storage elements are twofold: (1) They dissipate too much power, which limits the number of bits that can be put into a package module. (2) They occupy too much silicon area, which limits the number of bits that can be economically

FIGURE 11. Photomicrograph of a fully decoded 1024-bit charge-storage MOS memory chip.

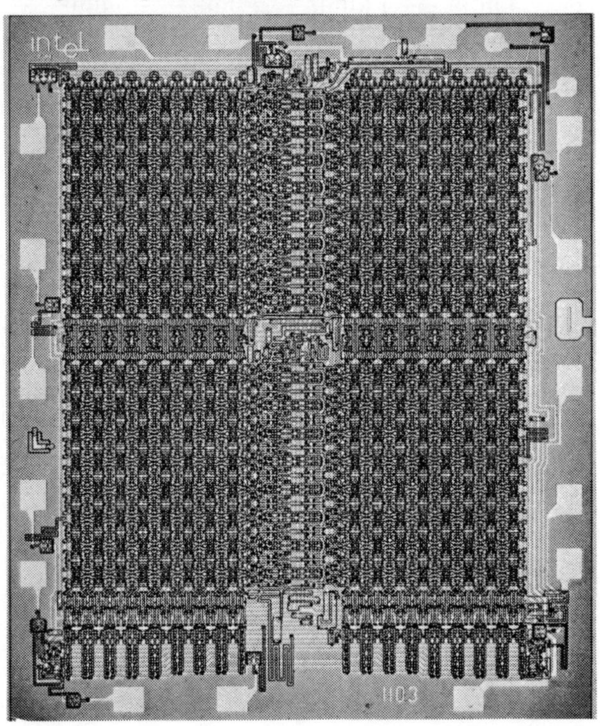

put on a monolithic circuit. Various techniques have evolved to combat these shortcomings of flip-flop-based semiconductor memories. For example, the simplest way to reduce the power dissipation of an MOS storage cell of Fig. 3 is to clock the V_{GG} supply line. While the V_{GG} supply line is negative the MOS transistors connected as load elements will be active. When V_{GG} is positive, the load elements will be turned off. Correct memory information is maintained by charge storage only. One side of the flip-flop was negative prior to the turnoff of V_{GG}. This side will maintain its potential by storing this negative charge on the gate capacitance of the other flip-flop side. Parallel with this gate capacitance, however, is the junction capacitance associated with the crosscoupled MOS devices. These junctions will provide a parasitic leakage path—through the leakage current of a reverse-biased junction—for the stored charge and, therefore, will limit effective charge stôrage to a finite time period. One characteristic feature of dynamic charge-storage semiconductor memories, therefore, is that they all need a *periodic refreshing* of the charge stored in their cells. In the case of the memory cell of Fig. 3(D), this is accomplished by periodically turning V_{GG} negative, which activates the load elements and replenishes the lost charge through this load element. The minimum rate is determined by the quality of the process and the temperature range of operation. Power dissipation figures as low as a few microwatts per bit can be achieved in the standby mode.

This or similar methods—while achieving significant power reduction and thus removing one important limitation—do not effectively attack the most significant problem: packing density of memory cells. Economics dictate that, for minimizing bit cost, the number of bits per chip has to be maximized. The technological constraints are a given maximum chip size and a set of photolithographic tolerance rules.

Somewhat better packing density can be achieved using the cell of Fig. 8. This is a four-transistor cell—basically a flip-flop without the load devices. Unlike a conventional flip-flop cell, the word-enable transistors and the load devices (Q_1 and Q_3) are one and the same. If the proper device ratios are maintained between Q_1 and Q_2, and between Q_3 and Q_4, the flip-flop maintains its proper logic state. To read, one senses the currents in the bit line. If Q_2 is on, the bit line associated with this side will carry current, the other bit line will not. To write into the memory cell, one forces the bit lines into the proper states and then transmits these data into the cell by enabling the word line. To refresh such a cell, the bit lines change to a common negative potential (for the case of p-channel circuits) and turn on the word line. This

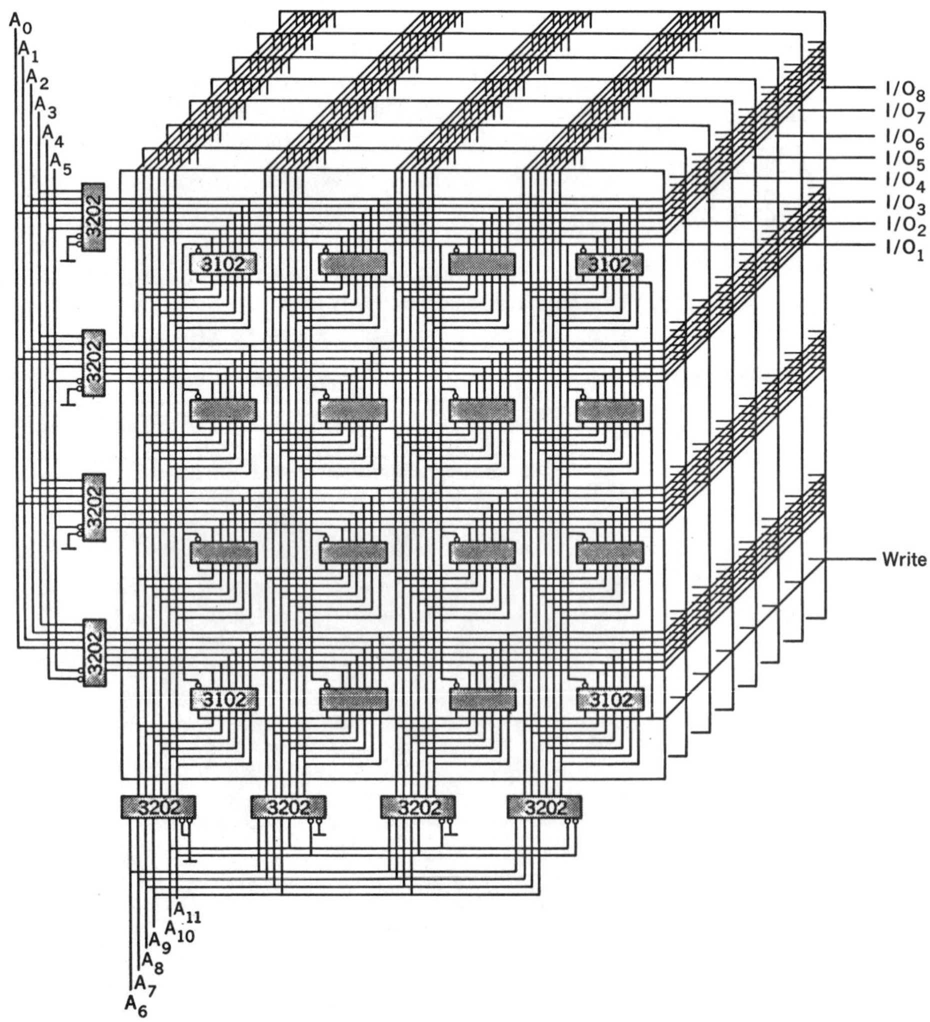

FIGURE 12. Organization of a 4096-word by 8-bit memory system that uses partially decoded 256-bit bipolar memory components. The memory chips are marked 3102, and the decoder-drivers are marked 3202. The diagram shows eight identical planes that each contain 4096 words of one bit.

FIGURE 13. Organization of a memory system that uses fully decoded 256-bit MOS memory components, which are marked 1101. All address inputs are connected in parallel, and the appropriate memory units are enabled by the chip-select inputs for input or output operation.

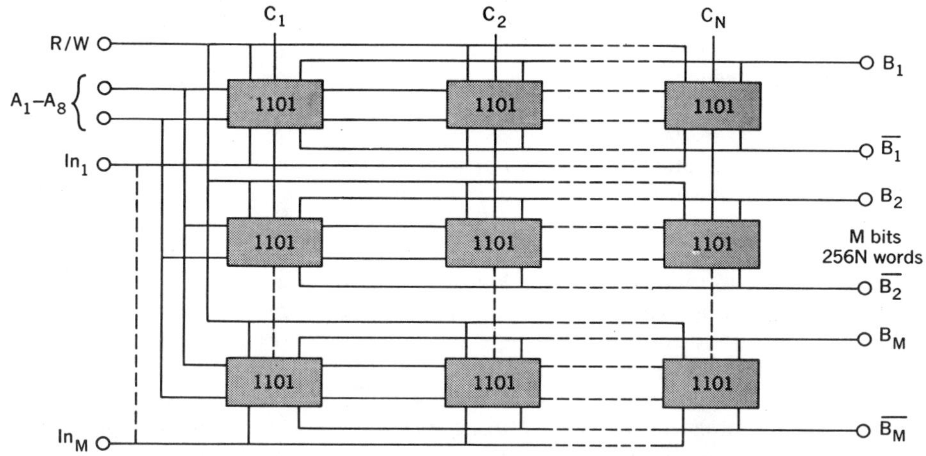

FIGURE 14. Organization of a memory system that uses fully decoded 1024-bit memory components. The memory components, marked 1103, can be connected in an array of 1024M words by N bits. System access times of under 250 ns and cycle times of under 400 ns are achievable with such memories. A—Organization of the memory plane. B—Block diagram of a 4096-word by 9-bit memory system.

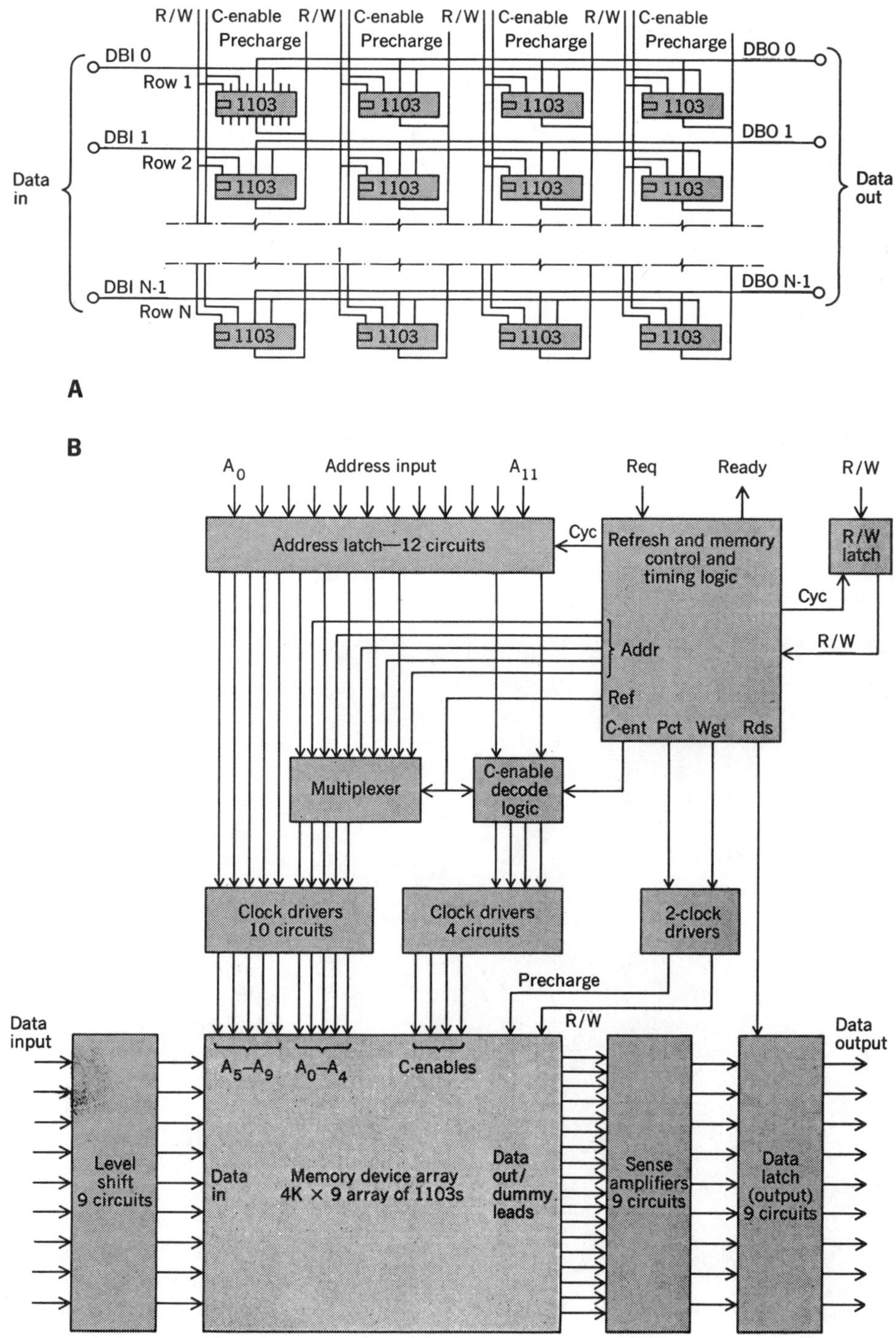

causes the regeneration of information in the cell. This type of cell is used in AMI's 512-bit dynamic RAM and in Cogar's 1024-bit memory chip.[8]

To store charge, however, a flip-flop is not needed; only one MOS transistor whose gate is the storage node is used.[9,10] For a *p*-channel MOS transistor the presence of a sufficient amount of negative charge on its gate will mean that the device is on. An insufficient amount of charge on this gate will mean that the device is off—an electrical analogy to the logic "1" and logic "0" conditions. A means for supplying charge or no charge to the gate of the MOS transistor (for writing into the cell) and a means for interrogating the cell for its content (for reading it) is needed. Figure 9(A) shows a configuration of such a charge-storage memory cell. Transistor Q_1 is the charge-storage element. Transistor Q_2 connects Q_1 to the *read bus* when it is activated by the *read-enable* signal. Transistor Q_3 provides a *write* path to the charge storage node when activated by the *write-enable* signal. Transistor Q_3 also provides periodic refreshing of the memory data. This process rewrites or reinforces the charge condition of the storage node from an on-chip refresh amplifier. In a typical memory array organization (Fig. 10) each column of *M* bits has its own refresh amplifier. There are *N* columns in the memory array. Activating any one of the *M* read-enable signals writes the content of the memory cells in that row into their respective refresh amplifiers. This, followed by the write-enable command on that same row, will refresh all memory cells in that row. Refreshing the entire memory plane requires going through *M* rows only because each refresh cycle will refresh *N* bits simultaneously.

The memory cell shown in Fig. 9(A) is used in Intel's 1103, a 1024-by-1 dynamic RAM, whose basic organization is shown in Fig. 10. A photomicrograph of this chip is shown in Fig. 11. Various other products (AMI's 6001 and Mostek's MK 4006P) use similar charge-storage concepts. There are some differences in cell connections, Fig. 9(B) and (C), or periphery designs, but all of these products are based upon the three-transistor memory cell.

Although these dynamic RAMs are only now available commercially, they already represent significant competition to any existing main-frame memory elements, such as cores or plated wire. They achieve high performance: less than 200-ns access time, less than 100-μW/bit power dissipation in the active mode, and less than 10-μW/bit power dissipation in the standby mode. Their small geometry allows extremely high packing density on a memory chip, as shown by the comparison in Fig. 4. A 1024-bit memory chip using three-transistor cells has about the same dimensions as a 256-bit memory chip using bistable flip-flop cells. As a result, these memories can be cost-competitive with either cores or plated wires.

Although all existing products based on charge storage use MOS transistors as storage elements, a bipolar equivalent of this principle has been proposed that uses diodes for storage.[11] This concept is in the development state at present.

FIGURE 15. The memory system. The two-layer board contains 4096 words of 18 bits and all driver, sense, and control circuitry needed to operate the memory.

System organization using semiconductor memories

Although semiconductor memories represent some of the most complex integrated chips in digital systems today, they are usually organized in such a manner that system design flexibility is maintained. The following simple examples illustrate how semiconductor memory components may be interconnected to form a variety of large memory systems.

Figure 12 shows a 4096-word by 8-bit system built of partially decoded 256-bit bipolar memory chips (marked 3102) and decoder–drivers (marked 3202). For purposes of illustration, the diagram is drawn to show eight identical planes. Each plane contains 4096 words of one bit. To increase the word length, one needs only to increase the number of planes. Word size may be increased by increasing the size of each plane in the *x* or the *y* direction, or both. Cycle time of such a system is 100 ns.

Fully decoded static MOS memories can be used to construct a memory system in a similar manner. Figure 13 illustrates the expansion of a 256-by-1 fully decoded memory component (marked 1101) into a 256*N* words by *M* bits memory system. In Fig. 13, the outputs can be OR-tied to expand the number of words. All address inputs are connected in parallel and the appropriate memory units are enabled by the chip-select inputs for input or output operation. The input and output parts are directly interfaced to commonly used bipolar logic circuits such as TTL integrated circuits. System performance is in the 1-μs cycle time range.

In larger memories (4 to 8 kilobytes, or more) dynamic MOS memories based on the charge-storage concept provide the most economical approach.[12] In Fig. 14, the memory components (marked 1103) can be connected in an array of 1024*M* words by *N* bits very much like the smaller system of Fig. 13. The interface circuitry, however, is not as simple as that used in a memory system based on fully decoded static memory elements. Address and clock inputs require a level shifter that converts a transistor–transistor logic (TTL) level to the higher voltage level required to drive the charge-storage memory array, Fig. 14(B). Instead of feeding into a TTL gate, the outputs will require a sense–amplifier interface.

MOS memories that use charge storage require periodic refreshing of information in the cells. For the 1103 this is accomplished by cycling through the 32 states of the address inputs A_0 to A_4. This has to be done within a certain time period. For the 1103 this time is 2 ms.

Special refresh cycles may be executed by the controller to guarantee this condition for a random-access memory. For a 600-ns cycle, the refresh operation requires less than one percent of available memory cycles. Although refresh control is a special requirement inherent in the use of dynamic MOS memory components, it adds only a negligible cost to the total system.

The performance of a system such as Fig. 14 equals or exceeds that of most core memory systems. System access times of under 250 ns and cycle times of under 400 ns are achievable with such memories. Figure 15 shows such a memory system.

The use of semiconductor memory components provides great system flexibility. Memory systems built of semiconductor memory components are easily expandable. Using the wide variety of commercially available semiconductor memory components permits systems of various sizes and speeds.

Conclusions

Semiconductor IC memory elements whose availability was announced in the past year or so herald the existence of a new memory element: the semiconductor memory component. They are the product of a fast developing field that is based upon the ten-year history of IC technology.

The evolution of the IC business has been very rapid. Circuit yields have improved continuously as production costs have decreased; unit prices reaching less than five cents per gate in 1970 for TTL elements, less than one cent per bit for MOS shift-register elements. These economics were made possible by the continuous technology improvements characteristic of the IC industry. These technology improvements are responsible for the existence of LSI arrays, such as semiconductor memory elements that contain thousands of devices on a monolithic chip, and will lead to further cost reduction of these components in the future.

REFERENCES

1. Ayling, J. K., Moore, R. D., and Tu, G. K., "A high-performance monolithic store," *ISSCC Dig. Tech. Papers*, pp. 36–37, Feb. 1969.
2. Rice, R., Sander, W. B., and Greene, F. S., Jr., "Design consideration leading to the ILLIAC IV LSI process element memories," *IEEE J. Solid-State Circuits*, vol. SC-5, pp. 174–181, Oct. 1970.
3. Dunn, R. S., and Jeansonne, G. E., "Active memory design using discretionary wiring for LSI," *ISSCC Dig. Tech. Papers*, pp. 48–49, Feb. 1967.
4. Bryant, R., Tu, G., Kwei, T., and Robinson, R., "A high-performance LSI memory system," *Computer Design*, pp. 71–77, July 1970.
5. Schmidt, J. S., "Integrated MOS random-access memory," *Solid-State Design*, pp. 21–25, Jan. 1965.
6. Pleshko, P., and Terman, L. M., "An investigation of the potential of MOS transistor memories," *IEEE Trans. Electronic Computers*, vol. EC-15, pp. 423–427, Aug. 1966.
7. Dumesnil, M., and Schroeder, J., "A large-scale integration hybrid substrate," *ISSCC Dig. Tech. Papers*, pp, 66–67, Feb. 1969
8. Lund, D., Allen, C. A., Andersen, S. R., and Tu, G. K., "Design of a megabit semiconductor memory system," *Proc. Fall Joint Computer Conf.*, pp. 53–62, Nov. 1970.
9. Boysel, L., Chan, W., and Faith, J., "Random-access MOS memory packs more bits to the chip," *Electronics*, vol. 43, pp. 109–115, Feb. 16, 1970.
10. Regitz, W. M., and Karp, J. A., "Three-transistor cell 1024-bit 500-ns MOS RAM," *IEEE J. Solid-State Circuits*, vol. SC-5, pp. 182–186, Oct. 1970.
11. Waaben, S., "Electronically variable semiconductor memory using two diodes per memory cell," *ISSCC Dig. Tech. Papers*, pp. 46–47, Feb. 1970.
12. Hoff, M. E., "Design an LSI memory system that outperforms cores—economically," *EDN*, pp. 6–15, Jan. 15, 1971.

Part II
Bipolar Transistor Memories—Technology and Circuits

Editor's Introductory Comments

New processes and devices have had an impact on bipolar memory performance and cost. The Schottky process (II,1) improves performance by eliminating minority carrier charge storage delays. Collector diffusion isolation [1], isoplanar isolation (II,2), and air isolation by means of vertical anisotropic etching (known as the AIM or V-ATE process) all reduce the silicon chip area required for realization of a given circuit. Chip area is a key measure of cost of integrated circuits.

The major improvements in memory density, performance, and power consumption over the past five years have come via improvements in circuit design and chip layout, rather than via new processes or reduced device geometries.[1] Early memory cell circuits were unnecessarily complex [2]. Almost all bipolar memories now available use one of the two cell circuits described in II,3. Improved address decoding and data buffering circuits have had a major impact too. Examples of current commercial practice in bipolar memory circuit design are shown in II,4 and II,5. A fully decoded 256-bit bipolar memory is now available in addition to the partially decoded component of II,5. Fully decoded 1024-bit RAM's of similar circuit design will be announced in 1972.

One new direction in bipolar memory circuit design is indicated by II,6. Neither a schematic diagram nor a drawing of surface and cross-section geometries is sufficient to fully explain the compact low-power cell circuit. A full understanding of devices and circuits is necessary to pursue this sort of design approach.

Another new approach is dynamic bipolar memory circuits [3]-[5]. Such circuits require restoration of stored information every 1-10 ms. This approach is attractive only if the complete memory function, including address decoding and data refreshing, can be integrated at higher density than competing static circuits.

[1]The best illustration of this point is the main memory for the IBM System 370 Model 145 (see VI,1), which achieves outstanding system-level characteristics without resort to any new integrated circuit processes or reduced device dimensions.

The final paper (II,7) shows the complete design for a 1024-bit READ-ONLY memory. Newer designs offer 8192 bits in about the same silicon area [6]. The density of bipolar ROM is about eight times that of bipolar RAM; the cost per bit should be roughly eight times lower for the ROM at comparable points in the production cycle.

References

[1] B. T. Murphy, V. J. Glinski, P. A. Gary, and R. A. Pedersen, "Collector diffusion isolated integrated circuits," *Proc. IEEE,* vol. 57, pp. 1523–1527, Sept. 1969.

[2] G. B. Potter, J. Mendelson, and S. Sirkin, "Integrated scratch pads give a new generation of computers," *Electronics,* vol. 39, pp. 118–126, Apr. 4, 1966.

[3] H. Henn, "Bipolar dynamic memory cell," *IEEE J. Solid-State Circuits,* vol. SC–6, pp. 297–300, Oct. 1971.

[4] S. Waaben, "100 ns electronically variable semiconductor memory using two diodes per memory cell," *IEEE J. Solid-State Circuits,* vol. SC–5, pp. 192–196, Oct. 1970.

[5] J. Mar, "A two-terminal transistor memory cell using breakdown," *IEEE J. Solid-State Circuits,* vol. SC–6, pp. 280–283, Oct. 1971.

[6] Monolithic Memories, Inc., "8192 bit bipolar read-only memory MM5280," *Data Sheet,* Sept. 1971.

Schottky diodes make IC scene

With reproducibility problems licked, these devices make attractive elements; they permit unusual component combinations, save chip real estate, reduce power dissipation, and enhance production yields

By R.N. Noyce, R.E. Bohn, and H.T. Chua

Intel Corp., Mountain View, California

Schottky diodes, because they don't store charge, open new worlds to digital integrated circuits. Diode-transistor logic with Schottky diodes can be made as fast as transistor-transistor logic and emitter-coupled logic. What's more these devices reduce an IC's power dissipation and permit unusual combinations of components on a chip. They also make large-scale integration easier to achieve because more functions can be performed in less area. Not the least of their virtues is the fact that they can increase production yields.

For all these advantages, Schottky-diode IC's have not been commercially available. Intel Corp., however, has concentrated on their development and is now making Schottky diodes that are stable and reproducible. The company has committed Schottky IC's to commercial production, and has found that they live up to expectations. Their first Schottky IC will be introduced in the next few weeks.

Structurally, the Schottky diode is little more than a metal in contact with a semiconductor. As long as aluminum is used for the metal, formation of Schottky diodes in a monolithic IC is compatible with standard processing. Some provision is needed to prevent high-field effects at the edge of the metal contact, but this too is compatible with standard processing. In fact, once reproducibility problems have been solved, Schottky diodes actually simplify processing of high-speed digital IC's because gold doping is eliminated.

The Schottky diode has been around in one form or another for a long time—the crystal detector in early radios, for instance. The modern device, which gets its name from the German scientist who developed the first valid theory of metal-semiconductor rectification, can be used to enhance the performance of existing IC designs as well as to develop entirely new configurations with exceptional features.

In the forward-biased p-n junction, current is carried by holes flowing from the p-type material into the n-type material, and electrons flowing from the n to p material. Either case results in an excess of minority carriers near the junction. If the voltage is reversed these carriers will flow back across the junction, creating a high current until the supply is exhausted. In other words, the p-n junction can't be turned off immediately.

In a Schottky diode made of aluminum on n-type silicon, essentially all of the forward current is carried by electrons flowing from the semiconductor into the metal. They quickly come into equilibrium with the other electrons in the metal, so there is effectively no stored charge to prevent rapid switching. Another major point of difference between the Schottky and the p-n junction diode is that the former has lower forward voltage for a given current.

In a practical circuit the Schottky diode is placed in parallel with the base-collector junction of an npn transistor; the metal electrode is connected to the base and to the n region of the collector, where it forms a rectifying contact.

Since the Schottky diode has a lower forward voltage compared with the collector-base junction, the diode clamps the transistor and diverts most of the excess base current through the Schottky diode, preventing the transistor from saturating. There's no charge storage, either in the transistor or diode.

Clamping techniques have been used in the past to prevent charge storage. Most are variations of the Baker clamp proposed in 1956.[1] In this scheme, a germanium diode shunts the collector-base junction of a silicon transistor to prevent it from being forward-biased; some charge is still stored in the germanium diode. The Schottky IC employs clamping but charge storage is eliminated.

In the past, other efforts to improve switching

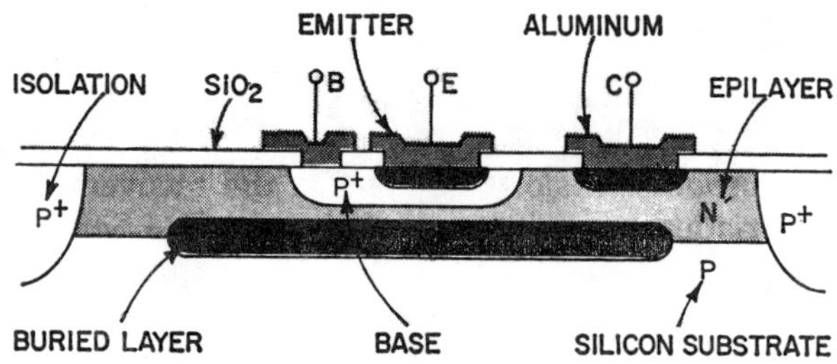

CONVENTIONAL NPN TRANSISTOR

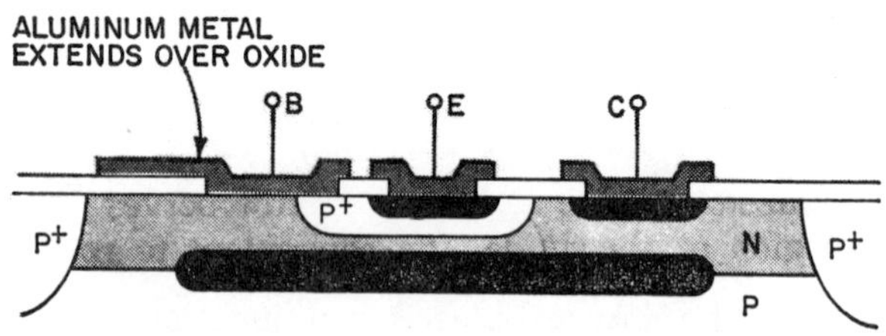

SCHOTTKY EXTENDED-METAL TRANSISTOR

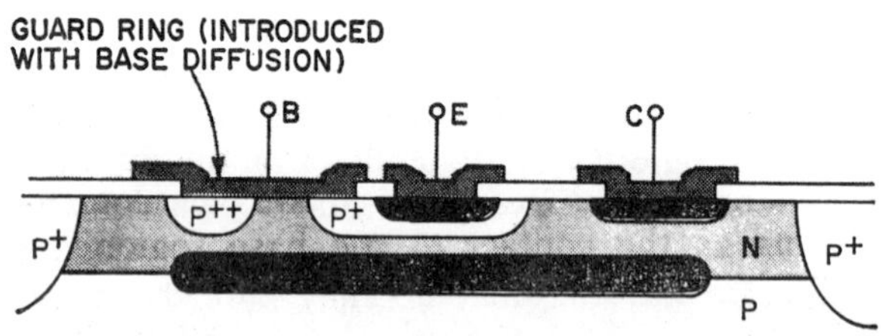

SCHOTTKY GUARD-RING TRANSISTOR

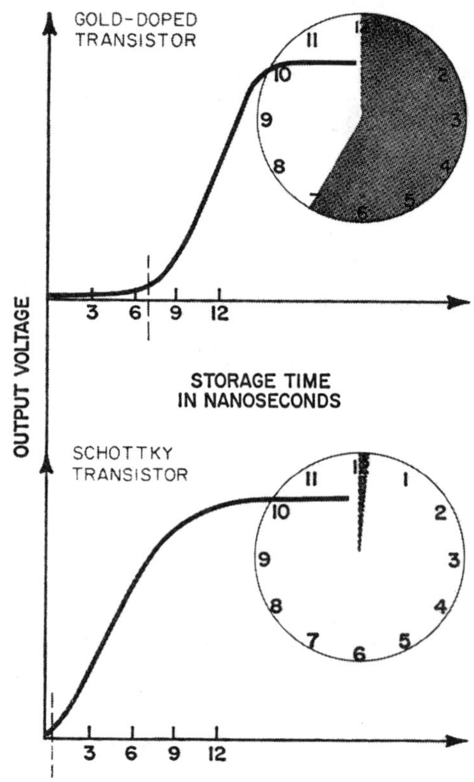

Double duty. In a Schottky transistor a Schottky diode is in parallel with the collector-base junction. The base metal is also the anode of the diode; the collector diffusion is also the diode cathode. Edge effects can be avoided with the extended-metal or guard-ring structure. The curves compare storage time of the gold-doped and Schottky transistors. The former uses up 7 nsec of recombination time before it can change state; the Schottky transistor responds almost instantaneously.

speed by reducing or avoiding stored charge have taken two approaches: process innovation and novel circuit design.[2, 3, 4]

Attempts to minimize storage time through novel circuit designs have had varying success. The most familiar is current-mode switching, which avoids saturation (and hence charge storage) by controlling the collector current so that the collector-base junction never goes into forward conduction. Current-mode logic is fast—propagation time through a gate can be as low as 1 nanosecond—but when such IC's are assembled in a system, they will oscillate unless precautions are taken to control their input impedance characteristic.

Off the gold standard

The standard processing approach to charge-storage reduction in conventional IC's is gold doping. Since gold acts as a recombination center, it reduces the lifetime of minority carriers; that is, it shortens the time for recombination of holes and electrons. Intentional reduction of minority carrier lifetime is somewhat ironic. In the early days of transistors, much effort was expended in increasing the minority carrier lifetime to increase current gain and decrease junction leakage.

Unfortunately, the gold diffusion has some highly undesirable side effects. It tends to reduce gain (h_{FE}). If this value gets too low, the IC's won't function because the transistors either don't saturate or have excessive turn-on delay. The manufacturer can use a narrower base to get higher h_{FE}, but this can lead to excessive junction leakage.

The Schottky process sidesteps this problem. Without the need for gold doping, low storage time and high h_{FE} can be achieved simultaneously.

Another difficulty with gold doping is the lack of selectivity—if one transistor must be high speed, all the others on the chip have to be high speed too. But with the Schottky process, fast and slow devices can be mixed on the same chip. This freedom of choice in component characteristics is exploited

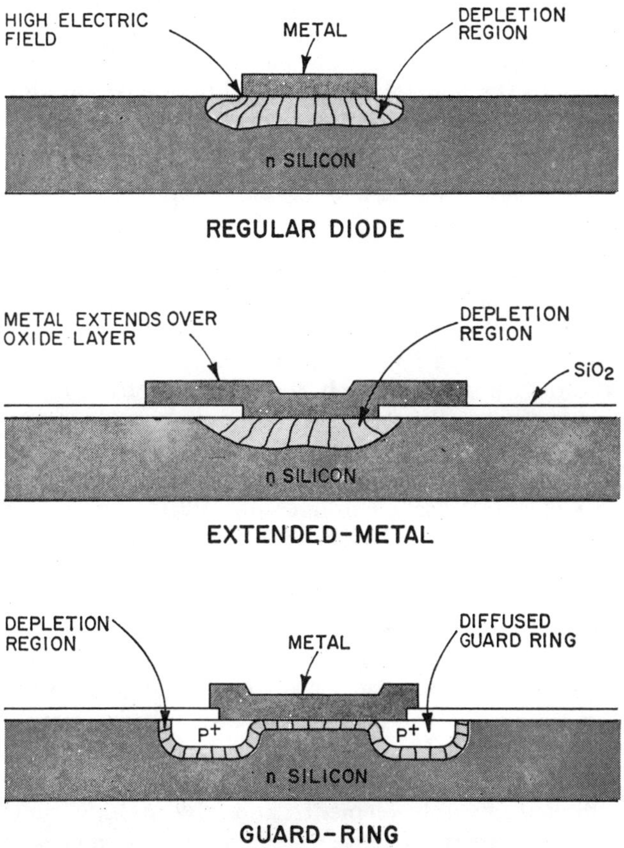

Edge effect. The high field concentration at the edge of the regular Schottky structure can cause spurious currents. These can be avoided by means of the extended-metal structure or the guard ring.

in some of Intel's new products.

But before Schottky diodes could be successfully integrated manufacturers had to resolve some dilemmas; diode characteristics were often grossly different from theoretical predictions and devices were not reproducible. The cause, it has since been found, is the high electric-field concentration at the periphery of the metal electrode, as shown above. This field, through the mechanism of tunneling, can lead to spurious currents that dominate the ideal diode characteristic. Because the shape of the edge varies from one device to another, the magnitude of these excess currents can vary greatly.

Two solutions to this problem were recently proposed.[5, 6, 7] In one, the metal electrode extends over the silicon-dioxide passivating layer. With this dielectric layer between the metal and the semiconductor, the electric field at the periphery of the electrode is greatly reduced, and the diode has ideal characteristics.

In the other solution, the edge of the metal extends over a diffused p+ guard ring. This also reduces the field, and moves it away from the edge of the electrode so that it can't affect the diode. Again, the characteristics are ideal.

Electrically, the aluminum-silicon Schottky diode differs somewhat from the diffused p-n junction diode in its d-c forward current-voltage characteristic, shown at right. The forward voltage of the Schottky diode at any given current is 200 to 300 millivolts less than that of the junction diode, making it an ideal low-voltage clamp.

Another major difference is the recovery characteristics, opposite page at bottom. The Schottky diode storage time is effectively zero, in contrast to typical values of 6 nsec for the gold-doped junction diode and 30 nsec for the junction diode without gold doping. In all three cases, the voltage across the diode decays with the same R-C time constant once the stored charge recombines.

The Schottky transistor is simply an npn IC transistor, produced by conventional photolithography, diffusion, and epitaxial growth techniques with an Al-Si Schottky diode in parallel with the collector-base junction. The base contact of the transistor serves as the metal contact of the diode and the collector region of the transistor serves as the n region of the diode. It's convenient to regard the npn-Schottky diode combination as a single device and to represent it by a single symbol:

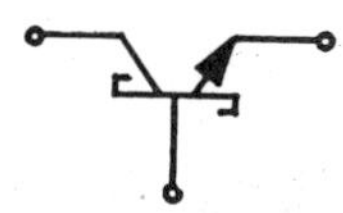

To form the Schottky transistor, the base-contact opening is extended beyond the base diffusion and over the collector region. When the aluminum metalization is deposited, it simultaneously functions as the contact to the base region and the anode of the Schottky diode. To prevent high field concentration, the metal can either be extended over the passivating oxide or terminated over a p+ guard ring, which can be diffused into the chip at the same time as the base, as shown on page 75.

One big difference between the Schottky transistor and the conventional gold-doped npn device is in offset voltage, which is lower for the latter—typically 120 millivolts vs. 250 mv. However, the saturation voltage need not be excessive because of this large offset; by controlling both device geometry and processing, $V_{CE(sat)}$ can be kept to 0.5 volt at 20 ma collector current and 1 ma base current. And because the Schottky diode has a lower temperature coefficient than the emitter-base diode, $V_{CE(sat)}$ decreases with increasing temperature.

This behavior of the Schottky diode amounts to partial temperature compensation of the emitter-base diode and hence of $V_{CE(sat)}$. It helps to maintain a high worst-case logic 0 noise margin at high temperature when the Schottky transistor drives fan-out circuits that operate at a $2V_{BE}$ logic threshold as standard DTL and TTL IC's do.

The storage time of a Schottky transistor is effectively zero, as shown on page 78, in contrast to typically 7 nsec for a gold-doped device and 34 nsec for a non-gold-doped unit. Moreover, storage time hardly varies with temperature. At 125°C,

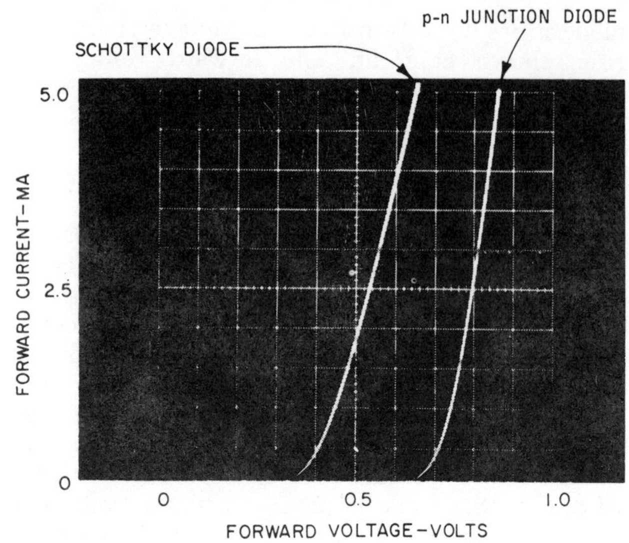

Forward characteristic. At 1 ma forward current, voltage drop is 0.45 volts for the Schottky diode, 0.74 volts for the p-n junction diode.

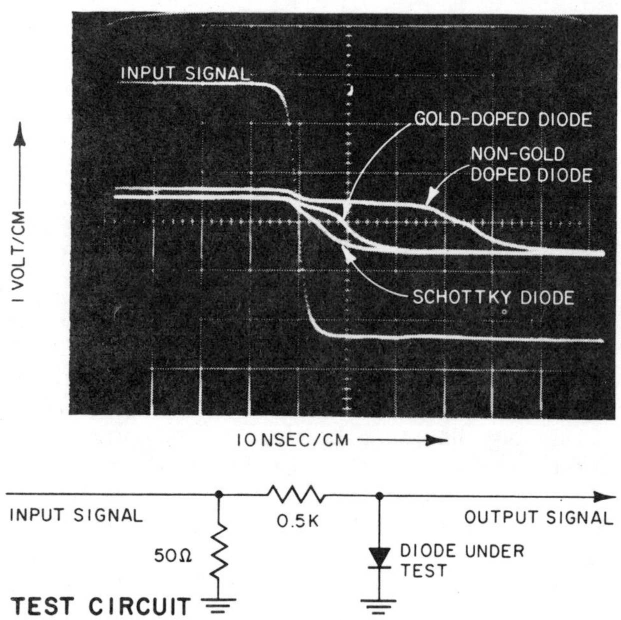

No storage. Transient characteristic shows that the Schottky diode, unlike p-n junction diodes, has essentially zero storage time.

it's still less than a nanosecond, whereas the gold-doped storage time doubles to 15 nsec at that temperature.

Aside from much faster switching speed, Schottky diodes offer the IC designer a greater range of devices to choose from. Because gold doping can't be done selectively, the designer has at his disposal only one active component: a fast switching npn transistor whose emitter-base and collector-base junctions can be used as diodes. But since the Schottky process doesn't require gold doping, the designer has seven active components to choose from. And he can mix them at will—fast- and slow-switching devices can be placed on the same chip. In both the Schottky and the gold-doped transistor, various levels of resistivity are available in the bulk material, as well as in the isolation, emitter and base diffusions for use as resistors.

Advantageous options

The Schottky diode across the base-collector junction of the npn transistor is an almost ideal active switch with less than 1 nsec storage time and high amplification. Without the Schottky diode, the device becomes a charge-storage npn transistor characterized by low saturation voltage, high inverse gain (h_{FE}), and lengthy recovery time.

Substrate pnp transistors can be diffused into the chip too; these can have h_{FE} of 10 or more and are ideal for input buffering. Lateral pnp transistors can be formed as well—with and without Schottky-clamped collectors to provide fast or slow recovery time. These devices are useful as curent sources and for voltage translation.

Then there are SCR's. Like transistors, these can be diffused into the chip in both charge-storage and Schottky-clamped versions, each of which can be optimized for different purposes. They make efficient bistable elements and give a high functional density to integrated shift registers and counters.

And, of course, there are diodes—emitter-base, charge-storage, and Schottky—each with different characteristics.

With just the Schottky diode and transistor, it's easy to upgrade old IC designs. Take DTL, for instance. If, without changing any of the resistor values, the basic DTL gate is modified with Schottky components as shown at top of page 80, the performance of the gate is improved as follows:

- Less sensitivity to power supply variations, since the circuit will continue to function at a lower minimum supply voltage. This is because, when all inputs are in the high state, the voltage at the base of Q_1 is typically 0.3 volts lower that it would be in the gold-doped counterpart.
- Less sensitivity to temperature variations because of the compensating effect of the Schottky diode temperature behavior.
- Faster switching; the gate turns on faster, and storage time is less than 1 nsec, remaining so throughout the operating temperature range.

In addition, modified DTL circuits can be fabricated at better yield (and lower cost to the user) because h_{FE} is higher and leakage is lower than in the gold-doped version.

The Schottky version of the TTL gate also has some important advantages over its gold-doped counterpart:

- Lower input inverse leakage current, typically less than 0.1 microampere. This is made up of the reverse-bias leakage of the emitter-base junction and the input clamp diode; there is no contribution from inverse h_{FE} since the collector base junction

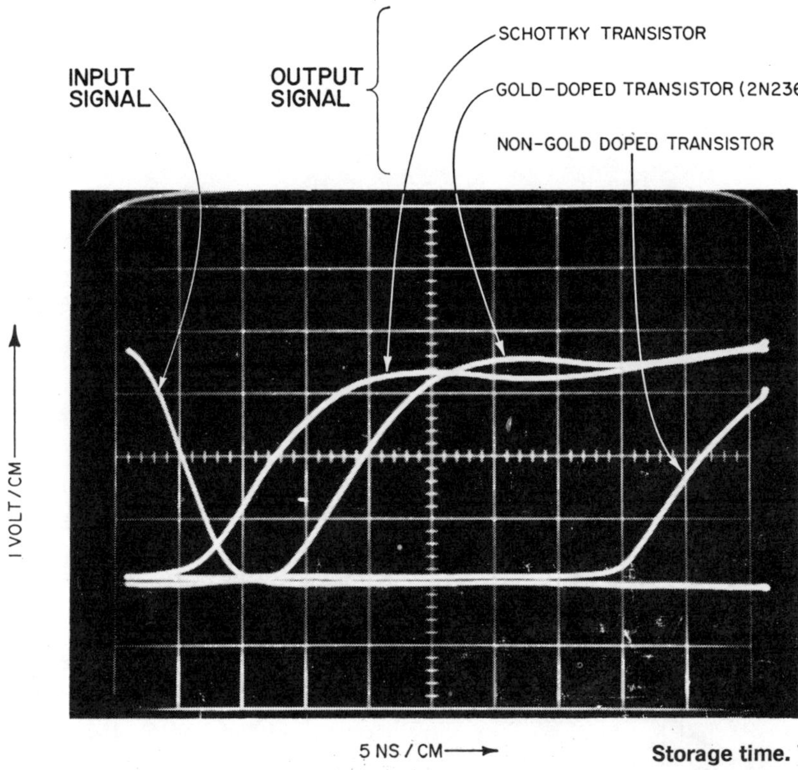

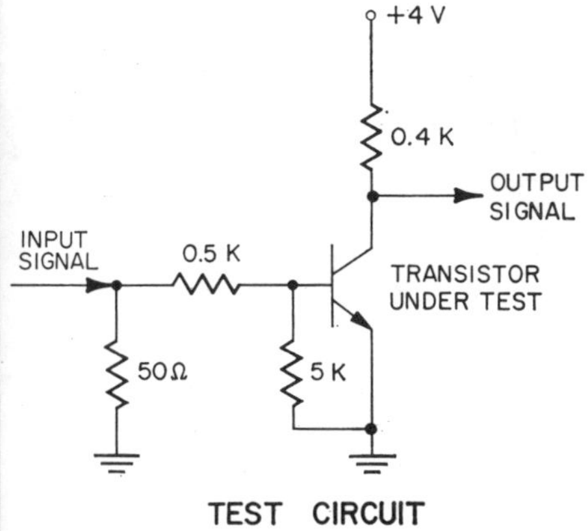

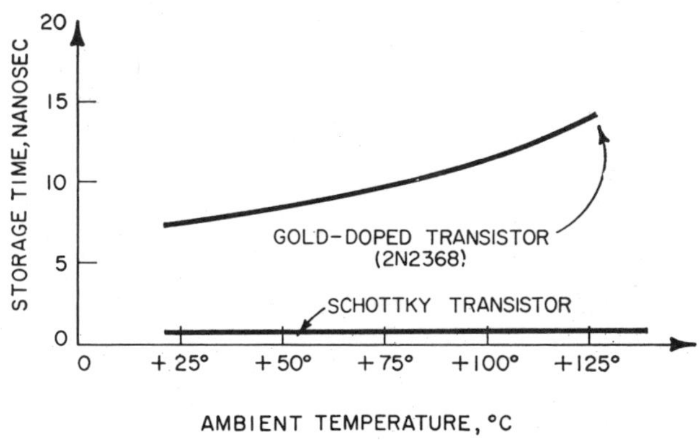

Storage time. The Schottky transistor, because it doesn't store charge, responds much more rapidly to an input signal (top traces, ambient temperature 25°C). This difference becomes even more pronounced at high temperatures (bottom curves).

of the input transistor never becomes forward biased.

- Less line reflection, since the input clamp diodes—now Schottky diodes—have a lower forward-bias voltage.
- Lower power dissipation because no charge is stored (and wasted) in the transistors.
- Higher output high level, by approximately 0.3 volt.

Although the Schottky process significantly improves many of the existing circuits, its real potential lies in entirely new designs. For example, by designing from scratch with the Schottky process in mind, a DTL gate can be made just as fast as TTL. An example of such a circuit is shown on page 80, bottom left.

When the circuit is loaded with two-200-ohm resistors, as shown, the switching speed is about 5 nsec, comparable to that of the fastest TTL circuits commercially available. The advantage of DTL over TTL, given equal speeds, is that the outputs of the DTL gates can be OR-tied. Moreover, DTL doesn't create heavy current transients in the power supply during switching.

The input current to this Schottky DTL gate is low, typically 250 microamperes. This presents a very small load to the driving gate. As a result, each gate can drive a large number of similar gates —fan-out can be 25 or more. If the two 200-ohm resistors are used as a terminating resistance for a 100-ohm line, there will still be enough current left to drive 10 gates.

Another novel circuit is the binary divider on page 80, bottom right. It combines the best of the slow and fast worlds: the Schottky transistors that comprise the flip-flop switch at high speed, and the slow charge-storage transistors serve as temporary-data-storage elements.

Bridge work

The special advantages of these circuits, and of the Schottky process that makes them possible, are particularly applicable to complex circuits. One such IC built at Intel is a 64-bit high-speed scratch-pad memory. It's organized as 16 words by 4 bits; each of the 16 words is addressable through its own binary decoder. It has four data inputs, four address inputs, and four outputs. If it's required, the outputs can be OR-tied with the outputs of other memory

Components compatible with the Schottky process

TYPE	SYMBOL	PROFILE	PRINCIPAL CHARACTERISTICS
SCHOTTKY NPN TRANSISTOR	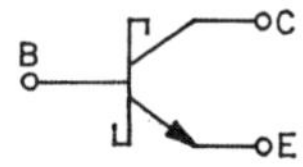	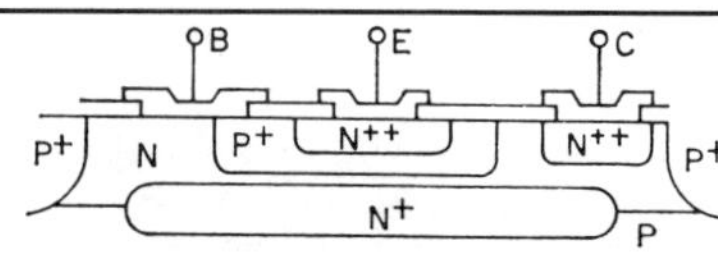	STORAGE TIME <1 NSEC $V_{CE(sat)}$ = 0.2 TO 0.4 VOLT h_{FE} = 60
CHARGE-STORAGE TRANSISTOR	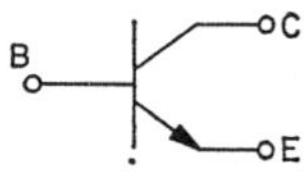	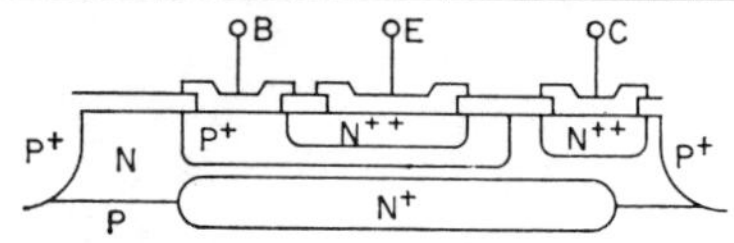	STORAGE TIME > 20 NSEC $V_{CE(sat)}$ < 0.1 VOLT h_{FE} INVERSE > 2
SUBSTRATE PNP	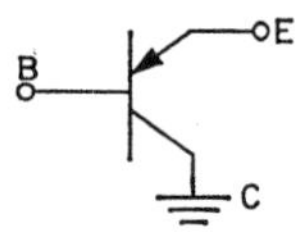	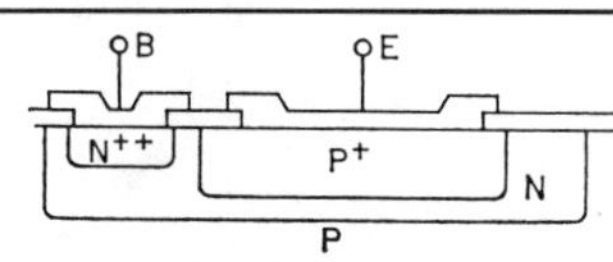	H_{FE} > 10
LATERAL PNP	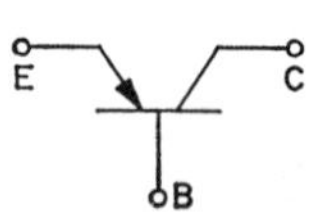	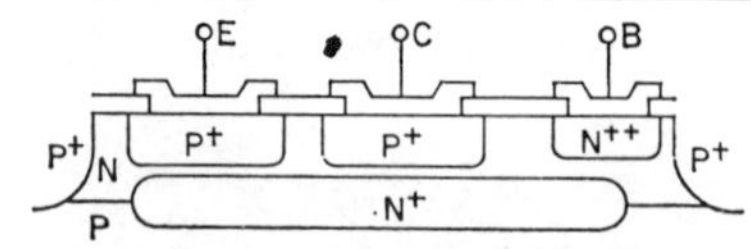	H_{FE} > 2
LATERAL PNP WITH SCHOTTKY COLLECTOR	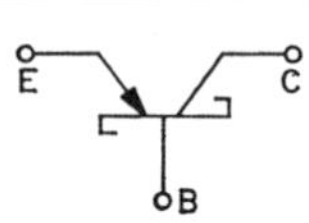	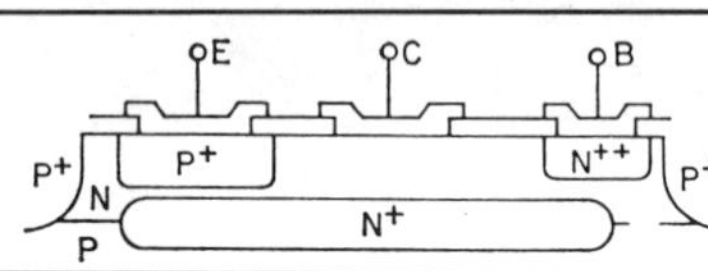	STORAGE TIME < 1 NSEC
CHARGE-STORAGE SCR	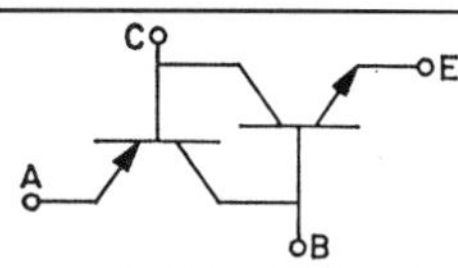	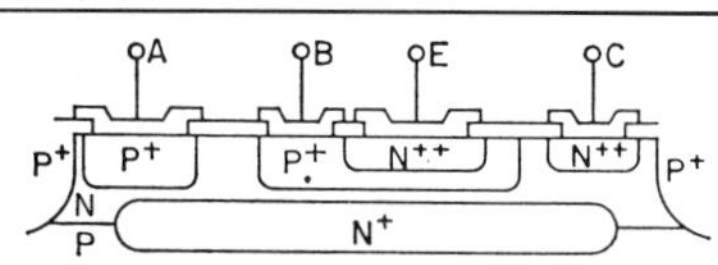	STORAGE TIME > 20 NSEC V_{ON} < 0.1 VOLT
SCHOTTKY-CLAMPED SCR	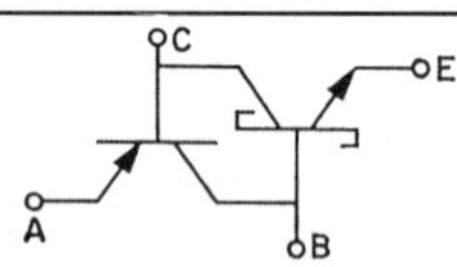	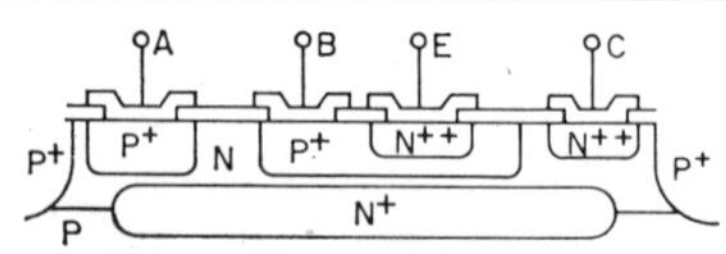	STORAGE TIME <1 NSEC V_{ON} = 0.2 TO 0.4 VOLT
EMITTER-BASE DIODE	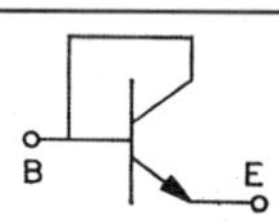	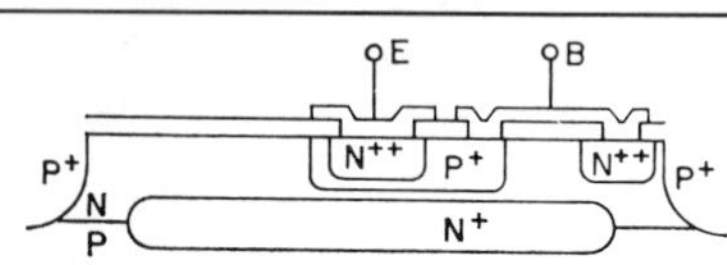	V_F = 0.6 TO 0.8 VOLT
CHARGE-STORAGE DIODE		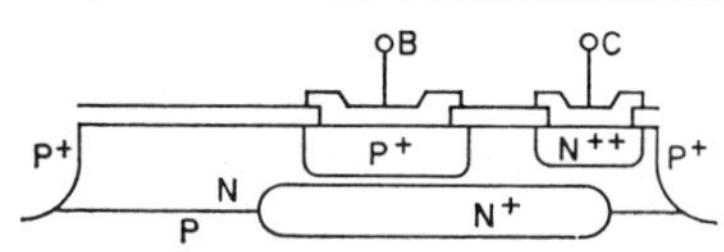	STORAGE TIME > 20 NSEC
SCHOTTKY DIODE		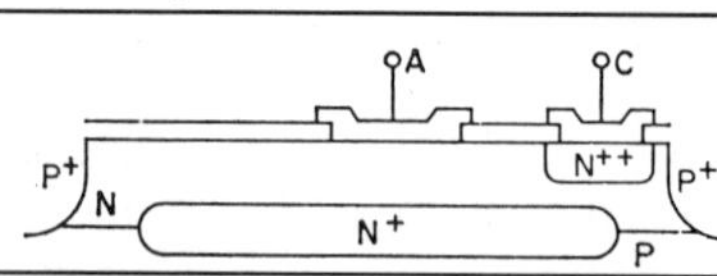	V_F = 0.3 TO 0.5 VOLT STORAGE TIME <1 NSEC
DIFFUSED RESISTOR		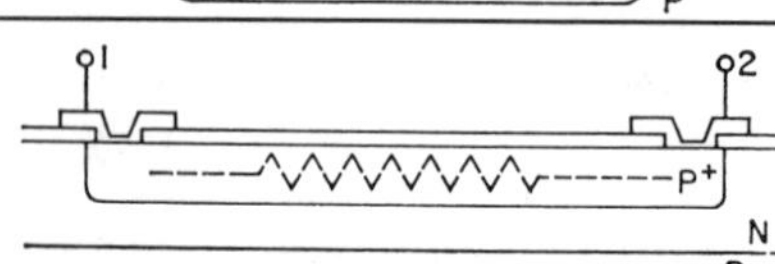	ρ_S = 100 TO 200 Ω/□
COLLECTOR-PINCH RESISTOR		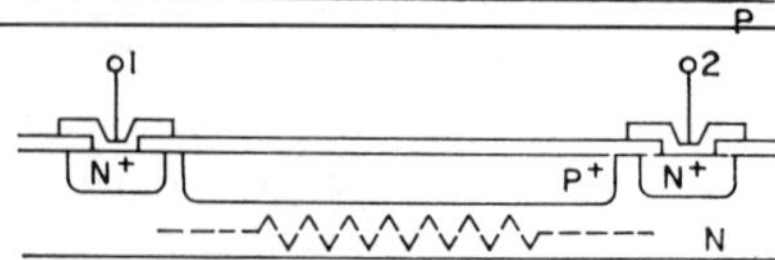	ρ_S = 400 TO 1000 Ω/□
BASE-PINCH RESISTOR		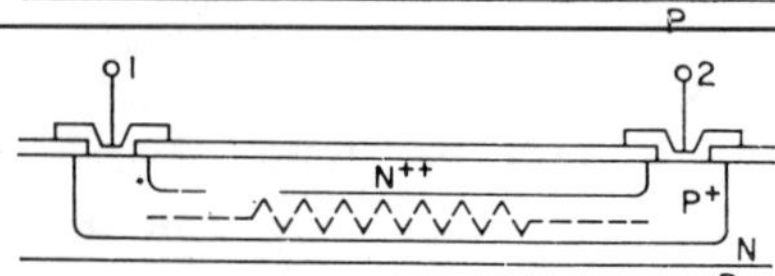	ρ_S = 2 TO 10 KΩ/□

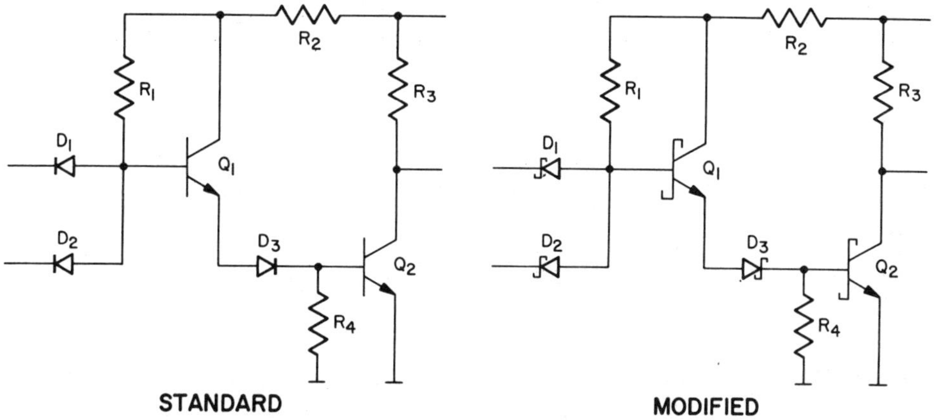

Updated. Modifying the standard DTL gate with Schottky diodes and transistors (top) affords significant performance improvement.

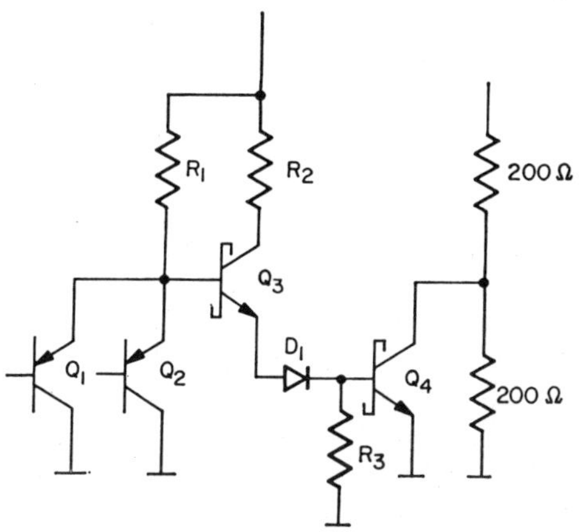

The new DTL. Without gold doping, many components become available to the IC designer. In the DTL gate at bottom left, input loading is reduced an order of magnitude by the substrate pnp's that replace the input diodes normally used.

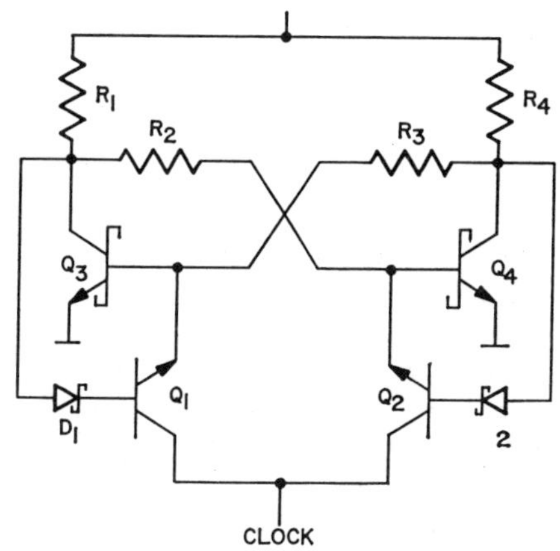

Simplified. With the Schottky process, a binary counter stage (bottom right) is simple and occupies a small chip area. This is because charge-storage transistors (Q_1 and Q_2, which are incompatible with the gold-doped process, can be used to store information.

integrated circuit chips.

The input load is equal to one unit TTL load, and the output can sink 20 milliamperes. The access delay—from address input to data output—is less than 50 nsec under a 20 ma resistive load. Power dissipation, including address buffering, decoding, sensing, and control logic, is typically 6 milliwatts per bit.

These specifications represent major improvements over gold-doped IC performance. Because storage time has been eliminated by the Schottky process, the input-to-output propagation delay has been reduced by as much as 40% for certain critical delay paths. And the variation in propagation delay due to temperature change is considerably less than in a gold-doped circuit. The higher h_{FE} values that result from the Schottky process also help keep propagation delay constant with temperature. This is particularly important for circuits that operate at low temperatures where large increase in delay, due to low h_{FE}, is a chronic problem in gold-doped IC's.

Another advantage of this higher-valued h_{FE} distribution is that transistors can be designed for higher forced-beta conditions, and power dissipation is therefore lower. For the bipolar logic-buffer chip, for example, power dissipation is reduced by about 30%.

Most significant, perhaps, the Schottky process has resulted in higher functional density and smaller chips. A comparison of the Intel 64-bit memory with an equivalent gold-doped circuit is revealing: Although the gold-doped IC is made with the same mask tolerances, the Schottky version uses a 30% smaller chip.

With all that they have going for them—high speed, greater functional density, and flexibility of design—Schottky-diode IC's certainly appear to have a bright future.

References

1. R. Baker, "Maximum Efficiency Switching Circuit", M.I.T., Lincoln Lab., Lexington, Mass., Report TR-110, 1956.

2. K. Tade et al., "Reduction of the Storage Time of a Transistor Using a Schottky Barrier Diode", PROC. IEEE, **55**, pp. 2064 - 2065, 1967.

3. E.R. Chenette and R.A. Pedersen, "Integrated Schottky Diode Clamp for Transistor Storage Time Control", PROC. IEEE, **56**, pp. 232 - 237, 1968.

4. A. Tarui, Y. Hayashi, H. Teshima and T. Sekigawa, "Transistor Schottky-Barrier-Diode Integrated Logic Circuit", Journal of Solid-State Circuits **SC-4**, pp. 3 - 12, 1969.

5. A.Y.C. Yu and E.H. Snow, "Surface Effects on Metal Silicon Contacts", Journal of Applied Physics **39**, 3008, 1968.

6. M.P. Lepselter and S.M. Sze, "Silicon Schottky Barrier Diode with Near Ideal Characteristics", Bell System Technical Journal, **47**, 195, 1968.

7. R.A. Zettler and A.M. Cowley, "The p-n Junction-Schottky Barrier Hybrid Diode", IEEE Trans. on Electron Devices, **Ed-16**, 58, 1969.

Isolation method shrinks bipolar cells for fast, dense memories

Passive isolation promises ICs that combine bipolar speed and MOS density, while simplified masking and a self-alignment feature presage higher yields and reduced costs

by Doug Peltzer and Bill Herndon, *Fairchild Semiconductor division, Mountain View, Calif.*

☐ A 40% savings in chip real estate is the promise of a process being developed for fabricating bipolar IC devices. This improvement on conventional bipolar techniques can be put to good use in producing LSI circuits, such as random access memories, that match the density of MOS devices while retaining the speed advantages of bipolar units. Moreover, in attaining its properties with simplified masking, the new process strongly indicates greater yields and lower costs. These factors could have a far-reaching effect in deciding whether MOS or bipolar devices will dominate the development of semiconductor memory technology.

The process is based on a new approach to circuit isolation called oxidized isolation, in which the active p-type diffusions that isolate conventional bipolar devices are replaced by passive insulator-oxide regions. With the isolation also serving as an insulator, there's no need to separate the isolation region from the transistor base. Hence the oxidized isolated device achieves a considerable size reduction over its diode-isolated counterpart.

Implementation of the oxidized isolation technique required development of a method to mask the active region of the transistor—where the buried collector and the base and emitter are located—during the oxidized isolation step.

The approach selected uses silicon nitride (Si_3N_4) layers to mask against thermal oxidation of the silicon in this region during the isolation step. Silicon nitride is ideal for this purpose because it remains practically inert during the oxide step. In fact, the nitride converts to silicon dioxide very slowly—1,000 angstroms of nitride will mask against oxidation for 10 hours at 1,150°C in steam.

Ironically, the basic silicon nitride mask technique was developed for fabricating MOS devices. This method, shown in Fig. 1, was devised by an Italian physicist, F. Morandi,[1] to circumvent the problems inherent in the standard practice of fabricating active and inactive circuit areas by using different oxide thicknesses. The result of using these different oxide thicknesses is weak spots in the metalization pattern as it runs over the steps.

In the Morandi process, a silicon nitride layer covers the device's active areas. When the wafer is oxidized, a thick thermal oxide grows in the field area; no oxide grows on the silicon surface beneath the nitride layer. For planar surfaces, the field oxide is removed by etching (the silicon nitride still is present) and a second oxide is grown, bringing the oxide level up to the silicon edge beneath the nitride layer.

Source and drain areas then are opened for diffusion by removing the nitride and conventional processing of the devices. Then the source and drain diffusions are performed, the nitride is removed from the gate, and the gate is oxidized.

However, Morandi's method of oxide isolation requires two oxidation steps. So another semiconductor specialist, J.A. Appels[2] of N.V. Philips Gloeilampenfabrieken in the Netherlands, directed his goal toward achieving a simplified process. Instead of double oxidation, he used a shallow silicon etch followed by a single oxidation to achieve a plane structure (Fig. 2). The principal advantage here is reduction of the total time required for oxidation: oxides 1 to 2 microns thick could be grown within 2 to 16 hours. (A planar MOS structure made this way is shown in Fig. 3.)

But more significantly, Appels extended his single oxidation method to fabricate discrete bipolar transistors; the structure is shown in Fig. 4. When the thick field oxide is formed, the nitride layer covering the top of the silicon mesa is removed with phosphoric etch. Since this etch attacks oxide slowly, the field oxide is relatively untouched during nitride removal. A p-type base then is diffused into the mesa and is covered with a thin layer of thermally grown

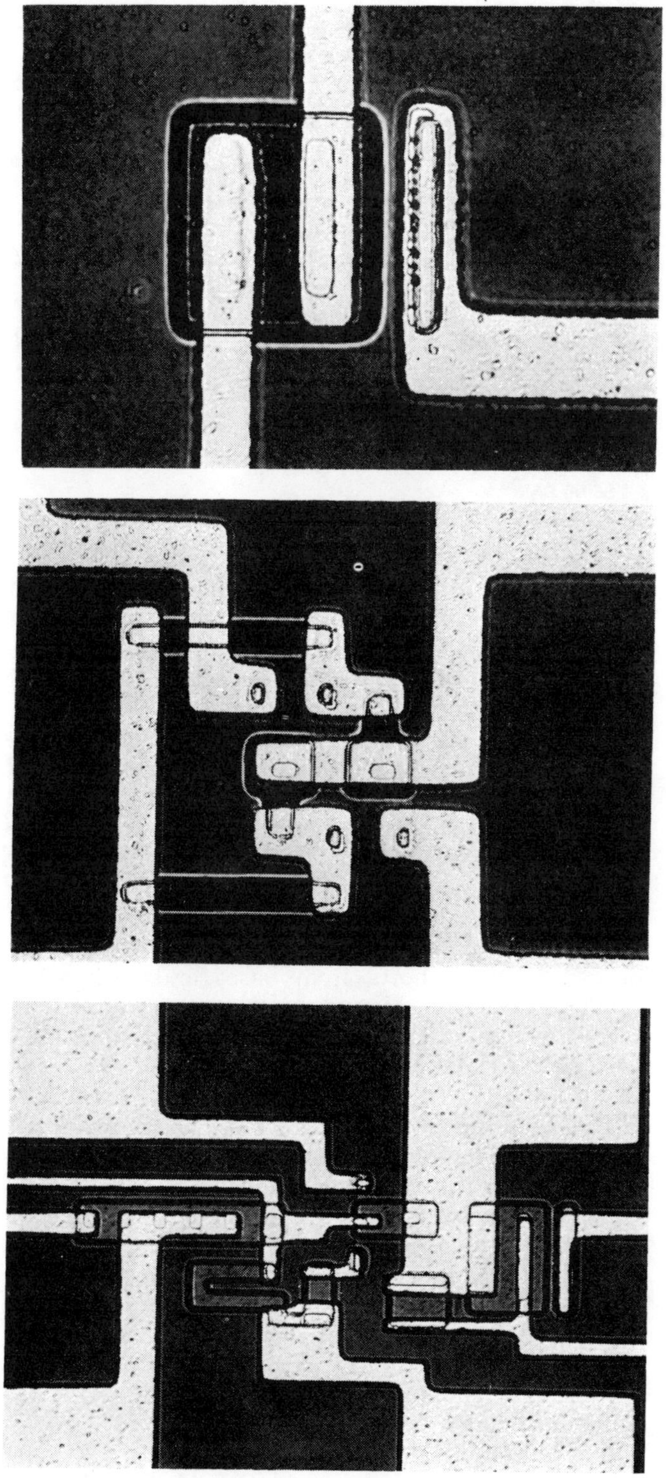

Their own devices. Using new oxidized isolation process, Fairchild Semiconductor has fabricated ICs intended for memory applications. The top photomicrograph shows a double-diffused transistor with a 1-mil by 0.2-mil emitter and a single base strip; it operates at bipolar logic levels and speeds but with greatly reduced size. Such devices comprise a 12.5-mill^2 diode-coupled cell (center) using the oxide isolation process with self-aligned emitters; conventionally processed cells would occupy 30 mil^2. Further area reduction is possible by reducing metal line widths and spaces. Decoder/word-driver (bottom), used to drive cells in memory, is only 22 mil^2.

oxide. Finally, an emitter region is cut into this oxide. The transistor is completed with an emitter deposition and diffusion, a contact masking step, metalization, and a metal masking step.

But important as Appels' work is, it's applicable only to discrete transistors: since the n-type collector also is the substrate, the device clearly cannot be integrated; a contact to one collector would be common to all. A process producing integrated transistors requires a method of isolating the collector while at the same time providing a means of contacting to the collector.

The answer is the new technique, called the Isoplanar process. To appreciate how it satisfies these requirements, consider the conventional diffused isolation npn transistor shown in cross-section in Fig. 5. In this structure, the collector is buried beneath an epitaxial layer, through which contact is made, and a diffused p-type region isolates the collector region of one transistor from the collector region of an adjacent device. With the oxidized isolation, the p regions are replaced by selectively grown thermal oxide regions formed to the depth of the buried collector, resulting in the cross-section shown in Fig. 6. The electrical contact to the buried collector is surrounded by an additional oxide region that is located between the collector sink area and the base region.

Analyses of the new structure makes it apparent that good isolation is obtained with the oxide regions without having to separate the isolation region from the transistor base. The region between the p+ isolation and the base in Fig. 5 can be eliminated entirely, bringing with it a 40% saving in valuable chip area.

But more efficient utilization of space is only part of the promise. The other part is the anticipated lower fabricating costs resulting from simplier masking processes.

To picture how the masking is simplified, remember that after the oxidation step, the nitride layer is still present on the wafer. It can be removed as required

Typical oxidized-isolation transistor characteristics

	Epitaxial base	Double diffused
Static forward current transfer ratio (h_{FE}) @ 1 mA	50-100	50-100
Emitter-to-collector breakdown voltage with base open (LV_{CEO}) @ 1 mA	7 V	10 V
Breakdown voltage, collector to base, emitter open (V_{BR}^{CBO}) @ 10 A	22 V	30 V
Saturation voltage (V_{sat}) @ 5 mA	0.13 V	0.27 V
Maximum frequency of oscillation (small signal) (ft_{max})	1-1.5 GHz	1.5-3 GHz
Carrier transit time (τ_t)	6-9 ns	13-20 ns

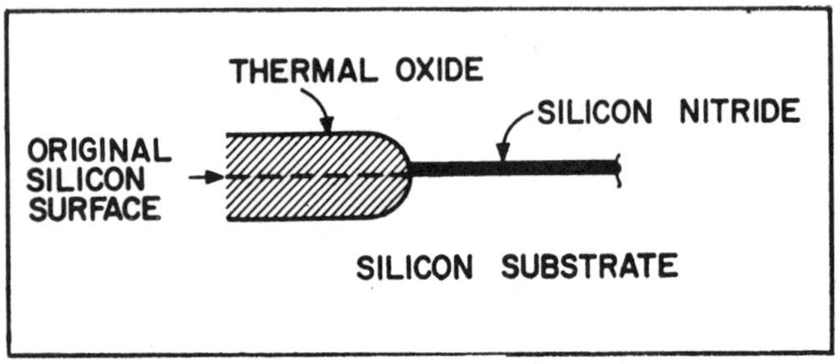

1. **New masks.** First developed for MOS technology, nitride layer deposited on silicon wafer lets thick oxide grow in field area, not beneath nitride.

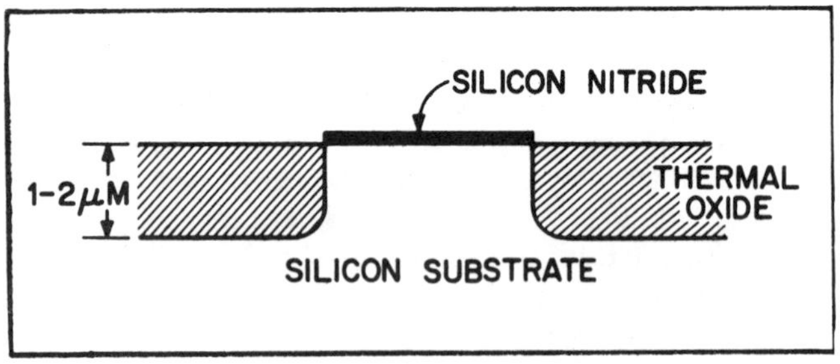

2. **Flat.** Early nitride masking required two oxide steps. To avoid this, a shallow etch first can be made in the silicon, followed by a single oxidation.

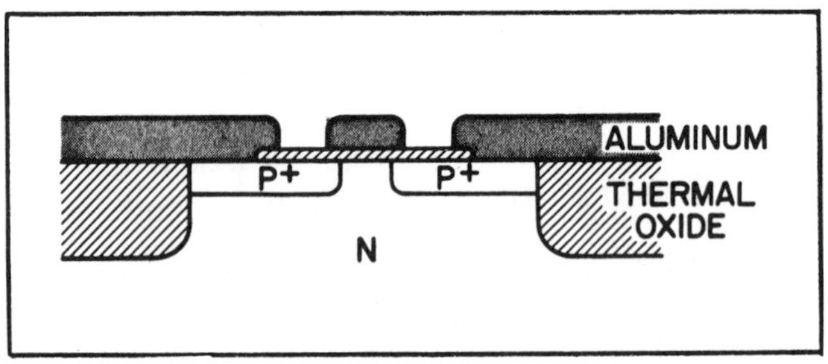

3. **Finished.** MOS device is completed by removing the nitride and performing the source and drain diffusions in the conventional manner.

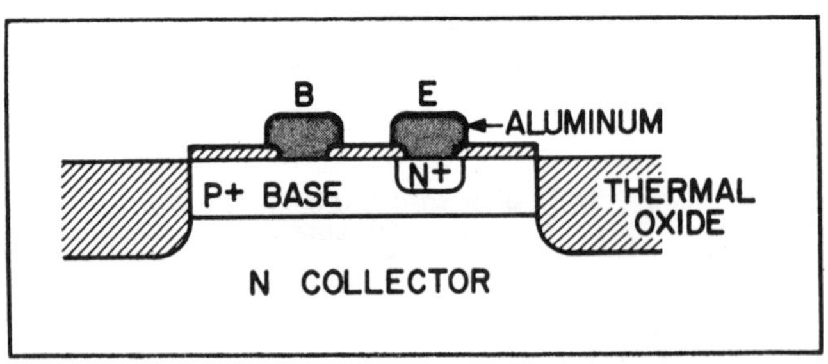

4. **Bipolar bound.** Applied to bipolar devices, once the thick field oxide is formed, the nitride layer covering the top of the silicon mesa is removed with phosphoric etch. A p-type base then is diffused into the mesa and covered with a thin layer of thermally grown oxide. Finally an emitter region is cut into this oxide and the transistor is completed with an emitter predeposition and diffusion, a contact masking step, metalization, and a metal masking step. Suitable for only discrete devices, a method was needed to isolate the n-type collector for IC applications.

Performance comparison

	Cell size	Total chip area per bit	Access time	Total power per bit
Standard isolation 256-bit RAM	30 mil²	$\frac{114 \times 146}{256}$ =65 mil²	80 ns	2 mW
Silicon gate MOS 256-bit static RAM	25 mil²	$\frac{112 \times 128}{256}$ =56 mil²	1 μs	exercised: 2 mW standby: 50 μW
Projected oxidized-isolation 256-bit RAM	12.5 mil²	$\frac{78 \times 84}{256}$ =25 mil²	80 ns	2 mW

with a selective phosphoric etch that attacks nitride and not oxide. Thus, subsequent masking steps defining the base, emitter, and collector sink diffusions, can be made with masks in sequential steps, with only minimal concern for errant diffusion in the unwanted areas, since these areas are protected by the previously applied nitride film or field oxides already grown in the isolation regions.

Yield also is improved because the new structures are less sensitive to the effects of pinholes formed during the isolation masking process. In standard structures, an oxide pinhole could cause low collector-isolation, or low collector-to-base breakdown voltage, or could cause shorts from bases and resistors to isolation regions. However, the same faults are minimized or eliminated in the Isoplanar process, because the isolator region does double duty by acting as an insulator.

The oxidized isolation process can provide still another advantage—a self-alignment feature. Since the nitride layers can be etched away selectively without harming oxide layers, the base and collector sinks can be fabricated right out to the oxide, so that the oxide, as an insulator, itself limits the extent of these diffusions. Thus, mask alignment is far less critical. Resistors can be aligned in the same manner, indicating the possibility of preregistration of both passive and active components with respect to the isolation regions.

In addition, a single base opening in a mask could form base regions for several individual transistors separated by properly located oxide isolations. In some cases it may even be possible to eliminate the base mask entirely simply by striping nitride across the entire surface.

To date, two types of oxidized isolation transistors have been fabricated—the double-diffused device shown in Fig. 6 and an epitaxial base transistor shown in Fig. 7. The epitaxial devices are similar to the collector diffused isolation transistors developed by Bell Laboratories, but thanks to the Isoplanar process they occupy only about half the area and have higher emitter-to-collector breakdown. They require 1-to 2-micron, p-type epitaxial layers and a collector

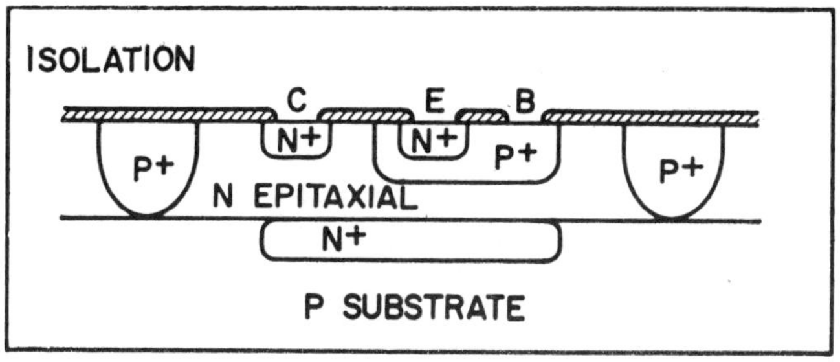

5. Old way. Conventionally fabricated npn transistors use p region for isolation, requiring space between isolation and base.

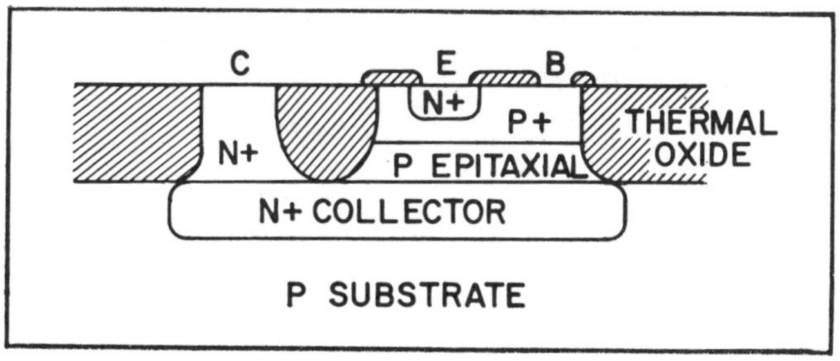

6. New way. By using thermal oxide in place of the p region, base can abut isolation region, saving 40% of chip real estate.

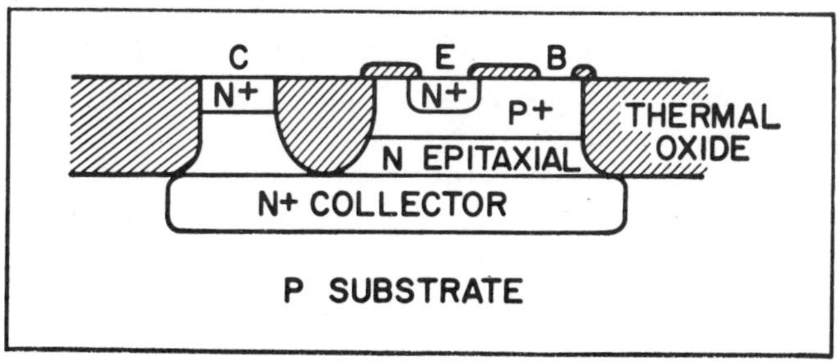

7. Versatile. Both double diffused and epitaxial structures can be made with oxidized isolation. This epitaxial version requires separate collector sink.

sink diffusion. With epitaxial structure it's possible to build oxidized-isolation npn transistors and epitaxial- and base-isolation isolated resistors. Also, the buried collector can provide a crossunder that in some circuits can provide still another layer of interconnection.

Double-diffused devices include npn and pnp transistors, Schottky barrier diodes for Schottky clamped circuits, base and epitaxial resistors, collector resistors —in fact, just about every circuit device that can be made with p-region isolation is possible with the Isoplanar process. Standard masking and diffusion steps can be used; for example, the double-diffused transistors require 1-to 2-micron, n-type epitaxial layers, and a shallow base diffusion, but do not need the separate collector sink diffusion required in the epitaxial-base structure.

Typical characteristics for both epitaxial-base and double-diffused transistors with 1-mil-x-0.2-mil emitter and single-base strips are summarized in table on page 53. Indications are that oxidized-isolation transistors are quite suitable in bipolar digital circuits, yielding devices that operate at standard bipolar logic levels and speed while occupying considerably reduced wafer area.

A large-scale memory best illustrates the chip density available with the oxidized-isolation technique. The table on page 54 compares a projected oxidized-isolation 256-bit random access memory with an available bipolar RAM using standard isolation, and with an available silicon gate MOS static RAM. The oxidized-isolation circuit's access time compares favorably with the other devices but it shows greatly reduced cell and die size. □

Fairchild's Isoplanar plans

Many at the Fairchild Semiconductor division feel the Isoplanar process is the first major development in isolation technology in a decade. And they expect the company to be putting its money into the process. Although it's still in the research and development stage, Fairchild expects to have an Isoplanar 258-bit random access memory by the summer; it will be followed by a 1,024-bit RAM, which Fairchild expects to pack in the same chip size used by Intel's 1,024-bit MOS RAM.

Memory is where the company thinks the Isoplanar device market is—RAMS, read-only memories, shift registers, anywhere chip size is a function of cell size. Because the Isoplanar process reduces cell size, more bits per square mil can be packed on a chip, and more chips on a wafer. Thus, yield is greatly improved, and after all, "yield is the name of the game," says co-author Doug Peltzer.

The reason the company feels that other bipolar integrated circuits, such as logic devices, may not benefit as much from the size reduction afforded by the Isoplanar process is that they already are metal limited; even if transistor size were greatly reduced, the metal needed to interconnect the devices covers almost all the surface area.

One limitation does apply to the Isoplanar process: the breakdown properties of the oxide insulator limit Isoplanar devices to operating voltages below 10 volts. However, Fairchild is quick to point out that most memory circuits operate below 10 V anyway.

References

1. Morandi, F., "The MOS Planox Process," IEEE International Electron Devices Meeting, session 18 (October 29-31, 1969).
2. Appels, J.A., Kooi, Faffen, Schatorje, and Verkuylon, "Local Oxidation of Silicon and its Application in Semiconductor Device Technology," Philips Res. Reports, Vol. 25, (1970) pp. 113-132.
3. Murphy, B.T., Glinski, Gary, and Pedersen, "Collector Diffusion Isolated Integrated Circuits," IEEE Proc. Vol. 57, No. 9, (September 1969) pp. 1,523-1,527.

Bibliography

V.Y. Doo, "Silicon Nitride, A New Diffusion Mask", IEEE Transactions on Electron Devices, Vol. 13, No. 7, (July 1966) pp. 561-563.
F.S. Green Jr. and N.V. Phan, "Megabit Bipolar Memory Systems," ISSCC Digest of Technical Papers, 1970, p. 41.
Rex Rice, Wendall B. Sander, and Frank S. Green Jr., "Design Considerations Leading to the Illiac 4 LSI Process Element Memories," IEEE Journal of Solid State Circuits Vol. SC-5, No. 5 (October 1970), p. 174-181.
Marcian E. Hoff, "Application Considerations for Semiconductor Memories," EEE, (June 1970), p. 62-69.
D.J. Lynes and D.A. Hodges, "A Diode-Coupled Bipolar Transistor Memory Cell," ISSCC Digest of Technical Papers, 1970, p. 44.

Memory Using Diode-Coupled Bipolar Transistor Cells

DENNIS J. LYNES AND DAVID A. HODGES, MEMBER, IEEE

***Abstract*—A diode-coupled bipolar transistor memory cell has been integrated on 64-bit beam-lead chips for use in random-access memory modules ranging up to 16 384 bits storage capacity. Other integrated circuits that perform decoding, driving, and sensing functions are used in the multichip store modules. Access time and READ or WRITE cycle time is 60 ns, with per-bit operating power of 400 μW and standby power under 75 μW. Complete stores manufactured in volume are expected to be cost competitive with magnetic memory systems.**

I. Introduction

NEW developments in bipolar transistor integrated-circuit process technology and memory-cell circuit design, described in this paper, reduce bipolar flip-flop memory-cell power consumption and silicon area to the ranges typical of p-channel insulated-gate field-effect transistor (IFGET) flip-flop designs. This is achieved without sacrificing the high-speed performance advantage of all-bipolar designs.

Considerable attention has been given recently to semiconductor memory designs employing IFGET memory cells in combination with bipolar transistor decoding, driving, and sensing circuits [1]–[4]. Relative to early bipolar transistor memory cells, IGFET memory cells require an order of magnitude less dc power, less silicon area per cell, and fewer steps in the fabrication process. These features have been taken to imply lower costs. Improved interconnection technologies and low IGFET threshold voltages make possible economical, mechanical, and electrically compatible combinations of IGFET memory-cell arrays with bipolar transistor interface circuits [2]–[4]. Such combinations can provide memory access and cycle times as short as 100 ns, about five times shorter than is feasible with all-IGFET memories and only perhaps two times slower than for economical all-bipolar memories. Dynamic memory-cell circuits offer substantial silicon area savings over flip-flop designs, but stored information must be refreshed at intervals on the order of 1 ms [5], [6]. The system reliability of such approaches remains to be demonstrated, and memory-cycle times are typically greater than for the flip-flop design described here.

Complementary IGFET (also known as CMOS) technology is of limited interest for most ground-based memory applications. Reductions in power consumption from the low levels achieved in the work described below appear to offer, from the overall system standpoint, advantages insufficient to justify the additional complexity and cost of complementary IGFET devices.

Manuscript received April 28, 1970; revised May 26, 1970.
D. J. Lynes is with Bell Telephone Laboratories, Inc., Murray Hill, N. J. 07974.
D. A. Hodges is with Bell Telephone Laboratories, Inc., Holmdel, N. J. 07733.

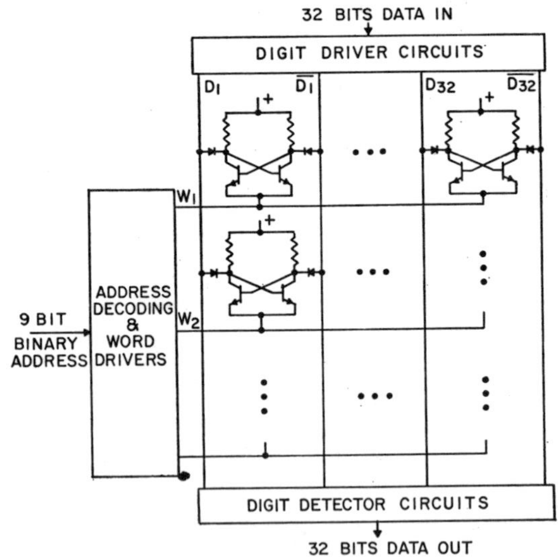

Fig. 1. Block diagram of a 512-word 32-bit-per-word memory plane.

A reexamination of the reliability question on the basis of equal interconnection characteristics and silicon areas per bit for the bipolar and IGFET technologies suggests that the ultimate reliability of all-bipolar memories is likely to be better. This conclusion is supported by the fact that bipolar devices are less susceptible to oxide defects and electrical threshold changes; these failure modes may dominate in a comparison of the two approaches on the basis indicated above. Finally, the potential cost advantage of IGFETs on a per unit silicon area basis has decreased as additional process steps have been added to improve IGFET performance, reliability, and reproducibility, and to simplify interconnections (e.g., by adding beam leads).

Our thesis is that the balance of the above performance, reliability, and cost considerations now favors an all-bipolar approach to memory system design, regardless of storage capacity and cycle-time objectives.

II. Memory-System Description

In order to indicate the type of application intended for the diode-coupled memory cell, Fig. 1 shows a block diagram of a proposed 16 384-bit memory module. It is word organized with a maximum capacity of 512 words, 32 bits per word. A 9-bit binary address (TTL signal levels) selects one out of 512 word lines through specially designed address decoding and word-driving integrated circuits. The data input and data output lines are also at TTL levels. Not shown on this simplified block diagram are the control signal lines, one each to the address decoder, digit driver, and digit detector blocks.

An optional feature, not shown, permits the addition of decoding and gating circuits to the digit driver and

Reprinted from *IEEE J. Solid-State Circuits*, vol. SC-5, pp. 186–191, Oct. 1970.

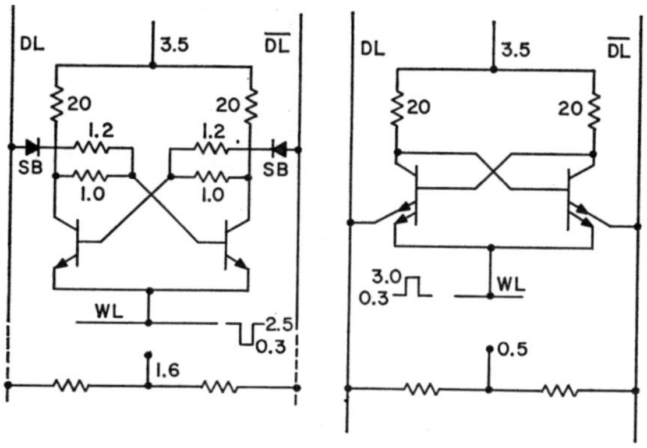

(a) (b)

Fig. 2. (a) Diode-coupled cell. Standby power, 65 μW; READOUT current 0.32 mA; WRITE delay, 5 ns; worst case margin, 0.2 volt. (b) Emitter-coupled cell. Standby power, 800 μW; READOUT current, 0.15 mA; WRITE delay, 20 ns; worst case margin, 0.3 volt. READ and WRITE operations involve differential signals on the digit-line pair (DL, $\overline{DL}$). Resistances are expressed in kilohms.

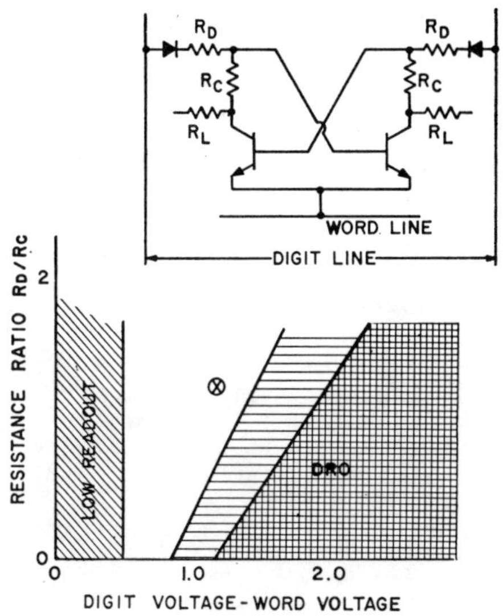

Fig. 3. Margin plot during a READ.

digit detector blocks. In this way, the complete module may be made to function as 16 384 words, 1 bit per word, as 512 words, 32 bits per word, or any intermediate binary configuration.

It is proposed that the complete memory system as shown in Fig. 1, comprising a total of about 350 beam-lead sealed-junction integrated-circuit chips, be assembled on six 1½- by 2-inch metallized ceramic substrates. These in turn will be mounted on a single two-sided printed wiring board about 5 by 6 inches overall. In operation, power dissipation for the complete module will be under 6 watts, less than 0.4 mW per bit. Standby power for the memory cells will be about 1 watt. An alternate approach being studied is a 4–8 kbit memory on a single, more complex ceramic substrate that becomes the system submodule.

III. Cell-Circuit Description and Comparison

The diode coupled cell is compared with the widely used emitter-coupled memory cell in Fig. 2. The cells were designed with similar ground rules, i.e., high values of load resistance for low standby power and sufficient readout current and margins to make them operational over a 0°–100°C temperature range.

The nominal operation of the diode-coupled memory cell (refer to Figs. 1 and 2) is as follows. V_{CC} is held constant at 3.5 volts. Assume that in the standby condition one of the transistors is on. The word lines (WL) are at 2.5 volts. This voltage, in conjunction with the 20-kΩ load resistors, gives a standby power of less than 75 μW. This word-line voltage also determines the back-bias voltage on the Schottky-barrier (SB) diodes connected to the digit lines. The quiescent digit line (DL) voltages are approximately 1.6 volts. During the READ operation, the word line is pulled down, forward biasing one of the Schottky-barrier diodes. A current of approximately 0.3 mA flows onto a digit line through each diode connected to the ON transistor. This current provides a 50-mV readout in 10 ns with a total capacitance per digit line of 60 pF.

During the WRITE operation, the digit line connected to the side of the cell to be written is raised to 2.8 volts by a low impedance driver. The word line is held near ground potential. This causes a current of 1 mA to flow through the cell if the cell transistor connected to this digit line is on. The voltage that is developed across the 1.0-kΩ resistor at the ON transistor is sufficient to turn on the OFF transistor. If the cell is in the desired state prior to the WRITE cycle, no change in state occurs as a result. When the word line is raised to the 2.5-volt standby potential, the diodes become back biased and the WRITE operation is complete.

The multiple-emitter cell is read by raising the word line from 0.3 to 3.0 volts. This action transfers the standby current from the WL to the DL connected to the ON transistor. The cell is written by raising the ON digit line to a positive voltage while the word line is high. When the word line is lowered to 0.3 volt, the cell regeneratively switches to the desired new state.

An important advantage of the diode-coupled cell is that readout current is substantially greater than standby current; this is helpful in obtaining short READ access time. With the multiple-emitter cell, readout current is limited to a value determined by the supply voltage and the load resistor. Furthermore, because the diode-coupled cell is written by overdriving rather than by regeneration as in the multiple-emitter cell, WRITE time can be very short. The emitter-coupled cell consumes a great deal more standby power and is fundamentally slower for both reading and writing.

Fig. 3 shows a dc margin plot for the diode-coupled cell. The vertical axis represents the resistance ratio R_D/R_C, the critical design parameter for this cell. The horizontal axis shows the voltage difference between digit line and word line during a READ operation. The margin (200 mV) is calculated for the case in which manufacturing variations (of resistor values and transistor parameters) and temperature effects combine in the worst way. This margin is considerably greater than the maximum noise amplitude anticipated on any word or digit line in a module assembly [7]. The clear area of the plot shows the safe operating region for the memory cell during a READ

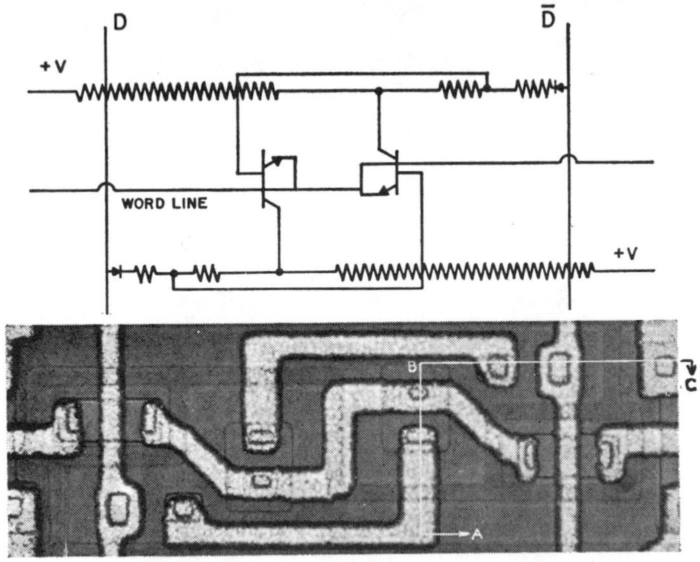

Fig. 4. One diode-coupled cell, 3 by 10 mils. Vertical digit lines are continuous metal; horizontal lines use diffused crossunders.

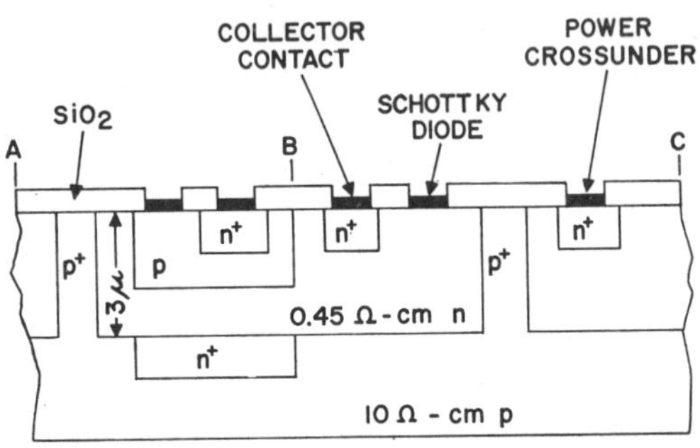

Fig. 5. Section view; interconnecting metallization not shown. All resistors are formed in 3-micron n-epitaxial layer.

operation. The worst case operating point is shown by a circled *X*. The slanted line that is closest to the operating point is the worst case minimum digit line minus word-line voltage as a function of the internal resistance ratio (R_D/R_C) at 100°C. The slanted line further to the right is for 25°C. The region of the plot labeled DRO (destructive readout) is the region into which the operating point must be shifted to write the cell. This is easily done when desired by raising the digit-line voltage. The straight line to the left of the operating point shows the voltage below which the readout current is too low to get operation in a 60-ns cycle time. This line is for 25°C; it shifts to the left for increasing temperature.

IV. Integrated-Cell Layout and Process Technology

Fig. 4 shows a top view of a single memory cell, and Fig. 5 shows a cross section with the important structural features. Minimum metal linewidth is 0.3 mil. Schottky-barrier diodes (PtSi on n-type silicon without guard rings) are used to provide low charge storage, low capacitance (0.03 pF) coupling to the digit lines. Leakage per diode is under 1 nA at 25°C with 3.5 volts reverse bias. These high-quality Schottky diodes are obtained without extra steps in the standard beam-lead process. The total capacitance per bit on each digit line, including the Schottky diodes, all metallization capacitance, and parasitics, is about 0.1 pF; this capacitance is far smaller than for other memory cells in which transistor emitters or collectors provide coupling to the digit lines. Low capacitance on the digit lines is helpful in obtaining high-speed operation of large arrays with low power consumption.

Epitaxial resistors are feasible and attractive because the silane epitaxial process used holds a tolerance of ±10 percent on epitaxial layer thickness and resistivity. The sheet resistance of the 3-micron-thick n-type layer is about 1500 ohms. Thus a total resistance of 44 kΩ is incorporated in each 30-mil^2 cell; cell area would be at least 60 mils^2 if standard diffused resistors were used. The temperature coefficient of resistivity for these epitaxial resistors is about +0.6 percent/°C. Because all resistors are formed in the same epitaxial layer and thus track with temperature, it is feasible to design for operation under worst case conditions over at least the range 0°–100°C. Low-barrier Schottky diodes as load elements would reduce cell area approximately 20 percent, but at the expense of some increase in processing complexity and some sort of temperature stabilization for the memory in operation [8].

It is interesting to compare the area of the diode coupled memory cell with alternatives. When the emitter-coupled cell is designed with the same tolerances, using epitaxial sheet resistors, it too occupies about 30 mils^2. Similarly, the common six-transistor IGFET flip-flop cell laid out with the same tolerances also occupies about this same silicon area.

A 64-bit beam-lead chip is illustrated in Fig. 6. Chip size is limited to 50 by 60 mils to minimize the total cost of a complete module, as discussed below. The chip is organized as 16 words of 4 bits each. Digit lines go off the chip at both sides; alternate word lines, power, and ground go off at top and bottom. Note that the digit lines are continuous metallization on the chip, while word lines and power lines employ emitter-diffused crossunders. The digit-line beam leads are 2 mils wide on 4-mil centers, in contrast to the 6–10-mil interconnection centers common with flip-chip or wire-bonded techniques. Experience to date suggests that the reliability of beam-lead devices is not primarily limited by interconnections [9]. Hence, the large number of leads per chip poses no problem.

Sixty-four of the beam-leaded chips of Fig. 6 are face-down bonded to one 1.5- by 2-inch alumina substrate with titanium-gold metallization as shown in Fig. 7. Two rows of eight decoder-word driver chips are added at top and bottom, forming an assembly of 128 32-bit words. Chips are attached to the substrate with a single alignment and bonding operation per chip. Chips damaged during bonding (about 1 out of 50) are readily replaced.

Single-level substrate metallization within the array of chips is adequate because word and power lines on the substrate can pass under the memory chips. (Digit-line continuity is maintained via metallization alternately on substrate and chips.) Outside the area occupied by the chips, crossovers are necessary; these will be provided by the Lepselter technique [10]. Patterns are defined in the

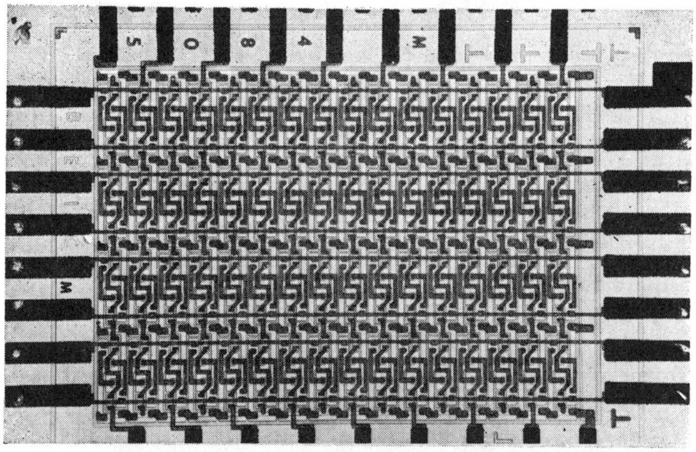

Fig. 6. 64-bit memory chip, 50 by 60 mils. Digit lines go off chip at both sides; alternate word lines go off chip at top and bottom.

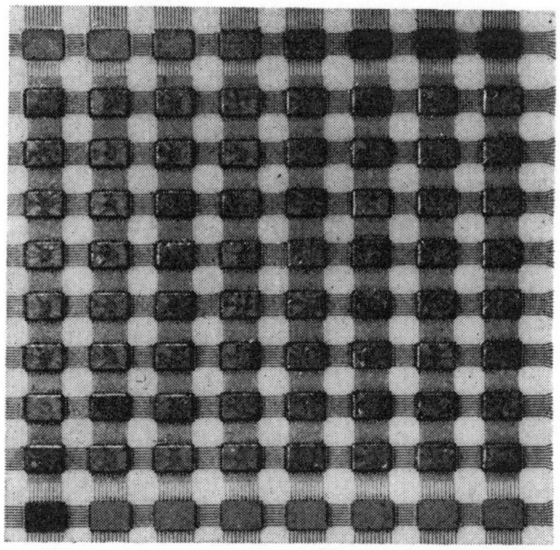

Fig. 7. 4096-bit memory array. Vertical word lines on 3-mil centers run under chips. Top and bottom rows are word decoder-driver chips.

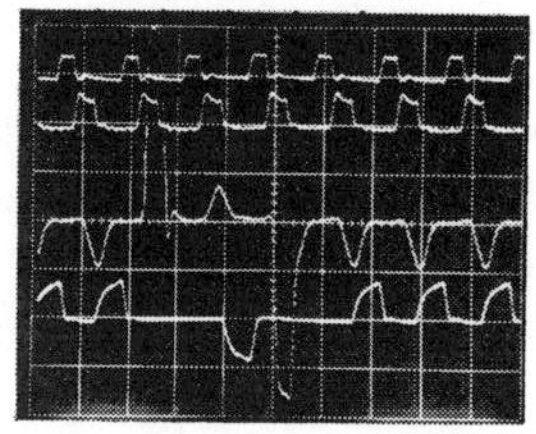

Fig. 8. Signals in a skeletonized 16 384-bit module.

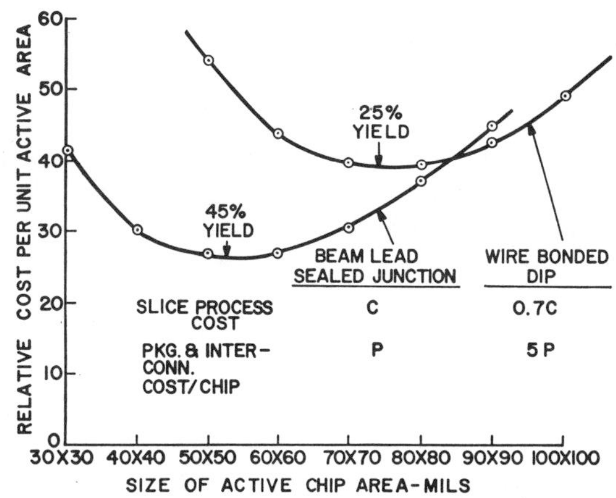

Fig. 9. Comparison of costs for completely assembled, tested memory modules; costs all stated on per-unit silicon area basis.

substrate metallization using conventional photolithographic and chemical etching techniques. Four of the memory assemblies of Fig. 7 are to be interconnected with two assemblies of driving, sensing, and control circuits on a single printed wiring board to form a complete 16 384-bit memory module.

V. Memory-System Performance

Fig. 8 shows the operating waveforms for a fully loaded skeletonized version of the 512-word 32-bit-per-word memory module. The address register gate input, shown at the top, combined with an address input, selects one of the 512 word lines. (The address register is included within the address decoding block of Fig. 1.) The digit-line unclamp signal frees the digit line so that the differential voltage from the memory cell may be coupled to the detector. The differential signal on the digit-line pair is shown next. The pattern shown is the READ of a "0," followed by the WRITE of a "1," followed by the READ of a "1," followed by the WRITE of a "0" and subsequent READ cycles. The bottom trace is the two-rail detected output, displayed as a differential signal for convenience. The detector used is described elsewhere [11]. Cycle time is seen to be 60 ns. Access time, measured from the 50 percent point on the address register gate input to the 50 percent point of the detector output, is under 50 ns as shown here. Under worst case conditions, access time may increase to 60 ns.

VI. Memory Module Cost

Fig. 9 is a plot of total memory module cost as a function of chip size for both beam-lead sealed-junction and conventional (wire-bonded chips in dual in-line packages) interconnection techniques. The horizontal axis represents active chip area, that is, the contiguous area within the borders of a chip that is occupied by circuit elements, essential clearances, and interconnecting metallization. The vertical axis represents relative cost per unit active area for a completely assembled memory module. Typical specific costs, and the method used to calculate the curves shown here, are shown elsewhere [12], [13].

An average effective spot defect density of 300 per square inch, determining chip yield as a function of chip area, is assumed in each case. Chip test cost is assumed equal for the two cases. The slice process cost for beam-lead technology is assumed to be C, while the corresponding figure for the somewhat simpler standard process with aluminum metallization is assumed to be 30 percent less. The largest difference arises in the cost of packaging, interconnection, substrates, and repairs, all reduced to a per-chip basis. In this respect, we estimate that conventional technology is at least 5 times costlier than beam-lead technology.

Fig. 9 shows that beam-lead chips should be designed with an active area of about 55 by 55 mils. At this size, a proven chip design should be manufacturable at a chip yield of 45 percent. In contrast, the dual in-line package technology results in an optimum chip size of about 75 by 75 mils, corresponding to a chip yield of about 25 percent. At the respective optimum points, the beam-lead approach is about 35 percent cheaper per unit active area than the conventional approach.

Cost per unit active area is not the sole factor determining memory module cost. Also important is the total active area per bit required to provide the complete memory function. The design described here, which removes all decoding, driving, and sensing functions from the memory chip, saves a substantial amount of silicon area because these interface circuit functions may be shared over a large number of memory bits. This design consumes only 30 percent of the total silicon area per bit in interface circuits. Designs that include most or all of the interface functions on the memory chip typically consume 50 percent or more of the total silicon area per bit in providing these functions.

Considering both cost per unit area and total area required, the memory design described above is expected to cost only $\frac{1}{3}$ to $\frac{1}{2}$ as much per bit as a functionally equivalent memory made up of fully decoded chips in dual in-line packages. The actual cost per bit in volume production should be equal to or less than that for 1-μs core memories in 10^6-bit modules.

VII. Summary and Conclusions

A diode-coupled bipolar transistor memory cell has been described. It is superior to other bipolar transistor flip-flop memory cells in power consumption and high-speed performance. It is competitive with p-channel IGFET flip-flop memory cells in packing density and power consumption, while offering appreciably better performance and probably better reliability.

The proposed 16 384-bit memory module employing the diode-coupled cell is word organized, and is partitioned so that no decoding, driving, or sensing functions are included on the memory chips. This approach achieves excellent high-speed performance, while reducing significantly total silicon area requirements and power consumption relative to other designs. A large number of interconnections within the module are inevitable with this approach. Experience with beam-lead sealed-junction technology indicates that overall module cost and reliability are not primarily determined by the number of beam-lead interconnections. Thus it appears that the chosen form of organization and partitioning will result in a highly desirable combination of cost, reliability, power consumption, and performance characteristics for memory modules of 10^3 bits or greater capacity.

Acknowledgment

W. C. Slemmer, J. D. Heightley, and J. H. Wuorinen contributed greatly to the circuit and store design. Chip design and fabrication have been carried out by E. N. Fuls, T. E. Saunders, and N. C. Wittwer. T. R. Finch, M. R. Barber, and many other colleagues provided indispensable suggestions, assistance, and encouragement.

References

[1] P. Pleshko and L. M. Terman, "An investigation of the potential of MOS transistor memories," *IEEE Trans. Electron. Comput.*, vol. EC-15, pp. 423–427, August 1966.

[2] D. A. Hodges, "Large-capacity semiconductor memory," *Proc. IEEE*, vol. 56, pp. 1148–1162, July 1968.

[3] E. J. Alexander, "P-channel IGFET memories," *1969 IEEE Internat. Conv. Digest*, vol. 69C, pp. 260–261, March 1969.

[4] T. W. Hart, Jr., D. W. Hillis, J. Marley, R. C. Lutz, and C. R. Hoffman, "A main frame semiconductor memory for fourth generation computers," *Fall Joint Computer Conf., AFIPS Proc.*, vol. 35, pp. 479–488, November 1969.

[5] W. M. Regitz and J. Karp, "Three-transistor-cell, 1024-bit, 500-ns MOS RAM," *IEEE J. Solid-State Circuits*, this issue, pp. 181–186.

[6] S. Waaben and H. Waggener, "100-ns electronically variable semiconductor memory using two diodes per memory cell," *IEEE J. Solid-State Circuits*, this issue, pp. 192–196.

[7] B. T. Murphy, S. M. Neville, and R. A. Pedersen, "Simplified bipolar technology and its application to systems," *IEEE J. Solid-State Circuits*, vol. SC-5, pp. 7–14, February 1970.

[8] D. A. Hodges, M. P. Lepselter, D. J. Lynes, R. W. MacDonald, A. U. MacRae, and H. A. Waggener, "Low-power bipolar transistor memory cells," *IEEE J. Solid-State Circuits*, vol. SC-4, pp. 280–284, October 1969.

[9] D. S. Peck, "Reliability of beam-lead sealed-junction devices," presented at the Ann. Joint Symp. on Reliability, Chicago, Ill., January 21–23, 1969.

[10] M. P. Lepselter, "Air-insulated beam-lead crossovers for integrated circuits," *Bell Syst. Tech. J.*, vol. 47, pp. 269–271, February 1968.

[11] W. C. Slemmer, "High-speed low-power strobed comparator," *IEEE J. Solid-State Circuits*, this issue, pp. 215–220.

[12] R. N. Noyce, "Making integrated electronics technology work," *IEEE Spectrum*, vol. 5, pp. 63–66, May 1968.

[13] J. H. Wuorinen and D. A. Hodges, "Component technologies for future systems," in *Information Processive 68*. Amsterdam: North-Holland, 1969, pp. 787–795.

Bipolar LSI Memory **3101A**

INTEL CORP. 365 Middlefield Road, Mountain View, California 94040 • (415) 969-1670

MAY 1971

HIGH SPEED FULLY DECODED 64 BIT MEMORY ---35 nsec MAX.

PACKAGE OUTLINE CERAMIC (C3101A)

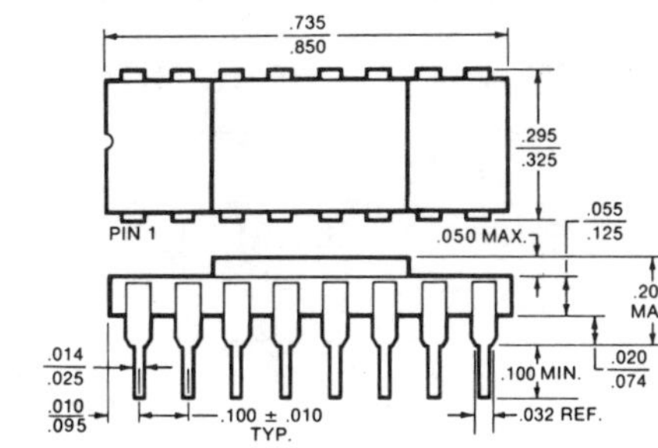

PACKAGE OUTLINE PLASTIC (P3101A)

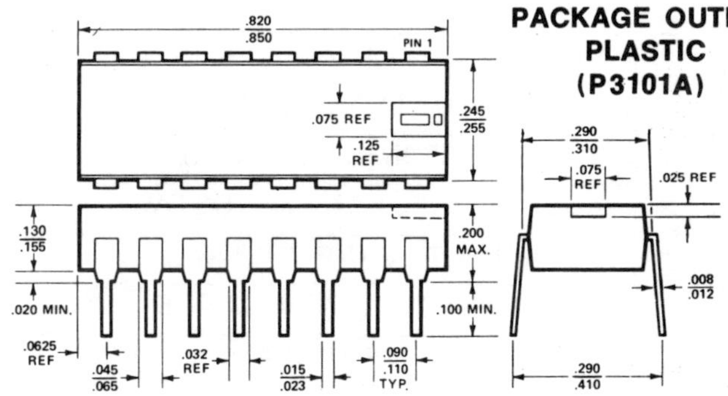

- **Fast Access Time -- 35 nsec. max. over 0-85° C Case Temperature.**
- **DTL and TTL Compatible -- Low Input Load Current: 0.25mA. max.**
- **OR-Tie Capability -- Open Collector Outputs.**
- **Simple Memory Expansion through Chip Select Input -- 17 nsec. max. over 0-85° C Case Temperature.**
- **Fully Decoded -- on Chip Address Decode and Buffer.**
- **Minimum Line Reflection -- Low Voltage Diode Input Clamp.**
- **Ceramic and Plastic Package -- 16 Pin Dual In-Line Configuration.**

PIN CONFIGURATION

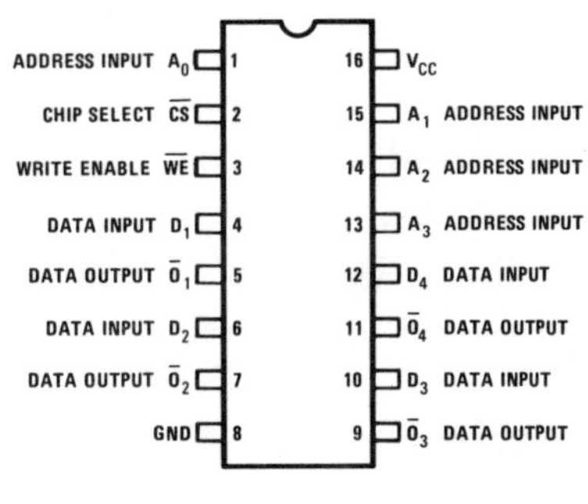

LOGIC SYMBOL

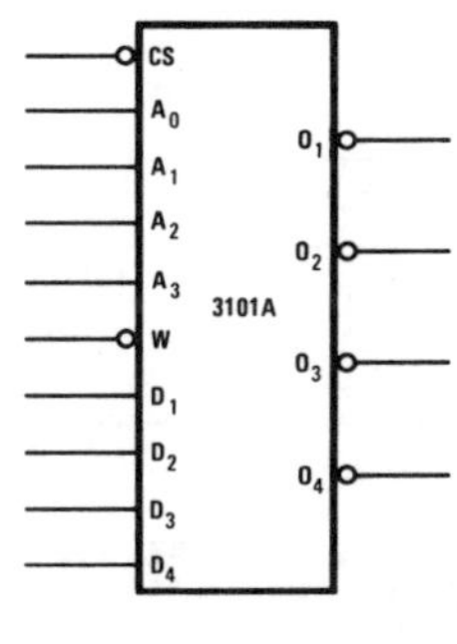

The Intel 3101A is a 64 bit random access memory. Its high speed makes it ideal in scratch pad applications. It is a pin for pin replacement with the Intel 3101. An unselected chip will not generate noise at its output during writing of a selected chip.

The use of Schottky barrier diode clamped transistors to obtain fast switching speeds results in higher performance than equivalent devices made with gold diffusion processes.

The Intel 3101A is packaged in either a hermetically sealed 16 pin ceramic package, or in a low cost silicone package, and its performance is specified over a case temperature range from 0° C to 85° C.

This device is organized as a 16 word by 4 bit array.

The storage cells are addressed through an on chip 1 of 16 binary decoder using four input address leads.

A separate Chip Select lead allows easy selection of an individual package when outputs are OR-tied.

In addition to the address leads and the Chip Select lead, there is a write input which allows data presented at the data leads to be entered at the addressed storage cells.

Maximum Guaranteed Ratings

Case Temperature Under Bias:	Ceramic	−65°C to +125°C
	Plastic	−65°C to +85°C
Storage Temperature:	Ceramic	−65°C to +160°C
	Plastic	−65°C to +150°C
All Output or Supply Voltages		−0.5 to +7 Volts
All Input Voltages		−0.5 to +5.5 Volts
Output Currents		100 mA

COMMENT:

Stresses above those listed under "Maximum Guaranteed Rating" may cause permanent damage to the device. This is a stress rating only and functional operation of the device at these or at any other condition above those indicated in the operational sections of this specification is not implied. Exposure to absolute maximum rating conditions for extended periods may affect device reliability.

D.C. Characteristics T_{CASE} = 0°C to 85°C, V_{CC} = 5.0V ±5% unless otherwise specified.

SYMBOL	TEST	MIN.	MAX.	UNIT	CONDITIONS
I_{FA}	ADDRESS INPUT LOAD CURRENT		−0.25	mA	V_{CC}=5.25V, V_A=0.45V
I_{FD}	DATA INPUT LOAD CURRENT		−0.25	mA	V_{CC}=5.25V, V_D=0.45V
I_{FW}	WRITE INPUT LOAD CURRENT		−0.25	mA	V_{CC}=5.25V, V_W=0.45V
I_{FS}	CHIP SELECT INPUT LOAD CURRENT		−0.25	mA	V_{CC}=5.25V, V_S=0.45V
I_{RA}	ADDRESS INPUT LEAKAGE CURRENT		10	μA	V_{CC}=5.25V, V_A=5.25V
I_{RD}	DATA INPUT LEAKAGE CURRENT		10	μA	V_{CC}=5.25V, V_D=5.25V
I_{RW}	WRITE INPUT LEAKAGE CURRENT		10	μA	V_{CC}=5.25V, V_W=5.25V
I_{RS}	CHIP SELECT INPUT LEAKAGE CURRENT		10	μA	V_{CC}=5.25V, V_S=5.25V
V_{CA}	ADDRESS INPUT CLAMP VOLTAGE		−1.0	V	V_{CC}=4.75V, I_A=−5.0 mA
V_{CD}	DATA INPUT CLAMP VOLTAGE		−1.0	V	V_{CC}=4.75V, I_D=−5.0 mA
V_{CW}	WRITE INPUT CLAMP VOLTAGE		−1.0	V	V_{CC}=4.75V, I_W=−5.0 mA
V_{CS}	CHIP SELECT INPUT CLAMP VOLTAGE		−1.0	V	V_{CC}=4.75V, I_S=−5.0 mA
V_{OL}	OUTPUT "LOW" VOLTAGE		0.45	V	V_{CC}=4.75V, V_W=V_S=0V V_D=2.5V, I_{OL}=15 mA
I_{CEX}	OUTPUT LEAKAGE CURRENT		100	μA	V_{CC}=5.25V, V_{CEX}=5.25V V_S=2.5V
I_{CC}	POWER SUPPLY CURRENT		105	mA	V_{CC}=5.25V, V_A=V_S=V_D=0V
V_{IL} *	INPUT "LOW" VOLTAGE		0.85	V	V_{CC}=5.0V
V_{IH} *	INPUT "HIGH" VOLTAGE	2.0		V	V_{CC}=5.0V

*This parameter is periodically sampled and is not 100% tested.

Typical Characteristics

OUTPUT CURRENT VS. OUTPUT "LOW" VOLTAGE

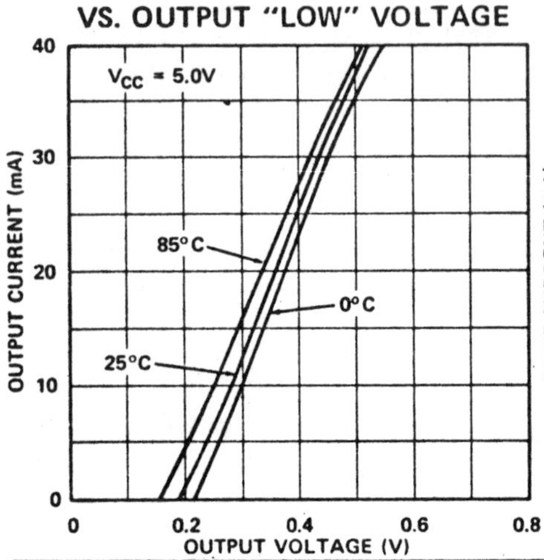

INPUT CURRENT VS. INPUT VOLTAGE

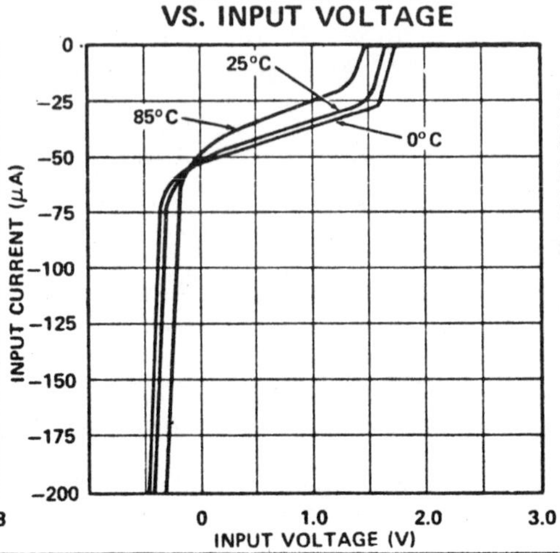

INPUT THRESHOLD VOLTAGE VS. CASE TEMPERATURE

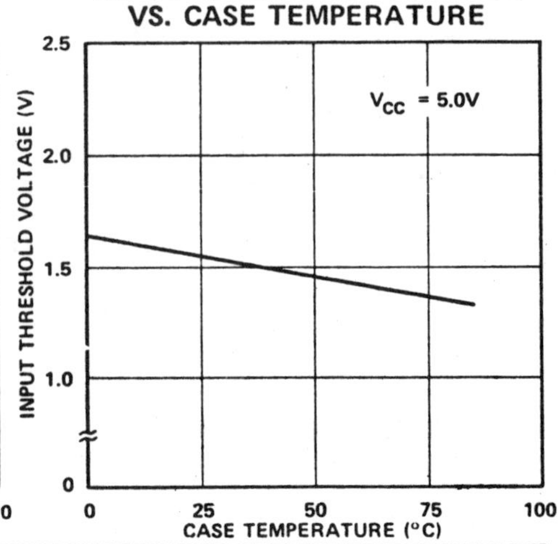

Switching Characteristics

Conditions of Test:

Input Pulse amplitudes: 2.5V

Input Pulse rise and fall times of 5 nanoseconds between 1 volt and 2 volts

Speed measurements are made at 1.5 volt levels

Output loading is 15mA and 30 pF

15 mA Test Load

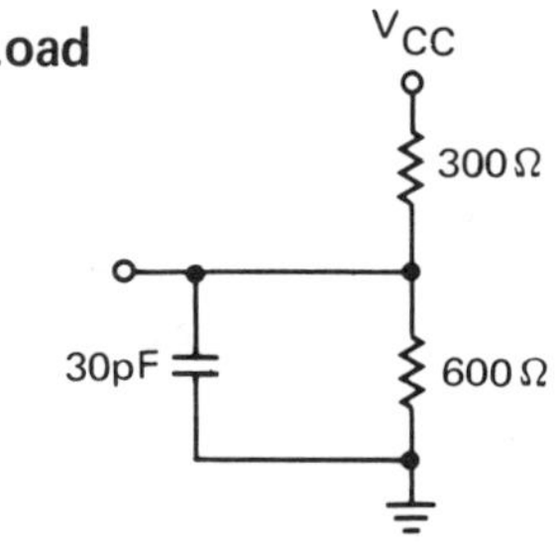

READ CYCLE

Address to Output Delay

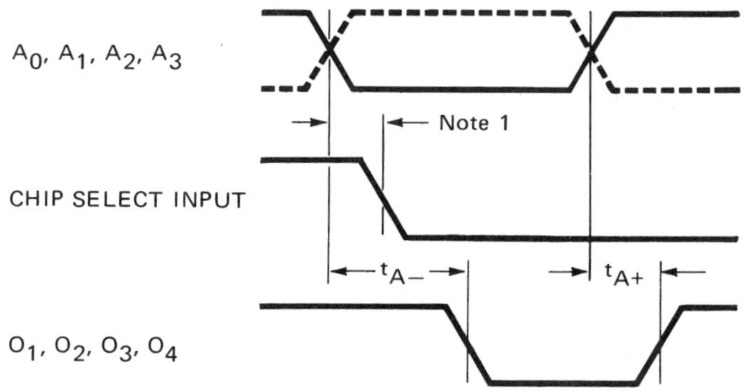

Chip Select to Output Delay

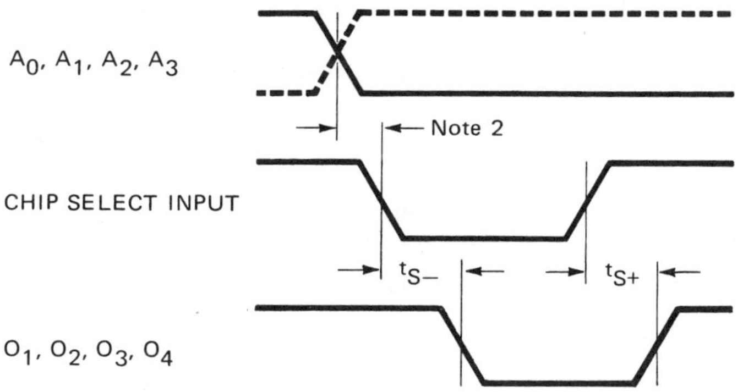

WRITE CYCLE

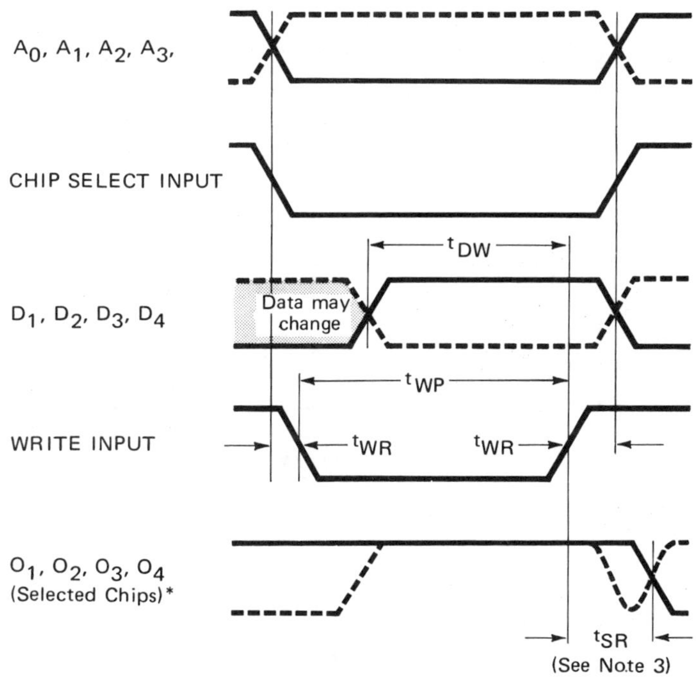

*Outputs of unselected chips remain high during write cycle.

NOTE 1: Chip select input may occur up to 18 ns after the address without affecting the maximum address to output delay.
NOTE 2: Addresses need only be stable for 18 ns before chip select to obtain the maximum chip select to output delay time.
NOTE 3: t_{SR} is associated with a read cycle following a write cycle and does not affect the access time.

A.C. Characteristics T_{CASE} = 0°C to +85°C, V_{CC} = 5.0V ±5% unless otherwise specified

READ CYCLE				WRITE CYCLE			
SYMBOL	TEST	LIMITS (ns) MIN.	LIMITS (ns) MAX.	SYMBOL	TEST	LIMITS (ns) MIN.	LIMITS (ns) MAX.
t_{S+}, t_{S-}	CHIP SELECT TO OUTPUT DELAY	5	17	t_{SR}	SENSE AMPLIFIER RECOVERY TIME		35
t_{A-}, t_{A+}	ADDRESS TO OUTPUT DELAY	10	**35**	t_{WP}	WRITE PULSE WIDTH	25	
C_{IN}**	INPUT CAPACITANCE C3101A (All Pins) P3101A	6(typ.) 5(typ.)		t_{DW}	DATA-WRITE OVERLAP TIME	25	
C_{OUT}**	OUTPUT CAPACITANCE C3101A P3101A	8(typ.) 7(typ.)		t_{WR}	WRITE RECOVERY TIME	0	

** This parameter is periodically sampled and is not 100% tested. Condition of measurement is f = 1 MHz, V_{bias} = 2V, and T_A = 25°C.

Typical Characteristics

ADDRESS TO OUTPUT DELAY
VS.
CASE TEMPERATURE

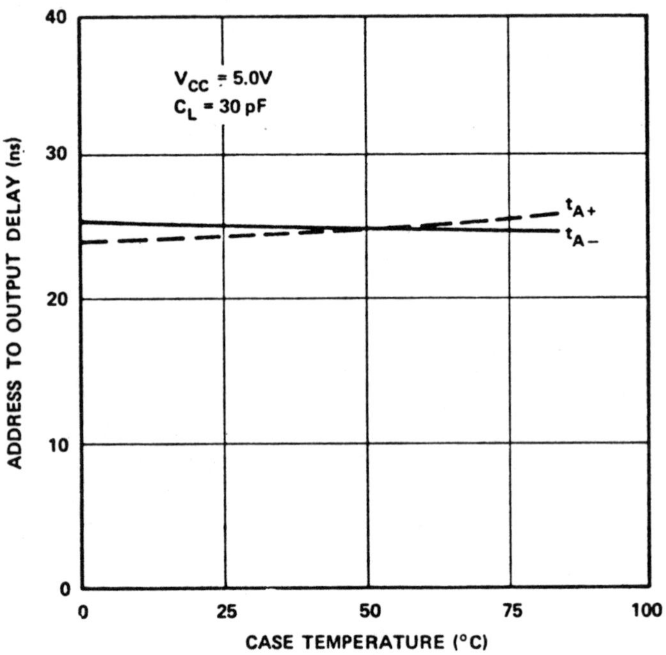

CHIP SELECT TO OUTPUT DELAY
VS.
CASE TEMPERATURE

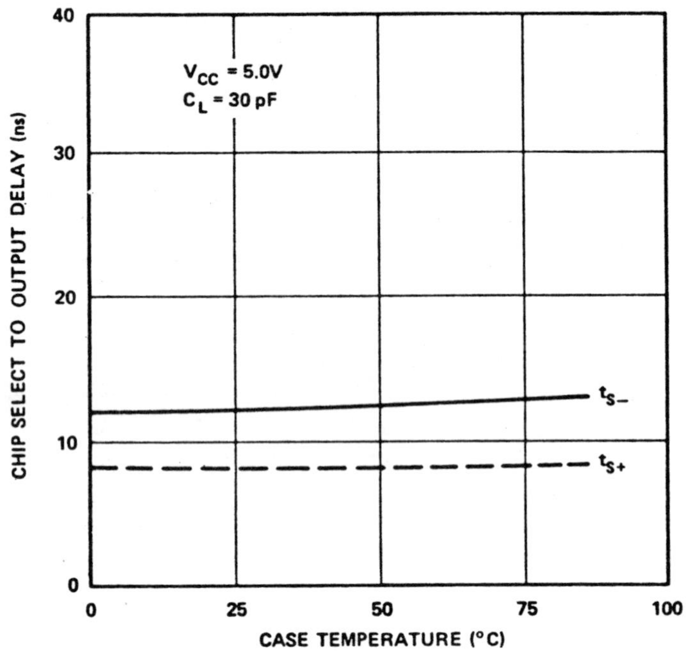

ADDRESS & CHIP SELECT TO OUTPUT DELAY
VS.
LOAD CAPACITANCE

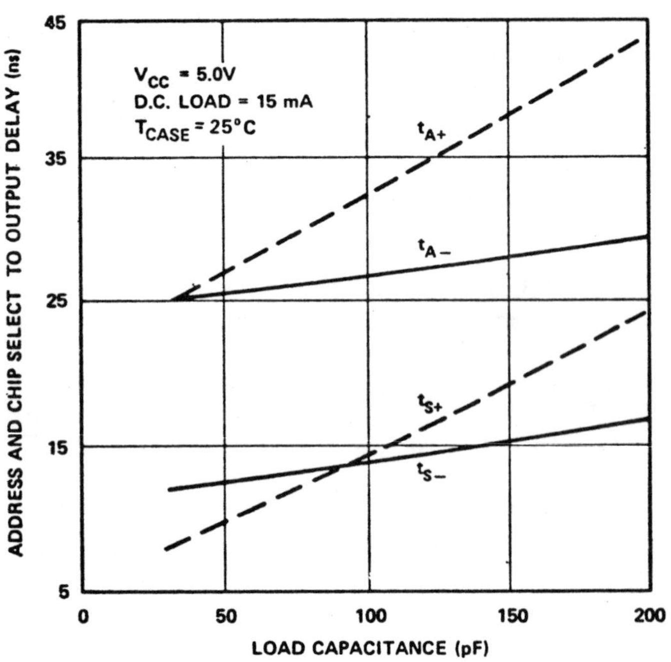

WRITE PULSE WIDTH & SENSE
AMPLIFIER RECOVERY TIME
VS. CASE TEMPERATURE

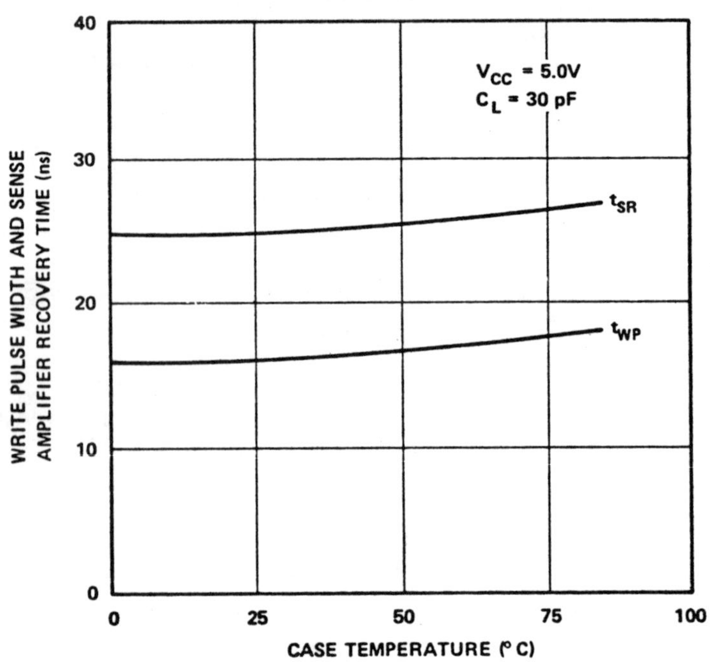

Circuit Schematic

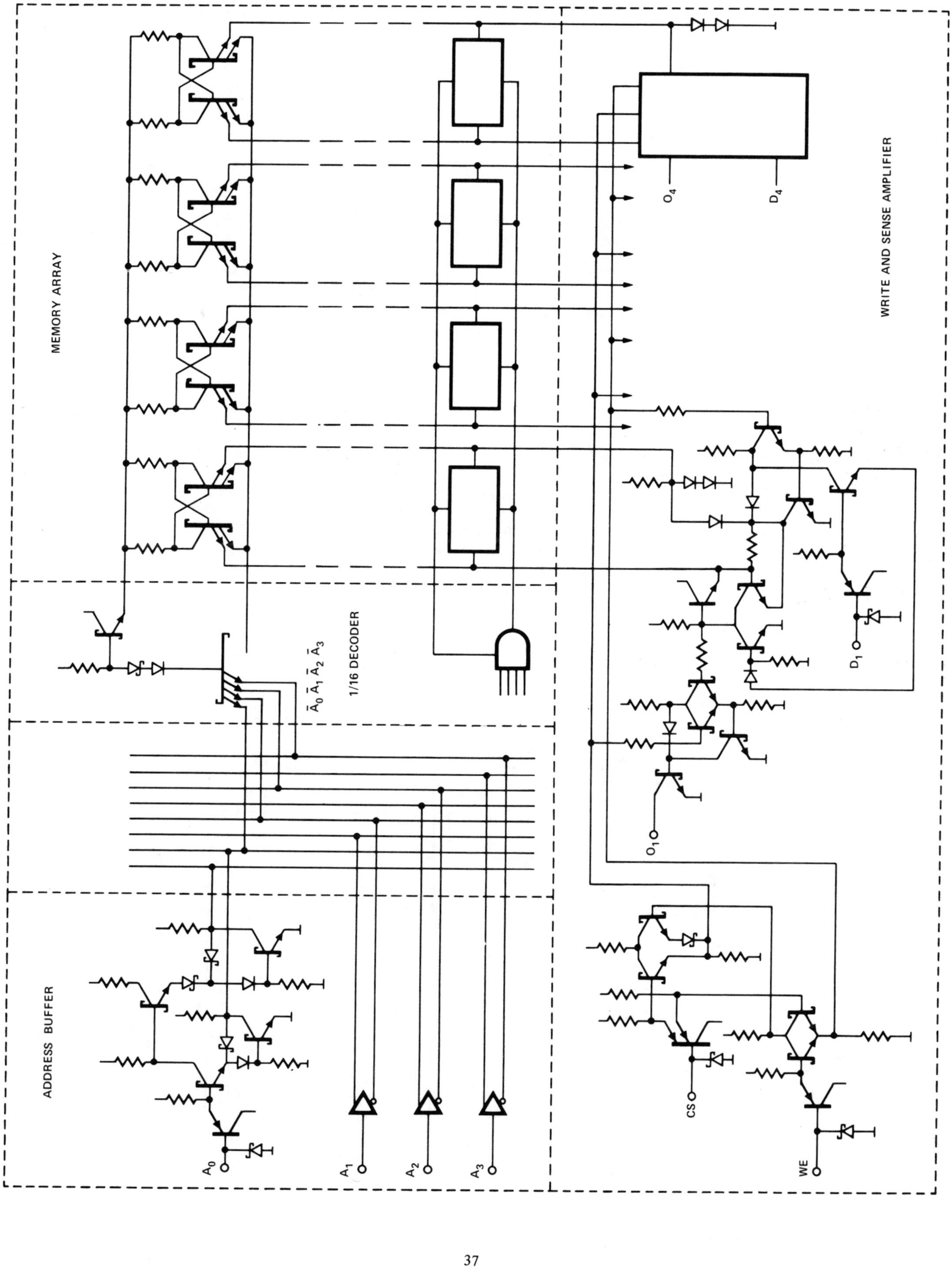

Application Information

Maximum as well as minimum delay times are specified in 3101A so that pipe lined systems may be designed with guaranteed race free conditions. Because the 3101A outputs go "high" when the WRITE pulse is enabled, the data input bus may be tied to the data output bus to create a single I/O port memory system. However, the WRITE pulse width may require an additional 15 ns in this configuration to compensate for the turn-off of data outputs after WRITE input is enabled.

Many of the requirements for small high-speed memories in digital computers and processors can now be more easily filled using Intel's 64 bit high-speed memory. Full 64 bit capacity, 16 word x 4 bit configuration, complete address decoding and wired-or output capability permit a wide variety of small to medium sized buffer memories to be constructed with a minimum of additional components.

Within the large digital processor, scratch pad memories, writeable control memories, and arrays of registers can be built quite easily.

During development of a machine, a writeable control memory constructed with 3101A's may be used to check out the bit patterns for subsequent use in high-speed read-only memories such as Intel's 3301.

In small signal processors, and fast Fourier transform calculators, the 3101A offers significant speed advantages over ferrite core and shift register memories. The random-access nature of the 3101A also removes some of the restrictions associated with serial memory components. Very high data rates are achievable with memories constructed from 3101A's permitting their use in such severe applications as pulse-height analyzers and queing buffers for video transmission lines.

Since the 3101A is DTL and TTL logic compatible, wiring precautions are no more stringent than those of a DTL or TTL system. Circuit diagram and logic representations are included in this data sheet to provide additional information for the user.

The following circuits are examples of the 3101A used as a building block in constructing memory systems of different size.

When expanding a RAM into a larger memory, additional levels of decoding are required. The example below shows the use of the 3205 (1 of 8 decoder) in a 128 word x 24 bit memory system. The new system access time would normally include the buffer delay (delay through the 3205) if the chip select to output delay of the 3101A was the same as its address to output delay.

However, the 3101A has a chip select delay of 17 ns and an address delay of 35 ns. Therefore, there are 18 ns available for decoding to occur before the access time of the system is affected. Since the maximum delay of the 3205 is 18 ns, the expanded memory system shown below has no additional access time delay.

128 Word x 24 Bit Memory System

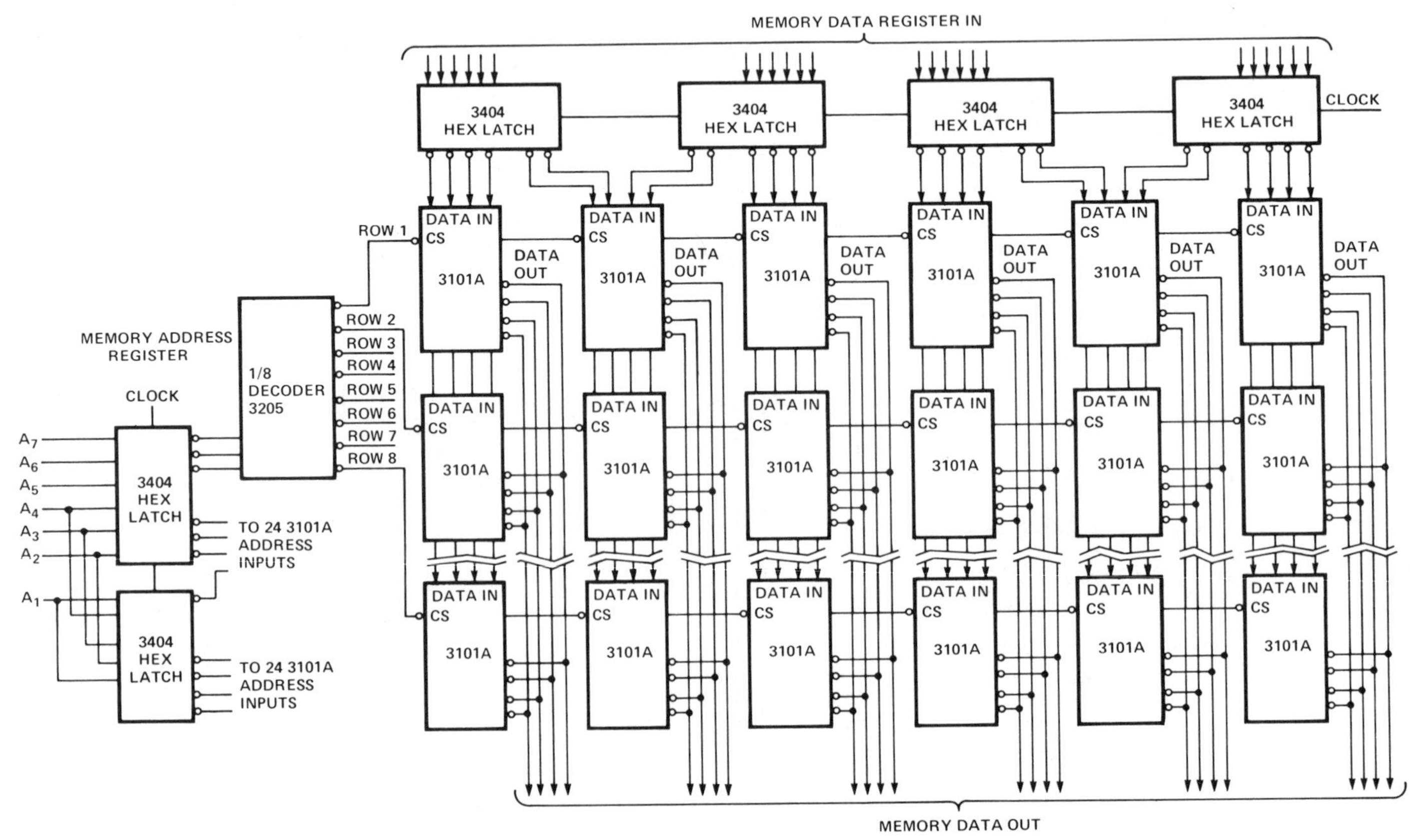

Application Information

16 Word x 16 Bit Memory

In this example, all chip select, write and address inputs are connected in parallel. A full 16 bit word is made up of 4 bits in each of the 4 packages.

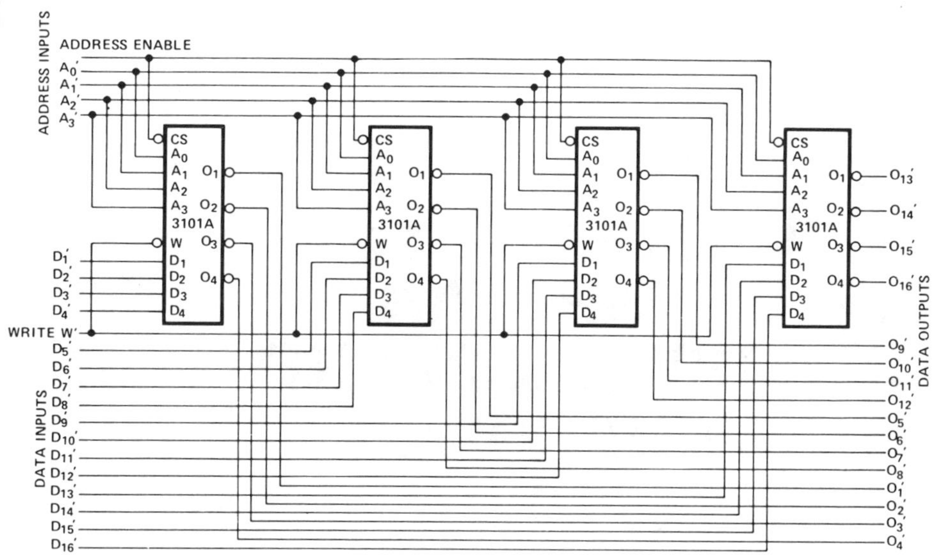

32 Word x 8 Bit Memory

Two 3101A devices are used to increase the number of bits per word to 8. This is accomplished by connecting the chip select and address inputs in parallel. To increase the number of words to 32, two of such parallel combinations are used.

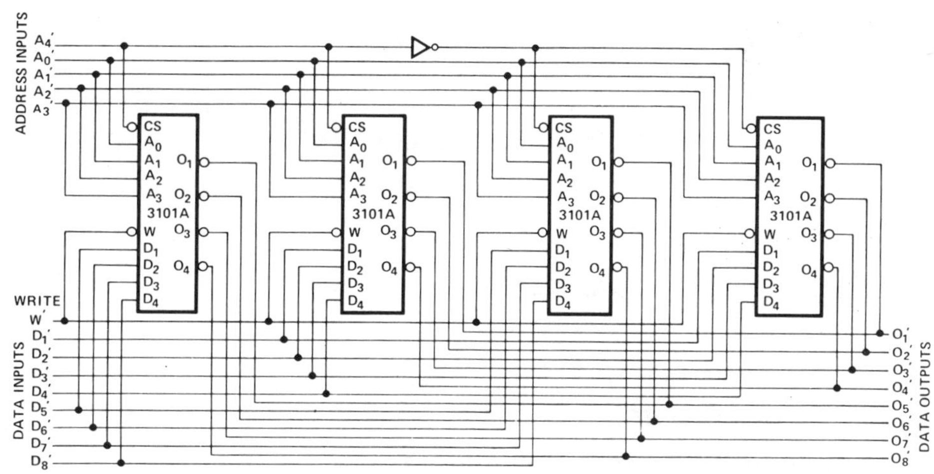

64 Word x 4 Bit Memory

The 64 word memory is made up of 4 3101A memory devices. Word expansion is made possible by utilizing the chip select input as an additional address line. A 1 out of 4 decoder is used to drive the chip select input and the outputs of each 3101A are OR-tied.

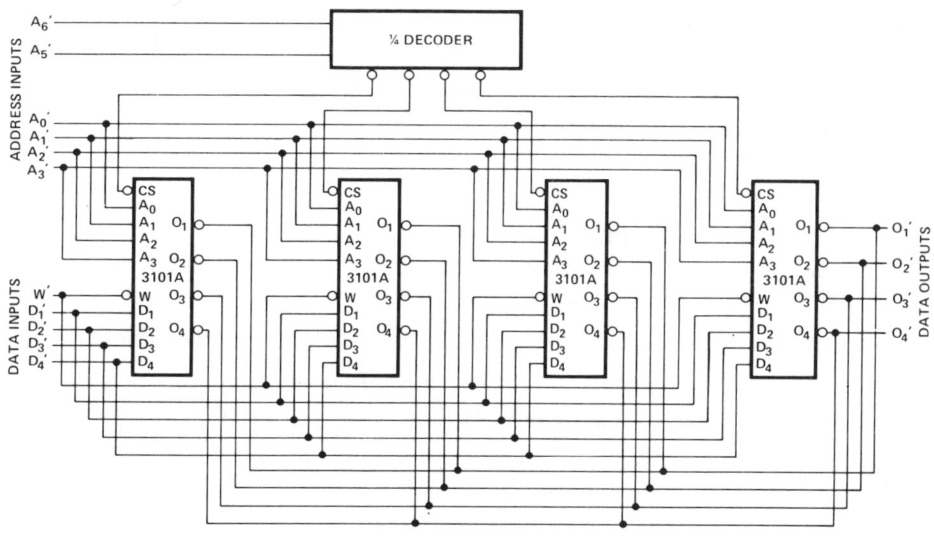

intel **Bipolar LSI Memory 3102, 3202**

INTEL CORP. 365 Middlefield Road, Mountain View, California 94040 • (415) 969-1670

DEC. 1970

PARTIALLY DECODED RANDOM ACCESS 256 BIT BIPOLAR MEMORY (3102) AND BINARY DECODER-DRIVER (3202)

- **Fast Access Time—120 ns total—including Decoder-Driver and memory**
- **Low Memory Power Dissipation—2.5 mw/bit**
- **Directly Compatible with DTL and TTL Logic circuits**
- **OR-tie capability**
- **Easy Memory Expansion through 4 chip select inputs on Decoder-Driver**
- **Minimum Line Reflection—low voltage diode input clamp**
- **Decoder-Driver (3202) Drives at least 32 memory elements, Memory (3102) Output Sinks 15 mA**
- **Low Input Load Current—.25 mA max., 1/6 Standard TTL Input Load**
- **Standard Packaging—Decoder and Memory both in 16-pin Dual-In-Line Configuration**

3102

PIN CONFIGURATION

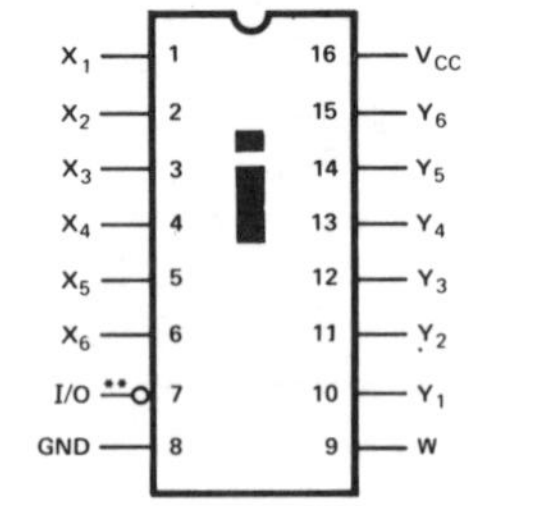

** DATA IN AND DATA OUT ARE OF THE SAME LOGIC LEVELS. FOR UNSELECTED CHIP, OUTPUT IS AT THE "HIGH" LEVEL.

V_{CC} = 16
GND = 8

BLOCK DIAGRAM

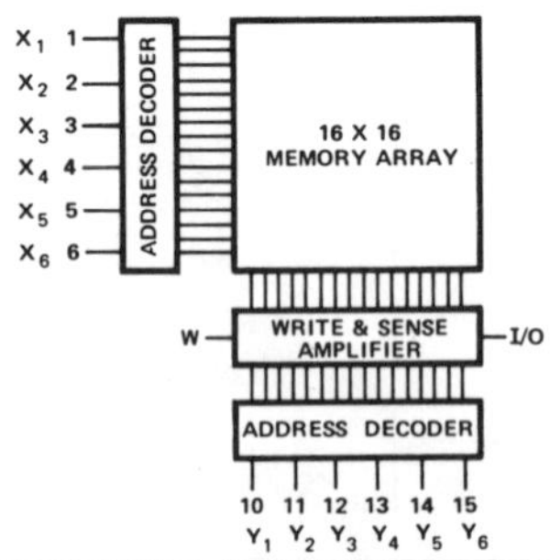

3202

PIN CONFIGURATION

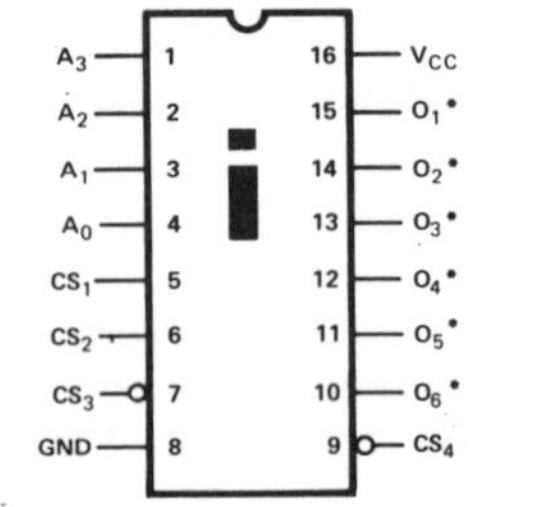

* O_1 THRU O_6 DRIVES X_1 THRU X_6 OR Y_1 THRU Y_6 ADDRESS INPUTS OF THE 3102 MEMORY CHIP.
CS_1 = CS_2 = HIGH, CS_3 = CS_4 = LOW TO SELECT A DECODER-DRIVER.

V_{CC} = 16
GND = 8

BLOCK DIAGRAM

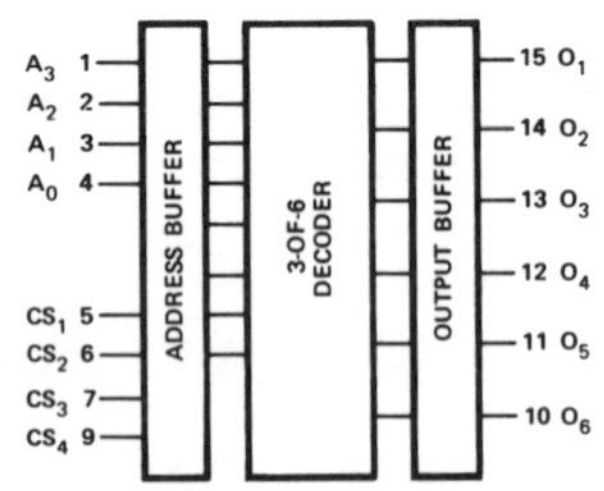

PACKAGE OUTLINE

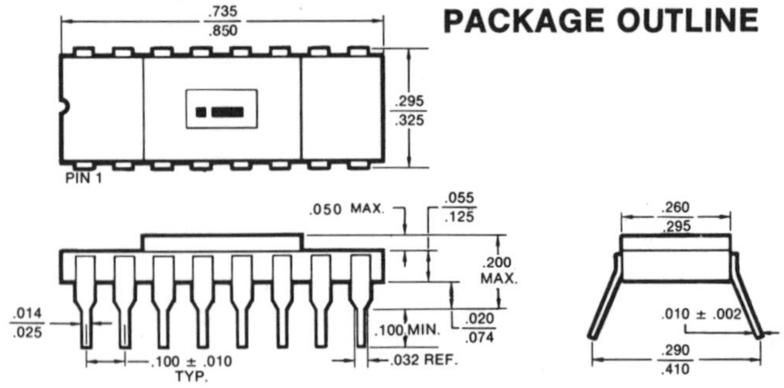

The Intel 3102 is a high speed 256-bit bipolar random access Memory, organized as 256 words by 1 bit. The address of the Memory is coded with a special 3-of-6 code which offers low power dissipation and high speed operation for the Memory chip.

The Applications section of this data sheet describes the special 3-of-6 code.

The Intel 3202 is a binary to 3-of-6 decoder-driver, capable of driving thirty-two 3102 Memory Chips. The use of 4 chip select inputs provides expansion of up to 4096 words without additional decoding.

Both the 3102 and 3202 are packaged in hermetically sealed 16 pin dual-in-line packages, and their performance is specified over the temperature range of 0°–85°C. The use of Schottky barrier diode clamped transistors to obtain fast switching speeds results in higher performance than equivalent devices made with a gold diffusion process.

3102 256 BIT RAM ELEMENT

Maximum Guaranteed Ratings*

Case Temperature Under Bias	−65° to +125°C
Storage Temperature	−65° to +160°C
All Input, Output or Supply Voltages	7 volts
Input Currents	20 mA
Output Currents	100 mA

*COMMENT:

Stresses above those listed under "Maximum Guaranteed Rating" may cause permanent damage to the device. This is a stress rating only and functional operation of the device at these or at any other condition above those indicated in the operational sections of this specification is not implied.

D.C. Characteristics (T_{case} = 0 to 85°C, V_{CC} = 5.0V ± 5%)

SYMBOL	TEST	LIMIT MIN.	LIMIT MAX.	UNIT	CONDITIONS
I_{FX}	X ADDRESS LOAD CURRENT		−500	μA	V_{CC}=5.25V, V_X=0.45V, V_W=0V, 2 Other X Lines =0V, 3Y Lines=0V
I_{FY}	Y ADDRESS LOAD CURRENT		-500	μA	V_{CC}=5.25V, V_Y=0.45V, V_W=0V, 2 Other Y Lines =0V, 3X Lines=0V
I_{FW}	WRITE INPUT LOAD CURRENT		-250	μA	V_{CC}=5.25V, V_W=0.45V, 3X, 3Y Line=0V
I_{FD}	DATA INPUT LOAD CURRENT		-250	μA	V_{CC}=5.25V, V_D=0.45V, 3X, 3Y Line=0V
I_{RX}	X ADDRESS LEAKAGE CURRENT		20	μA	V_{CC}=5.25V, V_X=5.25V, V_W=0V, All Other X Lines=0V
I_{RY}	Y ADDRESS LEAKAGE CURRENT		20	μA	V_{CC}=5.25V, V_Y=5.25V, V_W=0V, All Other Y Lines=0V
I_{RW}	WRITE INPUT LEAKAGE CURRENT		10	μA	V_{CC}=5.25V, V_W=5.25V, All X, All Y Lines=0V
I_{RD}	DATA INPUT LEAKAGE CURRENT		100	μA	V_{CC}=5.25V, V_D=5.25V, All X, All Y Lines=0V
V_{CX}	X ADDRESS CLAMP VOLTAGE		-1.0	V	V_{CC}=4.75V, I_X= −5.0mA, V_W=0V
V_{CY}	Y ADDRESS CLAMP VOLTAGE		-1.0	V	V_{CC}=4.75V, I_Y= −5.0mA, V_W=0V
V_{CW}	WRITE INPUT CLAMP VOLTAGE		-1.0	V	V_{CC}=4.75V, I_W= −5.0mA,
V_{CD}	DATA INPUT CLAMP VOLTAGE		-1.0	V	V_{CC}=4.75V, I_D= −5.0mA, All X, All Y Lines=0V
V_{OL}	OUTPUT "LOW" VOLTAGE		0.45	V	V_{CC}=4.75V, I_{OL}=15mA, V_W=0V, 3X, 3Y Lines =0V. Selected Location Stores "LOW"
I_{CEX}	OUTPUT LEAKAGE CURRENT		100	μA	V_{CC}=5.25V, V_{CEX}=5.25V V_W=0V, 3X, 3Y Lines=0V. Selected Location Stores "HIGH"
I_{CC}	POWER SUPPLY CURRENT		125	mA	V_{CC}=5.25V, V_W=0V, 3X, 3Y Lines=0V
V_{IL}	DATA WRITE INPUT "LOW" VOLTAGE		0.85	V	V_{CC}=5.0V
V_{IH}	DATA WRITE INPUT "HIGH" VOLTAGE	2.0		V	V_{CC}=5.0V
V_{ILA}	ADDRESS INPUT "LOW" VOLTAGE		.75	V	V_{CC}=5.0V
V_{IHA}	ADDRESS INPUT "HIGH" VOLTAGE	2.0		V	V_{CC}=5.0V
C_{IN}*	INPUT CAPACITANCE		20 (TYP)	pF	f=1MHz, V_{BIAS}=2.0V V_{CC}=0.0V
C_{OUT}*	OUTPUT CAPACITANCE		15 (TYP)	pF	f=1MHz, V_{BIAS}=2.0V V_{CC}=0.0V

*This parameter is only periodically sampled and is not 100% tested.

**INPUT CURRENT
VS
ADDRESS INPUT VOLTAGE**

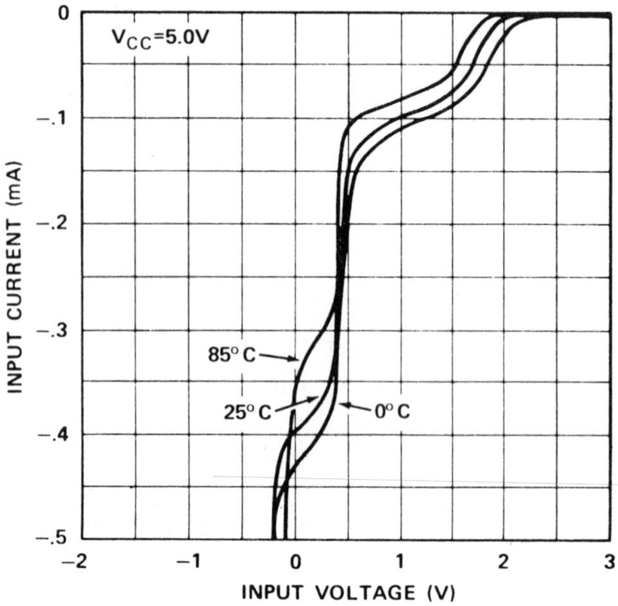

**INPUT CURRENT
VS
WRITE INPUT VOLTAGE**

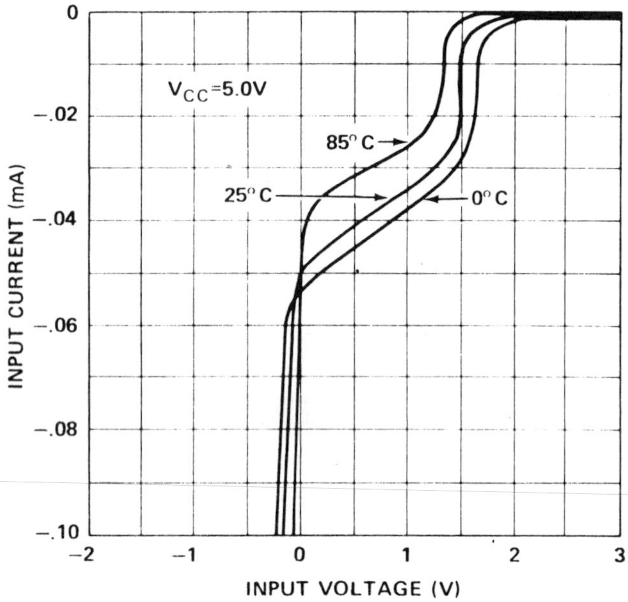

**OUTPUT CURRENT
VS
OUTPUT "LOW" VOLTAGE**

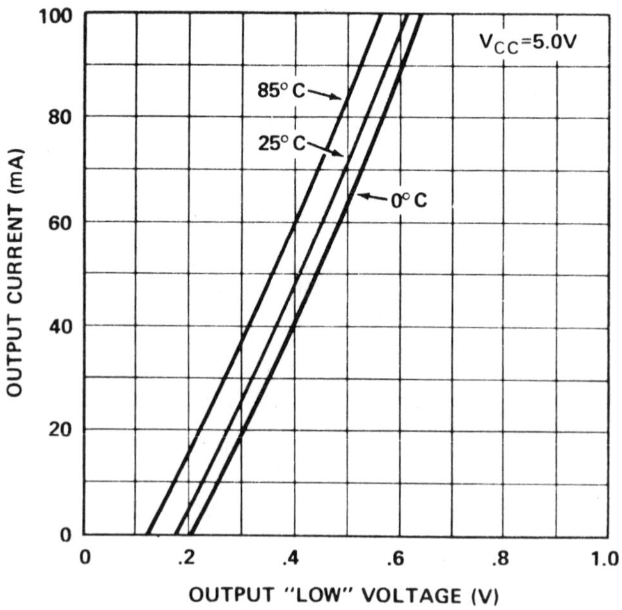

**INPUT THRESHOLD VOLTAGE
VS
TEMPERATURE**

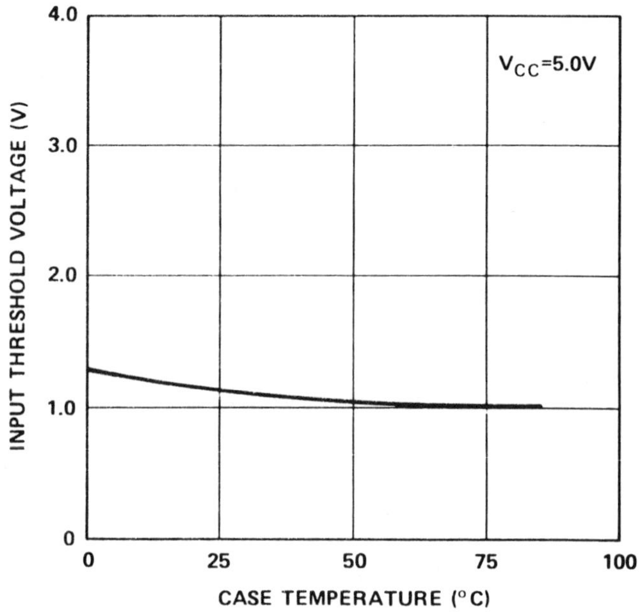

Y ADDRESS TO OUTPUT DELAY VS TEMPERATURE

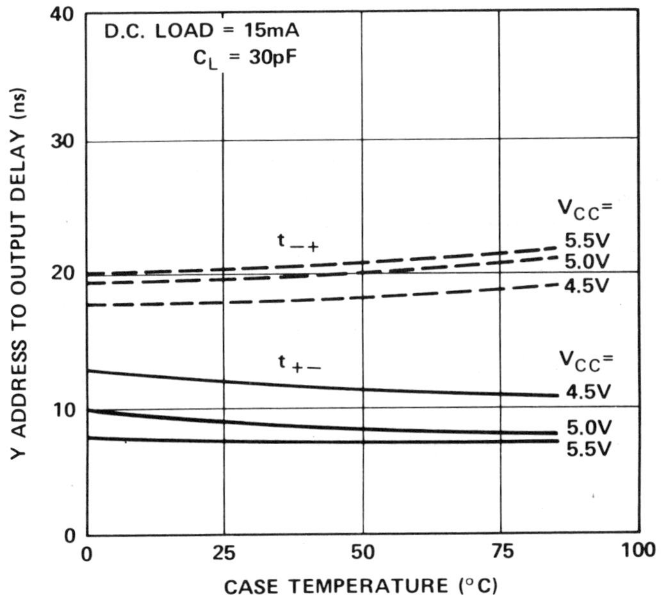

X ADDRESS TO OUTPUT DELAY VS TEMPERATURE

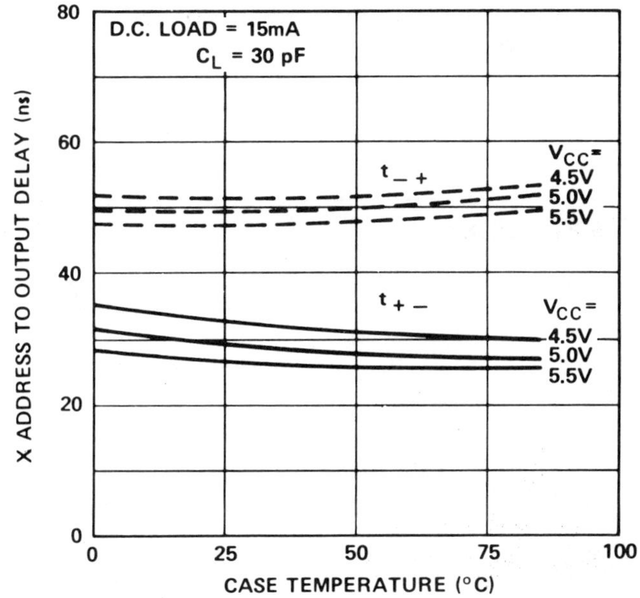

WRITE PULSE WIDTH VS TEMPERATURE

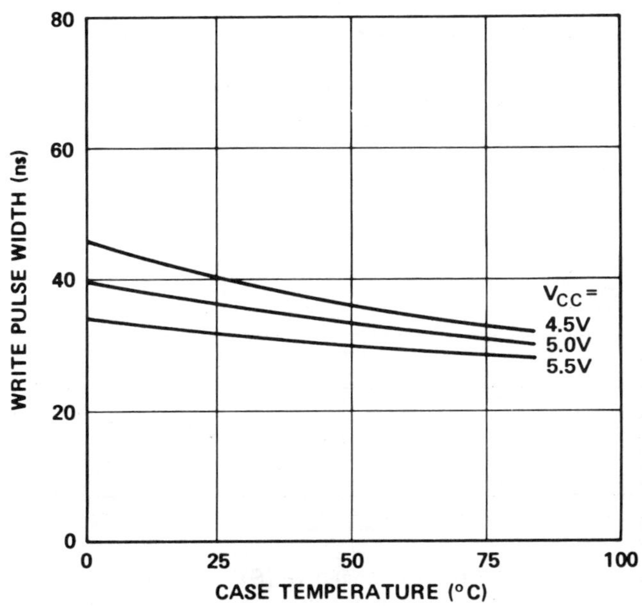

ADDRESS TO OUTPUT DELAY VS LOAD CAPACITANCE

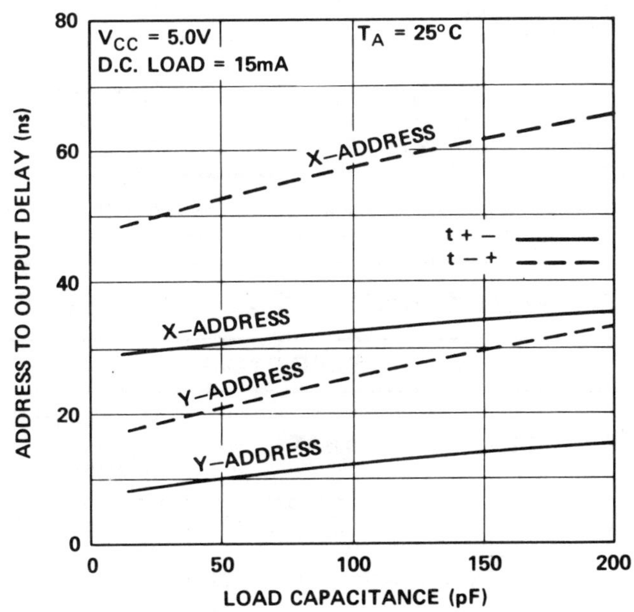

Switching Characteristics

Conditions of Test:

Input Pulse amplitudes—2.5V

Input Pulse rise and fall times of 5 nanoseconds between 1 volt and 2 volts

Speed measurements are made at 1.5 volt levels

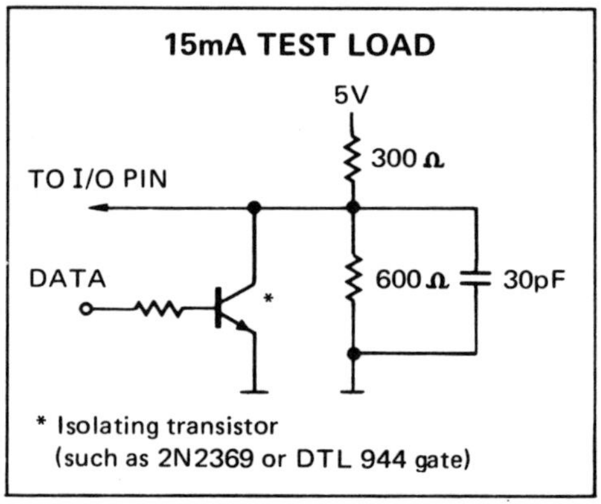

Read Cycle

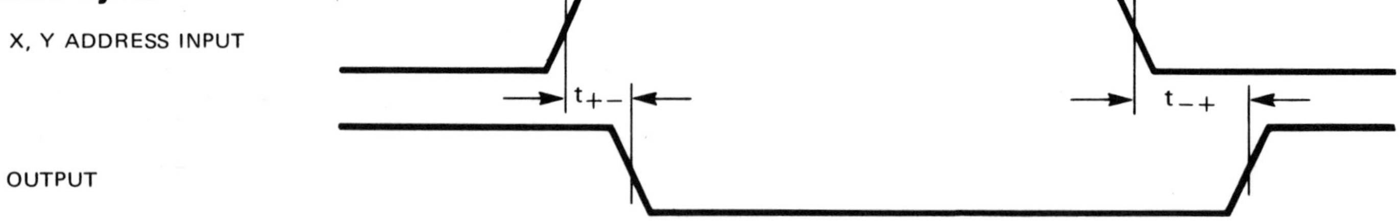

Write Cycle

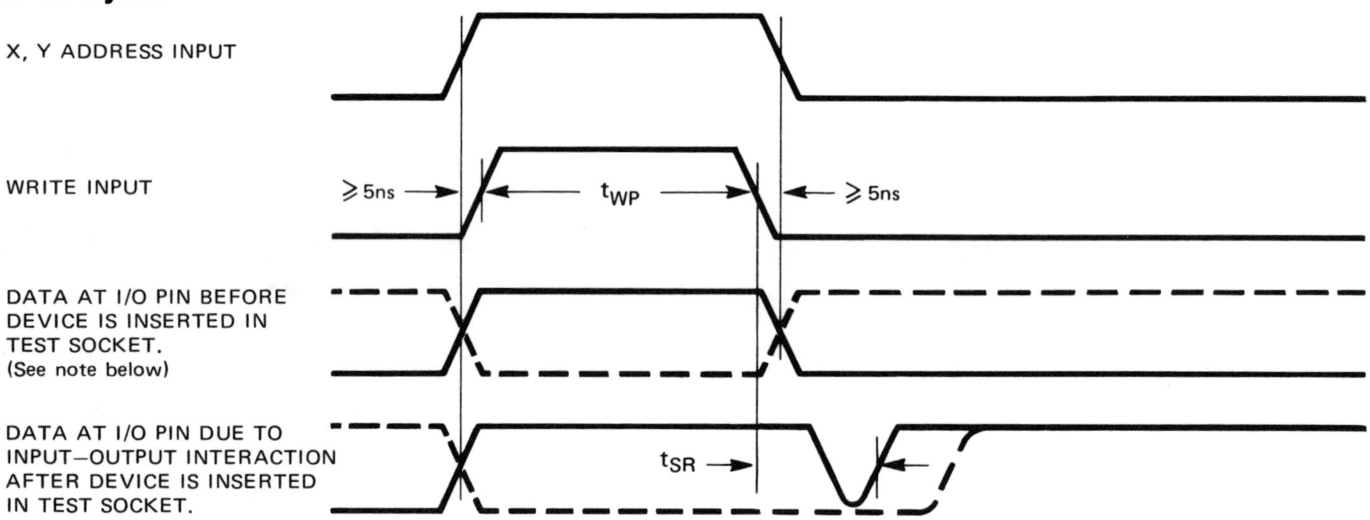

NOTE: Data input into I/O pin must be isolated with either an external transistor or a DTL gate as shown in circuit diagram for test load. Address may change ≥ 5 nsec at the end of the WRITE pulse.

Guaranteed Limits @ V_{CC}=5.0V, T_{case}= 0 to 85°C

READ CYCLE				WRITE CYCLE			
SYMBOL	TEST	LIMITS (ns) MIN.	LIMITS (ns) MAX.	SYMBOL	TEST	LIMITS (ns) MIN.	LIMITS (ns) MAX.
t_{+-}	Y ADDRESS TO OUTPUT		40	t_{WP}	WRITE PULSE WIDTH	80	
t_{-+}	Y ADDRESS TO OUTPUT		40	t_{SR}	SENSE AMPLIFIER RECOVERY TIME		60
t_{+-}	X ADDRESS TO OUTPUT		60				
t_{-+}	X ADDRESS TO OUTPUT		80				

3102 Circuit Schematic

3102–256 BIT RAM

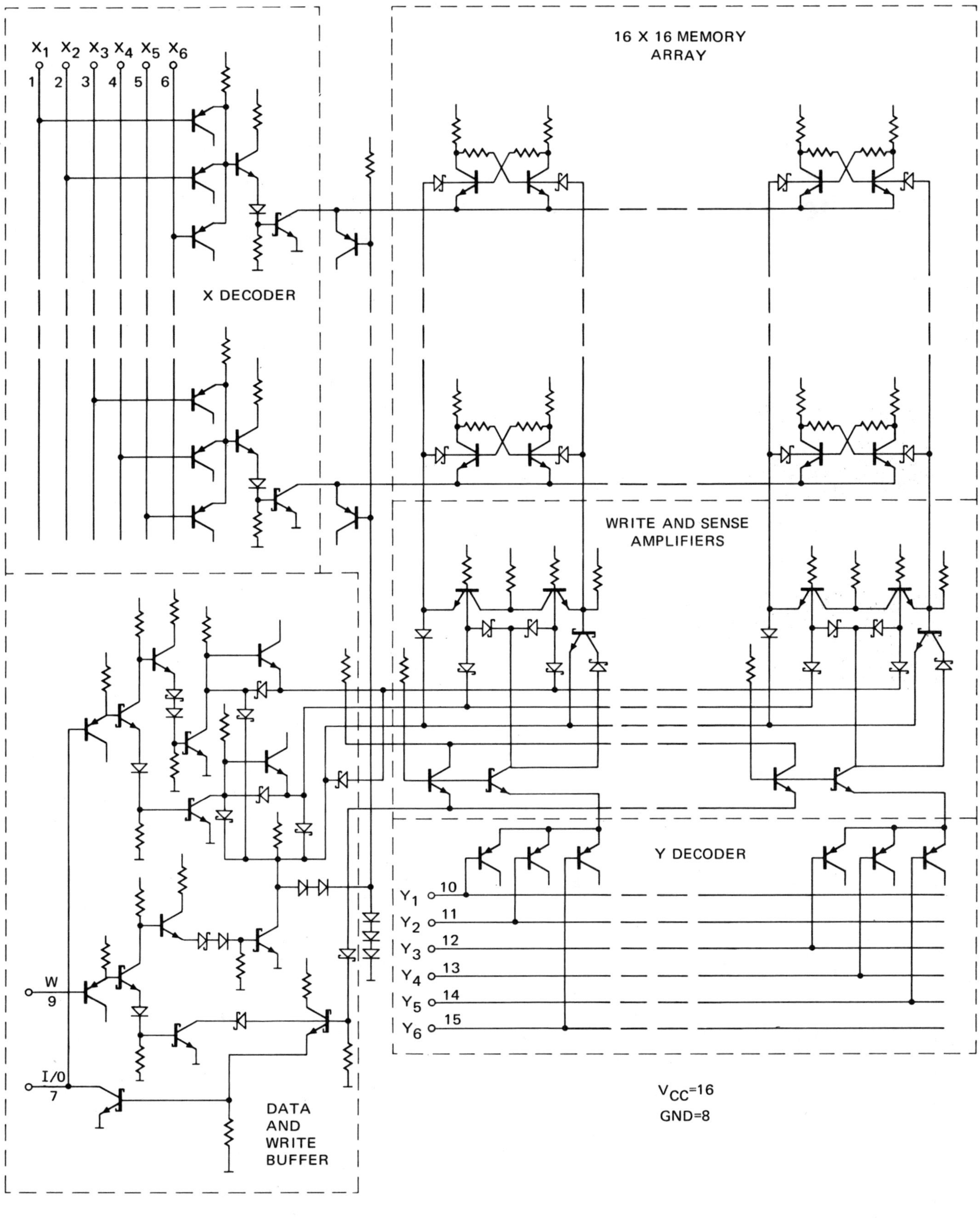

3202 DECODER DRIVER ELEMENT

Maximum Guaranteed Ratings*

Case Temperature Under Bias	−65° to +125°C
Storage Temperature	−65° to +160°C
All Input, Output or Supply Voltages	7 volts
Input Currents	20 mA
Output Currents	100 mA

*COMMENT:

Stresses above those listed under "Maximum Guaranteed Rating" may cause permanent damage to the device. This is a stress rating only and functional operation of the device at these or at any other condition above those indicated in the operational sections of this specification is not implied.

D.C. Characteristics (T_{case} =0 to 85°C, V_{CC}=5.0V ±5%)

SYMBOL	TEST	LIMIT		UNIT	CONDITIONS
		MIN.	MAX.		
I_{FA}	ADDRESS INPUT LOAD CURRENT		−250	μA	V_{CC}= 5.25V, V_A=0.45V
I_{FS}	CHIP SELECT INPUT LOAD CURRENT		−250	μA	V_{CC}= 5.25V, V_S=0.45V
I_{RA}	ADDRESS INPUT LEAKAGE CURRENT		10	μA	V_{CC}=5.25V, V_A=5.25V
I_{RS}	CHIP SELECT INPUT LEAKAGE CURRENT		10	μA	V_{CC}=5.25V, V_S=5.25V
V_{CA}	ADDRESS INPUT CLAMP VOLTAGE		−1.0	V	V_{CC}=4.75V, I_A=−5.0mA
V_{CS}	CHIP SELECT INPUT CLAMP VOLTAGE		−1.0	V	V_{CC}=4.75V, I_S= −5.0mA
V_{OL}	OUTPUT "LOW" VOLTAGE		0.45	V	V_{CC}=4.75V, I_{OL}=20mA
V_{OH}	OUTPUT "HIGH" VOLTAGE	2.4		V	V_{CC}=4.75V, I_{OH}=−1.5mA, V_{CS3}=V_{CS4}=0V
I_{SC}	OUTPUT SHORT CIRCUIT CURRENT	− 40	−120	mA	V_{CC}=5.0V, V_{OUT}=0V, V_{CS3}=V_{CS4}=0V
I_{CC}	POWER SUPPLY CURRENT		105	mA	V_{CC}=5.25V
V_{IL}	INPUT "LOW" VOLTAGE		0.85	V	V_{CC}=5.0V
V_{IH}	INPUT "HIGH" VOLTAGE	2.0		V	V_{CC}=5.0V
C_{IN}*	INPUT CAPACITANCE		5 (TYP)	pF	f=1MHz, V_{BIAS}=2.0V V_{CC}=0.0V

*This parameter is only periodically sampled and is not 100% tested.

INPUT CURRENT VS INPUT VOLTAGE

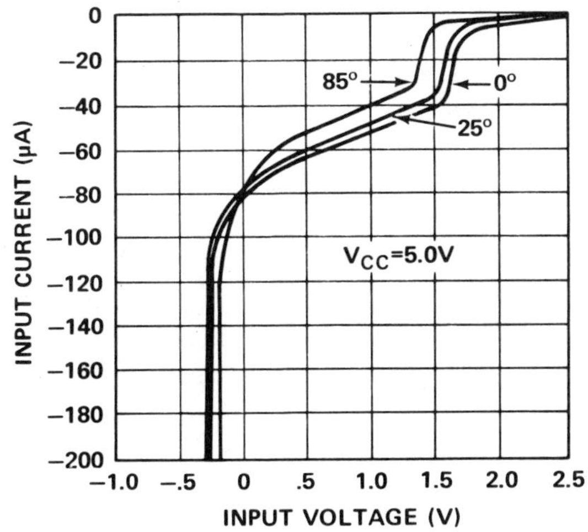

OUTPUT CURRENT VS OUTPUT "LOW" VOLTAGE

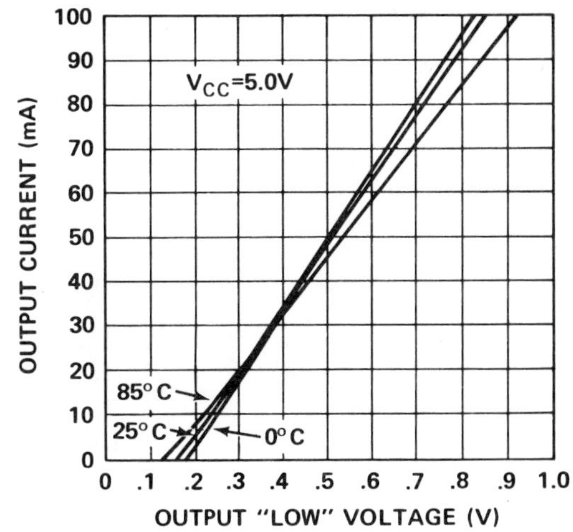

OUTPUT CURRENT VS OUTPUT "HIGH" VOLTAGE

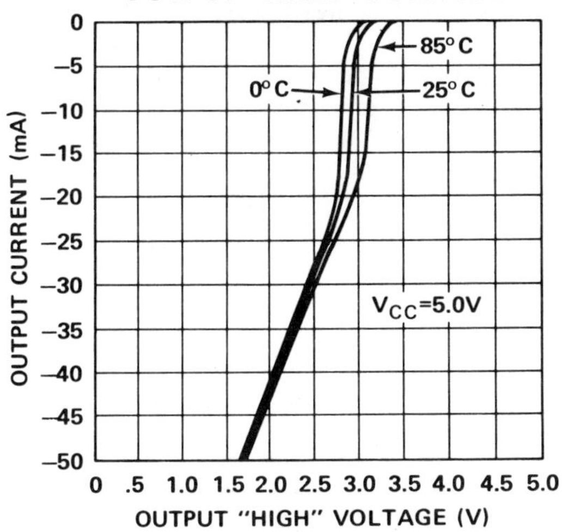

INPUT THRESHOLD VOLTAGE VS TEMPERATURE (°C)

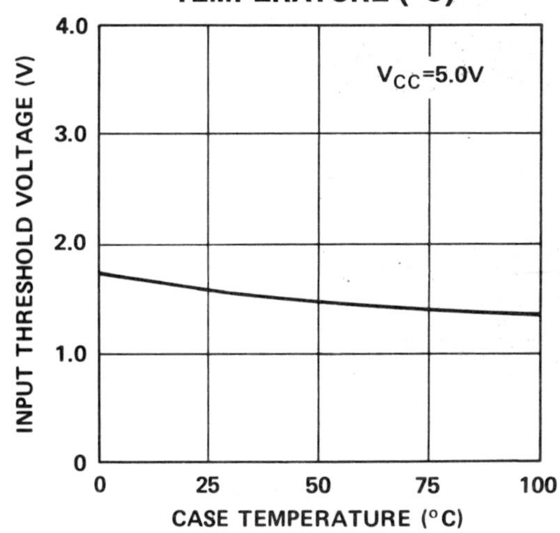

ADDRESS, CHIP SELECT INPUT TO OUTPUT DELAY VS TEMPERATURE

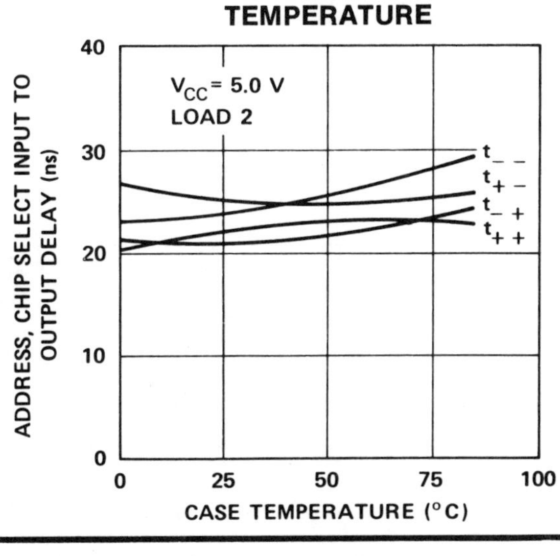

ADDRESS, CHIP SELECT INPUT TO OUTPUT DELAY VS LOAD CAPACITANCE

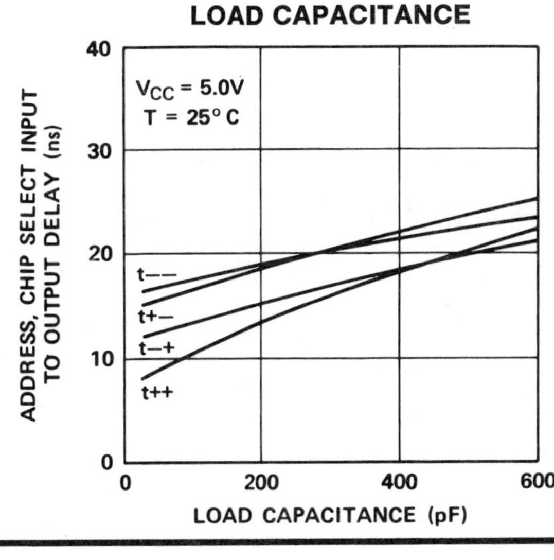

Switching Characteristics

Conditions of Test:

Input Pulse amplitudes—2.5V
Input Pulse rise and fall times of 5 nanoseconds between 1 volt and 2 volts
Speed measurements are made at 1.5 volt levels

3202 TEST LOADS*

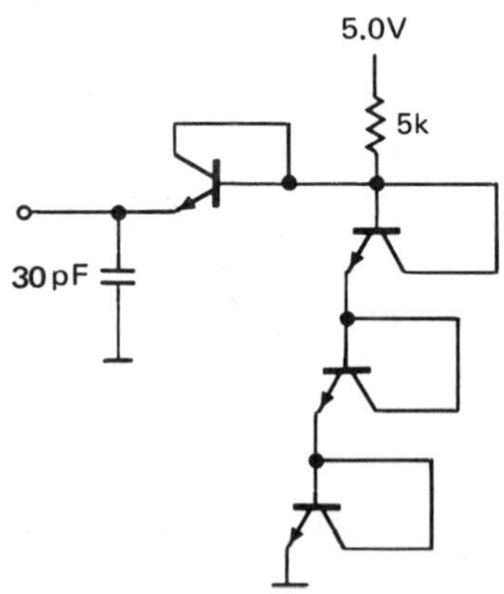

Load 1 (Fanout of one 3102)

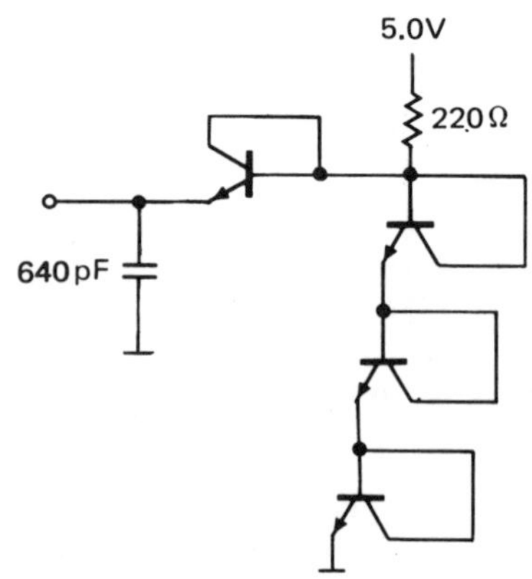

Load 2 (Fanout of thirty two 3102's)

* ALL TRANSISTORS 2N2369 OR EQUIVALENT

Through Delay (For Measurement Purpose Only)

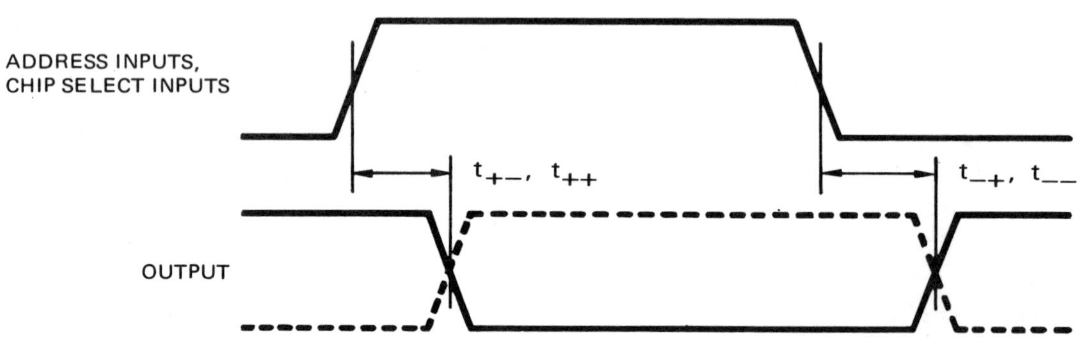

Guaranteed Limits @V_{cc}=5.0V, T_{case}=0 to 85°C

SYMBOL	TEST	MAX. LIMIT (ns)	
		LOAD 1	LOAD 2
t_{+-}, t_{-+} t_{--}, t_{++}	ADDRESS TO OUTPUT DELAY	25	40
t_{+-}, t_{-+} t_{--}, t_{++}	CHIP SELECT TO OUTPUT DELAY	25	40

Linear interpolation is permissible for computing worst case delay time for LOAD 1 ≤ LOAD ≤ LOAD 2.

3202 Circuit Diagram

3202 3/6 DECODER DRIVER

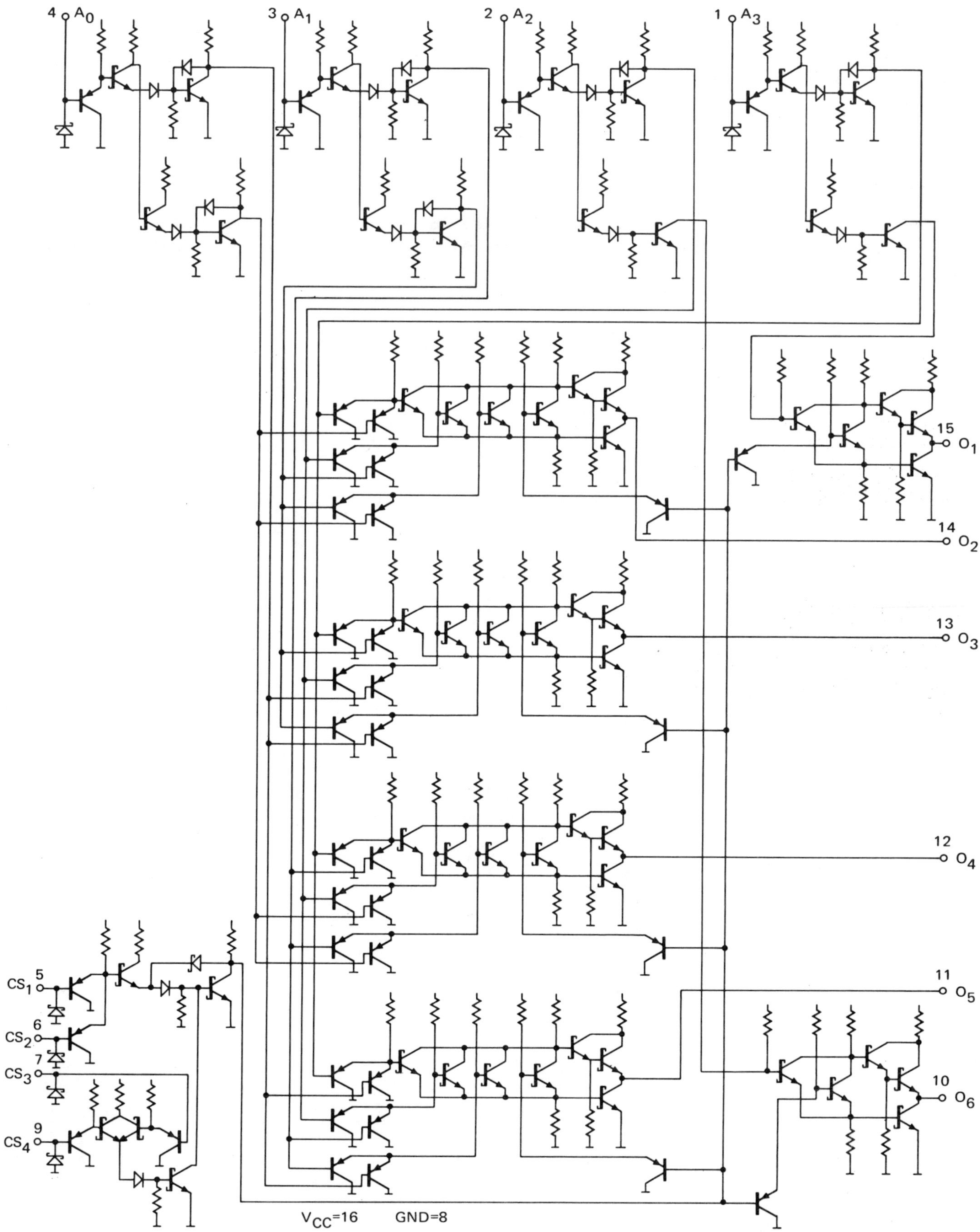

Application Information for 3102 and 3202

What is 3-of-6 code?

Each of the 16 binary codes from the 4-bit input of the decoder generates a unique set of 3 high and 3 low outputs. These six outputs select the X or Y coordinates of the 3102 RAM. With two 3202 decoders 256 unique locations are selected on up to 32 3102 memory chips.

Operation:

The 3102 has six X-address inputs, six Y-address inputs, one write enable pin and one data Input/Output pin. Both the X- and Y-address inputs accept a 3-of-6 code as shown below:

DECIMAL NUMBER	BINARY CODE				3-OF-6 CODE					
	A_3	A_2	A_1	A_0	X_1, Y_1	X_2, Y_2	X_3, Y_3	X_4, Y_4	X_5, Y_5	X_6, Y_6
0	0	0	0	0	1	1	0	0	0	1
1	0	0	0	1	1	0	1	0	0	1
2	0	0	1	0	1	0	0	1	0	1
3	0	0	1	1	1	0	0	0	1	1
4	0	1	0	0	1	1	1	0	0	0
5	0	1	0	1	1	0	1	0	1	0
6	0	1	1	0	1	1	0	1	0	0
7	0	1	1	1	1	0	0	1	1	0
8	1	0	0	0	0	1	0	1	0	1
9	1	0	0	1	0	1	1	0	0	1
10	1	0	1	0	0	0	0	1	1	1
11	1	0	1	1	0	0	1	0	1	1
12	1	1	0	0	0	1	1	1	0	0
13	1	1	0	1	0	1	1	0	1	0
14	1	1	1	0	0	1	0	1	1	0
15	1	1	1	1	0	0	1	1	1	0

"I" = HIGH LEVEL
"O" = LOW LEVEL

The Memory addresses may be driven by a decoder-driver (3202), DTL or TTL gates.

The connection order of address pins must be maintained between Decoder-Driver and Memory i.e., O_1 to X_1 or Y_1, O_2 to X_2 or Y_2 etc.

When a Memory location is addressed, stored data may be read out from the data Input/Output lead. To write into the Memory, data is entered into the data Input/Output lead and a write pulse (logical "high") is applied to the write enable pin.

Combined Operation of 3202s Driving 3102s

The binary address signal is applied to the inputs of the Decoder-Driver (3202). After the worst case delay through the Decoder-Driver (40 ns), the 3-of-6 code appears at the X and Y Address inputs of the memory chips (3102). This portion of the memory delay is identical for both the READ and the WRITE cycle.

During the READ cycle, output data is available at the I/O pin 60 to 80 ns after the X and Y Address of the memory chip receives the 3-of-6 codes. The worst case access time through the Decoder-Driver and memory chip is 120 ns total.

During the WRITE cycle, input data may be applied to the memory I/O pin on or before the 3-of-6 codes reach the X and Y Address pins. WRITE pulse is applied to the WRITE input after the worst case delay of the Decoder-Driver (40 ns) has elapsed. Data may leave the I/O pin after the write pulse ceases. New memory location may be addressed for either a new READ or a WRITE cycle at the end of the WRITE pulse. The total WRITE cycle time is 120 ns.

Memory expansion is possible through the use of 4 CHIP SELECT inputs in the Decoder-Driver. To select a Decoder-Driver chip, the CHIP SELECT inputs CS_1 and CS_2 must be "HIGH"; CS_3 and CS_4 must be "LOW." By applying proper signals to the chip select inputs, memory expansion of up to 4096 words is possible without any additional decoding element. If TRUE and COMPLEMENTARY signals are available as chip select signals, memory expansion of up to 65K words is possible without additional decoding elements.

The timing diagrams on page 13 show the READ and WRITE cycles of a high speed memory system using 3202s driving 3102s.

Read Cycle

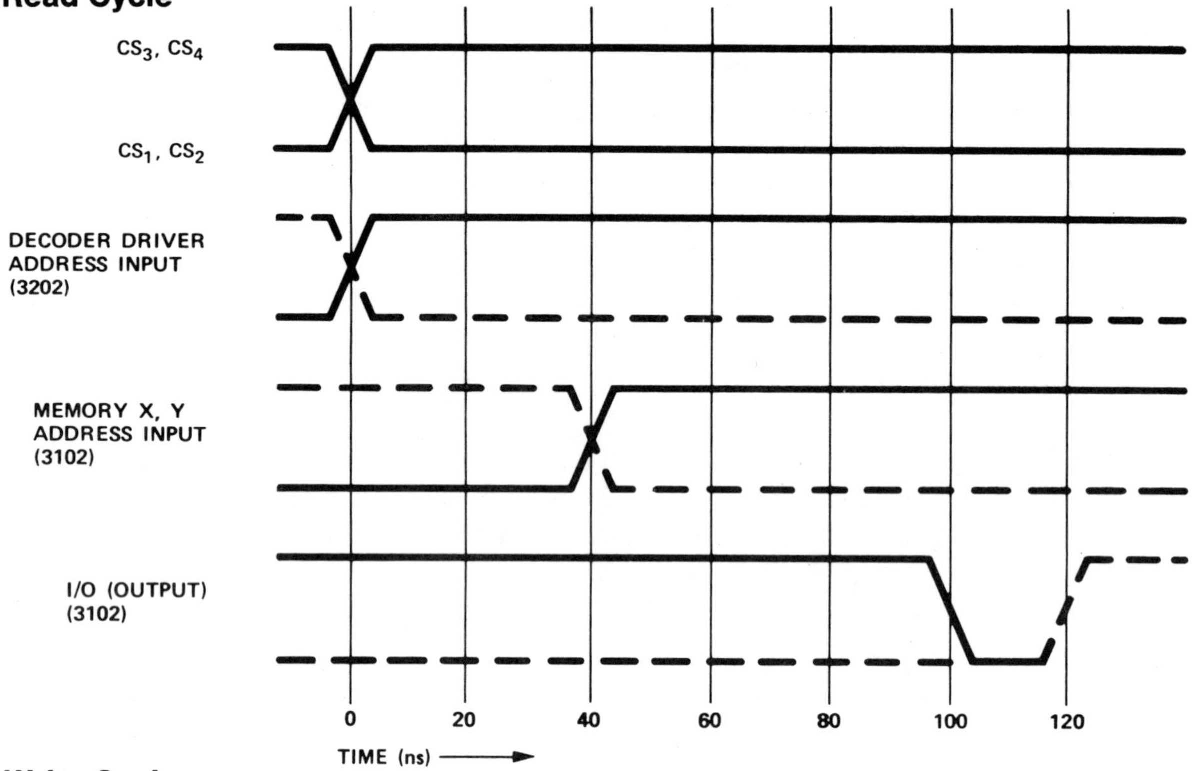

Write Cycle

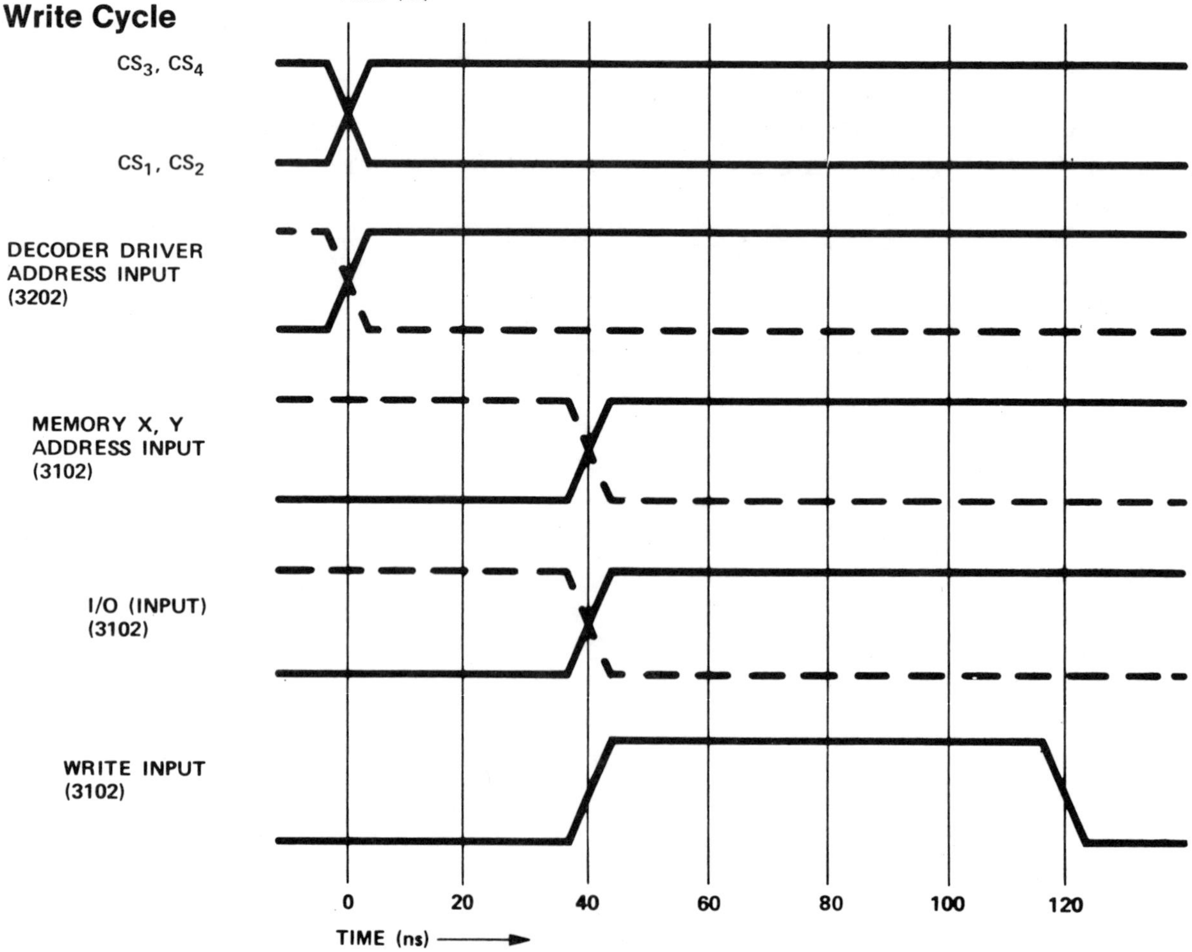

Typical Applications

LOGIC DIAGRAM
256 Word x 32 Bit Memory

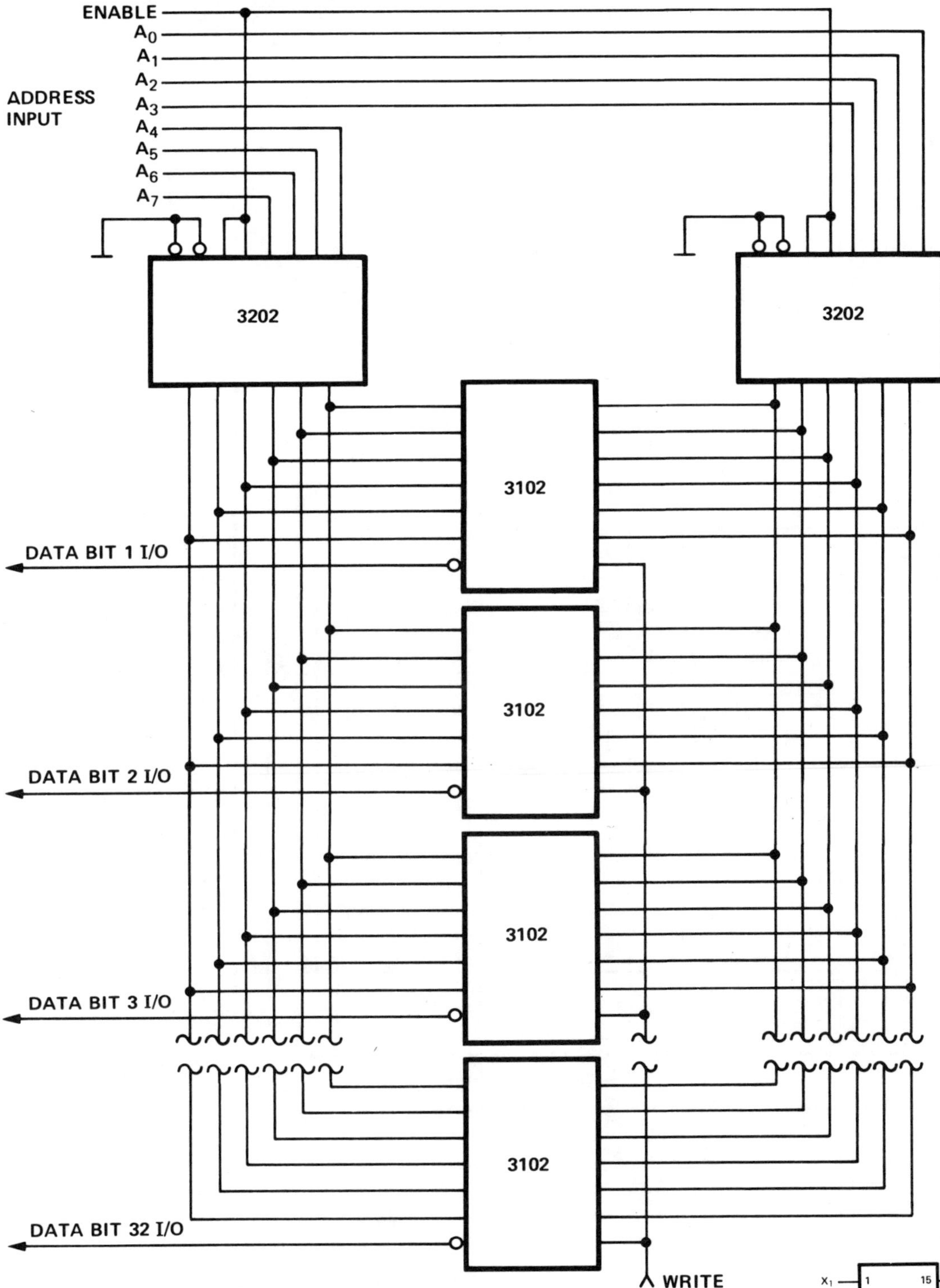

NOTE:
THE CONNECTION ORDER MUST BE MAINTAINED BETWEEN DECODER AND MEMORY CELL, i.e., O_1 TO X_1 OR Y_1, ETC.

NOTE: For actual physical location of pins, see page 1 of the data sheet.

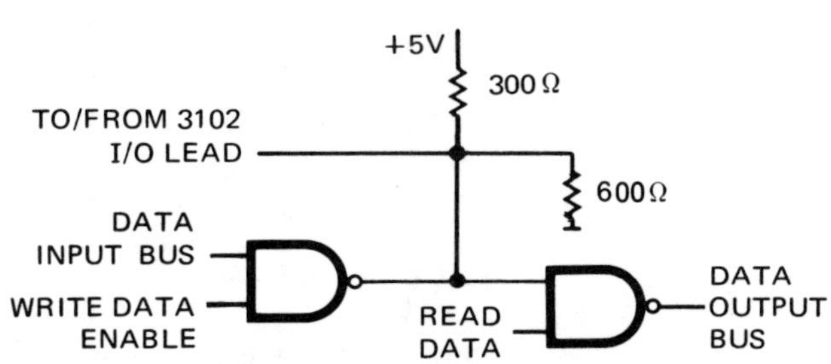

Typical Interface Circuit

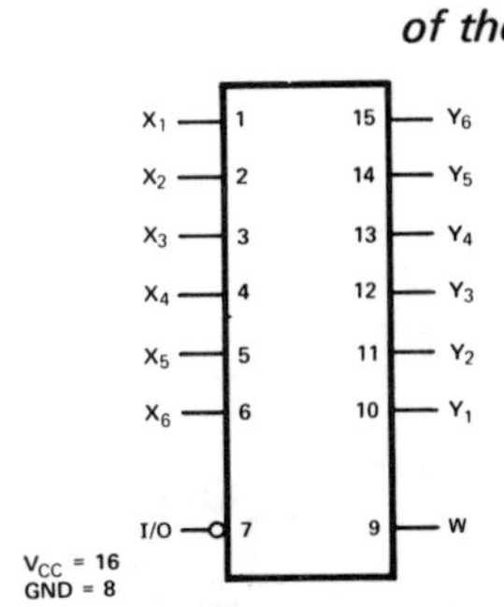

256 Word x 1 Bit Memory (3102)

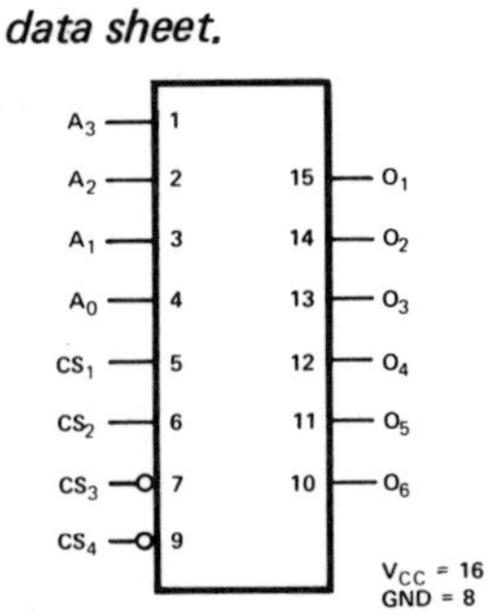

Decoder-Driver (3202)

Typical Applications

LOGIC DIAGRAM

4096 Word x 8 Bit Memory

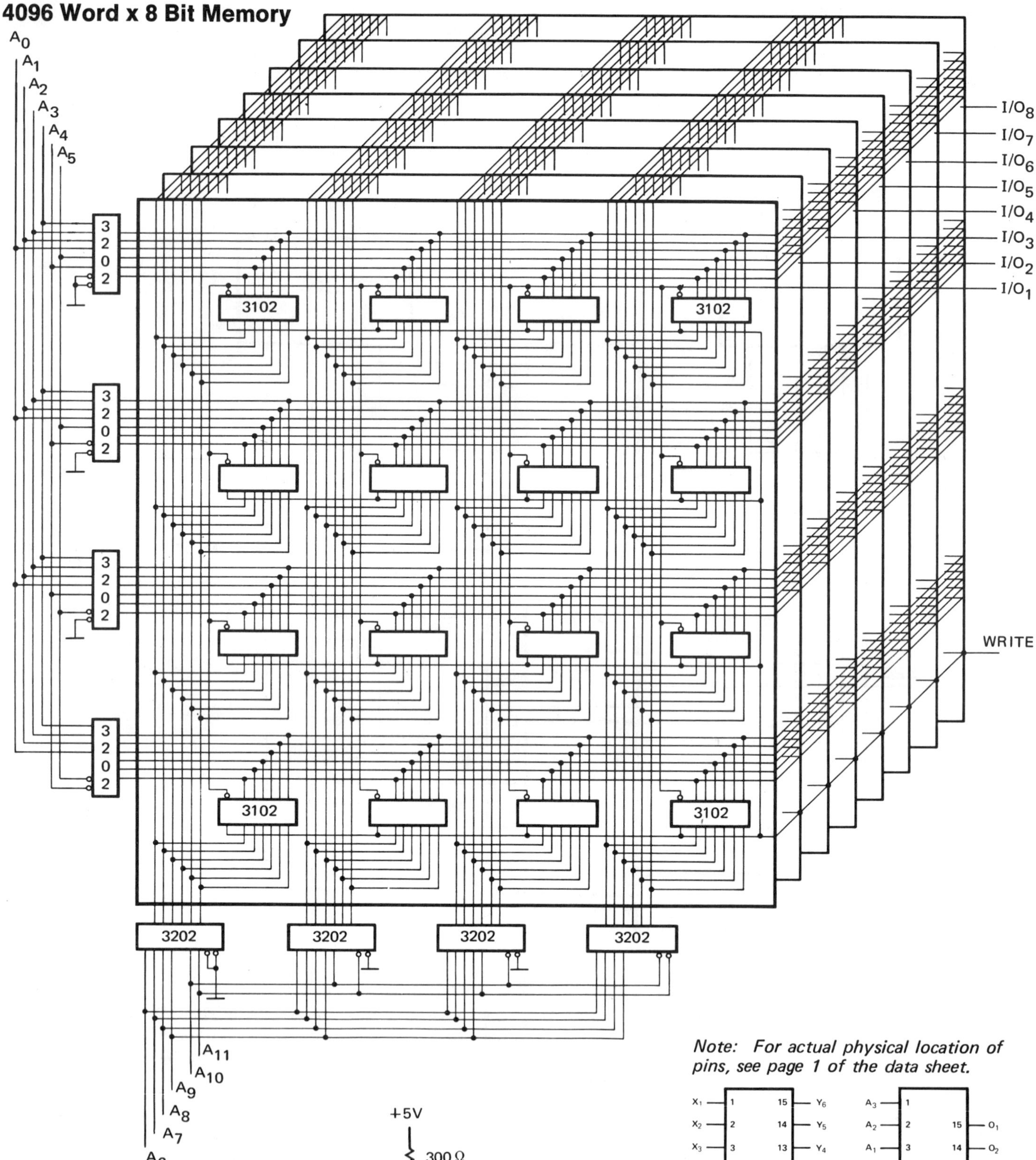

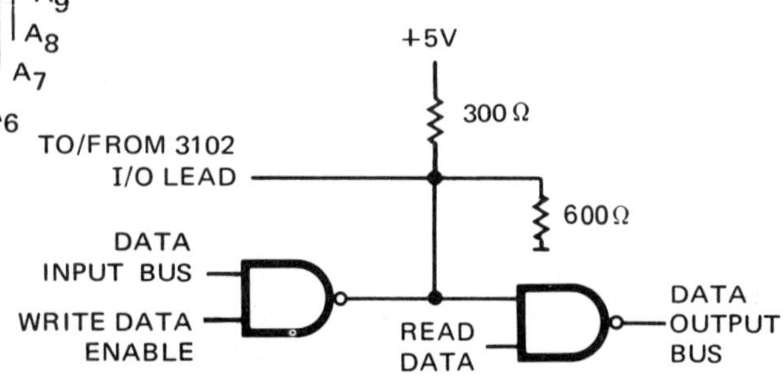

Typical Interface Circuit

Note: For actual physical location of pins, see page 1 of the data sheet.

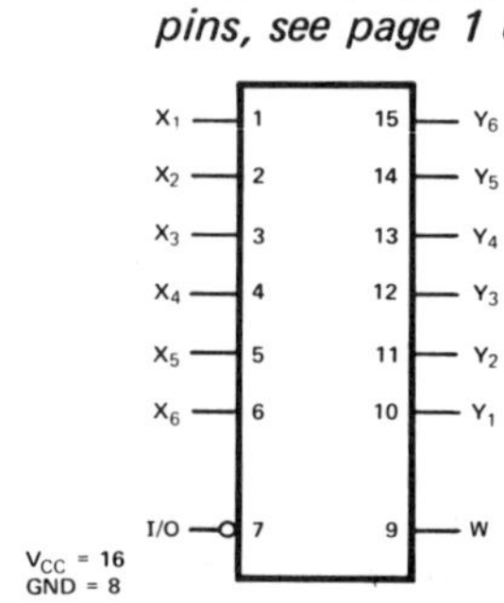

256 Word x 1 Bit Memory (3102)

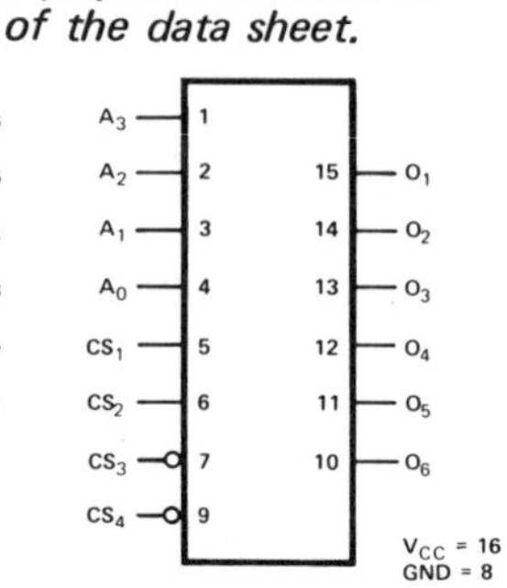

Decoder-Driver (3202)

Small-Size Low-Power Bipolar Memory Cell

SIEGFRIED K. WIEDMANN AND HORST H. BERGER

Abstract—**A dc-stable random-access memory cell employing n-p-n and p-n-p transistors has been designed in a concurrent circuit-layout approach. Test chips with 2 × 3 arrays have been processed in a standard bipolar technology using present manufacturing tolerances. Due to the merging of devices, the area required for a cell is only 14 mil². Furthermore, the cells have been operated at an extremely low dc standby power of less than 0.1 μW/cell. In spite of this low standby power, an array access time of 10 ns has been measured on a simulated 512-bit array in a pulsed power mode.**

Manuscript received March 10, 1971.

The authors are with the IBM Laboratories, Boeblingen, Germany.

I. Introduction

MAIN criteria in appraising monolithic memories are bit density, power dissipation, and performance.

Low power dissipation and high-speed performance have traditionally been incompatible. The superiority of the bipolar transistor technology for high-performance applications is unquestioned, but its potential for storage chips with high bit density and low power dissipation has been doubted in comparison with IGFET technology [1]. This belief is related to the experience that usual bipolar

Reprinted from *IEEE J. Solid-State Circuits,* vol. SC-6, pp. 283–288, Oct. 1971.

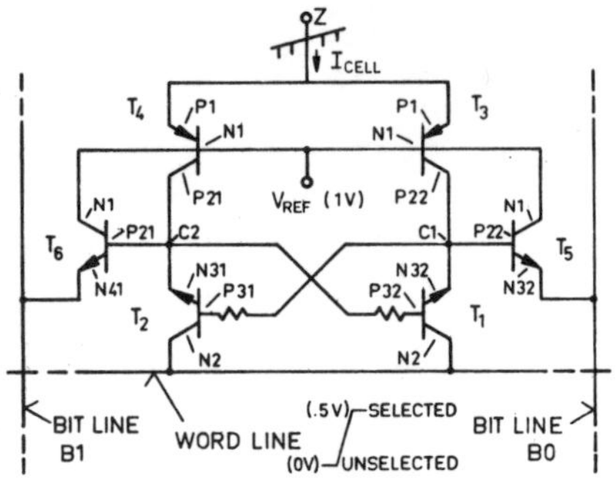

Fig. 1. Circuit schematic of memory cell.

circuitry needs a relatively large silicon area for the device isolation and for the high ohmic load resistors required particularly for low power consumption. In spite of this problem, some attractive bipolar memories have already been proposed [2]–[8].

In the following, a novel dc-stable bipolar flip-flop memory cell is described that simultaneously solves the problem of both the power dissipation and the storage density on the chip [8]. This is achieved by extensively merging devices in a combined circuit-layout design, which employs lateral p-n-p and inversely operated n-p-n transistors. It will be shown that the p-n-p load devices not only contribute to the small cell size but also enable operation of the cell in the submicrowatt power range. Furthermore, the use of inverse n-p-n transistors substantially simplifies the cell structure. Results obtained from 2×3 array chips processed in a conventional bipolar transistor technology fully verify the features of the novel cell, which are

1) extremely low dc standby power;
2) small size;
3) fast access time.

II. Basic Memory-Cell Structure and Operation

A. Cell Circuit

The electrical circuit schematic of the novel memory cell is shown in Fig. 1, as it finally emerged from the combined circuit-layout design. In Fig. 2, the cell arrangement in an array is depicted.

The flip-flop transistors T_1 and T_2 are preferably fabricated as inversely operated common-collector n-p-n transistors. In this case, the usual emitters N_{31} and N_{32} actually work as collectors. As will be shown later, such an approach offers decisive advantages in the layouts of the cell itself and the memory array. The normal n-p-n transistors T_5 and T_6 provide the decoupling from the bit lines $B0$ and $B1$.

Instead of ohmic resistors, p-n-p transistors T_3 and T_4 in common-base configuration have been employed as load devices. This measure is the key for the extremely low dc power achievable, and it also contributes to the small cell size. The low power feature will be ascertained by the following considerations. A flip-flop circuit loses its bistability when the loop gain drops below unity. Assuming that both flip-flop transistors conduct equal currents at the verge of bistability, and that the transistor output resistance is much larger than its input resistance, the following approximate condition for a loop gain larger than unity is valid

$$\frac{\beta}{1 + r_{e0}/r_c} > 1 \tag{1}$$

where

- β small-signal current gain of flip-flop transistors in common-emitter configuration;
- r_{e0} differential input resistance;
- r_c differential load resistance.

From this equation, the theoretical minimum current of a flip-flop memory cell can be derived by considering mainly the current dependencies of β and r_{e0}.

The differential input resistance r_{e0} may be expressed by

$$r_{e0} \approx (1 + \beta) \cdot \lambda \cdot v_T / I_E \tag{2}$$

where

$\lambda = 1 \cdots 2;$

$v_T = k \cdot T/q = 26$ mV at 300°K;

I_E emitter current.

Two approximations are of special practical interest. In cases where $\beta \gg 1$, it follows from (1) and (2) that

$$I_E > (\lambda \cdot v_T / r_c) = I_{E\,\min} \tag{3}$$

assures stability. Since the minimal cell current $I_{z\,\min} = 2 I_{E\,\min}$ (both transistors drawing equal currents), we arrive at

$$I_{z\,\min} \approx \frac{2\lambda \cdot v_T}{r_c} \approx \left. \frac{50\ \mu\text{A}}{r_c / k\Omega} \right|_{\lambda = 1}. \tag{4}$$

It is the load resistance r_c that determines the theoretical stability limit in this case.[1] Hence, we may call it the r_c-limited case.

However, in the cell circuit of Fig. 1, the p-n-p load transistor represents a much larger load resistance r_c than input resistance r_{e0} of the n-p-n flip-flop transistor. Therefore, (1) is reduced to the simple condition

$$\beta > 1. \tag{5}$$

In this β-limited case, the minimum cell current is determined by that current at which the current gain β_I of the inversely operated transistors T_1 and T_2 has dropped to unity. This is typically at a few nanoamperes. Of course, the current I_{cell} supplied to the cell has to be somewhat larger due to the common-base current gain α (p-n-p) < 1.

[1] For noise immunity reasons, the practical minimum cell current has to be approximately 5–10 times larger to assure a sufficient voltage difference between the two collector nodes.

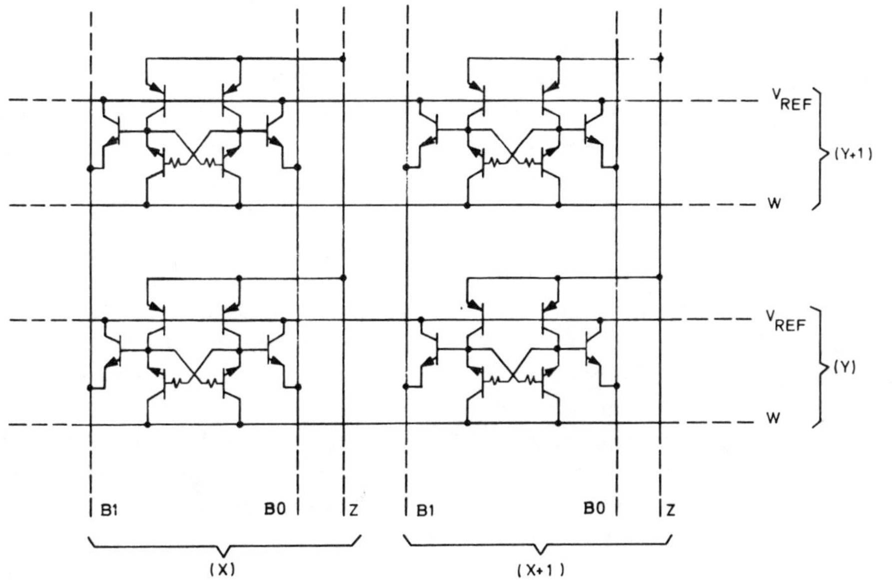

Fig. 2. Word-organized array of memory cells.

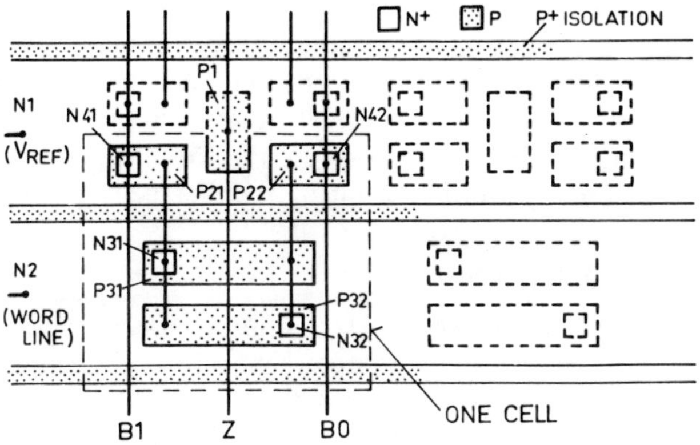

Fig. 3. Principal layout of memory cell (Fig. 1).

B. Principal Layout

The principal layout in Fig. 3 shows how the devices are arranged to achieve a small-size structure. Since the flip-flop transistors T_1 and T_2 in Fig. 1 are inversely operated, a common n bed N_2 can be used for all those transistors of a common word. As this n pocket is, in addition, utilized as a word line, an n$^+$ buried layer is preferably applied to achieve lower series resistances. The n$^+$ layer also enhances the inverse current gain of T_1 and T_2. The diffused p and n regions P_{31} and N_{31} as well as P_{32} and N_{32} of T_1 and T_2, respectively, are cross connected to provide the cross coupling of the flip-flop transistors. The p regions P_{31} and P_{32} have been prolonged so that the metal line Z can cross the devices. The resulting resistance appears in the cross-coupling path of T_1 and T_2 in Fig. 1.

A second n stripe N_1 provides the bases of the lateral p-n-p load transistors of two adjacent word rows. In order to reduce series resistances of the N_1 pockets, additional metal lines running in parallel with the bit lines and interconnecting these N_1 stripes can be inserted (not shown in the layout of Fig. 3). A common emitter P_1 is used for the collectors P_{21} and P_{22} of two cells. The n-p-n decoupling transistor T_5 can be merged with the N_1 base and the P_{22} collector of the p-n-p load transistor T_3. Accordingly, this applies to transistor T_6. Therefore, only the additional n$^+$-diffused areas N_{42} and N_{41} are necessary to implement these transistors. The n$^+$ emitters are tied to the bit lines $B0$ and $B1$. Thus, in total, the layout yields a truly integrated circuit with no need for double-layer metalization or area-consuming crossunders. Of course, the series resistance of the n-pocket lines have to be considered in the array design, depending on operating conditions.

C. Operation

In the word-organized array of Fig. 2, the power lines Z have been chosen to run parallel to the bit lines. This leads to an optimum layout in single-layer metalization, as has been shown in Fig. 2, but the Z lines can be also laid parallel to the word lines. The currents for all cells of a common Z line are supplied by a common current source, and the voltage V_{ref} for the common-base load transistors has to be chosen equal or larger than the word-line voltage. Adequate current sharing is provided because of the excellent base–emitter voltage tracking of the p-n-p devices, as will be demonstrated later by the experimental results.

When a cell is selected we raise the corresponding word-line voltage so that the bit-line potentials are determined by the potentials of the cell on the selected word line. Then reading is performed by simply sensing the potential difference on the corresponding bit-line pair. Since the bit-line transistors operate as emitter–followers, a multiple of the cell current is available to charge up the bit-line capacitances. This results in a fast READ operation even at relatively low cell currents, which is one of the outstanding features of this novel memory cell.

For writing, the word-line voltage is raised as for a

READ operation and the corresponding bit line is held at a slightly lower potential (typically 100–300 mV below word-line voltage). For example, if T_1 in Fig. 1 is conducting, the bit line $B1$ of transistor T_6 is held on this fixed potential causing T_6 to conduct. Thus, transistor T_1 is switched off via the base of T_6, and T_2 can turn on. The peak current in bit line $B1$ is considerably larger than the current to be pulled off the base of T_1, due to the current gain of T_6. However, the average WRITE current is much smaller than its peak value, since the memory cycle time is much larger than this period of WRITE current flow. Obviously, this current can be reduced by reducing the effective current gain of the bit-line transistor at least at higher currents. Introduction of a collector series resistance to obtain saturation at higher currents is also possible—however, at the expense of some additional cell area. Another measure would be the insertion of special WRITE switch transistors; such devices can be integrated with minor area sacrifice. In any case, the influence of the n-bed series resistances on the WRITE current flow also has to be taken into account.

It should be mentioned that the cell switching time is mainly determined by the time required to charge up the capacitance at node C_1 (Fig. 1) by the p-n-p collector current. As soon as transistor T_2 turns on, the current in bit line $B1$ is switched off.

D. Array Power Modes

The p-n-p transistor load offers additional significant benefits. The current of the p-n-p transistors—and thus the cell current—can be controlled over a few orders of magnitude by small voltage changes across the emitter–base junction. Furthermore, different from ohmic load flip-flops, there is no discharge path through the load for the capacitances of the cross-coupled transistors. This permits complete turnoff of the cell current supply for a certain period of time that depends mainly on the leakage currents. These features can be utilized for various array powering modes.

A straightforward mode is to operate the memory cells at a constant cell current throughout standby and addressing. In this case, the switching speed of the p-n-p transistor has obviously no impact on the memory performance.

Usually, a good tradeoff between power dissipation and performance of a memory system can be achieved by a pulsed power operation of the memory cell. This means that relatively few memory cells of the selected chip are powered up during addressing in order to increase their READ/WRITE speed. In this power mode, it is conceivable not only to boost addressed cells, but also to turn off unselected cells during the address period. Such an operation may offer advantages for some applications, e.g., to simplify the selection scheme.

The choice of a specific power mode heavily depends on the memory system objectives and process technology. Therefore, general guidelines cannot be given.

III. Experiments

A. Actual Layout

The actual layout of the cell for a certain technology depends on many parameters: layout rules; desired power/performance range and operation mode of the cell; basic device parameters given by the technology (particularly parasitic effects); and finally parameters such as size, shape and organization of the array, chip size, and peripheral circuits. Hence, based on the principal layout scheme of Fig. 3, many somewhat different layout schemes can be devised, each with its special advantages.

The experimental cell designs have been primarily intended for very low power applications with cell standby currents in the order of 1 μA. A test chip with fully wired 2×3 arrays, single cells, and single devices has been designed to be processed with a conventional technology (no gold doping). Fig. 4 shows the actual layout of this approach; the subcollector areas are not shown in the drawing for the sake of clarity. The cell size is only 13.7 mils^2 in a conventional 6-μ epitaxial layer process with a minimum metal linewidth of 0.4 mil and a minimum spacing of 0.2 mil on mask level. A photomicrograph of a test chip section is depicted in Fig. 5 showing a cell array, where a 2-word $\times$ 3-bit matrix has been wired.

B. Device Measurements

As pointed out in Section II-C, a uniform current sharing of the p-n-p load transistors connected to a common supply is desired for the memory-cell array. There was no suitable chip metalization pattern for measuring this tracking parameter on chip level. As a substitute, the p-n-p load devices from eight randomly picked chips of the same wafer have been analyzed. Although this is a rather pessimistic selection, the total deviation of the individual collector currents at a fixed base–emitter voltage has been less than ± 20 percent in the current range of $I_c = 1 \cdots 100$ μA. This includes the variation between the four collectors of a particular device.

Fig. 6 shows the common-base current gain of the four-collector p-n-p transistor versus the total collector current. With rather conservative emitter–collector distances of approximately 6 μ (9 μ on mask level) an alpha of 0.75 has been obtained in a current range between 1 and 100 μA. This means that 75 percent of the current supplied to the cell reaches the cross-coupled flip-flop transistors, so that only 25 percent of the cell current is not utilized. Because of the large basewidth the cutoff frequency is 3 MHz at 100-μA emitter current. This latter value can be substantially improved by decreasing the basewidth in order to achieve a better performance in a pulsed power operation.

The inverse current gain[2] of the n-p-n flip-flop tran-

[2] For comparison, the forward current gain of these devices is about 50 at 100-μA emitter current.

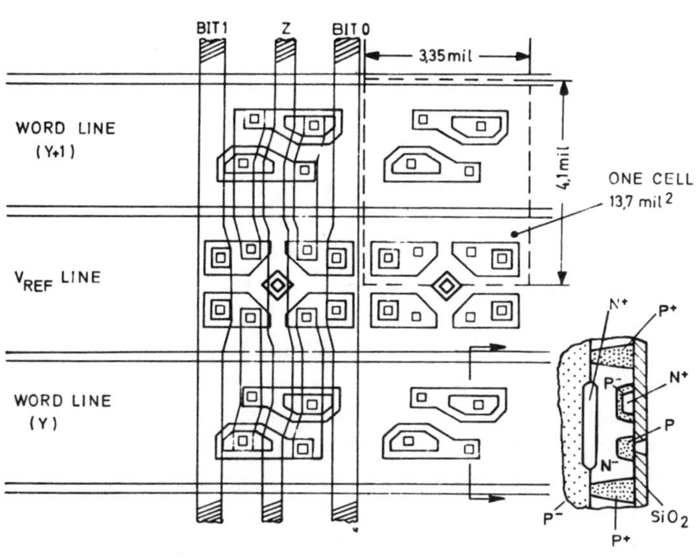

Fig. 4. Actual layout of memory cell.

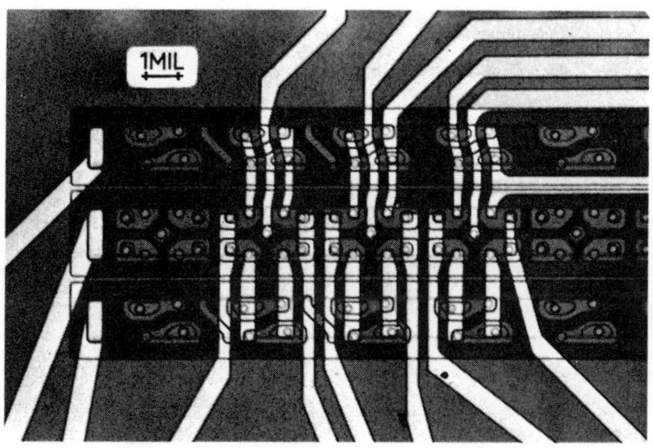

Fig. 5. Chip microphotograph of memory-cell array (2 × 3 cells are wired).

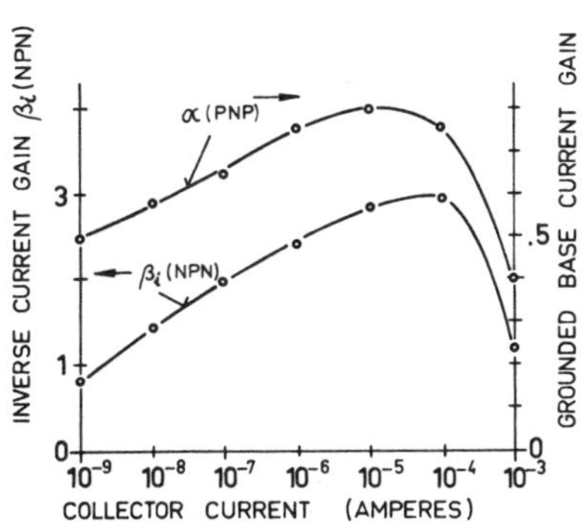

Fig. 6. Grounded-base current gain $\alpha_{p\text{-}n\text{-}p}$ of four-collector p-n-p transistor and grounded-emitter-current gain β_I of inverse n-p-n transistor.

sistor is also depicted in Fig. 6. It remains well above unity down to currents in the order of 10 nA. Since the current gain α (p-n-p) of the load devices also drops little, the cells retain their information at these extremely small dc currents. This has been verified by measurements: the cells proved to be stable at currents of 10–50 nA, corresponding to a power dissipation of 0.02–0.1 μW. Similar to the dc gain, the cutoff frequency $f_{T\,\mathrm{inv}}$ in the inverse mode is also much smaller than in normal operation ($f_{T\,\mathrm{inv}} \approx 15$ MHz at 100 μA). As in the case of the dc gain, this is not essential for the cell operation.

Because of the highly integrated cell structure one has to pay attention to parasitic transistors, especially as no gold doping may be employed. Gold doping would deteriorate both the lateral p-n-p and the inverse n-p-n transistor. If the lateral p-n-p transistor is kept out of deep saturation there is only a very small parasitic current to the p^+ isolation walls and substrate, which does not harm the cell operation.

The inverse n-p-n transistors require more attention. Fig. 7 depicts a schematic cross section showing the two cross-coupled flip-flop transistors of one cell and the devices of an adjacent cell. Here, transistor T_1 has been assumed conducting. Since it is inversely operated, the p-n junction of the p base and the n epitaxy is heavily forward biased. Hence, the p base can also act as an emitter and inject holes into the n region, where they can be partially collected by the other p regions. The numbers at the arrows in Fig. 7 indicate the measured currents. 20 percent of the total base current I_B has been found to flow into the substrate, mainly via the p^+ isolation walls (not drawn in the figure). Another 20 percent reaches the base of the other cell transistor T_2, but only 0.2 percent is collected by the p region (base) of a transistor belonging to the adjacent cell.

The effect of the substrate current is a slight reduction of the effective inverse current gain of the flip-flop transistor. The current injected into the base of T_2 has to be taken over by the collector of T_1 via the cross connection, thus requiring a minimum inverse current gain β_1 somewhat larger than unity to assure the cell bistability. Finally, the current injected into an adjacent cell might cause this to switch, if it is operated at a much lower current level (pulsed power operation). To reduce this coupling, e.g., an n^+ stripe can be laid between the devices to enhance the minority carrier recombination. This has been done in some parts of the cell array layout in Fig. 5. A decrease of the parasitic current by a factor of three has been observed, but even without this measure it is sufficiently small. In total, the parasitic transistors do not impose real problems, but have to be considered in the design.

C. Cell Performance

The measured READ and WRITE delay times have been plotted in Fig. 8 versus the constant cell current I_{cell} (no pulse powering). In these measurements a capacitive bit-line load corresponding to 16 cells has been provided in order to simulate a 16 word × 32 bit array. The READ delay time in Fig. 8 has been defined as the time period between the beginning of the word select pulse and the appearance of a 100-mV differential signal on the bit line pair.

According to Fig. 8, the cell exhibits a good READ performance as expected from the emitter–follower opera-

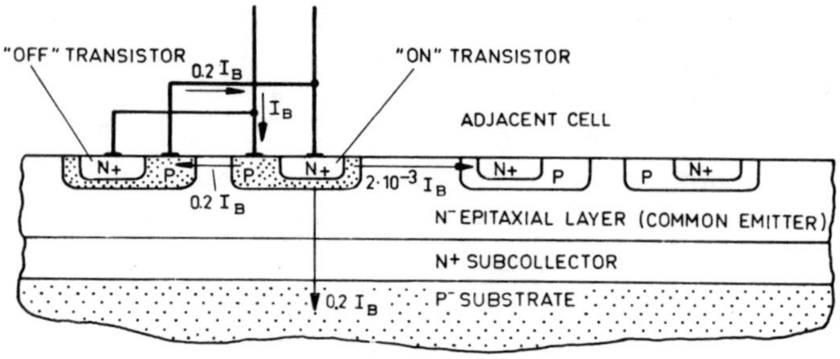

Fig. 7. Schematical structure of inverse n-p-n flip-flop transistors, showing parasitic current flow.

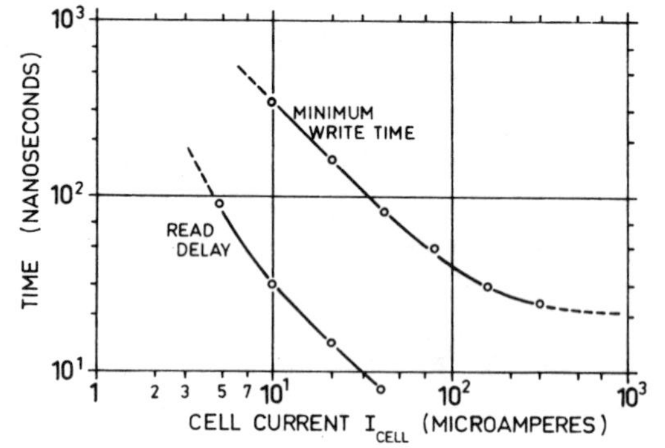

Fig. 8. READ delay and minimum WRITE time versus constant cell current I_{cell}.

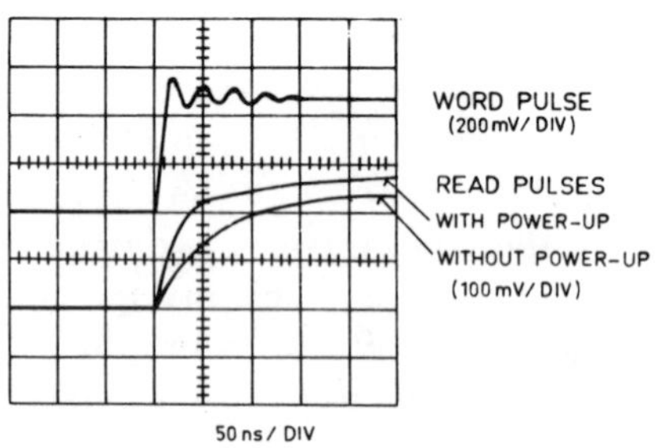

Fig. 9. READ signal for 10-μA cell current with and without powering-up to 250 μA.

tion of the bit line devices. The WRITE delay time, however, is considerably longer. This is not so critical in many applications; on the contrary, it offers the advantage of an excellent cell stability.

At low currents, a powering up of the cell during addressing significantly decreases the delay times. As an example, the oscillogram of Fig. 9 demonstrates how the READ speed is improved by pulsing the cell current from a 10-μA standby current to 250 μA synchronously with the word-line selection pulse. A 100-mV output signal is obtained after 10 ns, as compared to 30 ns without power-up. The WRITE delay time has been reduced from 400 to 80 ns at the same powering conditions.

IV. Conclusions

The novel dc-stable flip-flop memory cell investigated and described in this paper is characterized by a small-size structure and concurrently by an extremely low power dissipation. The mergence of devices, the replacement of ohmic resistors with active devices, and the utilization of isolation pockets for interconnections have led to the highly integrated cell structure. In particular, inversely operated n-p-n flip-flop transistors and lateral multicollector p-n-p load transistors are employed.

The memory cells have been successfully operated in the submicrowatt power range due to the p-n-p load devices. These load transistors also permit a variety of pulsed power operation modes to provide a wide range of power/performance tradeoffs. Remarkable in this context is the excellent READ performance even at a low cell power.

The feasibility of the novel approach has been fully verified by 2 × 3 array test chips processed in a conventional 6-μ epilayer bipolar process. With conservative dimensions (e.g., 0.4-mil minimum metal linewidth on mask level) the cell needs only 14 mil^2. At a total standby power of 10 mW, an array access time of 10 ns and a WRITE time of 80 ns has been achieved for a simulated 16 × 32 bit array.

From these characteristics (summarized below) it is concluded that monolithic READ/WRITE memories can be built in *bipolar* transistor technology with much higher density and much smaller power dissipation than expected so far. The memory cell characteristics are as follows.

1) Lateral p-n-p transistors as current–source load devices.

2) Inversely operated common-collector n-p-n transistors.

3) Cell size—14 mil^2 (6-μ epitaxial, 0.4-mil metal lines).

4) Standby power dissipation < 0.1 μW.

5) Access time 10 ns (array only) and write time 80 ns for 512-bit array and power-up of 0.5 mW per cell.

Acknowledgment

The authors wish to thank the members of the Applied Physics Department at the IBM Laboratories, Boeblingen, Germany, for their excellent support.

References

[1] L. L. Vadasz, A. S. Grove, T. A. Rowe, and G. E. Moore, "Silicon-gate technology," *IEEE Spectrum,* vol. 6, Oct. 1969, pp. 28–35.

[2] D. A. Hodges *et al.,* "Low-power bipolar transistor memory cells," *IEEE J. Solid-State Circuits,* vol. SC-4, Oct. 1969, pp. 280–284.

[3] W. Jutzi and C. Schuenemann, "Ein Symmertisches Bistabiles Thyristor Flip-Flop," in *Mikroelektronik,* vol. 3. München-Wien, Germany: Oldenburg, 1969, pp. 277–294.

[4] S. Waaben and H. A. Waggener, "100-ns electronically variable semiconductor memory using two diodes per memory cell," *IEEE J. Solid-State Circuits,* vol. SC-5, Oct. 1970, pp. 192–196.

[5] J. Mar, "Two-terminal transistor memory cell using breakdown," *ISSCC Digest Tech. Papers,* pp. 10–11, 1971.

[6] K. Jannigudin *et al.,* "A switched collector impedance memory," *ISSCC Digest Tech. Papers,* pp. 14–15, 1971.

[7] P. T. Panousis, "A trim memory employing both n-p-n and high-gain unijunction transistors," *ISSCC Digest Tech. Papers,* pp. 16–17, 1971.

[8] S. K. Wiedmann and H. H. Berger, "Small-size low-power bipolar memory cell," *ISSCC Digest Tech. Papers,* pp. 18–19, 1971.

Design Considerations for a High-Speed Bipolar READ-ONLY Memory

JOHN C. BARRETT, MEMBER, IEEE, ARNDT BERGH, MEMBER, IEEE, TOMAS HORNAK, SENIOR MEMBER, IEEE, AND JOHN E. PRICE, MEMBER, IEEE

***Abstract*—This paper describes the design evolution and performance of a high-speed integrated-circuit READ-ONLY memory fabricated with all-bipolar technology. Two circuit configurations are discussed. The first is based on a 1024 × 1 memory organization and the second building on the first with the addition of variable memory formats, three-state output circuitry, and a chip enable function. Performance figures cited verify that a ROM with complete on-chip decoding, input and output compatibility to standard IC logic specifications, and 1024 bits of memory achieves access times of less than 50 ns within the framework of proven IC technology.**

INTRODUCTION

THE application of integrated-circuit techniques to READ-ONLY memories (ROM) has made practical the use of ROMs in a number of diverse applications. These include character generation, code conversion, table lookup, arithmetic logic, and control logic.

Although the first-named examples probably represent higher volume usage, an application with far-reaching implication to the computer industry is in the control-logic area. ROM basically implements a combinatorial logic network with an arbitrary logic function defined for each input word or ROM address. Sequential logic can be accomplished by establishing the proper sequence of inputs at successive clock intervals. Complex sequences of logic control functions can be accomplished easily by the proper structuring of the contents of the ROM.

The advent of very fast inexpensive bipolar ROMs has made practical their use in control applications and expecially in central processors of computers. It has become feasible to implement hardware instructions that require a complicated sequence of data manipulations. Higher level language instructions become reasonable with attendant impact on system architecture, especially in the area of operating system data handling.

In the control-logic application the ROMs are required to have speeds compatible with the logic speeds used in the processor. They also must have a high packing density to keep the ROM size small so that signal path delays are minimized. Finally, they must be price competitive with alternate control approaches.

GENERAL DESIGN CRITERIA

In optimizing the design of a high-speed integrated-circuit READ-ONLY memory, there are a number of criteria that must be considered. Four major criteria are listed in Table I.

The first two criteria conflict with each other (the well-known power/delay tradeoff) and must usually be considered together during circuit optimization (e.g., reduction of RC time constants by reduction of resistance results in increased power consumption). However, circuit configurations can be optimized to achieve fast switching action and good capacitive driving capability without

Manuscript received April 27, 1970; revised June 22, 1970.
The authors are with Hewlett-Packard Company, Palo Alto, Calif. 94304.

Reprinted from *IEEE J. Solid-State Circuits,* vol. SC-5, pp. 196–202, Oct. 1970.

TABLE I

1) Minimize circuit power dissipation
2) Minimize signal propagation delay
3) Minimize circuit area (minimum cost)
4) Minimize pad count (maximize reliability and minimize package cost)

incurring excessive power consumption. Power-gating techniques can be utilized (especially in the address decoders) to effect power reduction without speed degradation. Minimization of active and passive device area (subject to adequate current-handling capability) leads to reduced capacitance and higher speed. Use of Schottky diode for storage time reduction in saturated transistors also serves to improve high-speed performance. Thus, a variety of circuit, component, and processing techniques may be employed to minimize power dissipation and propagation delay time.

The third criterion, that of minimizing circuit area, not only serves to minimize fabrication cost but also to improve high-speed performance, due to reduced capacitance at switching nodes. Reduced circuit area can be achieved by straightforward shrinkage of component area (consistent with masking and diffusion tolerances) and also by considerations of logic organization and circuit design optimized for minimum area. For example, using a coincident select structure rather than a linear select structure allows a considerable reduction in the number of address decoders, and hence circuit area. Simple, yet effective, circuit designs can be used to minimize area. For example, the memory-cell array circuit, to be described later, is designed to occupy a single common-collector isolation, hence minimizing area to about 6.25×10^{-6} cm^2 (1 mil^2) per cell. The bit pattern for the memory-cell array is specified by programmable contact openings in the silicon dioxide, rather than by using a programmable metal mask. It will be shown later that this technique almost halves the area of the memory-cell array.

The fourth criterion, that of minimizing pad count, implies that on-chip rather than off-chip address inversion and decoding be utilized, in direct conflict with previous considerations of minimum power and minimum circuit area. However, it is felt that the advantages of a fully decoded ROM justify the extra power and circuit area consumed. For example, with on-chip decoding, the ROM is easier to test, since there are fewer pads to probe. Inexpensive standard 16-lead dual-in-line packages (DIP) may be used, rather than approaches involving multichip assembly on complex multilayer substrates; thus reducing current packaging costs, and minimizing the complexity of package-to-package interconnections.

In accordance with the above design criteria, the various circuit elements in the ROM were designed so as to minimize both circuit area and the power/delay product figure of merit. Topological considerations involved in laying out the ROM exerted a strong influence on its organization and circuit design. All of which says that a total design concept was applied that brought together all these integrated-circuit constraints to optimize the design wherein electrical and topological circuit design was perceived as a single entity.

Design Objectives

The electrical characteristics of the ROM should be such that they are compatible with the majority of user requirements. A list of pertinent ROM characteristics includes format, access time, supply voltages, power dissipation, input voltages and currents, output voltages and currents, and temperature range.

Optimum format is highly dependent upon the application but to be cost competitive a minimum of 1024 bits is required. Access time must be less than 100 ns to satisfy a majority of high-speed applications. Supply voltages should be compatible with normal digital integrated-circuit supplies of 5 volts $\pm$ 10 percent. Power dissipation is perhaps less critical to the user than to the designer, but should be minimized for enhanced reliability. Input and output signal levels should be compatible with standard digital integrated-circuit types, such as TTL, DTL, and CTL. Buffered inputs minimize input current requirements and permit many ROMs to be addressed in parallel. A chip select capability enables ROM outputs to be tied in parallel for expanded total word capacity.

The operating temperature, ranging from 0 to +85°C, covers most environmental conditions likely to be encountered.

Circuit Components

While a variety of components are capable of realization with integrated-circuit structures, it is felt that only three (n-p-n transistors, Schottky diodes, and diffused resistors) are required for the fabrication of high-performance bipolar ROMs. p-n-p transistors, both lateral and vertical, generally exhibit characteristics that result in degraded circuit performance, due to their limited current gain and bandwidth. Also, the tighter control required of epitaxial film thickness and resistivity results in a lower yield process and hence higher cost READ-ONLY memory.

Circuit Design

Our ROM design effort was accomplished in two cycles. The first led to the realization of a 1024 bits of memory in a bit-oriented configuration that provided a single bit output, which gave it the descriptor "1 × 1024." It is the first design specifically discussed here. A second design that offered significant improvements over the first is then described. The tradeoffs between will become obvious during the design discussion that follows.

1 × 1024 Circuit Design

A block diagram of the logic organization of the 1 × 1024 ROM is shown in Fig. 1. There are a number of functionally distinct circuits contained in the ROM, con-

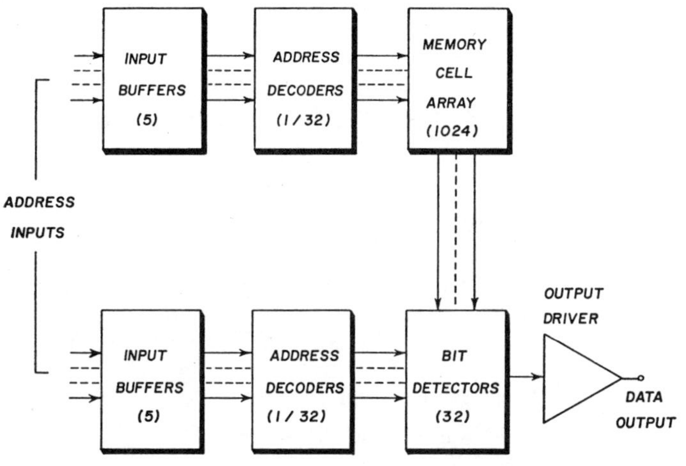

Fig. 1. 1 × 1024 ROM block diagram.

sisting of input buffer inverters, address decoders, memory-cell array, bit detectors, and output drivers.

Buffer-Inverter Circuit

An example of a high-performance input buffer inverter is shown in Fig. 2. The use of an active load transistor T_3 results in a low output impedance both in the high and low output states, giving high fan out and capacitive driving capability. The input resistor ratios $R_1 : R_2$ and $R_3 : R_4$ are designed so that transistor T_1 turns on before T_2, and hence the base of T_3 is pulled low before T_2 turns on, thus minimizing supply current transients that normally occur with active load output stages. With signal propagation delays of about 5 ns and average power dissipation of 10 mW, this circuit offers high-speed low-power performance in a small area (only two collector-isolation areas).

Address Decoder

Fig. 3 shows a circuit schematic of a memory address decoder that combines high performance with circuit simplicity. True and complement address signals are fed from the input buffer inverter via metal connecting lines to the emitters of the multiple-emitter transistor T_1. Programmable contact openings define emitter connections with either true or complement input signals to generate the appropriate function at the decoder output. The emitter–follower T_2 provides high pull-up drive without requiring resistor R_1 to be low value. Pull-down drive is supplied by the input buffer through the emitter–base junction of T_1, resulting in fast switching action with reduced power dissipation. The operation of the circuit is such that both T_1 and T_2 operate only in their active region, thus minimizing delay and storage time effects. Also, since the collector–base junction of T_1 is always reverse biased, the inverse current gain effect, which results in undesirable interactions between multiple-emitter inputs, is eliminated.

Typical performance figures for the address decoder circuit are propagation delays of about 5 ns at 4 mW

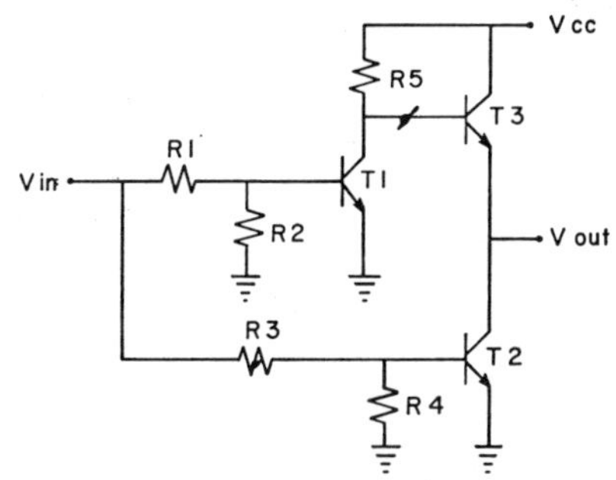

Fig. 2. Buffer-inverter circuit.

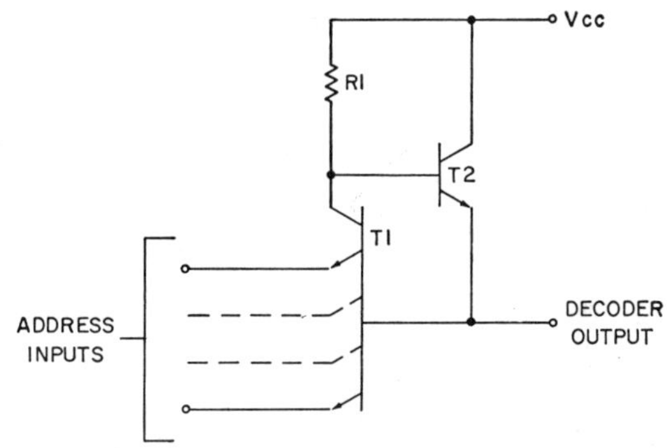

Fig. 3. Address decoder.

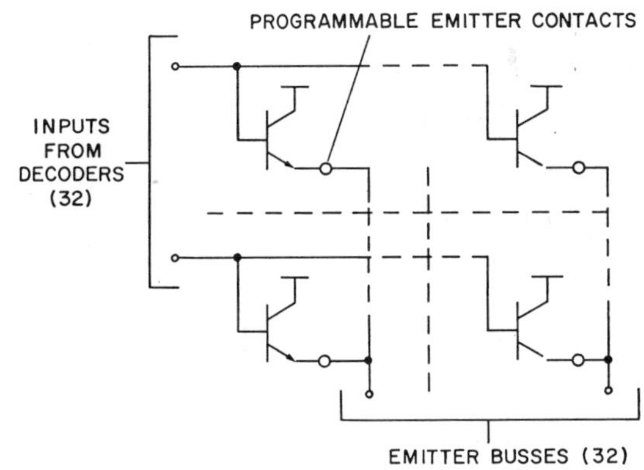

Fig. 4. Memory-cell array.

power dissipation with the circuit layout occupying about 40-mils^2 area.

Memory-Cell Array

Fig. 4 shows a portion of the memory-cell array, in which the orthogonal conducting lines are formed by base-diffused stripes in the horizontal direction and metal lines in the vertical direction, thus requiring only a single-layer metal technology. In order to prevent debiasing along the long base stripe and minimize the distributed RC network time constant an n^+ emitter diffusion was placed in the base stripe region and electrically connected to the base at intervals of 4 or 5 emitter inserts by a metal shorting bar (Fig. 5).

Fig. 6 shows how memory-cell area is minimized by programming the emitter contact openings rather than

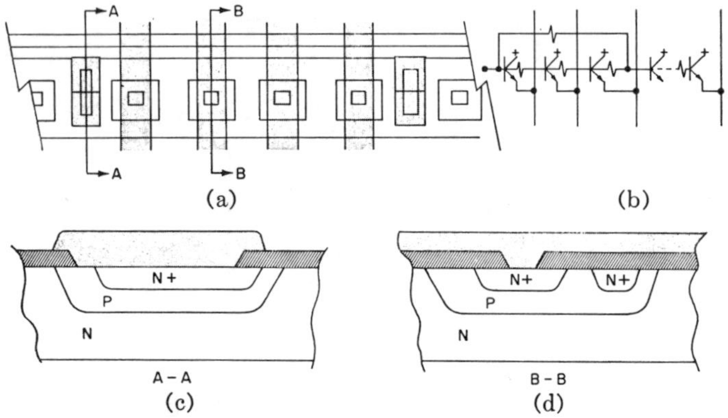

Fig. 5. Memory-cell device structure.

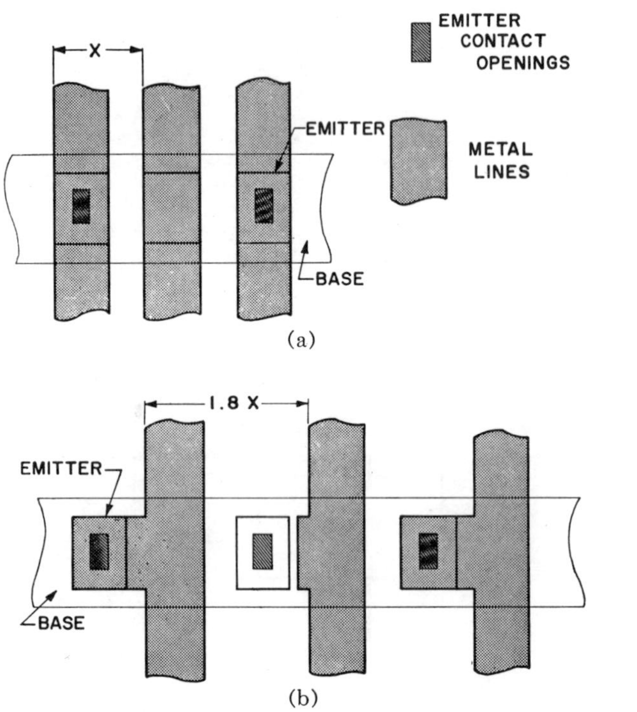

Fig. 6. Memory-cell contact programming. (a) Programmable emitter contact openings. (b) Programmable metal connections.

the metal connections. Center to center bit spacings differ by a factor of almost 2 to 1 for identical tolerances on diffusions, metal lines, and contact openings.

In operation for a 32 × 32 memory-cell array, the row select address decoders hold 31 out of 32 of the base stripes at a low voltage: the selected base stripe is pulled high and forward biases the base–emitter memory cells in that row. The outputs of these 32 emitter–followers are each fed into the 32 separate collectors of the bit detectors.

Note that 31 out of 32 of the bit select transistors will be held off, so that current is allowed to flow in only one of the 32 columns of the memory-cell array, thus minimizing power dissipation. Since the entire array of 1024 bits is encompassed in a single collector-isolation area, and the programmable emitter contact opening technique is used, a high-density memory-cell array is realized, occupying about 6.25×10^{-6} cm^2 (1 mil^2) per cell.

Bit Detectors

Bit detection consists of sensing the presence or absence of a high bit current in a selected vertical bus, and is accomplished as shown in Fig. 7. The column select address decoders hold the bases of 31 out of 32 of the bit select transistors ($T_1 - T_{32}$) at a low voltage: the selected base is pulled high, turning on the bit select transistor (say T_1). Then the presence (logic "1") or absence (logic "0") of bit current I_C is summed with the base current I_B to give two distinct current levels of $(I_{B_1} + I_{C_1})$ for a bit presence or I_{B_0} without a bit on the common-emitter output bus.

Output Driver

The circuit schematic of the output driver is identical in design to the input inverter circuit (Fig. 2).

The input signal is derived from the output bus of the bit detector. The input resistor values are chosen so that the threshold value of the input current I_{INTHR} (where the output changes state) is given by

$$I_{B_0} < I_{\text{INTHR}} < I_{B_1} + I_{C_1}.$$

I_{B_0} and $(I_{B_1} + I_{C_1})$ are the two distinct current levels on the output line of the bit detector. The presence or absence of an emitter contact opening at the addressed location in the memory-cell array will result in a low or high logic voltage level at the output of the output driver.

The input resistor ratios are also designed (like the input inverters) to minimize supply current transients during turn on and turn off. The totem-pole output results in a low output impedance both in the low and high input modes so that the effect of output loading on output rise and fall times is minimized.

1 × 1024 Performance Characteristics

The 1 × 1024 ROM was fabricated using standard IC processing techniques without gold doping or Schottky diodes. Access times were typically less than 100 ns, with a total power dissipation of 350 mW. The circuit was laid out on a 2.8 × 2.8 mm chip with basic masking tolerance of approximately 3 μm. The completed circuit is shown in Fig. 8.

4 × 256 Circuit Design

From the feedback gained in the evaluation by users of the 1 × 1024 configuration we established additional design goals for a second ROM design.

In addition to those previously discussed there were

- buffered inputs with $I_{in} < 100\ \mu$A
- access time less than 50 ns
- variable format (1 × 1024, 2 × 512, or 4 × 256)
- input and output compatible with CTL, DTL, and TTL
- WIRED OR output capability
- chip enable capability.

Obviously the inclusion of these features required a complete redesign that incorporated the best of the 1 ×

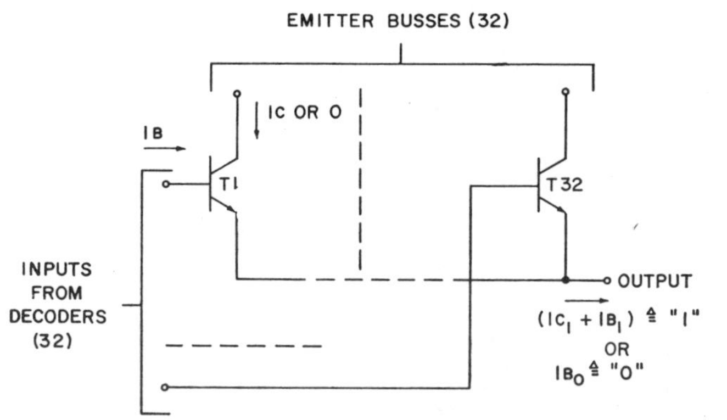

Fig. 7. Bit detectors.

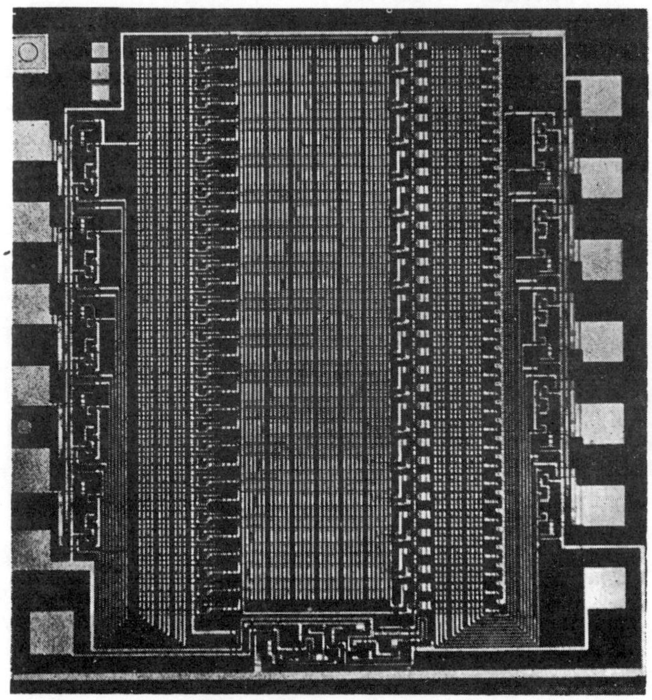

Fig. 8. 1024 × 1 ROM IC.

1024 design with significant additions. It was also decided to include the Schottky-barrier diode processing since our experience had clearly shown no yield loss and an increase in speed for the same power level of approximately 2 : 1.

Logic Organization 4 × 256

The block diagram in Fig. 9 shows the functional organization of the 4 × 256 ROM. The functions are similar to the 1 × 1024 organization but a third dimension of addressing in the output circuits has been added to permit the format variations specified. The chip enable function is incorporated in the third-dimension Z decoders. The memory array is partitioned into four 256-bit segments, each assigned its own output circuits. Coincident select addressing of one bit in each of the four segments is done by the X and Y decoders. These bits are further processed via the Z decoders to select the desired output combination. Dependent upon final metal option, the Z addresses select either one memory segment at a time for the 1 × 1024 format, two segments at a time for the 2 × 512 format, or all four segments in at a time for the 4 × 256 format.

Input Buffer Inverter

The input connection is through a special high-impedance front end circuit that allows the ROM to interface *directly* with either TTL or CTL logic. Two inverters provide both true and complement logic variables to the decoder inputs. The buffer front end and the first of the two inverters is shown schematically in Fig. 10. Schottky clamps are only placed across the transistors that would normally be driven into hard saturation.

The input buffer operates in the following manner. When V_{in} is low, T_2 is driven into saturation by the base drive current through R_2 and the emitter of T_1 is clamped at approximately 500 mV so that only leakage current flows across the nonconducting emitter–base and collector–base junctions. When V_{in} goes high, both T_3 and T_4 are driven into saturation and T_2 is pulled out of saturation and acts as a current source. The low input current is then independent of input voltage level.

X and Y Decoders

The decoder circuits are the same as utilized in the 1 × 1024 design. However, the upper node of the resistor R_1 and the collector of transistor T_2 (see Fig. 3) of the decoders $Y_0 \cdots Y_{15}$ are connected to the inverted output of the Y_{16} buffer instead of to V_{cc}. The corresponding points of the decoders $Y_{16} \cdots Y_{31}$ are connected to the "true" output of the Y_{16} buffer. Since one of the outputs of the Y_{16} buffer is always in the "low" voltage state, half of the Y-decoder power is saved.

Bit Detectors

Bit detection is accomplished through an emitter-coupled scheme (Fig. 11) that permits parallel selection of 4 bits within the four memory segments. The transistors ($T_{00} - T_{1023}$) in the four common-collector memory segments that are driven high by one of the 32 Y decoder lines perform as input transistors to four emitter-coupled switches each of which is assigned to one of the four memory segments. The X decoder (one out of eight format) drives transistors ($T_0 - T_7$), which through a resistor network ($R_0 - R_7$) selects one of eight possible vertical columns in each memory segment. The second half of the emitter-coupled switches is made up of four multiple-emitter bit-detector transistors ($TA - TD$), each assigned to one of the four memory segments.

One input to each of the 4 output drivers is controlled by the collector current of the bit-detector transistor. If a collector current is flowing (i.e., the emitter in the 1024-bit memory is disconnected), the driver output is in the "high" level, i.e., at logical "1." With no collector current the output is at logical "0."

To achieve a high switching speed the voltage swing of the X, Y, and Z decoder output nodes are limited by clamping. The clamping levels and also the dc base voltage

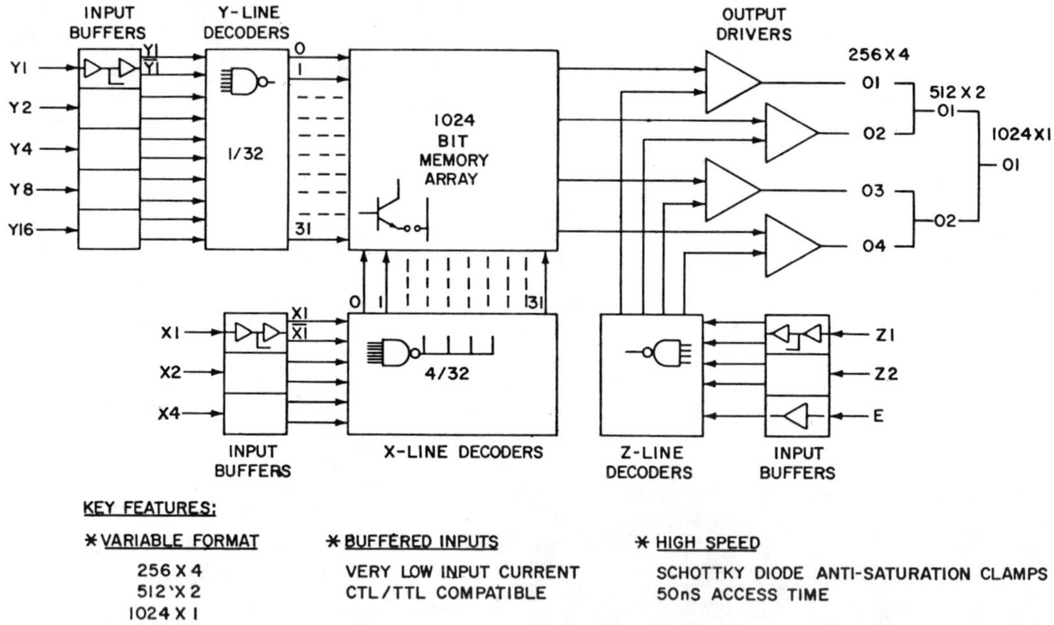

Fig. 9. 4 × 256 ROM block diagram.

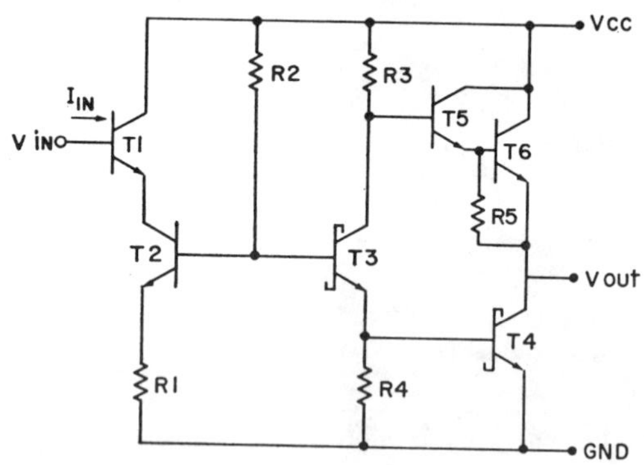

Fig. 10. Input buffer with Schottky-clamped transistors.

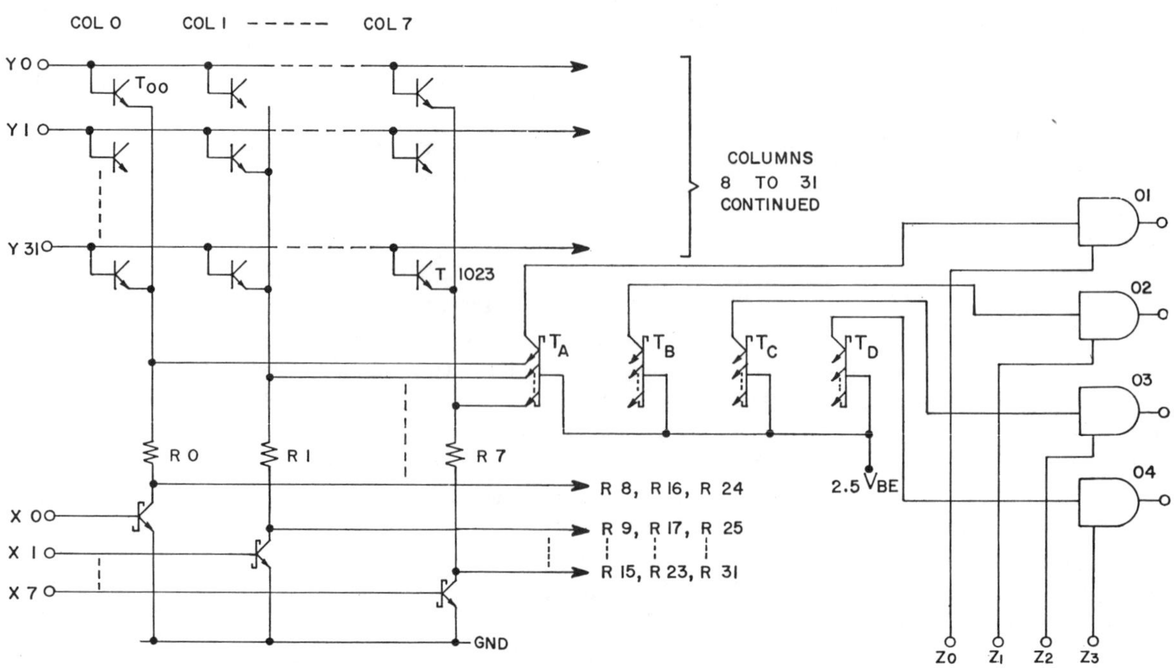

Fig. 11. Bit detection schematic.

of the bit-detector transistors ($TA \cdots TD$) are designed to be constant multiples of the base–emitter voltage V_{BE} and thus track with changing temperature. A second consequence of this is that the total chip power varies with temperature, with minimum power occuring at maximum chip temperature!

Output Driver

The output driver (Fig. 12) is an inverter with transistor T_1 operating as the enabling control device. With T_1 off the driver operates normally.

With T_1 in saturation, the base voltage of transistors T_3 and T_5 are pulled low, turning off T_3, T_4, and T_5 independent of the logic input state. The output then becomes a free floating node whose impedance is a function only of the capacitance and leakage currents of the node. This feature allows a parallel interconnection of the outputs of several drivers while still preserving an active pull up and pull down of the output node.

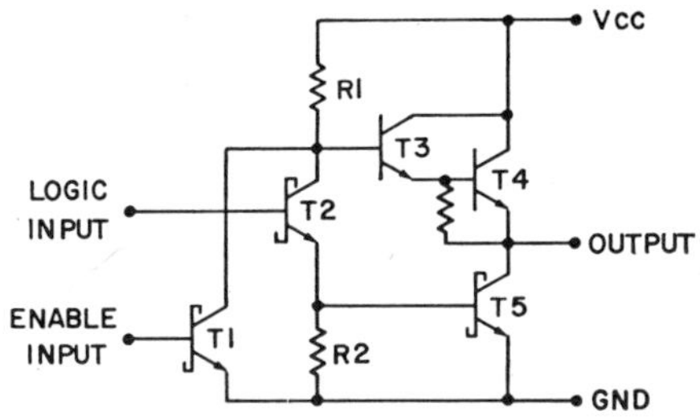

Fig. 12. Output driver and enable circuit schematic.

256 × 4 Performance Characteristics

Typical 256 × 4 ROM characteristics are shown in Fig. 13. Note that the ouput is specified to both sink and drive considerable output current, so that both current sinking logic types (TTL, DTL) and current sourcing logic types (RTL, CTL) can be driven from the ROM outputs.

Input current is very low over the range of input voltages, allowing the ROM to be driven by current sourcing or current sinking logic types. Access time (or signal propagation delay) is a function of the various signal path delays through the ROM that occur as the input address changes, so that for all the possible combinations of address pair changes, the spread in access time ranges from 15 to 40 ns.

The completed integrated circuit is shown in Fig. 14. Chip size is 2.75 × 2.9 mm.

Typical Characteristics (V_{cc} = 5 volts: T = 25°C)
"0" level output voltage = 0.4 volt at I_0 = +10 mA
"1" level output voltage = 3.3 volts at I_0 = −10 mA
"0" level input current < ±1 μA at V_{in} = 0 volts
"1" level input current = 50 μA at V_{in} = 5.0 volts
supply current = 120 mA
access time = 15–40 ns
format = 4 × 256
2 × 512
1 × 1024

Fig. 13. 4 × 256 Performance data.

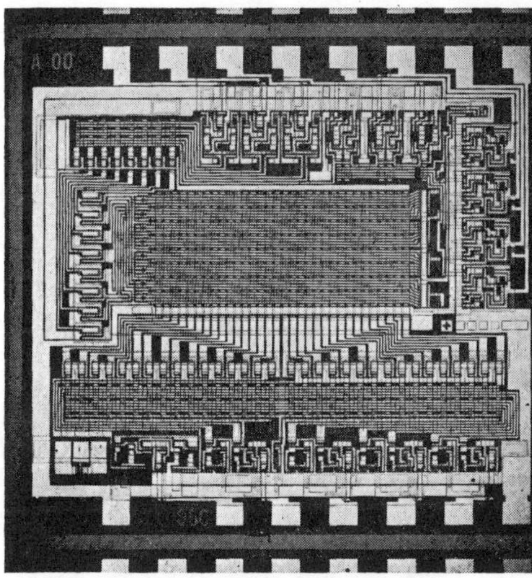

Fig. 14. 4 × 256 ROM IC.

Summary

Novel integrated-circuit design and logic organization have been applied to the design of bipolar READ-ONLY memories for minimum access time, maximum utility, and minimum cost.

A variety of integrated-circuit ROMs have been designed and fabricated in accordance with the design criteria previously described. Selection of a metal mask option allow various ROM formats (4 × 256, 2 × 512, and 1 × 1024) to be fabricated on a standardized chip whose area is about 7.8 mm². The combination of stable high-performance bipolar integrated circuitry with flexible functional specification of READ-ONLY memories promises to contribute to the designer an additional functional element that can be quickly modified to his custom requirements.

Acknowledgment

The authors gratefully acknowledge the contribution of many coworkers to the design, fabrication, and evaluation of the ROMs described here. In particular the significant contributions made by H. Hofmann and Z. Skokan are deeply appreciated.

Part III
MOS Transistor Memories—Technology and Circuits

Editor's Introductory Comments

A very wide range of alternative metal-oxide–semiconductor (MOS) (sometimes called MIS for metal-insulator-semiconductor or IGFET for insulated-gate field-effect transistor) processes have been applied or proposed for application to semiconductor memory. Silicon-gate p-channel MOS (III,1), aluminum-gate p-channel ion-implanted MOS (III,2), and aluminum-gate silicon-on-sapphire complementary MOS (III,3) techniques are described here. Other prominent techniques include the original aluminum-gate p-channel MOS, aluminum-gate n-channel MOS, silicon-gate n-channel MOS, and silicon substrate complementary MOS with either aluminum or silicon gate.

As in the case of bipolar memory, circuit design and chip layout seem to have had more impact on overall memory performance and potential cost than the choice of fabrication process. The diversity of processes is troublesome in the sense that complete reliability data and multiple sources of functionally identical components probably will not develop for all these technologies. At present, p-channel aluminum-gate and silicon-gate processes (PMOS) are well established. In the long run, silicon gate will likely displace aluminum gate for most p-channel applications. Silicon gate offers better performance, easier interfacing with bipolar circuits, and more freedom in chip layout due to the availability of the deposited silicon layer for crossovers and interconnections. Silicon-gate p-channel components are described in III,4, III,5, and III,6. An aluminum-gate p-channel memory is described in III,7. This offers performance superior to that obtained with the nearest silicon-gate equivalent (III,6), illustrating the point that good circuit design may be more important than advanced technology.

Silicon-gate n-channel devices (NMOS), just being introduced, offer better performance, easier interfacing with bipolar devices, and/or lower power consumption than p-channel devices. Unless unexpected reliability or manufacturability problems arise, the n-channel silicon-gate process likely will receive growing emphasis. A 4096-bit n-channel silicon-gate memory is described in III,8. This design is aimed at moderate performance, very high density, low power consumption, and ease of interfacing with bipolar circuits. A high-speed device is shown in [1].

Part III, 5–8 illustrate dynamic circuit techniques. Circuit density is increased and power consumption reduced, each by a factor of four to eight, compared to static MOS memory designs. Dynamic techniques thus are likely to dominate in large memory applications where cost is a primary consideration. Good circuit and system design can minimize the system-level impact of the required refreshing operations.

The very high density, good performance, and low power consumption of new designs such as III,8 suggest that complementary MOS (III,3) will not have much appeal for large storage capacity memory applications. Dynamic CMOS memories are not practical, and static CMOS memories cannot compete with dynamic PMOS or NMOS in density. Reducing power consumption below the levels shown in III,8 offers little economic benefit for ground-based systems. CMOS is attractive for electronic wrist watches, spacecraft electronics, and possibly general-purpose logic, but not for large ground-based memories.

Part III,9 describes an electronically writable READ-ONLY memory that may be erased by irradiating with ultraviolet light. Other techniques, such as MNOS (metal-nitride-oxide-semiconductor) and MAOS (metal-aluminum-oxide-semiconductor) can provide electrical writing and erasing, but these schemes require more complicated address decoding and data buffering circuitry. The device described in III,9 is commercially available; MNOS and MAOS devices are not.

Reference

[1] T. R. Walther and M. R. McCoy, "A three transistor MOS memory cell with internal refresh," in *IEEE Int. Solid-State Circuits Conf., Dig. Tech. Papers*, 1972, pp. 14–15.

Silicon-gate technology

Low-cost, large-scale integrated electronics based on metal-oxide-semiconductor design benefits from the application of silicon-gate technology

L. L. Vadasz, A. S. Grove, T. A. Rowe, G. E. Moore **Intel Corporation**

Silicon-gate technology provides an advantageous approach for implementing large-scale integrated arrays of field-effect transistors. Its advantages—principally resulting from the low threshold voltage and the self-aligned gate structure buried under an insulator—ease the problem of interfacing these circuits to bipolar integrated circuits and increase both their performance and functional density, making MOS integrated circuits easier and more economical to use. This article reviews recent progress with this technology and shows its application to the construction of complex digital functions as illustrated by a memory circuit.

Much of the early enthusiasm for metal-oxide-semiconductor (MOS) integrated circuits stemmed from their promise as low-cost digital functions. Optimism was based upon the apparent simplicity of integrated circuits made completely with self-isolated enhancement-mode transistors that could be constructed with a single diffusion step. By comparison, conventional bipolar integrated circuits typically involve four such steps.

However, the growth of MOS circuitry has been slow—compared with early expectations—for many reasons.

1. Early devices often had reliability problems, principally relating to ion drift in the oxide films or to dielectric breakdown in the thin gate insulator resulting from static electricity.

2. There were device-availability problems, and systems designed for MOS structures were delayed. (This suggests that the production yields were not as high as expected.)

3. For economy and to overcome the high-impedance drawback of MOS, it's important to build large functions. This aim contributed to the yield difficulties.

4. The expected cost advantages have also been slow in materializing. Yield problems have kept device costs high; but extra cost was incurred interfacing MOS to bipolar electronics, an important requirement in many systems. Generally, this factor related to the incompatibility of power supply and of signal levels between the two types of circuits, and tended to force designers to all-MOS or all-bipolar systems rather than use each to its advantage.

5. Performance of MOS circuits has restricted their applicability. The speed of MOS circuits is limited—for quite fundamental reasons—below bipolar structures.

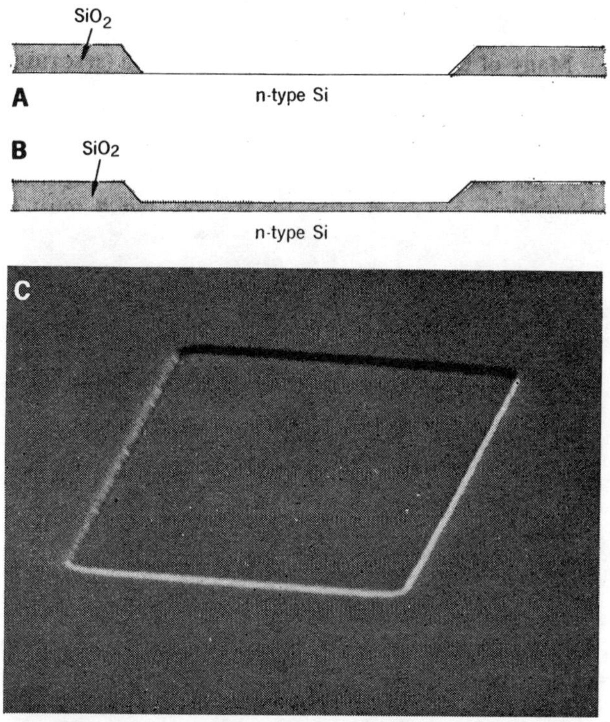

FIGURE 1. A—First oxide cut after initial oxidation for subsequent source, drain, and gate region. B—Second oxidation to define gate oxide thickness. C—Scanning electron microscope (SEM) photograph of a transistor fabrication at stage B.

On the other hand, MOS circuits clearly have certain advantages.

1. Their use permits logic circuits to be designed in a straightforward manner.

2. The ability to use such combinations as series gates greatly simplifies the circuitry, and the large fan-out and fan-in potential of a voltage-controlled device allows large margins for reliable operation.

3. New circuit forms, particularly multiphased-clocked circuits, are practical.

4. High-functional packing density can be achieved—probably two to ten times as much function per unit silicon area as with conventional bipolar circuits using the same geometrical tolerances. The cost of processing

Reprinted from *IEEE Spectrum,* vol. 6, pp. 28–35, Oct. 1969.

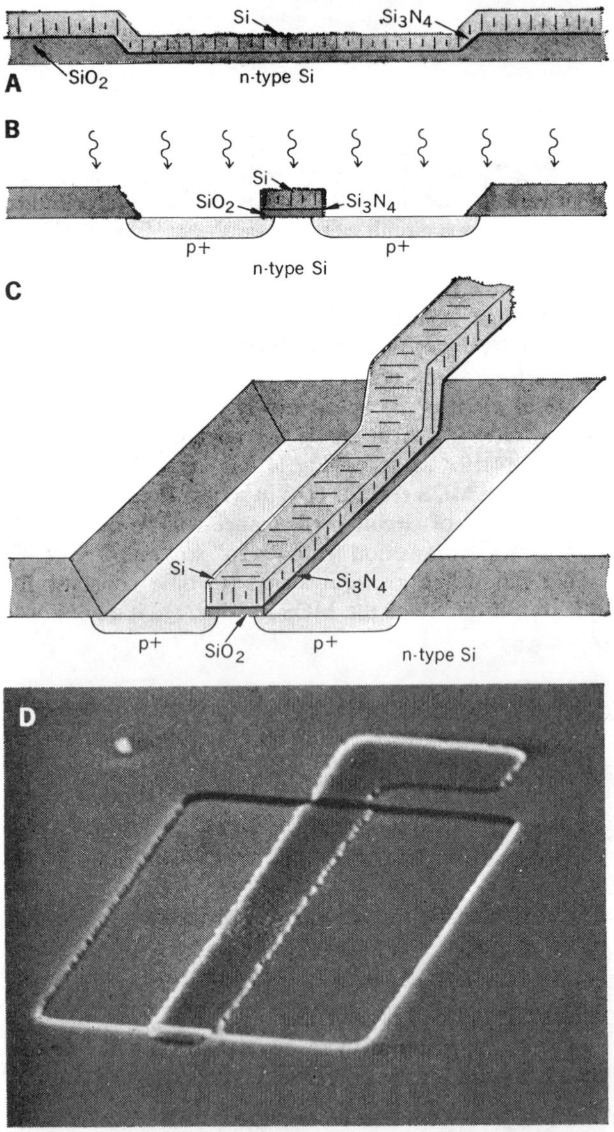

FIGURE 2. A—Device of Fig. 1(B) after deposition of nitride and amorphous silicon sandwich. B—Definition of gate insulator and gate electrode dimensions and subsequent boron diffusion. C—Perspective view of structure of 2(B). D—SEM photograph of the structure at end of this fabrication stage.

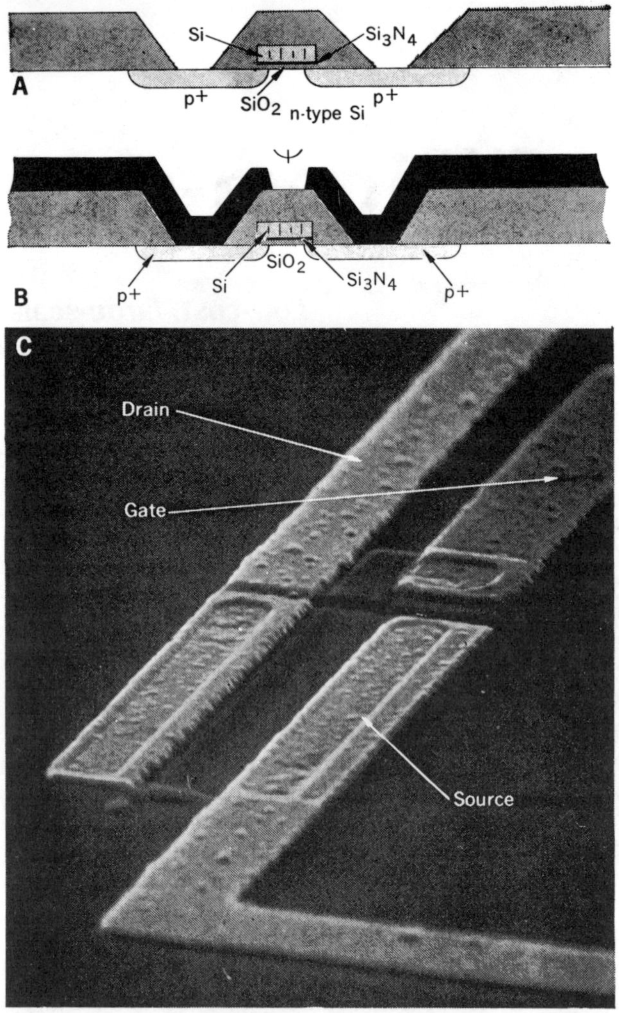

FIGURE 3. A—Device of Fig. 2(B) after deposition of silicon dioxide and definition of contact regions. B—Final structure after deposition and definition of metal interconnections. C—SEM photograph of the completed transistor structure.

is more nearly related to area than complexity, pointing to low cost per function in complex circuits.

5. Mask layout is simplified. No special isolation structures need be considered. The high impedance levels and resulting lower current densities greatly simplify the interconnection problems by allowing use of high-resistance signal paths. Fewer mask layers are generally required. The result is many fewer oversights.

Much effort has gone into improving MOS technology to remove the limitations while preserving as many of the advantages as possible.

In general, there are several ways that each limitation can be attacked. However, the use of polycrystalline silicon for the gate electrode affords a unique and powerful combination of properties that removes many of the objections that have been raised regarding metal-insulator-semiconductor (MIS) devices.[1,2]

A look at the technology

As in conventional MOS technology, the starting material for silicon-gate technology is n-type silicon. The wafer is first placed into an oxidizing atmosphere at high temperature and a relatively thick (about 1 μm) layer of silicon dioxide is formed on its surface. Next, regions for the source and drain of the final device *and* the eventual channel regions are defined by photomasking. The oxide is then etched from this area, as shown in Fig. 1(A). The slice is placed into an oxidizing ambient again, and a layer of silicon dioxide about 0.1 μm thick is formed in the window [Figs. 1(B) and 1(C)].

Subsequently, a thin layer of silicon nitride (Si_3N_4), another insulator, is deposited onto the entire surface of the wafer. This layer affects both the electrical characteristics and the reliability in a desirable manner.

Next, a thin layer of amorphous silicon is deposited on the wafer. The device structure at this point is as illustrated in Fig. 2(A).

The structure now is returned to photomasking for removal of the silicon and silicon nitride layers except where the gate area is to be or where the silicon film is intended for use as a circuit interconnection layer. The thin oxide is also removed by exposing it to an oxide etch, as shown in Fig. 2(B).

Finally, the wafer is placed into a diffusion furnace and boron—a p-type impurity that diffuses with relative rapidity in silicon, but very slowly in silicon nitride or in silicon oxide—is diffused into the surface of the structure. Consequently, after exposure, boron impurities will convert both the n-type silicon wafer, where exposed, and the deposited silicon layer into p-type silicon; but the n-type regions under the oxide-covered areas are unaffected [Figs. 2(B), 2(C), and 2(D)].

The device structure, except for the necessary interconnections, is now complete. A layer of silicon dioxide is deposited onto the entire surface—the exposed regions of the silicon wafer, the thick oxide layer covering most of the surface, and the deposited gate silicon. Openings are photoetched in this deposited silicon dioxide layer wherever a contact between the subsequent metalization and the underlying silicon wafer or deposited silicon is desired, as shown in Fig. 3(A). Aluminum is evaporated onto the surface so that it enters into these contact openings, and the desired interconnection patterns are defined by another photomasking operation. The device now appears as shown in Figs. 3(B) and 3(C).

It is desirable to protect the device both from mechanical damage to its intricate interconnection pattern and from contamination. For this reason, another layer of glass is deposited onto the wafer surface, and patterned by subsequent photomasking and etching to expose the pads where bonding wires are to make contact with the aluminum interconnection pattern.

It is interesting to note that, in fabricating this device, there is only one diffusion step, but there are five steps in which a thin film—an insulator or a conductor—is deposited. In the silicon-gate technology, yields, performance, and reliability—thus, success or failure—all depend on the manufacturer's ability to control the thin-film deposition steps.

Overcoming MOS's past, poor performance

Two key aims of MOS technological developments have been easier interfacing with bipolar circuitry, and higher component densities to decrease further the cost

FIGURE 4. A dc stable (static) inverter circuit. Figure of merit depends on operating voltages.

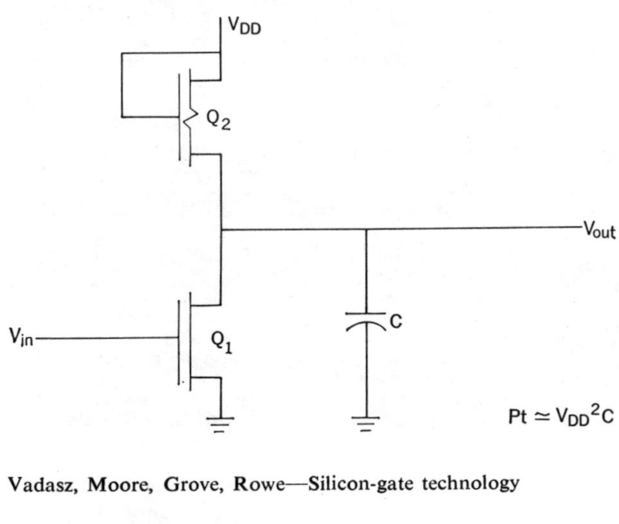

of electronic functions. The silicon-gate technology achieves significant improvements in both directions.

Low threshold voltage. The minimum operating voltages that must be used in an MOS circuit are determined by the gate voltage required to create a conducting channel between source and drain—called the *threshold voltage* (V_T). It is important that V_T be relatively low, not only to interface to bipolar circuits, but to improve circuit performance.

To demonstrate the improved performance attainable with the low threshold voltage compared with standard high-voltage technology, consider the use of a simple dc stable (static) inverter. The significant points of comparison are power dissipation and speed. A figure of merit is the speed–power product. An appropriate tradeoff can be made for any given application.

The inverter stage of Fig. 4 has an MOS transistor (Q_1) and an MOS resistor (Q_2) in series. It can be shown that the limit of circuit performance will be determined by the output transition from "zero" to "one."

We can define an equivalent *RC* time constant for this transition. Since the MOS resistor, Q_2, is a nonlinear resistor, a true time constant will not exist, but one can be approximated for the initial part of the transition. In terms of this time constant, the advantage of a low-voltage circuit is simply that, for the same power level, the resistor value can be significantly lower. In the power-dissipating state the dissipation of an inverter is given by

$$P \simeq \frac{V_{DD}^2}{R_{eq}} \tag{1}$$

The characteristic time constant is given by

$$t \simeq R_{eq}C \tag{2}$$

where *C* is the load capacitance.

The combination of these two equations will result in the expression for the power–speed product

$$Pt \simeq V_{DD}^2C \tag{3}$$

Equation (3) illustrates that, for a given capacitive load, lower-voltage operation gives a more meritorious figure.

In the silicon-gate technology, lowering the threshold voltage is achieved because a dielectric sandwich of silicon dioxide and silicon nitride is used under the gate and the gate electrode is made of heavily doped p-type silicon rather than of aluminum, as in conventional MOS technology. The threshold voltage depends on the *electric field* in the silicon reaching a certain critical value before a conducting channel appears. The higher the dielectric constant of the insulator, the lower the gate voltage has to be.

Silicon nitride has roughly twice the dielectric constant of silicon dioxide, so its use might be expected to lower threshold voltages. When silicon nitride is deposited directly onto a silicon wafer, however, spurious, undesirable charges appear near the interface between the nitride and the silicon and actually increase the threshold voltage. But if the layer of silicon dioxide under the layer of nitride is thick enough, these charges can be eliminated and one can still benefit from the higher average dielectric constant of the resulting sandwich.

The material from which the gate electrode is made also has a significant effect on the magnitude of the threshold voltage. The gate electrode and the silicon wafer act, in a way, like the two electrodes of a battery; an electromotive force is developed between them that can add to or subtract from the threshold voltage. Accordingly, the use of p-type silicon for the gate produces a threshold voltage for p-channel devices whose magnitude is about one volt lower than it would be in a comparable aluminum-gate MOS. Thus, the combined use of a silicon nitride–silicon dioxide sandwich and of a p-type silicon gate electrode results in threshold voltage values that are typically two volts.

Reducing parasitics. Reductions in parasitics attributable to silicon-gate technology are equally impressive. This advantage accrues from a basic feature of this technology—the gate electrode and the p-type diffused regions are "self-aligned" as shown in Fig. 2(B). The gate electrode, or rather the dielectric under the gate electrode, is the diffusion mask that defines the edge of the source and drain regions. Thus, they are automatically aligned. In conventional MOS technology, this alignment is done by the superposition of two sequential masking operations. The inherent inaccuracy of manually aligning two patterns causes a certain tolerance for misalignment that has to be included in the design. This tolerance wastes space, resulting in larger device structures, and, in particular, introduces relatively large interelectrode capacitances that degrade performance. Major advantages of the self-aligned gate structure result from these reduced parasitic capacitances.

1. *Reduced gate capacitance.* The thin gate oxide area is smaller. For a given channel length, the thin oxide region will extend over the source and drain regions only to the depth of the source–drain diffusion, typically <1 μm with the silicon-gate technology. Since standard technologies have as much as an 8–10-μm gate overlap above the source and drain regions, gate capacitance reduction can be as much as 50 percent.

2. *Reduced Miller capacitance.* From item 1, it is obvious that the gate-to-drain overlap capacitances have been reduced significantly by the use of a self-aligned gate. This reduction can be as much as an order of magnitude for equivalent device size.

3. *Reduced junction capacitance.* The reduction is mainly the result of reduced junction area. In a conventional metal-gate device, the size of the drain junction is larger primarily because a minimum separation has to be kept between gate metal and the line connecting to the drain. With the silicon-gate technology, the metal connected to the drain can overlap the gate regions since there is an insulating layer separating the gate from the metal. The only constraint is the proximity of the drain contact to the gate region; junction area reduction is about 30–40 percent.

High functional density. Smaller devices obviously contribute to higher packing density, but even more important is the circuit interconnection flexibility inherent in the silicon-gate technology. Figure 3 shows

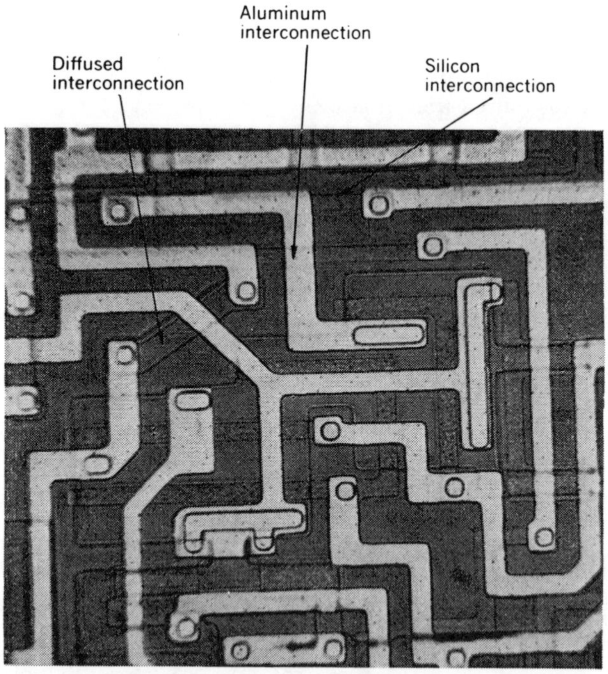

FIGURE 5. Photomicrograph of a small region of Intel's i-1101 256-bit RAM, showing the three possible modes of interconnection: diffused regions within the wafer, deposited silicon, and metal interconnecting lines.

FIGURE 6. Photomicrograph of static memory bits of three MOS memory products: (A) Intel's i-1101, (B) TI's 4003, (C) Fairchild's 3530.

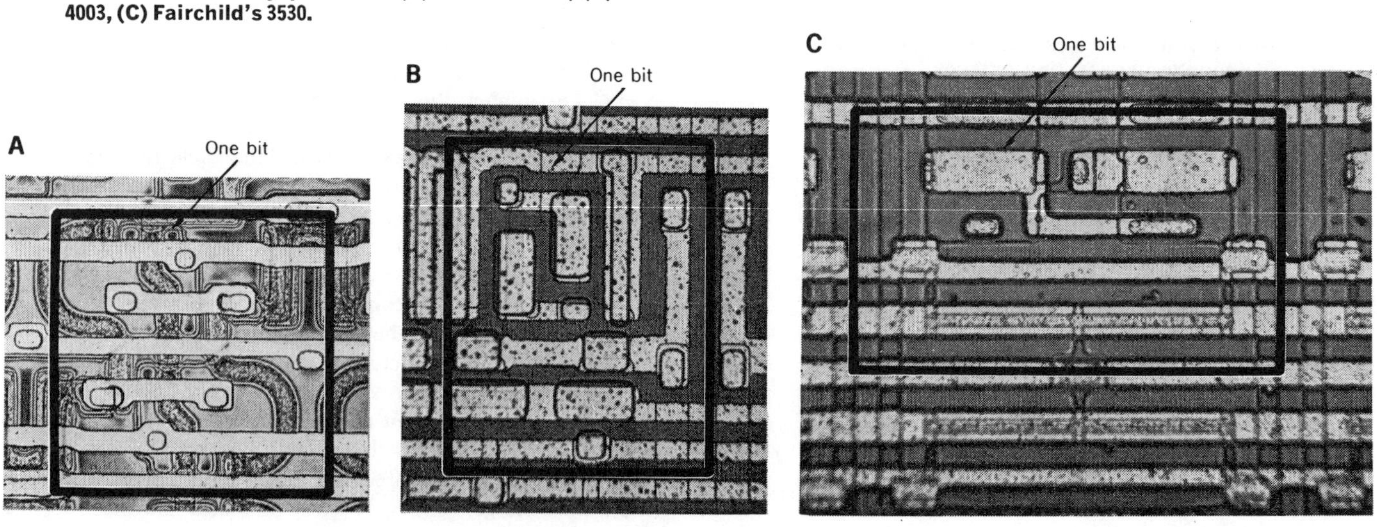

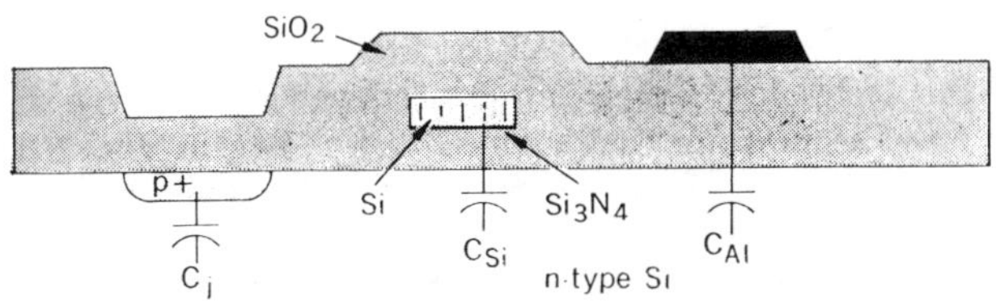

FIGURE 7. Capacitances associated with various types of interconnecting lines. Both metal and deposited silicon have a smaller capacitance per unit length than diffused region interconnections. For 10-μm-wide Ic lines, $C_j = 2.21$ pF/mm; $C_{Si} = 0.35$ pF/mm; and $C_{Al} = 0.28$ pF/mm.

that the aluminum interconnections are separated from the silicon-gate electrode by an insulator. Basically, this establishes all features necessary for metal patterns to cross the silicon film conductors, resulting in two layers of conductors above the substrate. The additional use of diffused conductors in the substrate, as in other MOS technologies, gives flexibility approaching that of three layers for interconnecting complex functions efficiently. These interconnections are illustrated in Fig. 5. By contrast, in the standard MOS technology, the only crossover possible is metal-crossing diffused regions.

An illustration of the higher packing density of this

FIGURE 8. Photomicrograph of Intel's i-1101 256-bit RAM. Block diagram is superimposed.

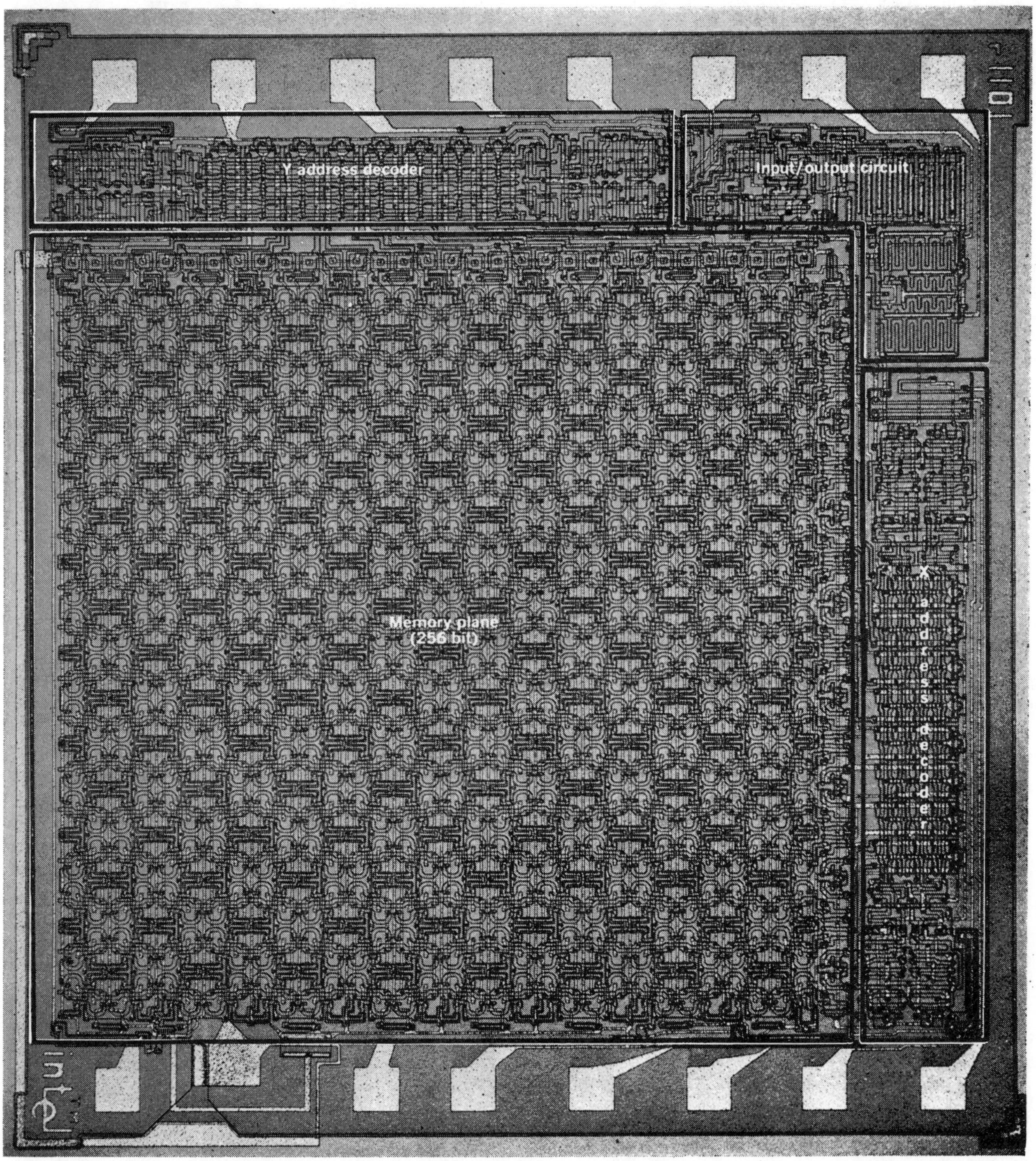

newer technology is given in Fig. 6, which compares a static memory bit from the silicon-gate memory array (described later) with a comparable bit from a circuit made using standard technology. The area saved approaches 50 percent. Similar area savings can be realized in other circuits, using static or dynamic schemes.

The use of deposited silicon for interconnection also reduces interconnection capacitances, further contributing to improved performance. Figure 7 shows the capacitance per unit length of three types of interconnection lines. Clearly, metal or silicon layers are associated with smaller parasitic capacitances than are diffused interconnections.

Lowering the failure rate. In addition to the device characteristic advantages and component density increase that silicon-gate technology provides, reliability benefits also result. The gate electrode and gate region in this device are buried under two layers of dielectrics. This shields the most sensitive region of the device from external contamination. In addition, the Si_3N_4 layer is a very effective barrier to ion migration. The use of lower voltages stresses the insulator less—also an advantage.

Manufacturability. It has been well established that yield is a strong function of integrated circuit area. Thus the higher packing density can be expected to produce higher yields for a given circuit complexity. The early protection of the sensitive, thin insulator region by the silicon-gate electrode minimizes the chance of damage during subsequent processing. It appears that the high yields and resulting low costs and product availability originally projected for MOS circuits will be realized with the silicon-gate structure.

Theory is one matter—practice proves

All of the advantages of the silicon-gate technology can be illustrated by Intel's 1101, a 256-bit, random-access memory element, fully decoded on the chip (i.e., including the circuitry necessary to select a particular bit from an eight-bit, binary-coded address) and with controls for read/write and inhibit modes provided. The functional block diagram superimposed on a photomicrograph of this element is shown in Fig. 8. The chip size is 3.10 × 2.8 mm. It is packaged in a 16-lead dual inline package (DIP). Operation and performance of this element are:

An 8-bit address code will select any one of 256 bits for either read or write operation. All address input logic levels are compatible with standard bipolar TTL or DTL logic levels. The mode of operation ("read" or "write") is determined by the R/W input control. In the "read" mode, the information from the memory will be available on the outputs, typically, less than 1 μs later than application of an address code. Note that there is no need to rewrite the data into the memory after a read operation since the read is nondestructive.

The "chip-select" function is accomplished by the "inhibit" input. This renders both *R/W* and *data input* leads ineffective and stops information transfer through the output buffer. The address decoder, however, is not inhibited—creating an effective increase in memory speed. The output leads are open while the memory array is inhibited, allowing OR-tying of many memory arrays.

Typical access times—defined as the lapse between application of an address code and the output of the TTL load—measured on the circuit are shown in Fig. 9.

The difference between positive-going output delay and negative-going output delay is due to the difference in driving capability of the push–pull output stage from the memory: A positive output transition from the TTL gate (negative transition from the MOS output) will always be 200–300 ns slower.

The access time degradation as a function of capacitive loading is also shown in Fig. 9. Note that the degradation is about 1 ns/pF.

Direct driving of bipolar circuits is accomplished by the use of a push–pull output stage on the i-1101. This output stage will typically "sink" 8 mA at a +0.45-volt output voltage (worst-case TTL logic zero level). Although this drive capability is reduced at higher tem-

FIGURE 9. Typical access time of i-1101 as a function of temperature. Outputs are those of TTL load and thus inverted from the actual memory output. Access time of i-1101 at a function of capacitive loading is also shown. Measurements are at room temperature.

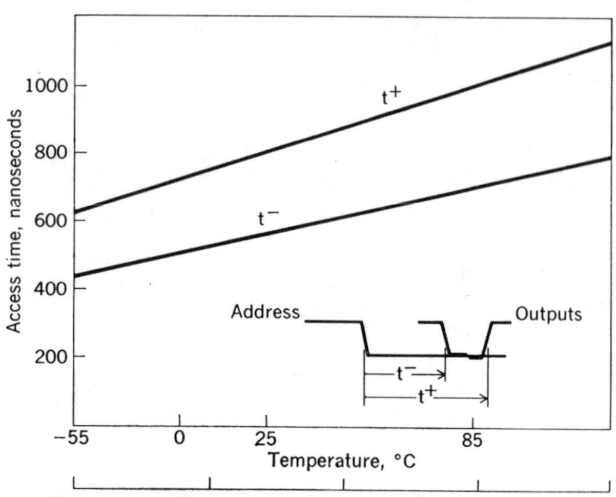

FIGURE 10. Typical current sinking capability of i-1101 at V_{out} = +0.45 volt (worst-case TTL, DTL logic zero level) as a function of temperature. Note that two worst-case TTL loads need only 2.8 mA, whereas the i-1101 can sink more than double this figure over the whole temperature range of operation.

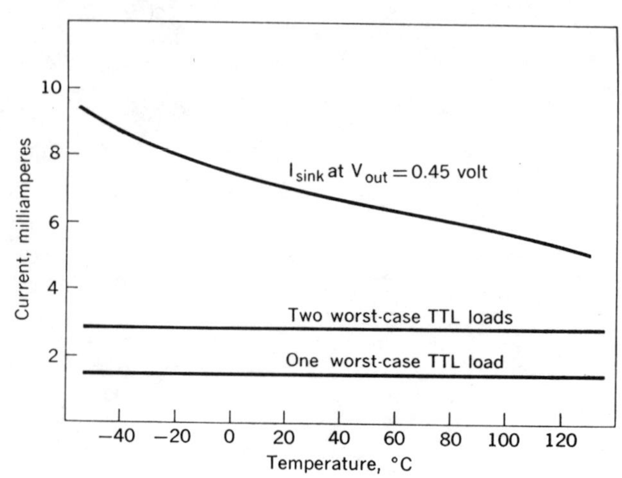

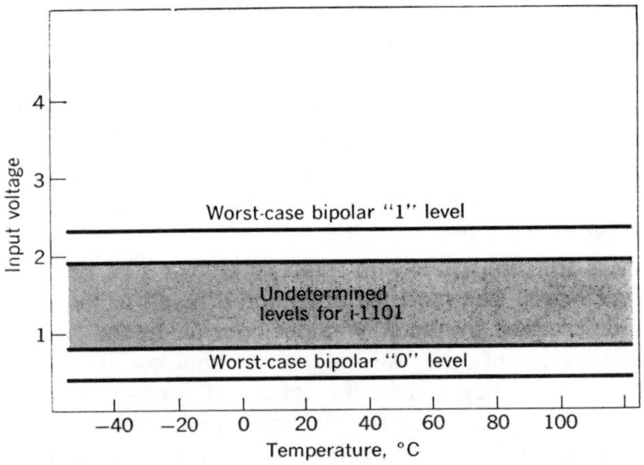

FIGURE 11. Invalid input voltage levels for i-1101 compared with worst-case logic zero and one levels of TTL gates.

FIGURE 12. Power dissipation in the cells and in the peripheral circuitry (I/O, decoder) for i-1101. Measurements are at room temperature.

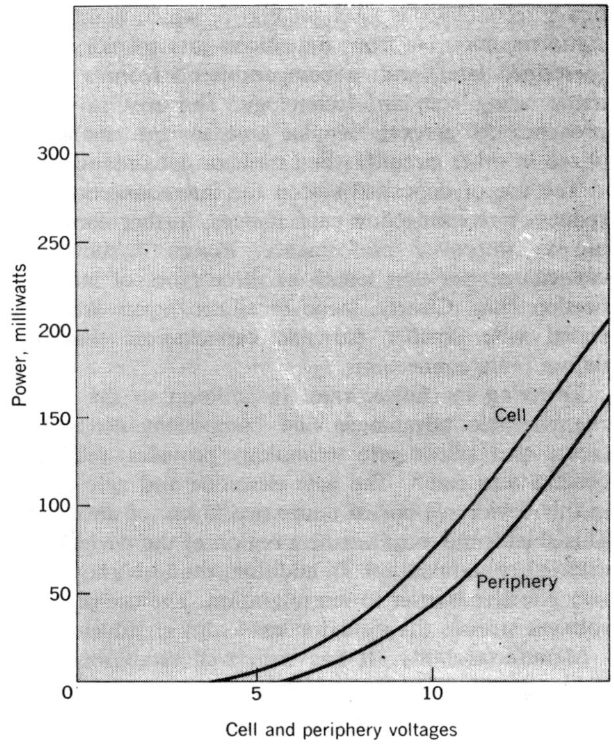

FIGURE 13. A 4096-word by one-bit RAM organization using i-1101 RAM elements.

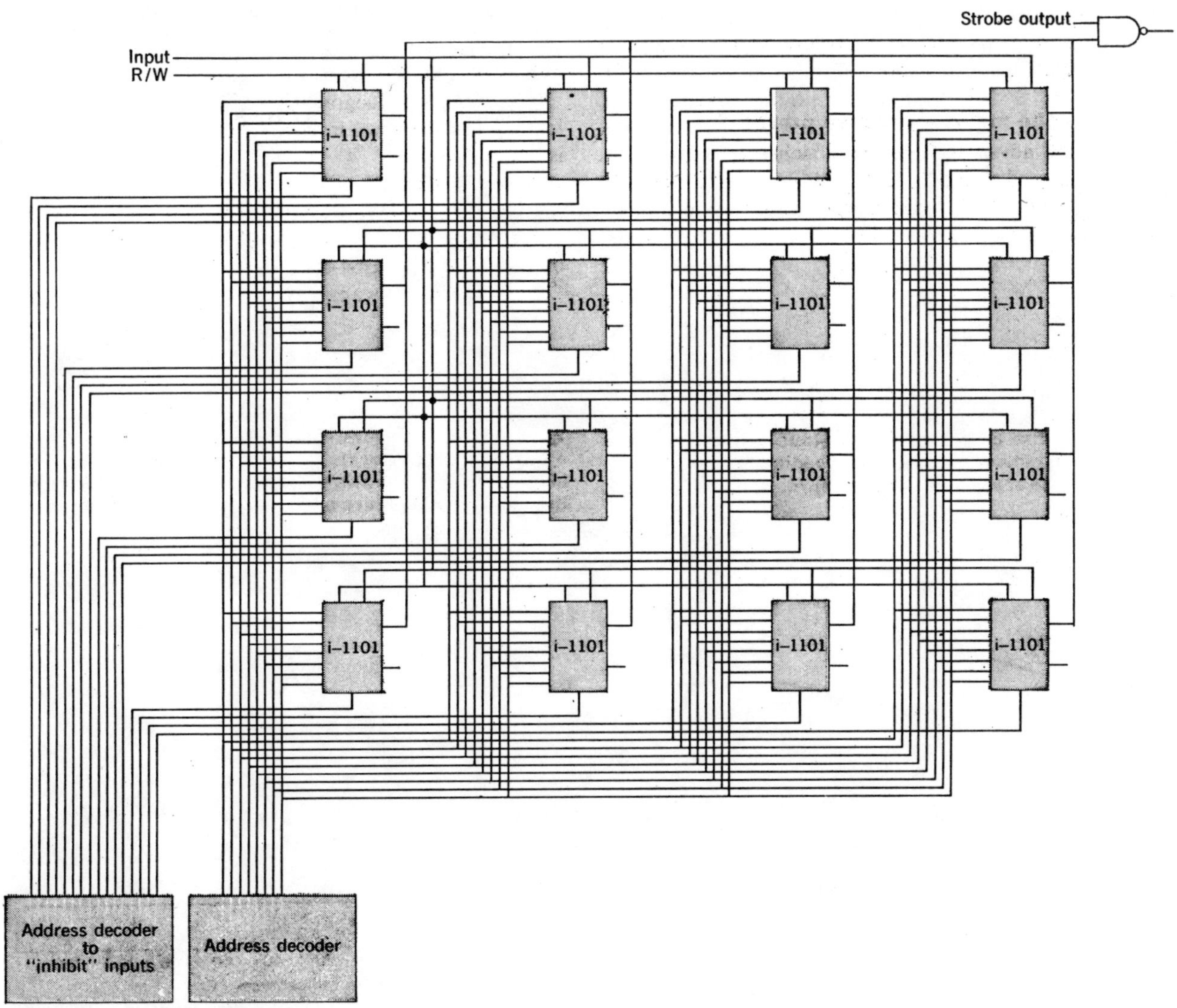

peratures, good fanout capability (>2 loads) is maintained even at 125°C. This is shown in Fig. 10.

Input voltage levels also are compatible with bipolar circuitry. Figure 11 shows the range of invalid input voltages at various operating temperatures. For comparison, worst-case logic 1 and 0 level specifications for bipolar circuits are also included. Ample margins exist at all temperatures for direct bipolar/MOS interfacing.

Power dissipation is typically less than 2 mW per bit for normal operation. This, however, can be lowered to below 50 μW. In standby mode—when the chip will only store information, but does not need to be accessed—the peripheral power supply can be completely shut down. This "idle" cuts the total power drain by a factor of 2. Furthermore, the cell power can be reduced considerably by reducing the cell voltage to -2 volts. This generates only -7 volts across the memory bits and is adequate for holding information in the memory cells. Figure 12 shows typical power dissipation in both the cells and peripheral (decode, I/O) circuitry. In this mode of operation, the total power dissipation is less than 12 mW, and corresponds to less than 50 μW/bit.

Larger memory arrays can easily be built with this memory element. Sixteen packages can make a 4096-word × 1-bit memory by simply OR-tying the outputs, paralleling all address lines, and decoding to each inhibit lead. This is shown in Fig. 13.

An interesting feature of an array of i-1101 memory elements is that the inhibit lead becomes part of the address inputs. This will allow improved performance for certain modes of operation. Since the inhibit/enable operation is much faster (~200 ns) than the address decoding, it is possible to address all 16 elements of the array of Fig. 13, at the same time inhibiting 15 of the 16 packages. Once the first element is accessed (say in 1 μs), 15 more bits of data at 200-ns intervals are obtainable only if the inhibit leads are switched. This gives a total of 4 μs for accessing 16 bits of data, or 250 ns average access time per bit.

The combination counts

To be sure, other MIS processes possess some of the advantages of the silicon-gate technology. Lowered threshold voltages can be obtained using different crystal orientation, or by utilizing multiple dielectric layers under the gate electrode.[3] Use of a molybdenum gate electrode[4] can also achieve a self-aligned structure and the use of ion implantation[5] can give the same result with conventional aluminum electrodes. However, the combination of desirable features available with the silicon-gate structures appears to be truly unique.

REFERENCES

1. Sarace, J. C., Kerwin, R. E., Klein, D. L., and Edwards, R., "Metal-nitride-oxide silicon field-effect transistors with self-aligned gates," *Solid State Elec.*, vol. 11, pp. 653–660, July 1968.

2. Faggin, F., Klein, T., and Vadasz, L., "Insulated gate field effect transistor integrated circuits with silicon gates," presented at the IEEE Internat'l Electron Device Meeting, Oct. 1968.

3. Nigh, H. E., Stach, J., and Jacobs, R. M., "A field-gate IGFET," presented at the Solid-State Device Research Conf., 1967.

4. Brown, D. M., Engeler, W. E., Garfinkel, M., and Gray, P. V., "Refractory metal silicon device technology," *Solid State Elect.*, vol. 11, pp. 1105-1112, Dec. 1968.

5. Bower, R. W., Dill, H. G., Aubuchon, K. G., and Thompson, S. A., "MOS field-effect transistors formed by gate masked ion implantation," *IEEE Trans. Electron Devices*, vol. ED-15, pp. 757–761, Oct. 1968.

Ion implantation offers a bagful of benefits for MOS

Better high-frequency performance is but one virtue of this new doping process; it also results in a broad range of low threshold voltages and makes possible new types of IC devices

By John Macdougall and Ken Manchester, *Sprague Electric Co., North Adams, Mass.*
and Robert B. Palmer, *Mostek Corp., Worcester, Mass.*

● Ion implanted MOS devices have been best known for their improved high-frequency performance over conventional metal oxide semiconductor IC's. But now device designers are learning that the ion-implantation process can produce even bigger performance dividends.

Not only do ion-implants yield consistently reproducible low thresholds, ranging below the standard 4 volts down to a TTL-compatible 1.5 volts, but the process permits fabrication of previously unrealizable IC's with devices such as depletion-mode loads and enhancement-type drivers on the same chip. This combination provides logic gates with speed-power products twice those of circuits with only enhancement-type loads.

What's more, ion implantation simplifies fabrication of n-channel IC's, heretofore rather tricky to make through diffusion methods, yet valuable because they are compatible with bipolar circuits. The n-channel allows complementary MOS IC's to be built that feature higher speed and lower power dissipation.

All this is possible because under the ion implantation process, dopants are introduced into the silicon IC wafer over a greater range of concentrations and with more precise control than is possible with diffusion techniques.

So far, the Mostek Corp., using techniques developed by the Sprague Electric Co. R&D Laboratories, has used ion implantation to manufacture—and sell commercially—several low-threshold IC's, including a dual 128-bit static shift register, a dual 256-bit dynamic shift register, and a 2,240-bit column-output character generator. These circuits operate at speeds comparable to other MOS IC's, but because of these ion-implanted channels, the Mostek devices have lower threshold voltages than are possible with the three other "low-threshold" MOS fabrication processes [*Electronics,* April 13, p. 118]. The threshold voltage, V_T, of ion-implanted IC's is 1.0 to 2.0 volts. The 1-0-0 process yields a threshold voltage of 2.0 to 3.0 volts, whereas silicon gate and silicon nitride techniques produce 1.5 to 2.5 volts.

Moreover, with ion implantation, it's possible to choose the threshold voltage. The threshold distribution can easily be shifted to any range the circuit designer considers optimum; it is not dictated by the process restrictions as it is in the other technologies.

Ion implantation offers still other advantages over these processes. Consider the 1-0-0 process. A transistor fabricated on the 1-0-0 plane of silicon exhibits less fixed-surface-state charge density per unit area, Q_{ss}, than a transistor made on 1-1-1 silicon. Thus, all factors being equal, the 1-0-0 transistors would have a lower V_T as indicated by the equation

$$V_T = -\frac{t_{ox}}{Ke_o}(Q_{ss} + Q_B) + \phi'_{MS}$$

where t_{ox} is the thickness of the gate dielectric, K is the dielectric constant of the gate material, e_o is the permittivity of free space, Q_B is the charge per unit area within the surface depletion region at the onset of strong inversion, and ϕ'_{MS} is the effective difference between the work function of the bulk silicon and that of the gate metal.

Unfortunately, carrier mobility in a 1-0-0 transistor's channel is lower, resulting in slower switching. More seriously, the field-oxide threshold voltage—the value at which unwanted transistor action occurs between metal on the surface of thick portions of the oxide and the silicon substrate—is low. This means that there is a greater probability of parasitic turn-on, and yields therefore tend to drop. The field-oxide threshold typically is 15 to 20 volts in 1-0-0 transistors, against at least 30 volts in conventional 1-1-1 transistors and in silicon nitride, silicon gate, and ion-implanted transistors. On the credit side, the 1-0-0 approach to low

Before and after. Conventional load device in MOS inverters is an enhancement-mode transistor (top). Ion implantation permits fabrication of depletion-mode load transistors (bottom). For equal power dissipation, the depletion load switches at least twice as fast as the enhancement load. In addition, the depletion load requires only one power supply, freeing chip area for other functions and simplifying equipment design.

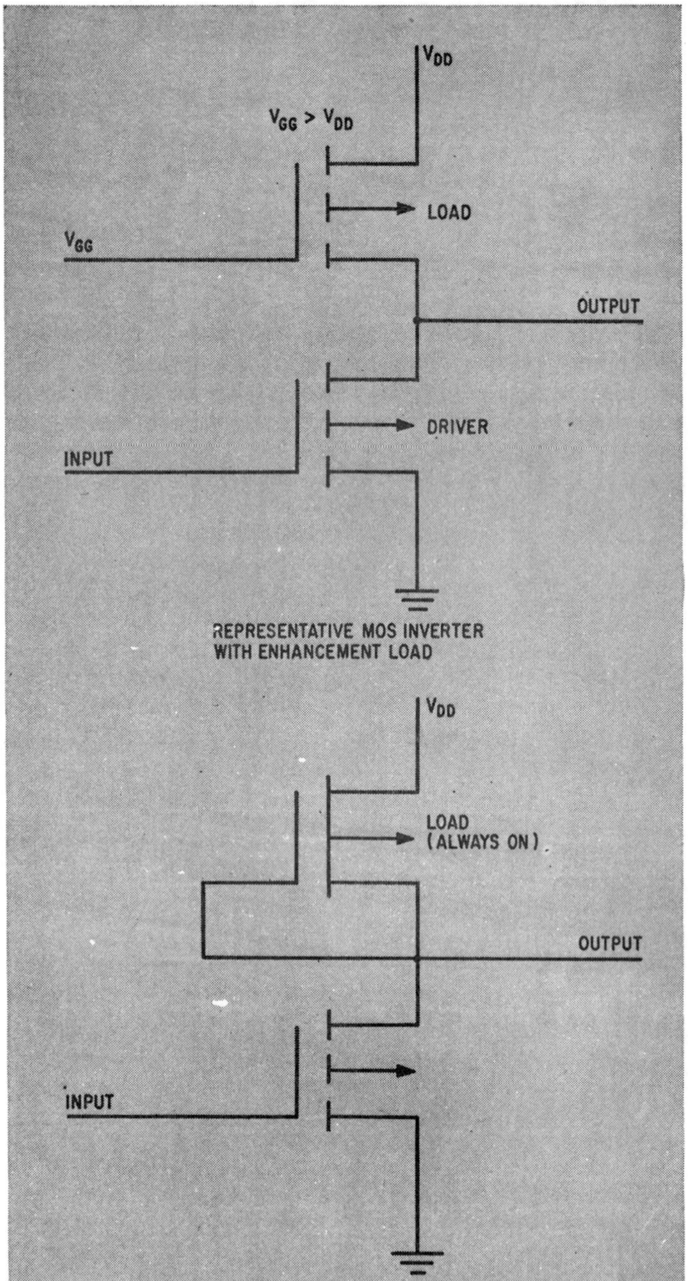

REPRESENTATIVE MOS INVERTER WITH ENHANCEMENT LOAD

DEPLETION-LOAD INVERTER

V_T does use standard, well-understood processes.

The silicon nitride approach, which employs a sandwich of Si_3N_4 and SiO_2, reduces V_T because the dielectric constant of the Si_3N_4-SiO_2 combination is greater than that of SiO_2 alone. This decreases V_T about 1.5 volts below that of the standard process. Moreover, the field-oxide threshold voltage remains high.

However the nitride process complicates fabrication: extra steps are required, and the two dielectric materials may etch at different rates. More seriously, some manufacturers have had difficulty because the threshold voltage of a transistor with a Si_3N_4-SiO_2 dielectric can shift, depending on the method used to form the gate structure. The disadvantage can be overcome, but at the cost of greater process complexity.

Then there's the silicon gate process. Here doped polycrystalline silicon is used as the gate electrode instead of the usual aluminum, thus reducing the work function between the two materials in the V_T equation. The reduction amounts to about 1 volt—more if a Si_3N_4 layer is interposed in the dielectric. The silicon gate also decreases parasitic capacitance and thus significantly speeds up circuit operation.

The principal disadvantages of the silicon-gate process are that it employs critical deposition and etching steps that are both more numerous and quite different from the standard sequence.

Ion implantation can reduce V_T by reducing the charge per unit area in the equation. This is accomplished by implanting ions into the MOS transistor's channel region, as shown on the following page. Virtually any threshold voltage below the 4 volts typical of the standard 1-1-1 process can be obtained in this manner. Moreover, IC's with ion-implanted channels have the highest ratio of field-oxide threshold to device threshold of any circuits available.

The implantation method does not require any additional deposition, photomasking, or etching steps that incur yield loss. And the same masks that are used in the standard MOS process also can be used in the ion-implanted channel technique; wafers that are to be fabricated for low-threshold voltage simply are diverted from the standard process flow after gate oxidation. Then they're ion-implanted and are returned to the normal process flow.

Sprague-Mostek's ion implantation techniques differ from those used at Hughes Aircraft Co., another firm that's closely identified with implantation. Mostek only implants the channel region [*Electronics,* May 25, p. 125], whereas Hughes employs ion implantation in conjunction with a self-aligning gate structure to form part of the drain and source regions. Since the gate doesn't overlap the source and drain, parasitic capacitance is reduced and switching speed is substantially improved.

The next step at Mostek will be to manufacture ion-implanted IC's with depletion-load transistors as well as the usual enhancement-mode driver transistors on the same chip. This is simply an extension of the ion-implanted channel process used to achieve low V_T: the channel of the load transistors on the IC are implanted with a slightly higher dosage of ions to make them operate in the depletion mode instead of the conventional enhancement mode. As a result, the IC's will operate with at least twice the speed of similar circuits that have enhancement-mode loads.

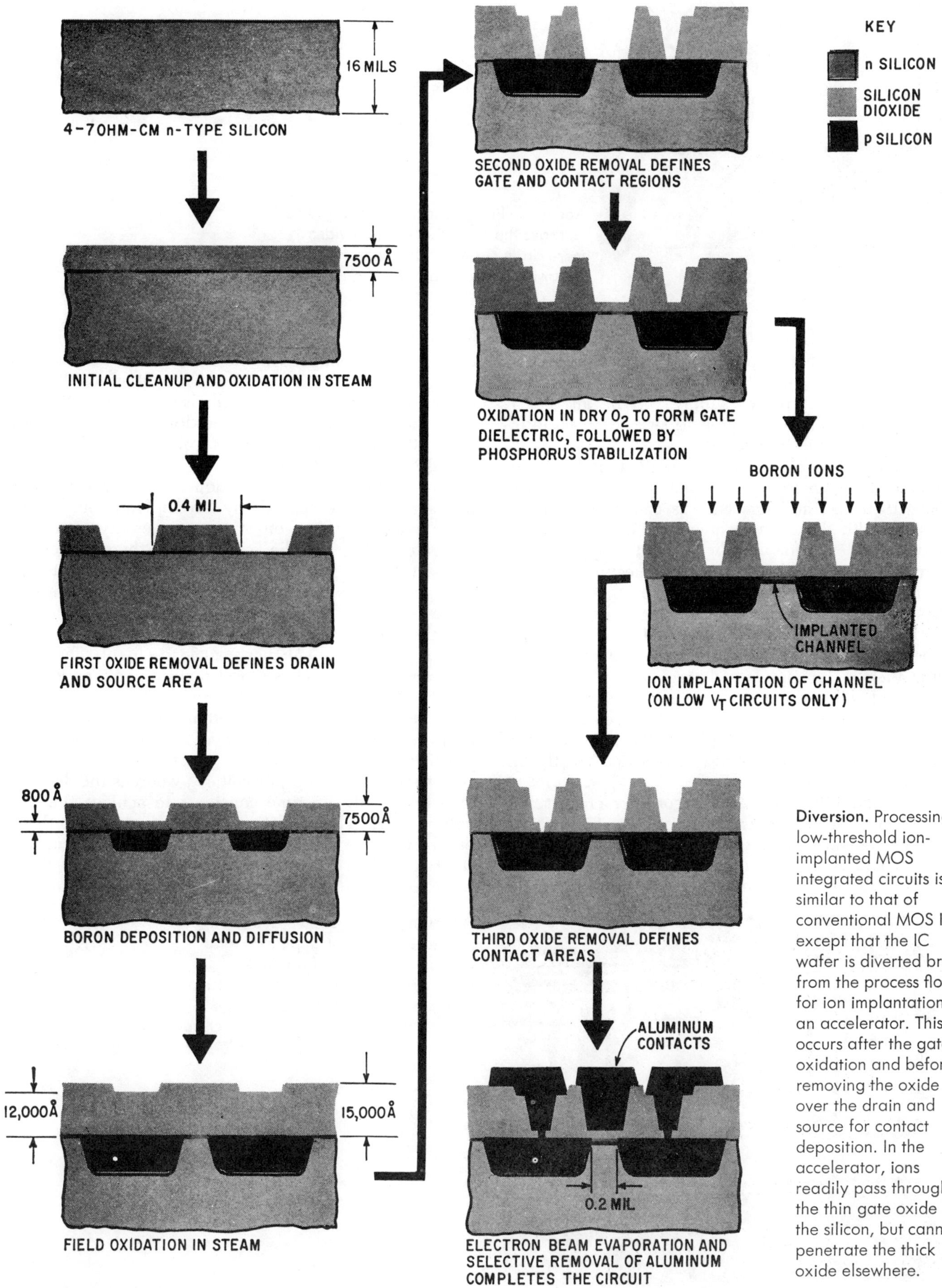

Diversion. Processing of low-threshold ion-implanted MOS integrated circuits is similar to that of conventional MOS IC's, except that the IC wafer is diverted briefly from the process flow for ion implantation in an accelerator. This occurs after the gate oxidation and before removing the oxide over the drain and source for contact deposition. In the accelerator, ions readily pass through the thin gate oxide into the silicon, but cannot penetrate the thick oxide elsewhere.

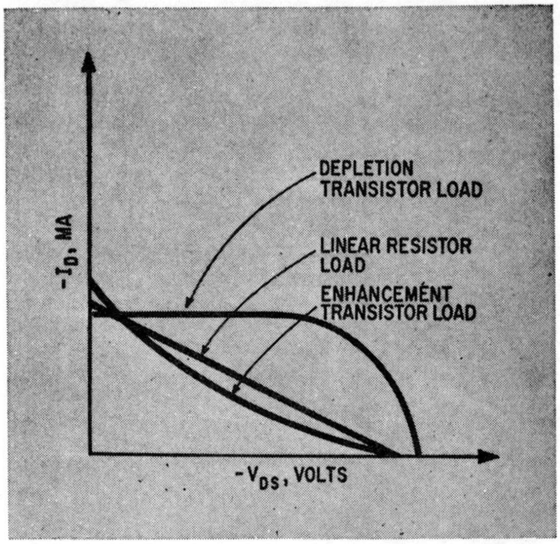

Above the line. For most drain-to-source voltages the depletion load provides more drain current than either the "ideal" linear resistor load or the enhancement load.

The first depletion-load circuit will be a 4,096-bit read-only memory that is completely transistor-transistor-logic compatible—even with the TTL power supply. A single TTL +5 volt power supply will be all that is needed to power the depletion-load MOS ROM; no interface components will be needed.

The desirability of using a depletion-mode device as a load has been recognized, but it was impractical to make them on the same chip as enhancement-mode transistors with diffusion techniques. For one thing, a depletion load calls for very light counterdoping of the load transistor channel. In fact, the concentration is so low that the dopant can be absorbed by the gate oxide at the high temperatures needed for growing the gate oxide. However, this doesn't occur under ion implantation because the gate oxide is grown before counterdoping; the ions are implanted through the gate oxide.

The main difference in operation between enhancement- and depletion-mode transistors, of course, is that the former is off at zero voltage on the gate, whereas the latter is on. The use of the new device greatly improves switching speed.

To understand the improvement, consider a representative MOS inverter circuit, as shown on page 87. This circuit, in its standard, commercially available form, invariably incorporates an enhancement-mode transistor as the load, as well as the driver. Suppose that the driver has just turned off. Then the load transistor will turn on if its gate voltage, V_{GG}, is greater than the threshold voltage of the load device, and the voltage at the node joining the driver and load will rise to a value V_{DD} (the drain voltage) minus the threshold voltage. Thus, if V_{DD} were 10 volts and V_T were 2 volts, the node voltage would rise to 8 volts.

Now consider the depletion-load inverter circuit on page 87, made possible by ion implantation. Because this device is always on, it's not necessary to apply a separate voltage to the gate, which therefore can be connected to the node. Since the gate-to-source turn-on voltage remains fixed during the switching transient, there is always a conduction path through the device, and the voltage at the node pulls up to V_{DD} much faster than it does in the enhancement load circuit. The depletion-load circuit reaches a useful output voltage in about half the time.

In terms of the time constant of the inverter circuit's resistance and capacitance, the enhancement-mode transistor load takes 18 time constants to switch from 10% to 90% of the output voltage, whereas the depletion load takes only 1.5 time constants to achieve the same output level. The depletion load is even faster than a purely resistive load, which would take 2.2 time constants to achieve this level. In practice, the 90% output voltage is not required, and the typical speed improve-

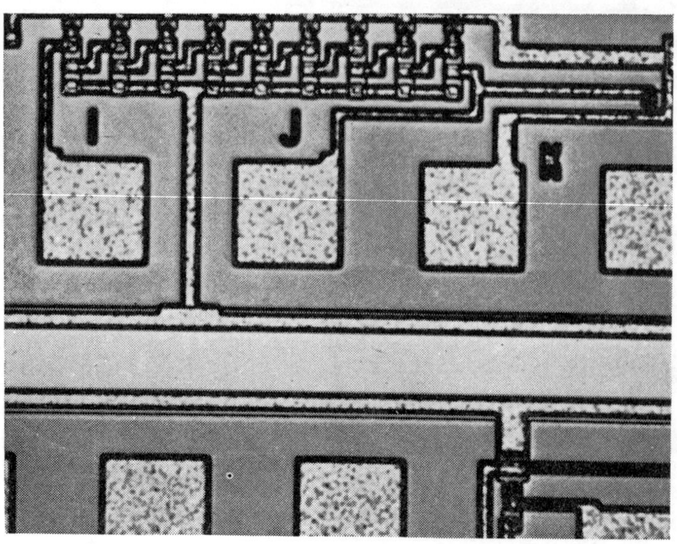

First string. String of nine depletion-load inverters on this test chip has a total propagation delay of about 200 nanoseconds, about half that of standard enhancement-load inverters. Large square areas at bottom are contact pads.

Short guide to MOS terminology

Enhancement-mode transistors show no channel conductance at zero gate bias. Forward bias increases the charge carriers in the channel and enhances channel conductance. **Depletion-mode transistors** are those that exhibit substantial channel conductance at zero gate bias. Application of a reverse gate bias depletes the number of charge carriers in the channel and reduces channel conductance.
Ion implantation is a technique for doping a silicon wafer. Ions of the proper dopant, such as phosphorus or boron, are accelerated to a high energy—40,000 to 300,000 electron-volts—and bombard the silicon wafer target. The ions' penetration depth in the silicon is determined by their energy. In those chip areas where ion implantation isn't wanted, an aluminum mask or a thick (12,000-angstrom) oxide absorbs the ions.

It's highly desirable to have enhancement-mode driver transistors and depletion-mode load transistors in p-channel MOS inverters. However, it wasn't feasible on a single chip until ion-implantation techniques came along. With precise control of the amount of dopant in the driver channels, depletion-mode devices now can be fabricated on the same IC chip as enhancement-mode transistors.

ment is more like a factor of two or three over enhancement loads.

As shown at top of preceding page, the load line for the depletion load always is above that for a purely resistive load, which in turn always is higher than the load line for the enhancement mode. Thus, the depletion load gets to a given voltage level faster.

Aside from speed, the depletion load has another very important advantage: it requires only one power-supply voltage, which can be the +5-volt TTL supply. This simplifies system design and also makes for more efficient utilization of chip area; since a substantial amount of metalization is omitted, additional area becomes available for logic functions. Physically, too, the depletion load device requires less space than an equivalent enhancement-mode device: the area of the depletion load channel is well under half that of the equivalent enhancement load.

In the future, Mostek plans to exploit ion implantation in conjunction with some of the other recent developments in MOS technology such as silicon nitride dielectrics, the silicon gate, n-channel circuits, and complementary circuits. The n-channel IC's appear particularly attractive because they are faster than p-channel devices and have greater functional density. Since the current carriers in the n-channel MOS transistors are electrons, and have roughly twice the mobility of holes at normal field intensities, the n-channel transistors have about half the on resistance of corresponding p-channel devices. This decreases the time constant and allows an n-channel device to turn on in about half the time.

In addition, both n-channel and bipolar devices operate at positive bias voltages. This similarity of voltage sign significantly reduces the MOS chip area required for output buffers to interface directly to TTL.

The advantages of n-channel over p-channel IC's as yet haven't been realized because n-channel devices fabricated on high-resistivity (10 to 15 ohm-centimeter) p-type silicon are inevitably, because of the low doping level, depletion-mode circuits: they're always on. This is fine for loads, but not for drivers. And if low-resistivity (1 ohm-cm or less) material is used to effect enhancement devices, mobility is so low that there is little speed advantage over p-channel.

However, ion implantation allows use of high-resistivity substrates to build enhancement-mode devices because the process can be controlled to give a shallow surface region in the channel of increased p-type dopant concentration.

Even these future applications don't exhaust the possibilities of the ion implantation process. Meanwhile, the low-threshold and depletion-load IC's made possible by this powerful technique provide an immediately useful introduction. ●

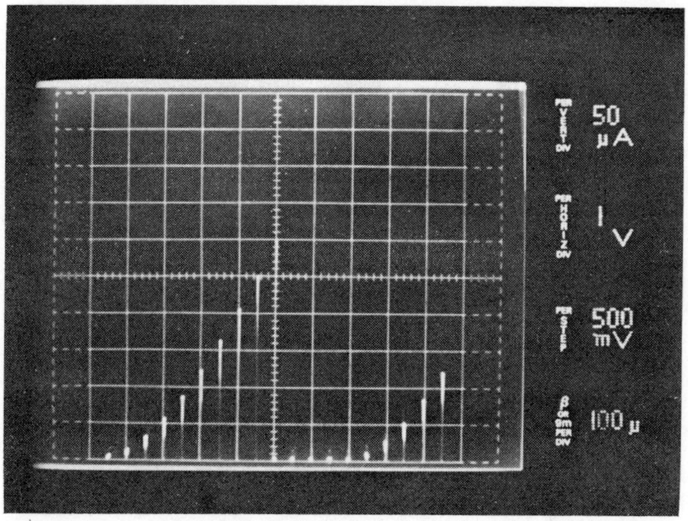

On and on. Unlike the enhancement-mode transistor (right), drain current versus gate voltage characteristic for the ion-implanted depletion-mode transistor load (left) is always in the on region in the voltage range that normally would be applied. (The horizontal center line represents 0 volts.) The depletion load behaves exactly like an enhancement load with a battery in series with the gate. This equivalent of an integrated battery both increases speed and eliminates the gate supply voltage normally required.

High-Performance Low-Power CMOS Memories Using Silicon-on-Sapphire Technology

EDWARD J. BOLEKY, MEMBER, IEEE, AND JOHN E. MEYER, MEMBER, IEEE

Abstract—**Complementary MOS/silicon-on-sapphire (SOS) inverters fabricated using silicon-gate technology and 5-μm channel-length devices have achieved nanosecond propagation delays and picojoule dynamic power-x delay products. In addition to high switching speed and low dynamic power, inverters with low leakage currents and therefore low quiescent power have been obtained.**

Two complex CMOS/SOS memories that realize the performance attributes of the individual inverters have been fabricated. An aluminum-gate 256-bit fully decoded static random-access memory features a typical access time of 50 ns at 10 V with a power dissipation of 0.4 μW/bit (quiescent) and 10 μW/bit (dynamic). The access time at 5 V is typically 95 ns. A silicon-gate 256-bit dynamic shift register features operation at clock signals of 200 MHz at 10 V and 75 MHz at 5 V. The dynamic power dissipation at 50 MHz and 5 V is typically 90 μW/bit.

Introduction

THE advantages of the MOS transistor as a basic element in large-scale digital arrays have been extensively described. The most important of these are processing simplicity, high packing density, and low power dissipation, all of which enable fabrication of complex arrays with high yields and low cost. It can be shown that on the basis of transit-time considerations, the MOS device is capable of high-frequency performance comparable to that of the bipolar transistor [1]. To date, however, MOS arrays have been characterized by relatively slow operating speeds because of harmful parasitic capacitance. High-speed performance, comparable to bipolar digital circuits at comparable supply voltages with less than one percent of the bipolar power dissipation, can be obtained from the MOS transistor only by the utilization of a technology that combines the best features of thin-film and monolithic silicon technologies in the environment of large-scale integration. The silicon-on-sapphire (SOS) approach comes closest to realizing these desirable features as MOS/SOS transistors with effective mobilities [2] comparable to those of bulk silicon MOS transistors can be fabricated in a thin single-crystal silicon film grown on an electrically insulating sapphire substrate [3]. The use of the thin-film silicon allows virtual elimination of the parasitic capacitance that seriously degrades the performance of bulk silicon MOS circuits. Fullest advantage of the low capacitance is realized in complementary symmetry (CMOS) circuit configurations that produce the highest speed with minimum power and circuit complexity [4]–[7].

The key electrical parameters that determine CMOS/SOS circuit performance are the switching speed and the leakage current of the CMOS/SOS inverters. These parameters have been investigated in detail; the results are presented and discussed below. After the characteristics of the individual CMOS/SOS inverters have been established, the design and performance of a 256-bit fully decoded CMOS/SOS static random-access memory and a 256-bit CMOS/SOS silicon-gate dynamic shift register are described.

Before proceeding to the detailed technical discussion, it is appropriate to describe briefly the basic concepts and advantages of MOS/SOS technology, especially in comparison with bulk silicon MOS technology.

Manuscript received June 30, 1971; revised December 1, 1971. This research was sponsored by the Air Force Avionics Laboratory, Wright-Patterson AFB, Dayton, Ohio, under Contract F33615-69-C-1499, and by NASA, George C. Marshall Flight Center, Ala., under Contract NAS12-2207, and by RCA Laboratories.

E. J. Boleky is with the Department of Electrical Engineering, Stanford University, Stanford, Calif., 94305.

J. E. Meyer is with the RCA Laboratories, Princeton, N. J. 08540.

Reprinted from *IEEE J. Solid-State Circuits*, vol. SC-7, pp. 135–145, Apr. 1972.

Silicon-on-Sapphire MOS Transistors

The switching speed of bipolar digital circuits is between one and two orders of magnitude faster than that of comparable bulk silicon MOS digital circuits. However, a comparison of the respective transit-time expressions for a representative n-p-n bipolar transistor and a NMOS transistor [1], as shown in Fig. 1, indicates that based solely on transit-time considerations, the NMOST can switch as fast as the n-p-n bipolar. This apparent discrepancy between the theoretical internal operating speeds of the two types of transistors and their actual switching performance in a circuit environment is due to the fact that the loading in a monolithic digital integrated circuit is primarily capacitive. Therefore, the important device parameter is the output current available per area for a given applied input voltage, or the transconductance, rather than the transit time or gain–bandwidth product. Due to the different physical mechanisms involved in the two devices, the bipolar transistor has a transconductance per area approximately two orders of magnitude higher than that of the MOS transistor. In addition to providing more current to charge load capacitances, bipolar transistors ordinarily have a lower threshold voltage for switching (0.7 V) than MOS transistors (1.5–3 V) so that lower operating voltages levels are normally used in the bipolar circuits. Therefore, bipolar digital circuits are faster because the transistors themselves provide more current to charge the circuit capacitances to lower voltages than those commonly used in MOS digital circuits.

Since the amount of current per unit area that can be provided by the MOS transistor is limited by the physical transport parameter involved (drift mobility) and by the state-of-the-art in photolithographic dimensional control (which determines the channel length L), the most direct way to increase the internal speed of MOS digital circuits is to remove all unnecessary circuit and device capacitance. This is easily accomplished by fabricating the MOS transistors and crossunders in a thin film of single-crystal silicon grown on an electrically insulating substrate, such as sapphire [7]. All unused silicon is removed from the substrate leaving perfectly isolated islands of silicon for transistors and crossunders, as shown in Fig. 2. No thick-field oxides for isolation are needed and the parasitic MOS problem is eliminated. Since the silicon is nominally 1 μm or less thick, the source and drain regions of the transistors can be "through diffused" to the silicon-sapphire interface [8] thereby reducing the capacitance of the source and drain regions by a factor (typically) of 25 over bulk silicon MOS devices. All metalization and degenerate silicon crossunders sit directly on the sapphire eliminating unnecessary wiring capacitance and diffused junction crossunder capacitance found in bulk silicon MOS circuits. In addition to the tremendous reduction in parasitic capacitance, there are also several less obvious advantages

$$T_{MOS} = \frac{4}{3}\,\frac{L^2}{\mu(V_{gs}-V_T)} \qquad T_{BIPOLAR} = \frac{W_B^2}{2.4 \times \mu\left(\frac{kT}{q}\right)}$$

FOR TYPICAL VALUES

$L = 8\mu$ $\qquad$ $W_B = 1\mu$

$\mu = 350\,\text{cm}^2/\text{v-sec}$ $\qquad$ $\mu = 600\,\text{cm}^2/\text{v-sec}$ for npn $\rho_B = 1\,\Omega\text{-cm}$

$V_{gs}-V_T = 10\text{v}$

$\frac{kT}{q} = 26\text{ mv @ } 300°\text{K}$

$T_{MOS} = 0.26\text{ns}$ $\qquad$ $T_{BIPOLAR} = 0.27\text{ns}$

Fig. 1. Transit time of MOS and bipolar transistors.

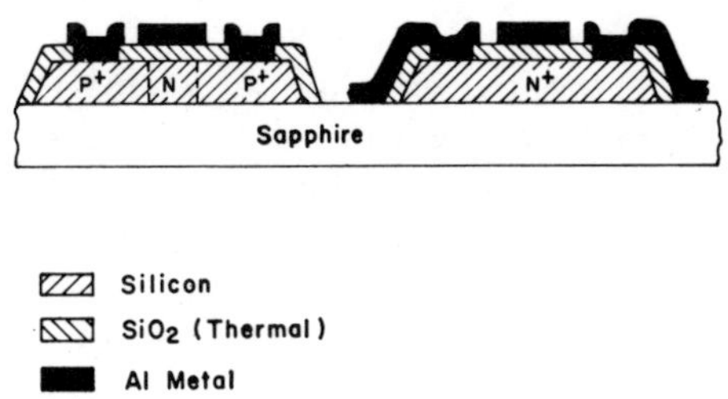

Fig. 2. Typical SOS transistor and crossover.

of MOS/SOS. Normally, each isolated channel region of the MOS transistors in the arrays is left floating, i.e., no electrical contact is made to it. The floating channel region results in two benefits. First, the back-gating effect on the threshold voltage is avoided because the channel always floats at a potential several tenths of a volt above the source potential. This unique feature further increases the speed advantage of MOS/SOS decoders in comparison with bulk silicon MOS decoders, so that large fully decoded MOS/SOS memories can be fabricated with very short access times. Second, all parasitic bipolar transistor action (in parallel with the MOS transistor) is virtually eliminated because the base (channel) region is floating with respect to the emitter (source). In addition, the current gain of the parasitic bipolar transistor is extremely small due to very low minority carrier lifetime ($\sim$ 1 ns) in the SOS films. Finally, SOS technology allows greater flexibility in reducing the carrier concentration of the channel silicon, since all field inversion problems have been eliminated. This feature of MOS/SOS allows the designer to adjust the device threshold voltage by simply varying the carrier concentration in the epitaxial SOS film. Consequently, lower threshold voltages, higher drain–source breakdown voltages, and higher effective mobilities can be obtained by the use of higher resistivity channel silicon without necessitating stopper diffusions, thick oxides, or electrostatic guard rings to avoid parasitic MOS transistors.

The above advantages of MOS/SOS technology apply to both single-channel (NMOS or PMOS) MOS circuits and CMOS circuits. Single-channel MOS/SOS circuits can be fabricated using process technologies similar to those currently in use on bulk silicon single-channel MOS circuits, with the exception that a silicon etch step is required. While NMOS/SOS and PMOS/SOS technologies have appeal as performance improvements to present bulk silicon MOS technology, the quiescent power dissipation of single-channel SOS digital circuits is comparable to that of the same circuit in bulk silicon, since the quiescent power is determined by the MOS load device rather than by reverse-biased junction leakage current. However, the lower stray capacitance of SOS devices greatly reduces the dynamic power dissipation of the circuit. The ultimate combination of low power dissipation (both quiescent and dynamic) and high performance is achieved only with a CMOS/SOS technology since the complementary circuitry reduces the quiescent power dissipation while increasing the switching speed. From a system viewpoint, CMOS circuitry is extremely attractive since it allows operation from a single supply voltage (which may vary over wide ranges), high noise immunity, and large fanout capability.

The combination of high switching speed and low power dissipation available through the use of CMOS/SOS technology has been realized for two CMOS/SOS digital circuits, a 256-bit static random-access memory fabricated using aluminum-gate technology and a 256-bit dynamic two-phase shift register fabricated using self-aligning silicon-gate technology. These circuits are currently being produced in a pilot-line environment at RCA Laboratories for yield evaluation. The next section will present detailed measurements of the switching performance and leakage current of CMOS/SOS inverters, the two parameters that primarily determine the performance of the CMOS/SOS arrays.

Leakage Currents in Silicon-on-Sapphire Diodes

It has been shown [9] that the quiescent power dissipation of complementary MOS logic circuits is due to the drain current of the off transistor (i.e., $|I_{DS}|$ at $|V_{GS}| = 0$ V and $|V_{DS}| = V_{DD}$). For bulk silicon CMOS circuits, this current (which can theoretically be of the order of picoamperes) can be due to either of two mechanisms: 1) the reverse current of the drain diode, and 2) the current in the surface channel (inversion layer) of the MOS transistor. It is known [10] that the drain–source current of an MOS does not become identically zero when the gate–source voltage $|V_{GS}|$ is lowered below the threshold voltage $|V_T|$[1] but decreases exponentially over many orders of magnitude [11] for $|V_{GS}| < |V_T|$ until the surface potential ϕ_S is less than the Fermi potential ϕ_F. To determine the micropower potential of CMOS/SOS circuits it is therefore important to distinguish between these two mechanisms. Measurements were made to characterize the total drain–source current of NMOS/SOS and PMOS/SOS transistors for values of $|V_{GS}| < |V_T|$. The current measurements were made using techniques to ensure that the currents were representative of the devices and not due to an uncontrolled surface leakage. Typical gate-transfer characteristics of CMOS/SOS transistors are presented in Fig. 3. These results demonstrate that the drain–source current near $|V_T|$ decreases exponentially until the channel component becomes less than the reverse current of the drain diode. The drain–source current then becomes approximately independent of gate potential until the gate potential is sufficient to cause surface avalanche. Therefore, for devices fabricated with sufficiently high threshold voltages (i.e., $|V_T| \gtrsim 1.5\ V$) the transistor off current (I_{DO}) will be the reverse-junction current of the drain-channel diode.

[1] The threshold voltage V_T corresponds to the condition that the surface potential ϕ_S equal $2\phi_F$ where ϕ_F is the Fermi potential of the channel silicon.

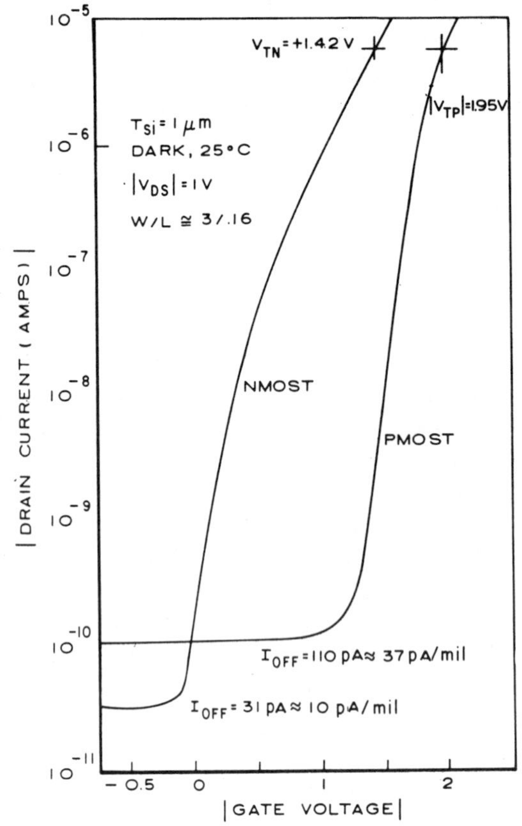

Fig. 3. SOS transistor drain current versus gate voltage.

In order to characterize and predict the standby power dissipation of CMOS/SOS inverters, a detailed study of the junction currents in SOS diodes has been made. Previous workers have reported the characteristics of vertical junction diffused diodes on SOS [12] and spinel [13]. The results of these investigations indicated that the forward characteristics of these diodes follow a voltage V dependence typical of the Sah–Noyce–Shockley (SNS) theory [14], i.e.,

$$I = I_0(e^{(qV/nKT)} - 1) \tag{1}$$

where q is the electron charge, k is the Boltzman constant, T is the absolute temperature, $n \approx 1.7$, and I_0 is a characteristic constant of the material proportional to the minority carrier lifetime. The currents measured under reverse bias, however, did not have the functional dependence on voltage as predicted by the

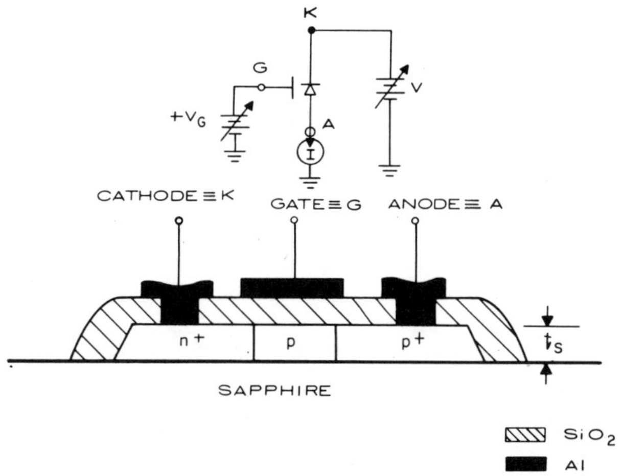

Fig. 4. SOS gate-controlled diode structure.

SNS theory. More recent results described below indicate that vertical junction SOS diodes do indeed have reverse currents indicative of step junctions according to the SNS theory.

Investigations of vertical junction diode behavior are most easily carried out using the gate-controlled diode structure shown in Fig. 4. The gate electrode affords positive control of the Si–SiO_2 interface potential in addition to allowing C–V measurements. The theory concerning gate-controlled p-n junction diodes has been described in detail. [15]. The diodes were usually measured at $V_G = {}_{CH} = 0$ (i.e., the surface potential just below the flat-band conditions). Under these conditions no surface inversion layer existed and the current being measured was just the reverse current of the vertical junction diode. Typical reverse current I versus voltage V characteristics, shown in Fig. 5, indicate that both the n^+-p and p^+-n diodes behave as step junctions according to the SNS theory of generation current in silicon diodes, i.e., both reverse currents are proportional to $V^{1/2}$ over several orders of magnitude. This voltage dependence of the current is due to the dependence of the depletion region width X_j, and hence the volume in which to generate hole-electron pairs on voltage. From the slope of the diode forward I–V characteristics, the constant n was determined to be ≈ 1.7 indicative of recombination current. It is not surprising that diffusion currents are not important in SOS structures because of the extremely low lifetime [12] in such material. From these forward characteristics the intercept value I_0 for a generation-recombination current was determined. One can than calculate a minority carrier lifetime τ from

$$I_0 = (qn_i/2\tau)X_jA \tag{2}$$

where n_i is the intrinsic carrier concentration, X_j is the depletion-region width, and A is the area of the junction. The calculated lifetimes of electrons in p-type SOS and holes in n-type SOS are 0.57 and 0.53 ns, respectively, for 1-μm thick films. These values are approximately three orders of magnitude lower than the lifetimes in comparable bulk silicon. Consequently, the current densities

$$J_0 = (qn_i/2\tau)X_j \tag{3}$$

obtainable on SOS diodes are higher by the same three orders of magnitude. The measured I_0 and, therefore, also the reverse leakage currents are not increased by that amount because the area of vertical junction SOS diodes is reduced by 20–50 times from that of a bulk silicon diode having the same width. This is so because the SOS diode area is simply the device width W times the silicon thickness t_s, which is typically only 1 μm. Both the forward and reverse I–V characteristics of vertical-junction SOS diodes are scaled by the device width as expected. The reverse currents of the n^+-p diodes were smaller than that of the p^+-n diodes because the depletion-region width in the more heavily doped p material is smaller. The temperature dependence of the reverse currents shown in Fig. 5 was measured over the temperature range of 25° to 225°C. The results indicate that the temperature variation of the reverse currents corresponds closely to the temperature variation of n_i, as predicted by SNS theory. From such data the band gap of thin-film SOS is determined to be 1.1 ± 0.2 eV.

From this discussion it is evident that NMOS/SOS and PMOS/SOS transistors can be fabricated with off currents of less than 80 pA and 150 pA per mil of width, respectively, at drain voltages of 10 V. Thus for the typical six-transistor CMOS memory cell of Fig. 10 with 0.5-mil-wide devices and $V_{DD} = 10$ V, a quiescent power dissipation of less than 1 nW is possible.

Switching Performance of CMOS/SOS Inverters

It has been shown [9] that the stage delay of a CMOS inverter can be expressed as:

$$\tau_{sd} = \frac{0.45C_L}{\frac{\mu_n C_{in}}{2L^2} V_0}\left[\frac{1}{\left(1 - \frac{V_{TN}}{V_0}\right)^2} + \frac{1}{\beta\left(1 - \frac{|V_{TP}|}{V_0}\right)^2}\right] \tag{4}$$

where

C_L	load capacitance of inverter,
C_{in}	input capacitance of NMOST $= \epsilon_{ox} W_N L/T_{ox}$,
L	channel length (assumed equal for NMOST and PMOST),
W_N, W_P	width of NMOST, PMOST,
μ_n, μ_p	effective mobility of NMOST, PMOST (electrons, holes),
β	$\mu_p W_P/\mu_n W_N$,
V_0	supply voltage,
V_{TN}, V_{TP}	threshold voltage of NMOST, PMOST,
ϵ_{ox}, T_{ox}	permittivity, thickness of gate oxide.

If the CMOS inverter drives an identical inverter then C_L is proportional to L. Therefore,

$$\tau_{sd} \propto L^2. \tag{5}$$

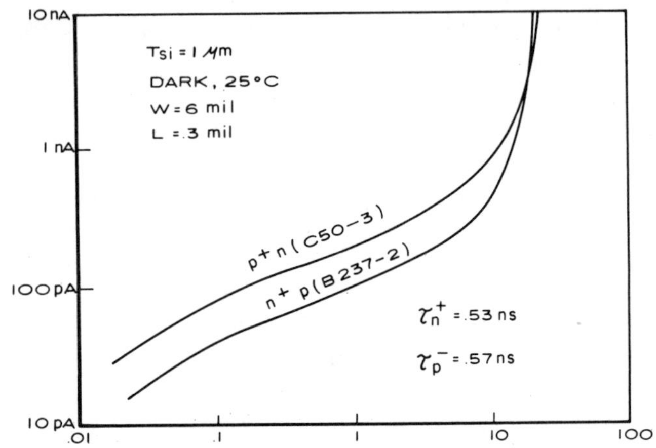

Fig. 5. Typical reverse current versus voltage characteristics.

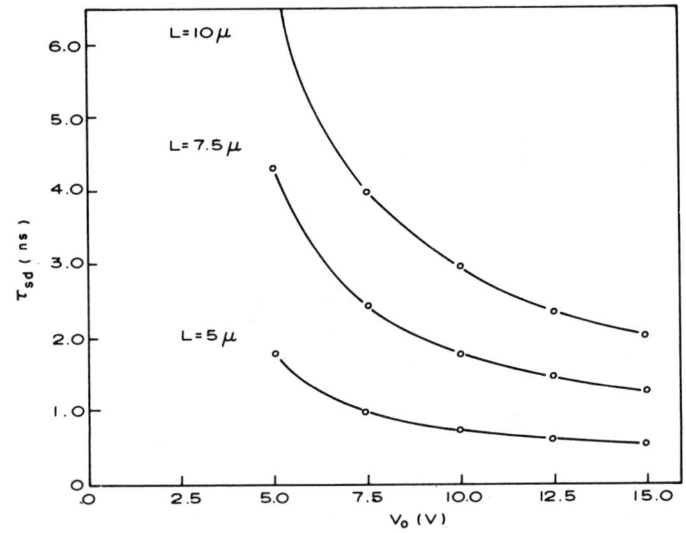

Fig. 6. Stage delay versus voltage CMOS/SOS silicon-gate inverter.

Also,

$$\tau_{sd} \propto 1/V_0 \tag{6}$$

assuming that $V_{TN}/V_0 \ll 1$, $|V_{TP}|/V_0 \ll 1$, and the effective mobilities (μ_p, μ_n) are independent of V_0. Equation (4) does not take into account the lower drain saturation voltage and increased back gating of the NMOST due to the more heavily doped p-type substrate, the nonlinear effect on C_L of the Miller (overlap) capacitance, and the dependence of the effective mobilities on V_0. However, (4) is useful for determining the functional dependence of τ_{sd} on L and V_0.

A 25-stage ring oscillator test array was used to accurately determine the stage delay and the power dissipation of the CMOS/SOS silicon-gate inverters as the supply voltage V_0 and the channel length L were varied [17]. The ring oscillator consists of 25 inverters connected end-to-end with the output of the last inverter brought back to the input of the first inverter. Each test array contained 3 ring oscillators, which differed only in the channel lengths of the devices, L = 5, 7.5, and 10 μm. The static CMOS/SOS inverter characteristics were determined from an identical inverter alongside each ring oscillator. Both n-type and p-type silicon-gate capacitors were included on the chip for stability measurements. The ring oscillators were fabricated using the CMOS/SOS silicon-gate technology described below in the discussion of the 256-bit CMOS/SOS dynamic shift register. The stage delay τ_{sd} is defined as the average time for the inverter to change state assuming a standardized signal waveform within the oscillator and is obtained by dividing the oscillator period by 50.

The measured stage delay as a function of supply voltage and channel length is shown in Fig. 6. The stage delay for a 5-μm channel length varies from 0.55 to 1.8 ns and decreases approximately as V_0^{-1}. The τ_{sd} versus V_0 curve flattens out at large values of V_0 due to the decrease in mobility with increasing gate voltage [2]–[3]. (Note that extremely fast switching (τ_{sd} = 1.8 ns) is possible with V_0 as low as 5 V.) By comparing stage delays at the same V_0 for different values of L (e.g., 7.5 V), it is seen that increasing L from 5 to 10 μm results in τ_{sd} increasing from 1 to 4 ns. Since all wiring capacitance and essentially all Miller capacitance is eliminated, the load capacitance C_L of each stage is proportional to the gate area of the next stage. Therefore, the stage delay increases approximately as L^2 [in agreement with (5)] and the stage delay is not affected by scaling W_N and W_P by the same factor. However, the power-x delay product of the CMOS/SOS inverter $P_{\text{inv}} \times \tau_{sd} = \frac{1}{2}C_L V_0^2$ can be reduced by scaling down the widths of the NMOST and PMOST.

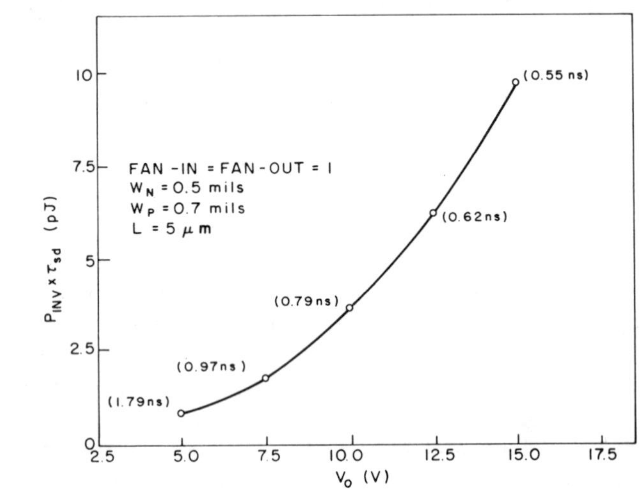

Fig. 7. Power-x delay product CMOS/SOS silicon-gate inverter.

Fig. 7 shows the experimental power-x delay product for the L = 5-μm ring oscillator inverters where W_N and W_P have been scaled (by a factor of 7) to sizes consistent with high-density CMOS/SOS integrated circuits. τ_{sd} is also shown in parentheses. It is to be emphasized that the power-x delay products shown in Fig. 7 are the product of the measured stage delay and the measured average value of the dynamic power dissipated during switching. The measured quiescent power dissipation

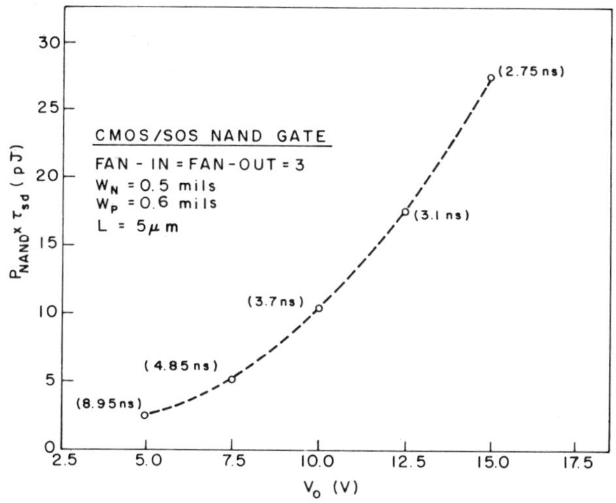

Fig. 8. Power-x delay product CMOS/SOS silicon-gate logic.

under these conditions is six orders of magnitude smaller than the dynamic power dissipated while switching at the maximum rate. It is seen that extremely low power-x delay products are available with SOS technology without sacrificing internal operating speed because of the very small internal capacitances. Operation at $V_0 = 5$ V results in $\tau_{sd} = 1.8$ ns, $P_{\text{inv}} \times \tau_{sd} = 0.81$ pJ and a standby power dissipation of 0.13 nW per inverter. Increasing V_0 to 15 V increases the power-x delay product to 10 pJ, increases the standby power dissipation to 2 nW, and decreases the stage delay to 0.55 ns.

The use of CMOS/SOS technology does not solve the usual MOS problem (discussed above) of inability to charge and discharge large load capacitances at high speeds. However, the following discussion illustrates the significant advantages of CMOS/SOS in terms of power dissipation (at comparable switching speeds) when compared with standard transistor–transistor logic (TTL) bipolar gates for the small fanouts normally encountered in decoding and memory circuits. To provide a basis for comparison with commercial high-speed logic gates, the predicted switching delay and power-x delay product for a 3-input 3-output CMOS/SOS NAND gate was calculated using an expression for τ_{sd} derived by Burns [9] in addition to the switching and power-dissipation results of the $L = 5$-μm ring-oscillator inverters. A worst case analysis [9] was used where it is assumed that two inputs are already high and the third input is then turned on. The results shown in Fig. 8 indicate that a switching delay of less than 10 ns is obtained at $V_0 = 5$ V with a power-x delay product of 2.20 pJ, and a standby power dissipation of 0.4 nW. If higher switching rates are desired, a stage delay of 2.75 ns at 27.5 pJ is available for $V_0 = 15$ V with a standby power dissipation of only 5.5 nW. Table I compares the predicted performance of the 3-input CMOS/SOS NAND gate operated at $V_0 = 5$ V with the performance of basic logic gates from the TTL family (low power, standard, high speed, and Schottky clamped) and the high-speed emitter-coupled logic (ECL) family of circuits. Note that CMOS/SOS is comparable to standard TTL in propagation delay (the TTL will be somewhat faster for equal fanout of $n = 3$), but the quiescent power dissipation of the CMOS/SOS gate is more than six orders of magnitude lower than TTL. This tremendous reduction in quiescent power is particularly important in memory applications where each circuit spends the majority of its time in the quiescent mode. In addition, the extremely low capacitance of CMOS/SOS technology manifests itself in the extremely low dynamic dissipation illustrated in Table I, e.g., at 50 MHz the total power of the CMOS/SOS gate is a factor of 40 lower than the power of the TTL gate. In the next section, the fabrication and performance of two complex CMOS /SOS memory circuits will be described to further illustrate the high-speed low-power potential of this technology.

256-Bit CMOS/SOS Random-Access Memory

One of the major potential applications of SOS technology is in the field of semiconductor memories. Because of higher performance and potentially lower cost both MOS and bipolar memory arrays are already serious competition to core memories. MOS serial memories presently dominate the market for storage in peripheral equipment and date terminals. SOS will compete with bipolar memories in the area of static high-performance buffer and cache memories where fast access and cycle times are required. CMOS/SOS memories, because of their inherent self-isolating and micropower characteristics, can achieve higher packing densities and therefore potentially lower cost than present TTL bipolar memory arrays. The 256-bit CMOS/SOS random access memory described below was developed for use as a high-performance micropower large-scale integration (LSI) memory array for use in space stations [18]. The circuit contains 1836 devices on a chip of 126 × 137 mils^2.

The 256-bit CMOS/SOS static random-access memory (RAM) is fabricated using an aluminum-gate process that requires seven masks including passivation [20]. The enhancement-type transistors are fabricated in 1-μm thick isolated blocks of n-type ($N_D = 3 - 5 \times 10^{15}$ cm^{-3}) and p-type ($N_A = 1 - 3 \times 10^{16}$ cm^{-3}) single-crystal (100) silicon, which are provided by two heteroepitaxial growth steps [19]. Typical aluminum gate CMOS/SOS transistor characteristics are shown in Fig. 9 for the device and material parameters listed. The I–V characteristics are virtually identical to those of similar bulk silicon MOS transistors with the exception that the effective carrier mobilities are slightly lower in the SOS transistors [2]–[3]. The threshold voltages (measured at 1-μA/mil width) of the NMOS/SOS and PMOS/SOS transistors are 1.0 and −2.0 V, respectively.

A schematic diagram showing the basic organization

TABLE I

IC Family	Low-Power TTL	Standard TTL	RCA CMOS/SOS	High-Speed TTL	Schottky TTL	High-Speed ECL
Logic gate	3-input NAND	3-input NAND	3-input NAND	3-input NAND	3-input NAND	4-input OR/NOR
Dc load factor	10	10	>50	10	10	10 (low Z) or 50 Ω
Propagation delay at n (fan-out)	33 ns at $n = 10$	10 ns at $n = 10$	9 ns at $n = 3$	6 ns at $n = 10$	3 ns at $n = 10$	1.1 ns at $Z_L = 50$ Ω
Average power dissipation at 50 per cent duty cycle (unloaded except where noted)	1 mW	10 mW	0.4 nW static 0.25 mW at 50 MHz $n = 3$	23 mW	20 mW	55 mW
Average power x propagation delay product (unloaded except where noted)	33 pJ	100 pJ	3.6×10^{-6} pJ static 2.2 pJ at 50 MHz $n = 3$	138 pJ	60 pJ	60 pJ
Power supply voltage	5.0 V ± 10%	5.0 V ± 10%	(3–15 V) 5.0 V	5.0 V ± 10%	5.0 V ± 5%	5.2 V ± 10%
Typical dc noise margin	1.2 V	1.2 V	2.25 V at V_{DD} = 5V	1.2 V	1.2 V	0.35 V

All values at 25°C.

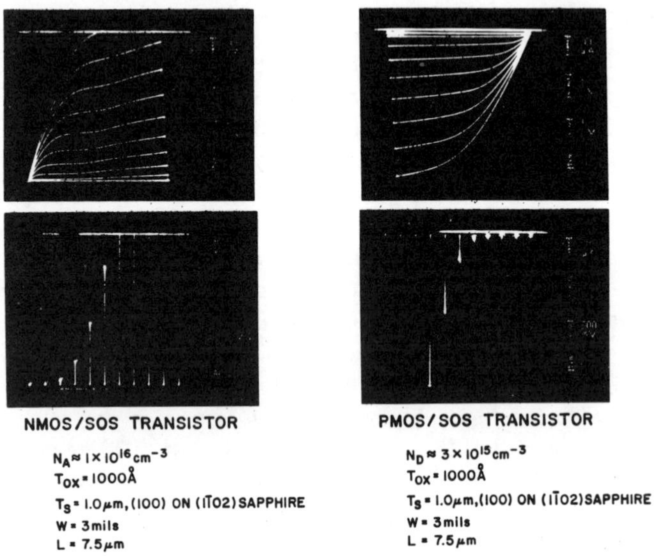

Fig. 9. Aluminum-gate CMOS/SOS transistor characteristics.

of the 256-bit RAM is shown in Fig. 10. The arrays are organized as 256 words × 1 bit and include full x- and y-binary address decoding. READ and WRITE commands are determined by the state of the two bit lines, D_0 and D_1. The address decoders are realized as dynamic NAND gates that are activated by a chip-enable (CE) input. The CE input also allows easier memory expansion when the chip is used in a large memory system.

The high-speed performance of the CMOS/SOS memory is best illustrated by the waveforms shown in Fig. 11. The access time shown here includes the 15-ns delay of the MOS–ECL sense amplifier-level shifter circuit. As indicated by Fig. 11(a) the CE-sense current delay of the RAM at V_{DD} = 10 V is 47 ns. Typical quiescent power dissipation is 100 μW or 0.4 μW/bit at V_{DD} = 10 V and in high-speed READ/WRITE operation the dynamic power is equivalent to 10 μW/bit. As further circuit and process enhancements are incorporated, even higher performance can be anticipated, thereby enabling the RAM to operate with the system timing shown in Fig. 12. While the performance of the CMOS circuits at lower voltages is obviously less than at higher levels, the ability to operate at lower levels eliminates the need for special high-power (i.e., expensive) TTL–MOS voltage-level shifters. This is a most important advantage for any CMOS technology. CMOS/SOS, however, allows the system designer the possibility of relatively high performance even at TTL levels. The access time of the CMOS/SOS 256-bit RAM is increased from 47 to 92 ns (Fig. 11(b)) at V_{DD} = 5 V. The static and dynamic power are decreased by typically 50 percent. A comparison of speed and other performance criteria with those of various commercially available 256-bit RAM (Table II) illustrates the unique low-power high-performance nature of CMOS/SOS technology. As the table indicates, the SOS RAM far exceeds the performance of other MOS technologies at power levels comparable to bulk complementary MOS. The cell area of the CMOS/SOS memory can be significantly reduced by the use of silicon-gate technology and less conservative design rules. The table also points out that CMOS/SOS technology is comparable in high-speed performance to standard TTL bipolar memories but dissipates two orders of magnitude less power. In contrast with static bipolar memories, no compromises between speed and power dissipation will be required as the packing density of CMOS/SOS static memories is increased by technological movements. Therefore, at densities of 1024 bits per chip, the CMOS/SOS memory high-speed performance should exceed that of the equivalent bipolar memory.

256-Bit CMOS/SOS Dynamic Shift Register

In order to demonstrate the high-speed capability of CMOS/SOS technology and the further improvements

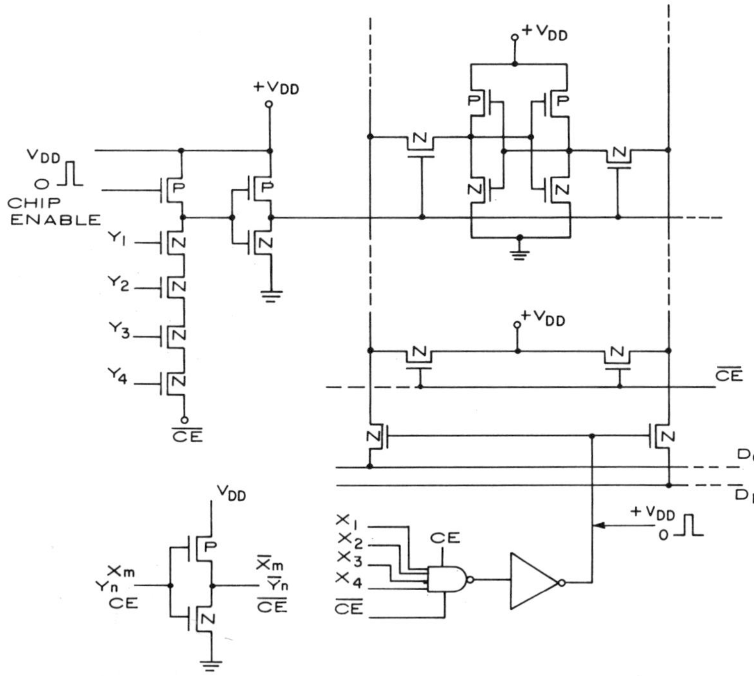

Fig. 10. Schematic diagram of 256-bit RAM.

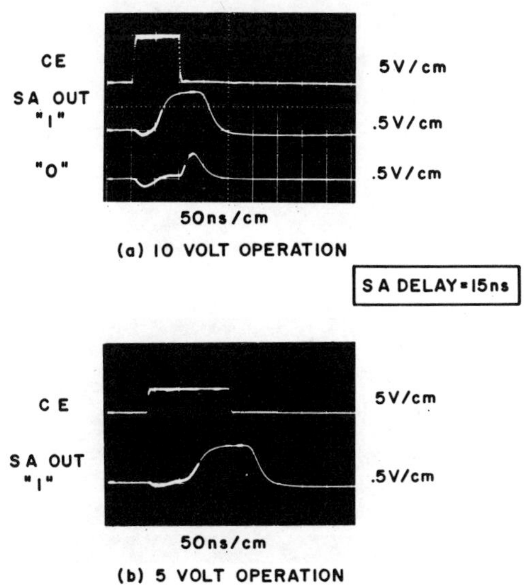

Fig. 11. 256-bit CMOS/SOS RAM performance.

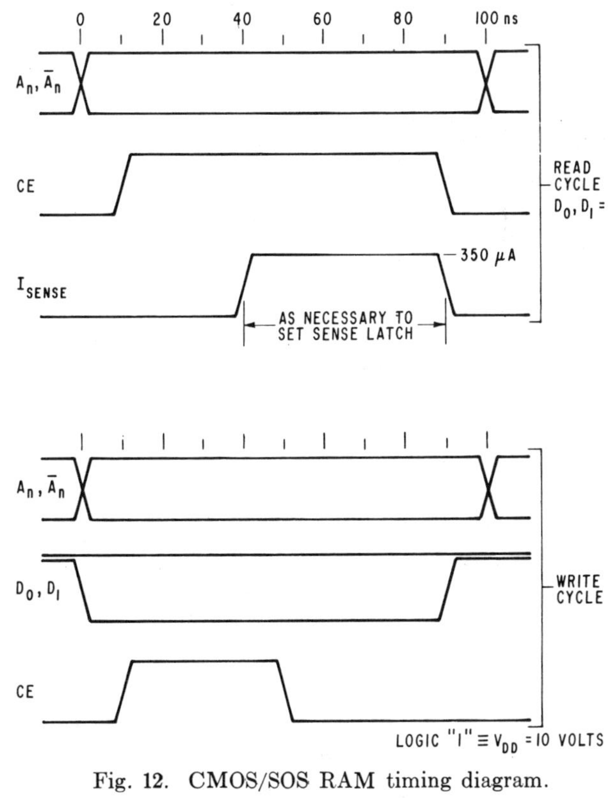

Fig. 12. CMOS/SOS RAM timing diagram.

in switching speed and packing density that are obtained by the use of a self-aligned polycrystalline silicon gate [21], a 256-bit CMOS/SOS silicon-gate dynamic shift register has been designed and fabricated [20]. Similar high-speed registers have already been used as the basis for a digital correlator for use in radar and communication system signal processors [22]. The schematic of a single cell is shown in Fig. 13. Data are shifted through the register by the clocked complementary transmission gates and storage is dependent upon the capacitance at nodes A and C. Computer-aided circuit simulations [23] indicated that in order to achieve the goal of 250 MHz operation, complementary transmission gates would be necessary. This was so for two reasons: 1) to achieve maximum speed it is necessary to avoid source–follower operation in any transistor, and 2) to maintain proper logical data on the gates during fast clock transitions, it is necessary to balance the ca-

TABLE II
Comparison of Static 256-Bit Fully Decoded RAMs

Type	TTL Compatibility	Cell Size	Access Time	Sense Current	Power/bit
PMOS	yes	15–20 mil²	0.7–1 μs	>3 mA	0.75–1.5 mW
COS/MOS (RCA TA6042)	10 V—input only	25 mil²	200 ns	300 μA	25 μW[a]
					1.2 μW
TTL	yes	20–30 mil²	60–75 ns	voltage	1–2 mW
CMOS/SOS (RCA TA6142)	5 V—input only	27 mil²	100 ns	100 μA	5μW
	(10 V—no)		50 ns	350 μA	0.2 μW[a]
					0.4 μW[a]
					10 μW

[a] Quiescent power.

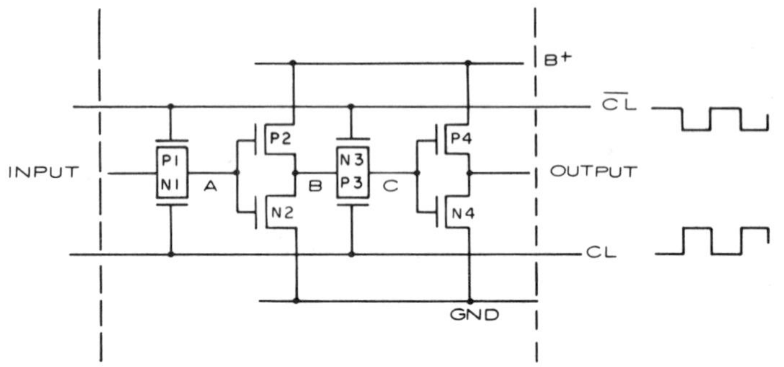

Fig. 13. Schematic of 256-bit CMOS dynamic shift-register cell.

pacitive feedthrough currents. The chip has an active area of 97 × 121 mils² including bonding pads. All inputs and outputs are protected against electrostatic charge. The entire chip contains 2072 transistors. The shift register is packaged in a 14-pin ceramic dual in-line package (DIP). The cell area is 4.4 × 6.0 mils², using state-of-the-art design rules.

The CMOS/SOS silicon-gate transistors [17] are fabricated using a process that requires seven masks, including passivation. The resultant NMOST has an n⁺-gate electrode and the PMOST has a p⁺-gate electrode. The threshold voltages are +1.5 and −0.75 V for the NMOST and PMOST, respectively, and under bias-temperature stress of ± 15 V and 150°C for 30 min, the device threshold voltage shifts are less than −0.2 V. The source–drain avalanche breakdown voltages (with $V_{GS} = 0$) are +25 and −40 V for the NMOST and PMOST with no evidence of punch-through currents. The transconductances of the NMOST and PMOST measured at $|V_{GS}| = |V_{DS}| = 10$ V are 300-micromho/mil and 200-micromho/mil width, respectively. The source–drain leakage currents measured at $V_{GS} = 0$ and $|V_{DS}| = 5$ V are less than 50-pA/mil width for both devices. All the transistors in the shift register, with the exception of buffer inverters, have channel lengths of 0.2 mil and channel widths of 0.5 mil.

Fig. 14 demonstrates the CMOS/SOS silicon-gate shift register performance at $V_{DD} = 5$ V and 50 MHz. The 1-0-1 data pattern is strobed into the register by the positive excursion of the clock or input phase as shown in Fig. 14 (a). The data is strobed out of the register on the positive excursion of the output or clock phase [Fig. 14 (b)]. The output suffers an additional delay that is due to two buffer inverters. The total shift-register delay is equal to 256 clock periods, or 5.12 μs, as indicated in Fig. 14(c) for a single data pulse. The shift register outputs were current sensed using a TI 751.7 line driver as a sense amplifier.

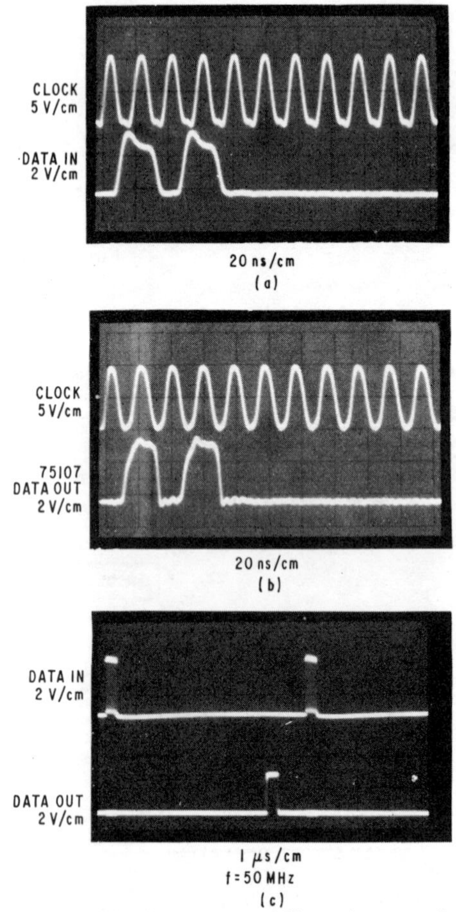

Fig. 14. CMOS/SOS silicon-gate shift-register performance. All verticals—2 V/cm.

The total dc power supply current drain of the shift register is equal to a quiescent or leakage component

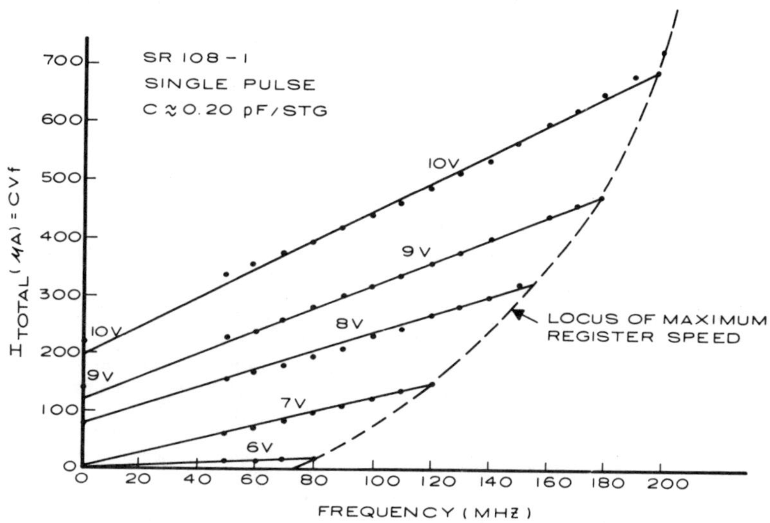

Fig. 15. Current drain of SOS dynamic shift register versus frequency.

I_{DO} plus the average value of the switching current I_{DYN}

$$I_{\text{total}} = I_{DO} + I_{\text{DYN}} = I_{DO} + nCVf \quad (7)$$

where

- n number of data transitions/clock cycle,
- C total node capacitance/stage,
- V supply voltage,
- f clock frequency.

Fig. 15 is a summary of shift-register total current drain versus clock frequency for the case of a single data transition per cycle with supply voltage a parameter. Both the quiescent and dynamic components of the current drain increase with the supply voltage as expected. The quiescent current drain at 10 V for these registers varied from 8 to 200 μA. The reason for the higher magnitude and wide variation of the leakage current I_{DO} is the low PMOST threshold voltage (−0.75 V). As noted above, a remnant channel current flows in low-threshold-voltage MOS transistors even at $|V_{GS}| = 0$. For this particular shift-register application, the highest possible operating speeds were required, which necessitated low threshold voltages. As indicated by the data in Fig. 3, a small increase in V_T will reduce the leakage current by several orders of magnitude. From the data a capacitance/stage C of 0.20 pF is estimated, which agrees very well with calculated values based on cell layout and oxide thickness. The measurements indicate that the maximum dynamic power (i.e., 128 transitions/cycle; square-wave data pattern at $f/2$) of the 256-bit CMOS/SOS register at 5 V and 50 MHz is only 22 mW or 86 μW/bit. This is equivalent to a dynamic power-x delay product of 0.86 pJ per inverter, which is in excellent agreement with the individual inverter measurements (Fig. 7). It is also interesting to note that the performance of the 256-bit dynamic shift register described here (using 5-μm channel lengths) is higher by precisely a factor of 4 over that of a 50-bit CMOS/SOS dynamic shift register (using 10-μm channel lengths) previously reported [5]. Therefore, the proportionality between τ_{sd} and L^2 (Fig. 6) is valid for complex CMOS/SOS arrays.

Conclusions

CMOS/SOS technology permits the realization of large-scale integrated arrays that combine the best feature of monolithic bipolar and MOS technologies. The perfect isolation and reduced capacitance of the SOS technology make possible static MOS circuits with nearly an order-of-magnitude improvement in speed and dynamic power dissipation over their monolithic counterparts. In fact, CMOS/SOS memory arrays have been fabricated with speeds comparable to TTL bipolar memory arrays even when operated at TTL compatible (5 V) levels. Quiescent power dissipation of the two arrays described here is typically less than 1 μW/bit.

The SOS/MOS technology is compatible with both aluminum and self-aligned silicon-gate processing. Further improvements in packing density and performance can be expected as the silicon-gate technology is refined.

Acknowledgment

The authors wish to thank their co-workers who contributed to various phases of this work. N. Kudrajashev and S. Connor were responsible for the mask layout and artwork generation. Test arrays for inverter switching speed and diode leakage were fabricated by J. Sokoloski, W. Salt, and Mrs. R. Gilchrist. A semiautomatic RAM tester was designed by W. Lile and built by L. Maurer. L. Maurer, A. O'Toole, and J. Kingsbury performed many of the electrical measurements throughout this effort. Both of the CMOS/SOS circuits were fabricated in the Integrated Circuit Technology Center, RCA Laboratories with the cooperation of J. Sarace, Miss D. Flatley, and S. Policastro. The high-quality SOS wafers and considerable materials support were provided by P. A. Crossely.

We greatly appreciate the valuable discussions, encouragement, and support of J. H. Scott and G. B. Herzog. The enthusiastic support of R. M. Werner, Air Force Avionics Laboratory, Wright-Patterson Air Force Base, and D. L. Anderson, NASA George C. Marshall Space Flight Center, are gratefully acknowledged.

References

[1] J. R. Burns, "High-frequency characteristics of the insulated-gate field-effect transistor," *RCA Rev.*, vol. 28, no. 3, 1967.
[2] E. J. Boleky, "The effective mobility of MOS transistors on insulating substrates," presented at the IEEE Int. Electron Devices Meeting, Oct. 28, 1970.
[3] ——, "The performance of complementary MOS transistors on insulating substrates," *RCA Rev.*, vol. 31, p. 372, 1970.
[4] J. R. Burns and J. H. Scott, "Silicon-on-sapphire complementary MOS Circuits for high speed associative memory," in *AFIPS Conf. Proc.*, vol. 35, p. 469, 1969.
[5] J. E. Meyer, J. R. Burns, and J. H. Scott, "High speed silicon-on-sapphire 50-stage shift register," *1970 ISSCC Dig. Tech. Papers*, p. 200.
[6] E. J. Boleky, J. R. Burns, J. E. Meyer, and J. H. Scott, "MOS Memory travels in fast bipolar crowd," *Electronics*, vol. 43, no. 15, p. 82, 1970.
[7] J. F. Allison, D. J. Dumin, F. P. Heiman, C. W. Mueller, and P. H. Robinson, "Thin-film silicon: Preparation properties and device applications," *Proc. IEEE*, vol. 57, pp. 1490–1498, Sept. 1969.
[8] E. C. Ross and C. W. Mueller, "Extremely low capacitance silicon film MOS transistors," *IEEE Trans. Electron Dev.* (Corresp.), vol. ED-13, pp. 379–381, Mar. 1966.
[9] J. R. Burns, "Switching response of complementary symmetry MOS transistor logic circuits," *RCA Rev.*, vol. 25, p. 627, 1964.
[10] T. Klein, "Technology and performance of integrated complementary MOS circuits, "*IEEE J. Solid-State Circuits*, vol. SC-4, pp. 122–131, June 1969.
[11] M. B. Barron, "Computer aided analysis of insulated gate field-effect transistors," Stanford Electron. Lab., Stanford Univ., Stanford, Calif., Tech. Rep. 5501-1, Nov. 1969.
[12] D. J. Dumin and R. S. Silver, "Diffused diodes in silicon-on-sapphire," *Solid-State Electron.*, vol. 11, p. 353, 1968.
[13] K. H. Zaininger and C. C. Wang, "MOS and vertical junction device characteristics of epitaxial silicon on low aluminum-rich spinel," *Solid-State Electron.*, vol. 13, p. 943, 1970.
[14] C. T. Sah, R. N. Noyce, and W. Shockley, "Carrier generation and recombination in p-n junctions and p-n junction characteristics," *Proc. IRE*, vol. 45, pp. 1228–1243, Sept. 1957.
[15] A. S. Grove and D. J. Fitzgerald, "Surface effects on p-n junctions: Characteristics of surface space-charge region under non-equilibrium conditions," *Solid-State Electron.*, vol. 9, p. 783, 1966.
[16] H. Kressel, "A review of the effect of imperfections on the electrical breakdown of p-n junctions," *RCA Rev.*, vol. 28, p. 175, 1967.
[17] E. J. Boleky, "Sub-nanosecond switching delays using CMOS/SOS silicon-gate technology," presented at 1971 ISSCC, Philadelphia, Pa., Feb. 17, 1971, paper 6.6.
[18] J. E. Meyer and W. R. Lile, "Design, processing, and testing of LSI arrays for space station." Quart. Tech. Rep. 4, NASA Contract NAS12-2207, Sept. 1971.
[19] G. W. Cullen, J. F. Corboy, G. E. Gottlieb, and J. H. Scott, "The epitaxial deposition of silicon on insulating substrates for MOS circuitry," *Trans. Met. Soc. AIME*, vol. 2, p. 653, 1971.
[20] E. J. Boleky and K. M. Schlesier, "Advanced CMOS circuit techniques," Interim Rep., Air Force Contract F33615-69-C-1499, Project AVT/4159, Aug. 1971.
[21] F. Faggin and T. Klein, "Silicon-gate technology," *Solid-State Electron.*, vol. 13, p. 1125, 1970.
[22] J. R. Burns, *et al.*, "Advanced CMOS circuit techniques," Interim Rep., Air Force Contract F33625-69-C-1499, Project AVT/4159, July 1970.
[23] J. E. Meyer, "MOS models and circuit simulation," *RCA Rev.*, vol. 32, no. 1, p. 42, 1971.

MOSFET Memory Circuits

LEWIS M. TERMAN, MEMBER, IEEE

Invited Paper

Abstract—Metal-oxide-semiconductor field effect transistors (MOSFETs) are currently being used in a variety of memory applications. The requirements of memory usage and the characteristics of MOSFET devices and technology have led to a number of unique circuits for these applications. Organization and design considerations of memory systems using MOSFET devices are reviewed, and examples of specific circuits are presented and analyzed. These include random access cells, shift registers, read only storage, and on-chip support circuits; both complementary and noncomplementary circuits are discussed.

I. MOSFET Memory Considerations

THE POSSIBILITY of using semiconductor circuitry for a large fraction of the memory requirements in computing and other digital systems has received increasing attention over the past several years [1]–[5]. Cost is a very important consideration in memory, and the advent of large scale integration holds the promise of drastically reducing the cost of an individual semiconductor circuit; appreciable reductions in digital circuit costs have already occurred. In addition, the use of circuitry as a storage technology offers a degree of flexibility unavailable with other memory technologies. Consider the block diagram of a memory system shown in Fig. 1. The basic elements of a memory are the array, in which the information is stored, and the supporting circuit blocks, which carry out the functions of selection, driving the array, and amplification and detection of sense signals. In a conventional magnetic memory system different technologies are used for the array and the support circuits and, as a result, there is a pronounced interface between them. This interface is both electrical and physical. Relatively complex circuitry is required to provide signals to drive the array, as well as to amplify and detect the sense signals coming out of the array, and different packaging is necessary for the circuits and the array. The use of circuitry in the storage array greatly reduces the interface. Clearly, the packaging can be done in the same technology, but more important, since the use of circuitry permits placing gain in the system anywhere it is desirable, the electrical interface can also be reduced. Conceptually, the entire memory system could be operated at logic levels, eliminating any need for special driver, sense amplifier, and detector circuits. In practice, this is not done because it is more efficient to simplify the storage circuits at the expense of support circuit complexity because of the far greater number of the former, but there exists considerable freedom in tradeoffs between the various circuit functions to achieve an optimum system, and, in fact, storage, decode, drive, and sense circuits may be intermixed as desired for this end.

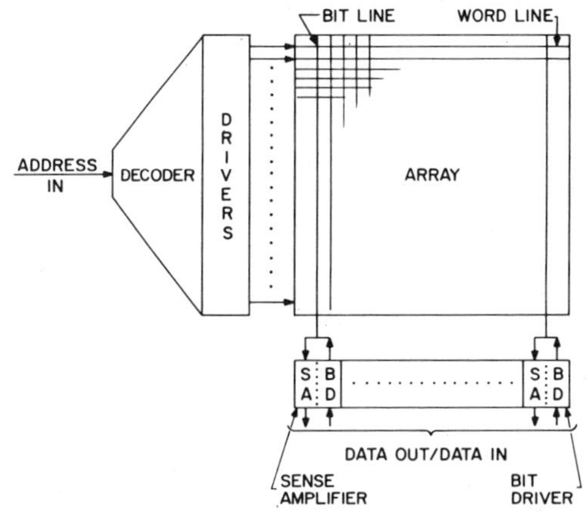

Fig. 1. Memory system block diagram.

There are additional advantages in a semiconductor memory system. Nondestructive reading is easily achieved, as is a dc base sense signal, and it is easier to obtain high performance because of the reduction of the array-support circuit interface, and the characteristics of the technology. On the other hand, there are also disadvantages. These include the power dissipation necessary to retain information, the volatility of storage which this implies if input power is lost, the reliability of a system having a great many chips and interconnections, and the question of whether or not sufficient circuit cost reduction can be achieved through large scale integration to allow semiconductor memory systems to be cost competitive with other memory technologies, especially ferrites, over a wide range of applications.

One important result of easing the interface between the array and support circuits is flexibility in choosing array size without affecting per bit cost. In magnetic systems, the support circuitry is comparatively expensive, while the cost of the individual storage bit is relatively low; thus, as the array is made larger, the support circuitry is spread over a larger number of bits, and the cost per bit declines. In contrast with a semiconductor memory the major cost is in the bit storage circuitry, and the cost per bit will remain essentially constant as the array size increases or decreases. These trends are sketched in Fig. 2. The ability to fabricate smaller systems at relatively low cost has given semicon-

Manuscript received September 21, 1970; revised January 20, 1971. *This invited paper is one of a series planned on topics of general interest—The Editor.*

The author is with the IBM Thomas J. Watson Research Center, Yorktown Heights, N. Y. 10598.

Reprinted from *Proc. IEEE,* vol. 59, pp. 1044–1058, July 1971.

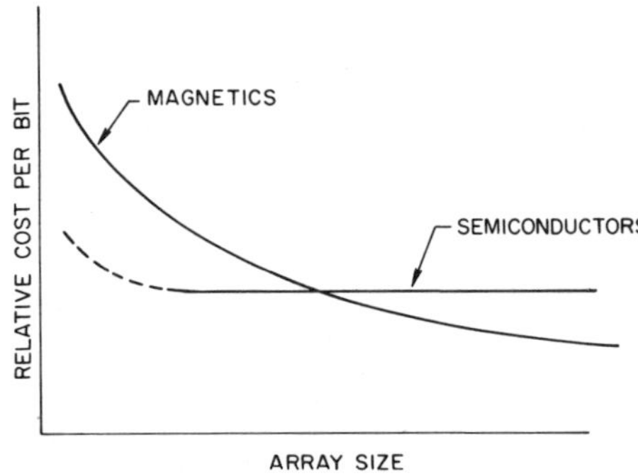

Fig. 2. Trends of cost per bit as a function of array size for magnetic and semiconductor memories.

ductor technology a toehold in the memory market, and, as larger systems become economically competitive, will result in the possibly advantageous flexibility that the memory does not have to be made in a single large system to keep the cost down, but rather can be segmented into a number of smaller units, which can be placed as desired through the system in order to optimize it.

The point at which the crossover occurs in Fig. 2 is very important, as it indicates the relative areas of application of the technologies (it is, of course, dependent upon the performance range being considered). Since with time the magnetic system curve will go to lower and lower costs, the semiconductor system cost must drop even faster in order to expand its area of application. Cost, therefore, is very important.

The circuit used to store information, the memory cell, is a prime element in determining semiconductor memory cost. Desirable characteristics are small area, allowing high bit storage density, light loading to the drive and sense circuitry, and the ability to be written and read with a reasonable delay. It also should be able to retain information with low power dissipation.

Metal-oxide-semiconductor field effect (MOSFET) devices are attractive for memory because the technology offers the potential of low cost and of fulfilling the previously mentioned requirements. The processing itself is simple, requiring four masking steps and only a single diffusion, which should result in higher yields and/or capability for larger chips. High circuit density can be obtained because the devices themselves are small, self-isolated, amenable to efficient layouts, and can be combined in simple dc coupled circuits without requiring additional resistors or capacitors. In addition, the devices themselves have a number of characteristics that are useful in designing the circuits. These include an absence of minority carrier storage effects, bidirection conduction between the source and drain of the device, the absence of any appreciable dc loading imposed by the gate circuit, low capacitive loading per device, and the high impedance level of the device. The last permits the fabrication of high values of resistance in relatively small areas which is helpful in reducing power dissipation, and allows the use of relatively long diffused lines without having their impedance

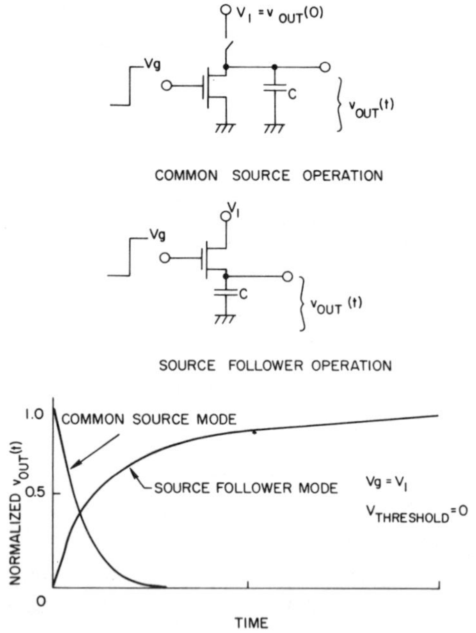

Fig. 3. Comparison of MOSFET switching transients. The effect of substrate voltage on threshold voltage has been ignored.

affect the circuit performance, which can be advantageous in chip layouts.

Performance has been considered a major drawback of MOSFET devices, and it is true that the best present MOSFET logic circuits operate at the low performance end. However, in the context of a memory system, the combination of MOSFET device characteristics and the requirements of memory usage permit operation of a memory system with sufficient performance to encompass a much wider range of applications. The access or cycle times of most core or film memories are much longer than the delay of the associated logic circuitry in the remainder of the system (a computer main frame memory is an example). Although a MOSFET system may have a long transient which limits performance, the system can still operate in conjunction with much higher speed processor logic.

The longest delays in MOSFET circuits occurs in driving off-chip, where the high impedance level of the devices is relatively slow in driving large interconnections and stray capacitances. With proper system organization the MOSFET transients can be kept on-chip, eliminating the need for off-chip FET drivers, and avoiding the resultant long transients.

The speed of a MOSFET switching transient depends upon the mode in which the device is operated. As shown in Fig. 3, the common source transient is much faster than the source follower transient; it also avoids the voltage attenuation of the latter.[1] Transient speed depends upon

[1] An exposition on MOSFET switching transients is given in [6]. The source–follower curves given therein are calculated for the substrate connected to the device source, which is impractical in an integrated circuit, and which eliminates the augmentation of the device threshold with increasing source to substrate voltage. A satisfactory qualitative correction can be made.

the voltages applied to the devices. Judicious use of clocking pulses to sequentially turn devices on and off, such as in multiphase logic schemes, [7]–[10], can eliminate dc paths and gain ratio requirements in switching circuits, permitting further improvement in performance. In addition, the MOSFET device exhibits essentially no delay in the source–drain current in response to a voltage pulse, applied either to the gate to turn the device on, or on the drain when the device is already on [11]. Neither of these modes of operation is applicable to logic circuits, since a current output is obtained for a voltage input, but either may be used to sense the state of a storage element without delay. This permits high speed reading of the state of a memory circuit.

For many applications, the optimum system is obtained by interfacing the MOSFET array chips with bipolar support circuit chips [12], [13]. This permits interconnection lines which have large capacitive loads to be driven by low impedance bipolar circuits, resulting in much faster transitions than if MOSFET circuits were used to drive the lines. Similarly, in sensing, transition time delay can be minimized by the use of a low impedance sense amplifier, which essentially senses current. The low impedance sense amplifier keeps the sense line time constant short, and, as outlined previously, current sensing results in minimum cell delay. The result is a hybrid system, the goal of which is to combine MOSFET cost advantages with bipolar performance capabilities. MOSFET chips are used in the array, which has the great majority of total circuits, and thus determines system cost. Bipolar chips are used for support circuits, which basically determine the performance, and are much more adept at driving larger capacitive loads. As a result, the best characteristics of both technologies are obtained simultaneously, and the approach is widely used where performance is a requirement.

The best performance in a hybrid system is obtained if the array chips contain no decoding. This permits the bipolar support circuits to have direct access to the on-chip word and bit lines, and in system operation the only delay involving MOSFET circuitry occurs in the cell itself. In reading, the delay can be essentially zero; in writing, there is a delay due to the transient of the cell as it switches, but since within the cell the device or devices causing the switching are comparatively lightly loaded with an inherent fan-out of 1, the delay can be made small compared to the system cycle time. The performance is largely determined by the bipolar support circuits.

The price paid for this organization is the number of connections to the chip required. Consider Table I, where the number of the chip address and information connections required for various organizations of hypothetical 256 and 1024 bit memory chips is given.[2] It is clear that the chip with full binary decoding has many fewer chip connections, which in turn should be advantageous from the standpoint

[2] A memory cell with 1 word line and 2 bit lines has been assumed in Table I.

TABLE I

	256	1024	Bits per Chip
Undecoded chip:	16 × 16	32 × 32	organization
	16	32	word line connections
	32	64	bit line connections
	48	96	total connections
X–Y selected chip:	256 × 1	1024 × 1	organization
	16	32	X connections
	16	32	Y connections
	2	2	bit line connections
	34	66	total connections
Full binary decoded chip:	256 × 1	1024 × 1	organization
	8	10	binary address lines
	2	2	bit lines
	10	12	total connections

of reliability and packaging yield, and the bit organization of a binary decoded chip has advantages of ease and flexibility in assembling a complete system. The decoded array chip is still interfaced with bipolar support circuit chips to obtain the best performance.

There are also a number of drawbacks to a decoded chip. The decode circuits require chip area and increase power dissipation on the chip. There is also more total capacitance in the system which must be driven. There is a significant effect on performance. On-chip decoding requires MOSFET logic circuits, and at least some decoding at MOSFET rather than bipolar speeds. Although the circuits do not have to drive off-chip (thus avoiding interconnection capacitances), they must drive large fan-outs and the associated capacitive loading, such as occur on the word line or the outputs of the address inverters.[3] The result is longer chip transients and a lengthening of the chip access and cycle times. This is quite apparent in present-day commercially available chips, as is discussed in subsequent sections.

There is still appreciable controversy over decoded versus undecoded chips, and currently there are examples of chips and systems using both approaches [15]–[18]. The key to the undecoded chip is the interconnection technology, which must have sufficient reliability and density. As the number and density of bits on a chip increase, there will be increasing requirements on the interconnection technology. Historically, improvements in chip size and density have occurred faster than corresponding interconnection capabilities, and it seems likely that the ability to put bits on a chip will outstrip the interconnection technology eventually, if this has not already occurred. Further, evolutionary im-

[3] Actually, the address inverters can be eliminated, at a cost of increased address connections, by presenting double rail address information to the chip, or by the use of combinatorial logic [14]. Neither approach has seen appreciable use.

provements in MOSFET devices should result in improved device performance, and reduce the performance difference between decoded and undecoded chips.

II. Random Access Read/Write p- or n-Channel Memory Cells

The first [19] and still most common MOSFET memory cell is the six-device circuit shown in Fig. 4.[4] Devices T_1–T_4 form a conventional bistable cross-coupled flip-flop circuit. T_1 and T_2 are the cross-coupled or "active" devices, T_3 and T_4 are the load devices. The interior nodes of the flip-flop are connected to the bit lines through two gating devices, T_5 and T_6, which can be switched on by the word line. Coincident selection versions of the cell have been shown in the literature [20]–[21]. The basic cell operation is similar.

In the quiescent state, the word line is at ground and T_5 and T_6 are off, isolating the flip-flop from the bit lines. Information is retained in one of the two stable states of the flip-flop circuit. The cell dissipates power due to current flowing through the load device in the on side of the flip-flop. The off-node load device conducts only the very small amount of current necessary to supply leakage current to the substrate from the off node, and essentially dissipates no power.

For the purposes of illustration, assume node 1 is high, and node 2 is low. Then T_2 is on and T_1 is off. To read the cell the word line is pulsed turning on gating devices T_5 and T_6. Both bit lines are kept at a supply potential. Current flows between the one bit line and ground through the on flip-flop device T_2 and the gating device T_6. No current flows in the zero bit line because T_1 is off.[5] The state of the cell is determined by detecting on which bit line the sense current occurs. There is essentially no cell delay in reading. The sense current will flow in the gating device in direct response to the word line voltage pulse (once the threshold voltage has been exceeded) as previously discussed. The drain of the gating device is connected to the bit line supply, and its source is connected to ground through the on flip-flop device, which is a comparatively low impedance. To the first order, the sense current does not charge or discharge capacitance, and it can flow to the sense circuitry without delay.

To write the cell, the word line is pulsed and one of the bit lines is grounded. In the illustrative state, if the one bit line is grounded, the cell is already in the state to which it was to be written, and no change occurs. If the zero bit line is grounded, node 1 is pulled down towards ground through the gating device T_5. As the level of node 1 drops below the threshold voltage of T_2, the device turns off, and node 2 is pulled up through T_6 and T_4. The writing process thus consists of two transients: discharging the initially high node through a gating device operating in common source mode, and charging the initially low node through the other gating device and a load device in source follower operation. The transients overlap somewhat, but it is the charging transient which largely determines cell writing delay in practice.

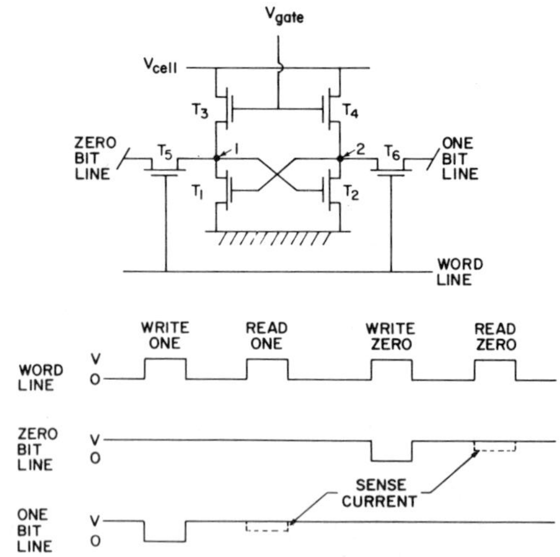

Fig. 4. Six-device memory cell.

There are two limiting cases for the writing operation. In the "low voltage" mode, the word line is kept on only until the cell node voltages have sufficient offset that threshold voltage variations and other circuit asymetries cannot cause the cell to lapse back to the initial state. At this point the word pulse is terminated, and the remainder of the pull-up of the high node occurs through the associated load device. In the "high voltage" mode, the pull-up transient occurs through the gating device, and the word pulse remains on until the final level of the high node is obtained. In the limiting case the contribution to the pull-up transient by the load device is negligible. Clearly, a spectrum of intermediate combination modes also exist.

The low voltage mode permits the use of a lower word line voltage pulse, but is unattractive because of poor performance. Low cell standby power requires a high impedance load device, while a fast writing pull-up transient requires a high gain (low impedance) device. These are conflicting requirements, and result in the performance and standby power of the cell being interrelated. To achieve a reasonable cell standby power, poor write performance must be accepted.

In the high voltage writing mode, the pull-up transient occurs through a gating device, which is turned off in the quiescent condition and which does not affect cell standby

[4] Most of the circuits shown (with the exception of the complementary circuits) will operate in the same general manner whether made in p-channel or n-channel technology, assuming the appropriate voltage supply polarity. Because of this, the device technology type and supply polarity are not explicitly shown on the circuit diagrams, although descriptive terminology as "high" or "low" level, "charge" or "discharge," etc., will be used as if n-channel devices and positive supplies were used. Enhancement mode devices have been assumed throughout except where contra-indicated in the text.

[5] Actually, there can be a current in the off side bit line through the gating device and the load device if the cell supply is larger than the bit line supply. This current will be in the opposite direction to the "1" sense current.

power. As a result, the performance and the standby power are independent, and may be designed for separately. Arbitrarily low standby power can be achieved by raising the load device impedance, since the only requirement on a load device is that it deliver enough current to the flip-flop off side node to compensate the node leakage current, which is quite small. In practice, a lower limit on cell power arises because higher impedance load devices take increasing area, and the cell becomes too large to be economically attractive.

Write delay is primarily determined by charging a node capacitance through the associated gating device. Since this capacitance consists mainly of the gate capacitance of the active cross-coupled device and some additional stray capacitance, write performance may be improved by increasing the gain of the gating device. However, there is an upper limit on the gain of the gating device. During reading, the cell can be disturbed if the current through the on-side active device is sufficient to raise the drain on the on-side active device above the threshold of the off-side active device to where the positive feedback of the circuit, and possible asymetries in device characteristics, can cause the cell to change state. This places an upper limit on the gain of the gating devices (or more precisely upon the ratio of gating device gain to active device gain).

Thus write delay and sense current are interrelated;[6] however, worst case write delay is associated with minimum sense current conditions, and the disturb condition occurs generally at or near the maximum sense current conditions. Thus tolerances on device characteristics and supply levels also affect the write performance. The looser the tolerances, the smaller the gating device must be made to avoid possibly disturbing the cell, and the worst case write delay is correspondingly increased. A limit on the minimum gating device size is imposed by the minimum sense current requirement and the necessity, during writing, of discharging the initially high node, which occurs through voltage division between the load and gating devices. An alternative, increasing the active device gain, generally increases cell area, which is undesirable.

A significant reduction in standby power dissipation of the cell may be achieved by pulsing the load devices off through the load device gate line, and occasionally pulsing them on. While the load devices are off, standby power is virtually eliminated. The read operation is unchanged and the write operation must be in the high voltage mode but is otherwise unaffected. This is termed "stored charge" operation, in that information is retained dynamically as charged stored on the capacitance of the off node of the flip-flop. There is effectively no leakage through the gate structure of the active device; however, since the nodes of the flip-flop are partially diffusion areas, the back biased

[6] One way to break the interrelation is to use a bilevel word pulse, with reading done at the lower voltage, and writing done at the higher. This requires additional complexity in the word driver circuits, and has not been used in practice.

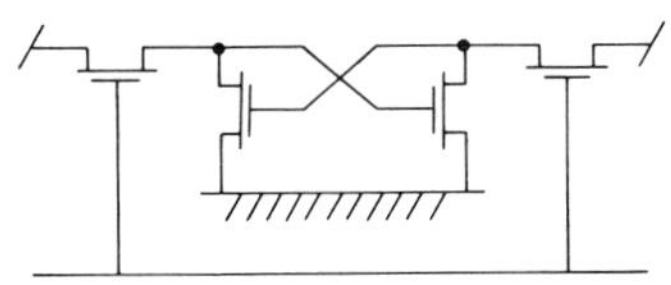

Fig. 5. Four-device stored charge cell.

p-n junction leakage currents potentially cause eventual loss of information. To overcome this, the load devices must be periodically pulsed on to "restore" or "refresh" the original charge condition. While the load devices are on, the cell does dissipate power. In practice, the interval between restore pulses is on the order of milliseconds, while the restore pulsewidth is in the microsecond range or less, and the duty cycle on the load devices is on the order of 10^{-2} to 10^{-3}. Average power dissipation is reduced by the same factor.

Reading and writing operations can be unaffected by stored charge operation, and can continue through the restore cycle. The cell design must take into account the droop in the off-node level due to leakage, and also due to the capacitive coupling to the trailing edge of the restore pulse, but otherwise the design criteria are essentially the same as for the conventional six-device cell. The possibility of the load devices being on during a read or write operation due to a simultaneous restore cycle must be included.

The load devices are not necessary for the read or write operations and, in fact, are also unnecessary for the restore cycle, and thus may be eliminated, resulting in the four-device cell [4], [22] shown in Fig. 5. Restoring the cell is accomplished by a read cycle (although no actual detecting of the cell state need take place). When the word line is pulsed, the gating devices turn on and act as load devices to the active devices, forming a cross-coupled flip-flop and the off-side node is restored by charge flowing from the associated bit line. Basic cell operation and design are similar to that of the six-device stored charge cell, except that the load devices are absent. The basic advantage this cell has over the six-device stored charge cell is in circuit area. The elimination of two load devices and two power supply lines result in a cell area about two-thirds to a half that of the six-device cell.

A flip-flop in stored charge operation is actually unnecessarily redundant, and information may be stored on the gate of a single device. Elimination of one cross-coupling device in the four-device cell leads to the three-device cell shown in Fig. 6 [17]. Information is stored on the capacitance of node 1. This is one of a class of memory cells in which information is stored on a node capacitance associated with the gate of a device, and which do not use a cross-coupled flip-flop circuit.

The cell is written by turning on the gating device T_1 with the word line, which establishes the condition of the write bit line on node 1. In reading the cell, the word line is turned on. Devices T_2 and T_3 form an AND gate. T_3 will be on or off, depending upon the charge condition of node 1,

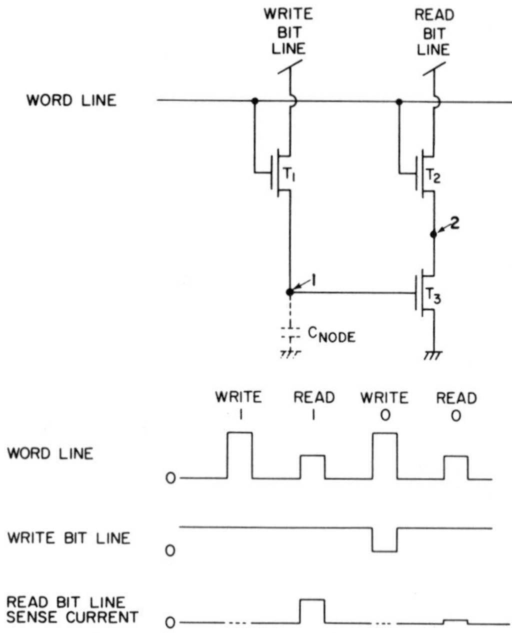

Fig. 6. Three-device stored charge cell.

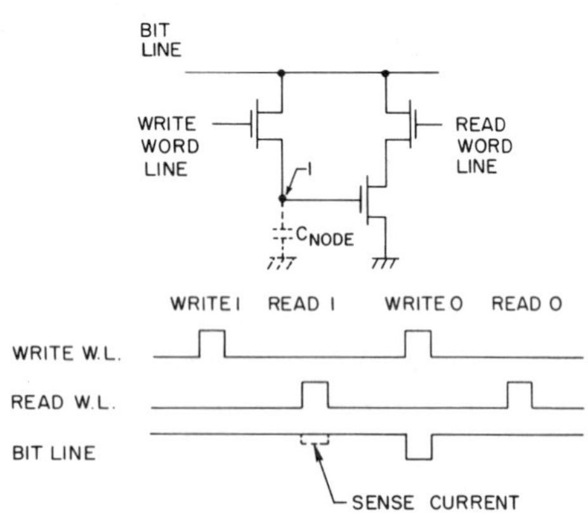

Fig. 7. Alternate three-device cell.

and the presence or absence of current in the read bit line can be detected to determine the state of the cell. A lower word line voltage is used in read to prevent inadvertent writing of the cell. This is aided by the fact that the word line must be at least two threshold drops above ground in order to turn on T_3, while it need be only one threshold drop above ground to turn on T_2 if T_3 is already on. An alternate version of the three-device cell which eliminates this possibility by having two word lines is shown in Fig. 7 [23].

The design of these cells is significantly different than that of the cross-coupled circuits. The absence of the cross-coupling device and the associated positive feedback paths results in the elimination of device gain ratio requirements. The cells cannot be disturbed by excessive sense current. As a result, the devices may be laid out for minimum cell area rather than being restricted to certain size ratios due to circuit requirements.

The write transient of the three-device cell can be inherently faster than that of the cross-coupled cells due to the absence of a gain ratio requirement. In both circuits, the

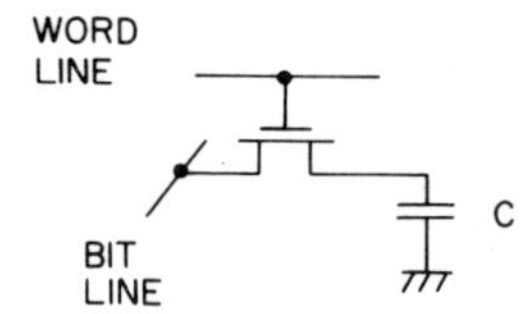

Fig. 8. Capacitor stored charge cell.

worst case write transient is a source–follower pull-up of an information node through a gating device. In cross-coupled circuits, the capacitance of the information node is largely determined by the flip-flop active device which must be larger than the gating device by some ratio. This results in a slower transient than for the three-device cells, where no such adverse ratio of device sizes is necessary. The three-device cell also has the advantage that only a single transient is required for writing, rather than the discharge transient, followed by an overlapping but still somewhat sequential pull-up transient that occurs in the cross-coupled circuits.

Read may be slightly slower in the three-device cells because (at the beginning of the read cycle) some sense current may be lost in charging node 2, or because a false sense current spike may charge node 2 even if T_3 is off. Such delays are generally minor from a system standpoint.

The three-device cells occupy much smaller circuit area than the six-device cell, not only because of fewer devices used, but also because they are ratioless circuits, and can be laid out for minimum area. In practice, an area advantage of perhaps a factor of four is obtained for the circuit of Fig. 6; that of Fig. 7 is somewhat less.

Perhaps the simplest stored charge random access cell is that using the circuit of Fig. 8 [24]. It consists of a capacitor, upon which information is stored, and a single gating device. The cell is written by establishing the bit line level on the capacitor through the gating device; reading occurs by connecting the capacitor to the bit line through the gating device, and sensing the charge on the capacitor, either as an ac base current pulse, or as a quasi-dc base voltage on the bit line capacitance. The cell has a number of drawbacks—the readout is destructive, the stored charge redistributes between the cell and bit line capacitances, and the writing speed is limited by the information storage capacitor which must be made large enough to yield a reasonable sense charge. On the other hand, the cell is simple, and the word and bit line loading can be about the minimum possible for any cell.

Each of the various stored charge cells must be refreshed periodically, but the procedure varies for different cells. The six-device cell in stored charge operation is refreshed through the load device gate lead, independent of read or write operations, and a chip, a block of the memory, or the whole memory system may be refreshed simultaneously at any time. The four-device cell is refreshed by pulsing the word line, similar to a conventional read operation, and consequently read and write operations must be inhibited during the refresh cycle. Block refreshing can be done by turning on all the word lines on a chip simultaneously, although there may be a problem with chip power dissipation and the difficulty of delivering sense current to all cells simultaneously.

Cells which are not cross-coupled, such as the three-device cells, will not automatically refresh themselves into the proper state when driven. The information to be refreshed must be read out first, and then be rewritten. This results in a longer refresh cycle, and also means that only one word line on a chip (or in any block of memory which shares the same sense amplifiers and bit drivers) can be refreshed at a time. A fully decoded bit organized chip will require an on-chip sense amplifier for each chip bit line, since writing one bit along a word line results in all bits along the word line being written into whatever state the respective bit lines are in, and in a bit organized chip, only one bit line is connected off chip at a time. The on-chip sensing can basically be a logic circuit, and can be relatively simple (as in Section VI), but additional chip circuitry is required. There will also be additional delay involved, and if the on-chip sensing is used in the read operation, the chip read access delay will be longer compared to using bipolar off-chip sensing.

The present state of the art, as reported in the literature, is 256 bits per chip using the six-device cell, and 1024 bits per chip using the three-device cell, indicating the density advantage of the latter. Chip and system performance depend upon chip organization. Using chips without decoding, cycle times of 150 ns or better for the full memory system have been reported using bipolar support circuits [1], [15]. Systems using fully decoded bit organized chips generally have cycle times in the 500-ns to 1-μs range [17], [18].

III. p- or n-Channel Shift Register Circuits

An alternative to the random access systems of the previous section is serial access systems using shift registers. The shift register is essentially an electronic delay line. As shown in Fig. 9, memory circuits are connected serially, so that the output of one feeds the input of the next. Under the control of clock pulses, all information bits are simultaneously shifted from one stage or cell to the next throughout the shift register. The output of the shift register may be fed back into the input to form a closed loop, around which information can be circulated.

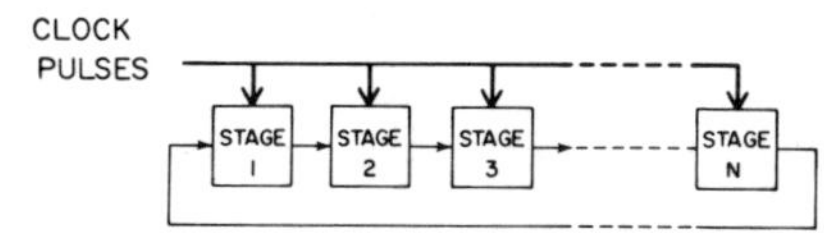

Fig. 9. Shift register block diagram.

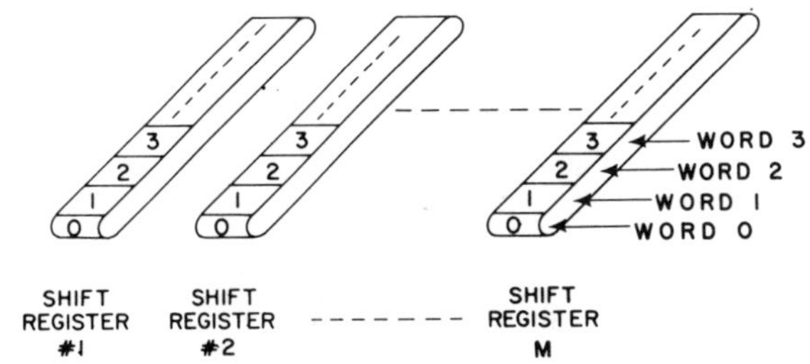

Fig. 10. Possible shift register memory system organization.

Once a bit enters a shift register, it must be shifted through all stages before it becomes available at the output.[7] To access a particular bit of information it may be necessary to shift it through every stage in the shift register, which results in a long delay, and for this reason shift registers are not attractive for random access applications. On the other hand, there are numerous applications where data are generated, manipulated, or transmitted serially. Also, if a sequential block of information is to be accessed, the long delay in obtaining the first bit of information may be outweighted by a high bit rate for transferring the rest of the information in the block. In this case a possible system organization is sketched in Fig. 10. Each shift register contains 1 bit of N words. To access a word, one shift register for each bit in the word are shifted synchronously so that all bits of the same word arrive at the shift register outputs simultaneously. The next word is obtained after the next shift. Worst case access time depends upon the length of the shift register; a shorter length results in faster access, but requires more decoding to select between the larger number of shift registers required in a system of a given size.

A shift register chip may require no decoding since any number of bits may be handled by a single set of input and output connections, and the number of connections to the chip needed is correspondingly low. The area of a single stage can be made quite small, and reasonable shifting rates can be obtained. A shift register stage is essentially a logic circuit which must drive another similar circuit with a fan out of 1; there is no large lumped capacitance which the cell must drive, such as the word or bit line of a random access chip. The shifting rate will be limited by the slowest cell in the shift register, and care must be taken to avoid having performance unduly limited by a single stage with an unusually large load, such as might occur at the output stage or as a result of nonuniform cell layout on the chip, because the shift register circuit must physically turn a corner or loop back. Interfacing the shift register chips with bipolar support circuits results in advantages similar to those obtained for random access systems; in particular, the clock pulse drivers must drive every cell on the chip, and thus see an appreciable capacitive load. Bipolar devices result in faster transition times on the clock pulses, particularly if a number of chips are to be driven in parallel.

The serial nature of the shift register has implications in reliability. If one cell in a shift register fails, the cell information passing through the shift register subsequently is lost, as all bits must pass through the bad cell. On the other hand, the low number of chip interconnections required per bit is favorable from a reliability standpoint; the net effect depends upon which failure mode predominates.

A shift register stage must be able to transfer an information bit to the next stage, while simultaneously receiving an information bit from the previous stage; subsequently, the received bit must be transferred within the stage, preparatory to the next stage-to-stage transfer. Thus a complete

[7] It is possible to have intermediate tap-off points.

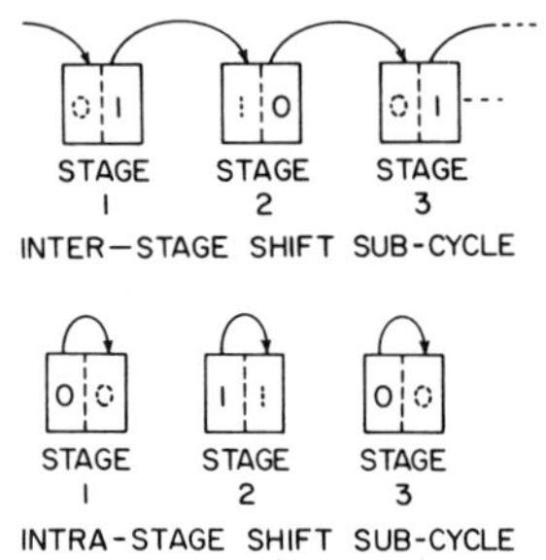

Fig. 11. Shift register subcycles.

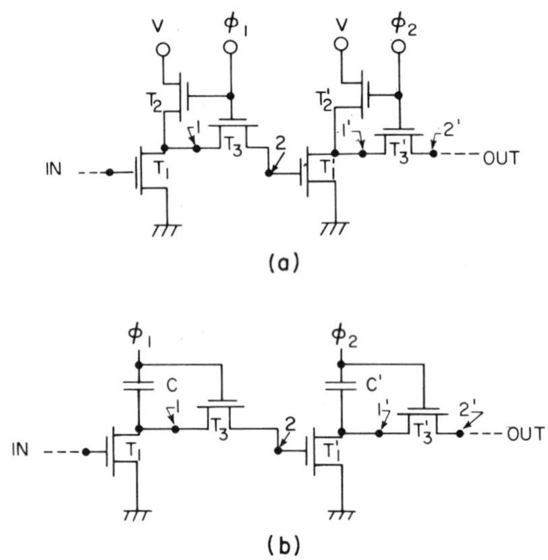

Fig. 12. Two phase shift register circuits.

shifting cycle actually requires two shifts, as shown in Fig. 11, and the stage is made up of two substages, which are usually identical. In a dynamic shift register, an information bit is generally stored on the gate capacitance of a device in an inverter and is transferred by pulsing the subsequent inverter properly, or through a series connected device used as a transmission gate which is pulsed on at the appropriate time. An inverter circuit provides gain, while the transmission gate may operate in the source–follower mode, resulting in voltage attenuation.

A two-phase shift register circuit is shown in Fig. 12(a) [25]. There are two identical substages; in each, T_1 and T_2 form an inverter, and T_3 acts as a series transmission gate connecting it to the inverter to the next substage. Prior to the application of phase (clock) pulse ϕ_1, information is retained as charge stored on the capacitance of the input node, as in the stored charge random access memory cells discussed previously. When ϕ_1 is applied, the inverter action establishes node 1 at the inverse of the input level, and node 2 follows through device T_3. When ϕ_1 terminates, device T_3 turns off, locking in the charge condition of node 2, and the information has been transferred in inverted form, from the input to node 2. In a similar manner, ϕ_2 transfers the information to the output node 2′ where, because it has undergone a second inversion, it has the same "sense" that it had at the input to the stage. To prevent interaction of successive bits of information, the phase pulses are kept from overlapping.

Performance is limited by the necessity of an impedance ratio between load and active devices in the inverters. Conventional dc inverter operation dictates this be around 1:10, which results in a comparatively low gain to capacitance ratio for the pull-up transient, where the load device charges the capacitances of nodes 1 and 2, the latter being primarily determined by the gate capacitance of the active device in the next inverter. The pull-down transient, occurring through T_1 and T_3 in the common source mode is much faster. The necessity of a gain ratio also results in a larger cell area. The transmission gate device has no dc requirement for a gain ratio with the load or active devices, but cannot be made too small or it will cause appreciable further slowing of the pull-up transient, and cannot be made too large because of the gate to source capacitive coupling on the trailing edge of the associated phase pulse. This coupling will reduce the amplitude of a high level (or "1") stored on node 2, and may cause the substrate diode to become forward biased if a low level (or "0") is stored there. In practice the gain of the transmission gate device is in the middle between those of the load and active devices.

The stage dissipates dc power while the phase pulses are on, since during that time there is then a possible path between the phase lines and ground. A modification of the two-phase circuit which eliminates this problem is shown in Fig. 12(b) [26], where the load device is replaced by a capacitor C. Operation is similar to the previous circuit. Node 2 is charged through C and T_3 if the input is low (T_1 is off), and is discharged as in the previous circuit if the input is high (T_1 in on). The capacitor eliminates the dc current path from the phase pulse. The charging transient of node 2 is faster than before because of the elimination of the load devices and dc gain ratio requirements. The discharge transient is longer than before, since the load capacitor C must be charged. This capacitor is made several times larger than the sum of the capacitances of nodes 1 and 2 to reduce loss in the phase line level coupled into node 2,[8] and a longer discharge transition results. An additional performance limitation arises from the possible injection of carriers into the substrate from the source of T_3, due to the coupling to the trailing edge of the phase pulse forward biasing the substrate diode. Overall, however, the performance is somewhat better with the capacitor load. Cell area is roughly the same, since the capacitor occupies appreciable area.

Both circuits in Fig. 12 are dynamic; i.e., similar to the stored charge random access cells, and the information is stored as the presence or absence of charge on a node capacitance. Leakage necessitates periodic refreshing of the information, which is accomplished when information is shifted to the next inverter. Thus the information is continually circulating in the shift register, with the leakages determining the minimum shift rate.

A widely used four-phase dynamic shift register stage with improved performance is shown in Fig. 13(a) [9]. The

[8] Reducing the value of load capacitor C can be compensated for by raising the phase pulse level, but this aggravates the problem of capacitive coupling to node 2 from the trailing edge of the phase pulse.

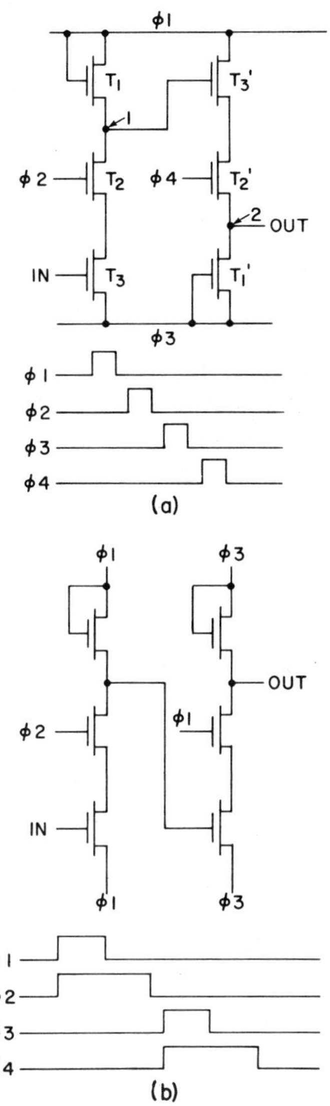

Fig. 13. Four-phase shift register circuits. (a) Nonoverlapping clock pulse circuit. (b) Overlapping clock pulse circuit.

stage consists of two identical pulsed inverters with the second drawn upside down to simplify the circuit schematic. Clock pulse ϕ_1 unconditionally charges node 1 through device T_1, irrespective of the conditions of the input line, since ϕ_2 is off. When ϕ_2 goes on, node 1 is conditionally discharged through the two-device AND gate formed by T_2 and T_3. At the end of ϕ_2, the input information has been shifted, in inverted form, to node 1. ϕ_3 and ϕ_4 perform the corresponding operations on the second inverter, and shift the information (now reinverted to true form) to the output node 2.

The clock pulses are kept from overlapping to avoid interference of adjacent information bits, and to prevent any dc current paths during shifting. This eliminates any requirements of a gain ratio between devices and allows a stage to be laid out for minimum area. The four-phase stage has a particularly efficient layout, and is similar in area to that of the three-device random access cell, despite having twice as many devices.

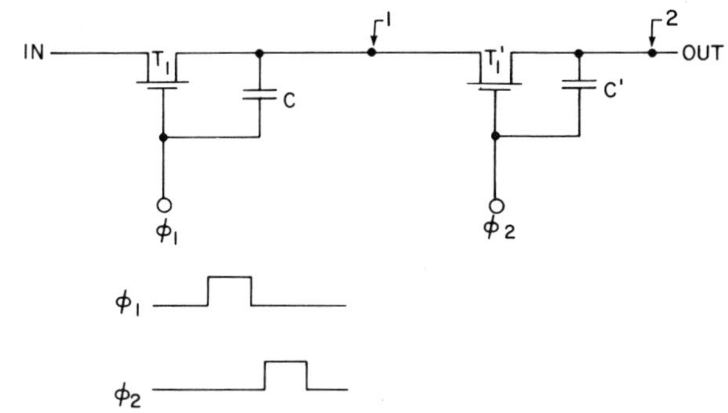

Fig. 14. Bucket brigade shift register circuit.

The elimination of any gain ratio requirements also results in improved performance because the possibility of a small device having to charge or discharge the gate capacitance of a much larger device is avoided and large capacitances, such as the load capacitance of the circuit of Fig. 12(b), are not present. The shift register performance can be improved by raising the amplitude of the clock pulses. This places extra strain on the supporting drive circuits and increases the power dissipation of a stage. Only dynamic power is dissipated in a stage, arising from the charge and discharge of the various capacitances. It is normally small but can become appreciable for high clock pulse amplitudes and shifting rates.

This four-phase shift register is widely used because of good performance and small layout area. One of the ultimate limitations to high speed shift register operation is the necessity of generating four separate nonoverlapping clocks. A four-phase shift register with somewhat simple clocking is shown in Fig. 13(b).[9] Operation is similar to the previous four-phase circuit, but the phase clocks can overlap without allowing a dc current path. Design considerations are similar to the nonoverlapping clock circuit, performance is slightly better, especially at low clock pulse amplitude, but the layout area is perhaps 25 to 50 percent larger.

The shift register stages shown so far have gain in each stage. This, in fact, is not necessary. Fig. 14 shows a shift register stage using the so-called "bucket brigade" technique [27]. It is a simple circuit requiring comparatively small layout area. A simplified explanation of the operation follows, assuming enhancement mode devices with a zero volt threshold. Initially, node 1 is charged to e_0 V and the phase lines are at ground. When ϕ_1 is pulsed to e_0 V node 1 is raised to 2 e_0 by capacitive coupling through C. Device T_1 now operates as a source follower in the saturation mode. Assume the input node is uncharged (i.e., at zero volts). Charge flows from node 1 to the input node until it charges up to e_0 V, at which point device T_1 turns itself off through normal source–follower action. At this point, node 1 is also at e_0 V. When ϕ_1 returns to ground, e_0 is coupled out of node 1, which ends up at ground potential. If the input node had been at e_0 V initially, no charge would have flowed through T_1, and node 1 would have ended up at e_0 V after

[9] The circuit is related to logic circuits shown in [9] and [51].

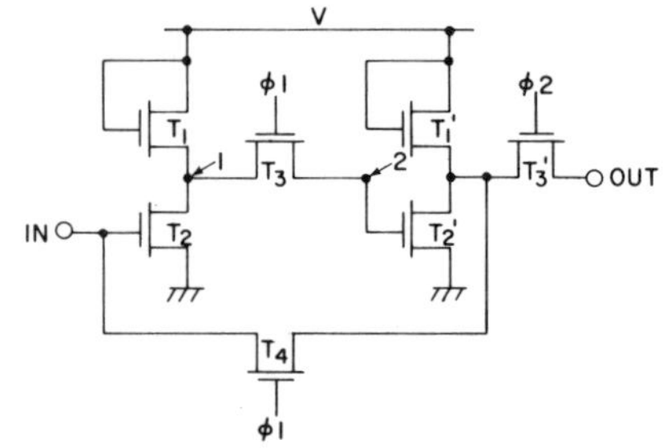

Fig. 15. Static shift register circuit.

ϕ_1 terminated. In this manner, the state of the input node is transferred to node 1 by ϕ_1. Clock pulse ϕ_2 performs a similar transfer of the information to the output node.

Since the devices operate in the source–follower mode, there is voltage attenuation at each transfer, and gain must be placed in the shift register periodically to reestablish the "1" and "0" levels. In order to avoid requiring gain too frequently, as little voltage attenuation as possible in each transfer is desired, which implies limited performance due to long source–follower transients. Capacitance C must be large in order to couple the phase pulse to the internal node, further limiting performance.

Information is stored dynamically on node capacitances, and must be periodically refreshed. The effect of leakage is cumulative between amplifying stages, and the low frequency clock rate must be adjusted accordingly.

A shift register can be made dc stable by incorporating a flip-flop circuit in each stage. An example is shown in Fig. 15 [28]. The stage consists of a pair of inverters and a pair of series transmission gates. Information is shifted by alternately pulsing ϕ_1 and ϕ_2 very much as in the two-phase shift register stage previously discussed. When ϕ_1 is on, a cross-coupled flip-flop circuit is formed, which statically retains the information. A potential race condition occurs in setting the flip-flop where ϕ_1 is turned on; it is overcome by designing the response time of the forward path through T_3 to be faster than that of the reverse path through T_4. The use of dc inverters result in gain ratio requirements with attendent power dissipation and performance limitations. The stage may be made bidirectional by adding a device between node 1 of each stage and node 2 of the previous stage [29], and altering the phase pulse pattern. Alternately pulsing the forward path series gates shifts information to the right; alternately pulsing the reverse connected gates shifts it to the left.

Static shift registers have found use because they do not require continual shifting of the stored information. However, in comparison to dynamic shift registers, they have generally lower performance, require more power, and occupy larger layout area.

For currently available shift registers, shifting rates run typically 1 to 5 MHz for p-channel devices, and up to 15 MHz has been reported using n-channel devices [4]. Chips are available with from several hundred bits of storage to as much as a thousand bits, operating in the megacycle range.

IV. Complementary Circuits

In complementary MOSFET circuits both n- and p-channel devices are used [30]–[33]. Such circuits have some unique characteristics and advantages. Consider the simple complementary inverter circuit in Fig. 16(a). When the input is low, the n-channel device is off and the p-channel device is on. The output node is high. If the input is high, the n-channel device is on and the p-channel device is off. The output node is low. The circuit acts as an inverter, but has the very desirable characteristic that in either quiescent condition, one of the two devices in the circuit is off. No dc current flows (other than leakages), and standby power dissipation is essentially eliminated.

When the input goes from a low to a high level, the output node is pulled down through the n-channel device in the common source mode of operation. Similarly, the upward output node transient occurs through the p-channel device, also in common source operation. Thus the source–follower transient is eliminated. This not only gives a faster mode of operation for the transients, but also eliminates the voltage loss between gate and source which occurs with enhancement mode devices in source–follower operation.[10] A further reason for improved transient performance is that the load device is turned off and is not pulling against the active device in the inverter. This also eliminates any requirement for gain ratios between the devices, which, as before, leads to better transient response. Complementary circuits are advantageous from the standpoints of having a firmer on state and better noise immunity.

The complementary circuit analog of a series transmission gate device is shown in Fig. 16(b) [34]. The polarity of the pulses on the clock lines are chosen so that both devices are either on or off. When the gate is on, there is always one device in the pair operating in the common source mode, and although the other device operates as a source follower, the limitations of source–follower operation are avoided.

Two complementary inverters can be cross-coupled to form a bistable memory element with essentially zero standby power. Addition of two gating devices leads to the memory cell of Fig. 17 [35], which is the complementary analog of the cell shown in Fig. 4. Operation is qualitatively similar. In writing, the complementary load device can be used to help the pull-up transient; however, initiating the writing cycle is more difficult because the initial pull-down transient of the high node must be accomplished against a common source load device that is turned on hard. If the load device is made low gain, the writing becomes easier to initiate, but the speed advantage gained in the upward transient is sacrificed.

Historically, complementary memory cells have been designed for high performance, and have been relatively complex and have required large layout areas to achieve that

[10] Augmentation of the device threshold as the magnitude of the source to substrate voltage increases due to the source–follower transient is also avoided.

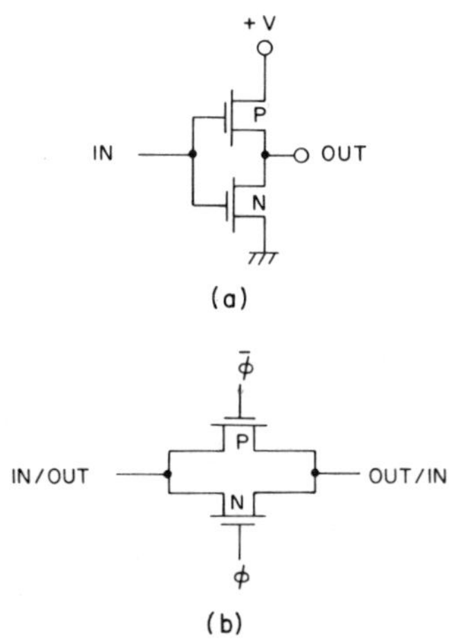

Fig. 16. (a) Complementary MOSFET inverter. (b) Complementary series transmission gate.

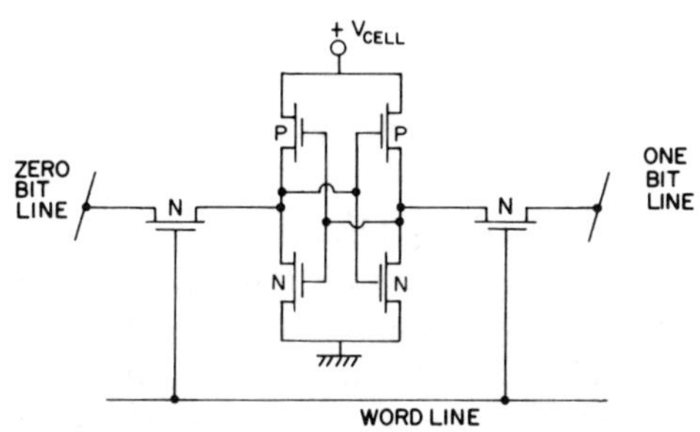

Fig. 17. Complementary six-device cell.

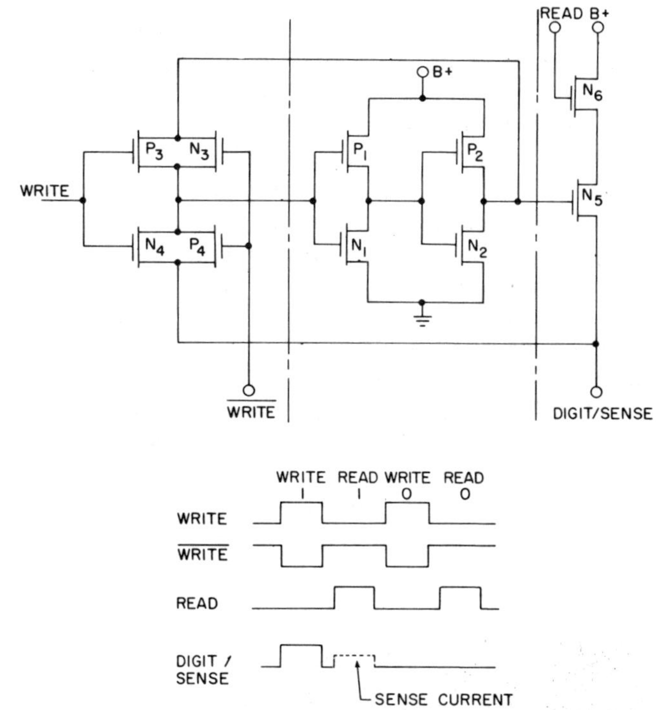

Fig. 18. High performance complementary cell.

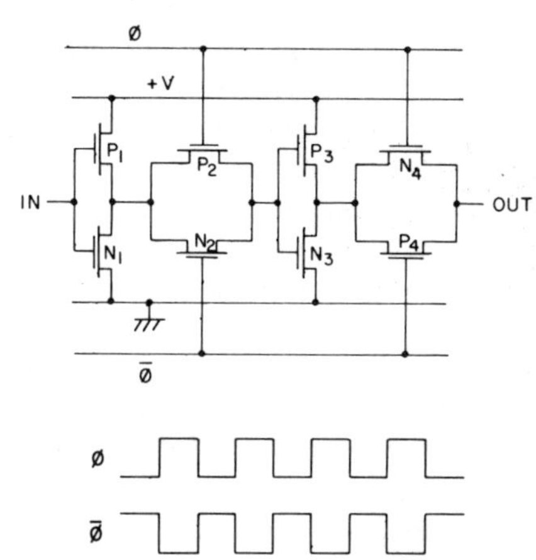

Fig. 19. Complementary shift register.

end. An example is shown in Fig. 18 [36]. In the quiescent condition, the transmission gate formed by P_3 and N_3 cross-coupled the inverters formed by P_1/N_1 and P_2/N_2, while the P_4/N_4 transmission gate is off. To read, sense current conditionally flows through the AND gate formed by N_5 and N_6, depending upon the state of the cross-coupled flip-flop circuit. To write, the polarity of the WRITE and $\overline{\text{WRITE}}$ lines are reversed. This turns off the transmission gate formed by P_3 and N_3, and breaks the feedback path in the cross-coupled circuit. At the same time, the transmission gate formed by P_4 and N_4 is turned on, forcing the information on the digit/sense line onto the gate of the P_1/N_1 inverter, from where it propagates through to the output node of the P_2/N_2 inverters in one pair delay. The WRITE and $\overline{\text{WRITE}}$ lines are then returned to their original polarity, reestablishing the cross-coupling of the cell. The fast write speed of the cell is obtained through the operation of the complementary circuits and the breaking of the feedback loop. The performance is obtained on the cost of circuit complexity, which is clearly greater than with the previous cells. A variety of complementary memory cells are shown in [31].

Quiescent standby power of both noncomplementary stored charge and complementary memory cells is determined by leakage currents. The stored charge cells require a periodic refresh cycle, during which power is dissipated. In practice, complementary systems dissipate only nanowatts per cell in standby, while a stored charge system will be two or three orders of magnitude higher. In both cases, power is low enough that volatility of stored information can be overcome by a battery back-up power supply.[11]

A high speed complementary shift register circuit is shown in Fig. 19 [37]. A stage consists of alternating two inverters and two transmission gates. Pulsing the clock lines to turn on the P_2/N_2 transmission gate allows the information at the input to propagate to the gates of the P_3/N_3 inverter in inverted form. Reversing the polarity of the clock lines locks the information in on the gates of the P_3/N_3 inverter, and allows it to propagate forward into the next stage through the P_4/N_4 transmission gate.

The performance of this shift register is better than that of noncomplementary circuits for a number of reasons. The

[11] A battery back-up is possible for dc cells where the system standby dissipation is not too high.

complementary inverter and transmission gate operation, which eliminates source–follower limitations and gain ratio requirements, is important. In addition, 1) the capacitive coupling from the clock lines to the information nodes is balanced and cancels; 2) information transfer requires only a single transient per shift, rather than an unconditional charge followed by a conditional discharge, as is done in the four-phase and similar circuits; and 3) the phase clock pulsing is much simpler.

At present, only a limited amount of complementary circuit hardware is available. Almost all of this has been aimed for the high performance end. Shifting rates of 75 to 100 MHz have been reported for the shift register circuit shown [37], and random access chips of 64 to 256 bits with cycle times in the 10–20-ns range have been reported for both undecoded and fully decoded bit organized chips [36], [39], [40]. These performance figures are significantly better than that reported for available p-channel counterparts.

The performance difference with n-channel circuits would be much less. Pulsed logic circuits are clocked to avoid dc current paths and gain ratio requirements, thus achieving low standby power operation and device geometry requirements similar to complementary logic, and the use of a p-channel common-source load device in complementary circuits is not inherently significantly faster than the n-channel source–follower transient of an n-channel load device, due to the higher mobility of electrons (approximately a factor of three [38]) and the greater capacitive loading of complementary circuits, where two gates must be driven per unit of fan-out. Also, electron mobilities reported in complementary integrated circuits tend to be lower than that obtainable in n-channel circuits. Thus it is not clear that complementary circuit transients are inherently faster than those of n-channel pulsed circuits. On the other hand, pulsed n-channel logic will suffer from the voltage attenuation of source–follower operation; the necessity of serial transients to do logic, which may be at least partially overcome by overlapping transients, and the generation, timing, and distribution of the clock pulses, which can add significantly to the complexity of a system. As devices improve, the clocking could become a major inhibitant to utilization of improvements in inherent device performance.

The eventual importance of complementary devices is still not definite. In the past, they have suffered from a more complex technology and larger circuit areas, which has resulted in higher costs and tended to restrict applications to areas where the very low standby power characteristics are of prime importance. Some recent efforts have been made towards achieving competitive cell density, and achievement of similar comparable density between the six device noncomplementary and complementary cells has been reported [39]. Complementary circuitry does not improve the density of high density storage arrays, such as those using the three-device cells, and noncomplementary circuits have an edge for low cost applications.

With effort now being directed towards a wider range of applications, predictions have been made that complementary circuits will approach noncomplementary circuits in cost, although perhaps lagging behind some time in development [41]. It remains to be seen whether design claims and laboratory hardware can be carried through to a wide range of product applications, or if they will continue to be used in special applications only.

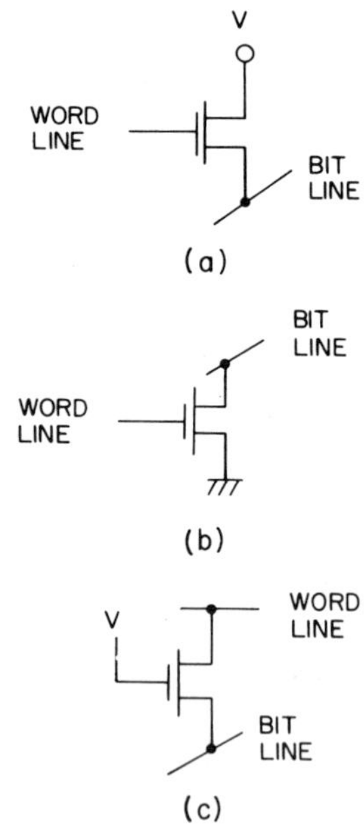

Fig. 20. Read only memory array circuits.

V. Read Only Memory and the MNOS Device

A frequent area of application for MOSFET devices is read only memories (or ROMs). In such a memory, the information is written in permanently during the fabrication process. In operation, the memory is read to retrieve the stored information, but the contents cannot be altered and there is no write cycle. There are many uses for a ROM; among them are control memories, code conversion, table look-up character generation, etc. [42].

Because of the elimination of the write requirement, only a single device is needed per bit. Circuits for the storage array are shown in Fig. 20. The information storage mechanism is the thickness of the gate oxide, which is determined during fabrication in the gate oxide mask step. If the device has a normal thin gate oxide, its threshold voltage will be low; if the gate oxide is thick, the threshold voltage is high. In response to a pulse on the word line, the low threshold device will conduct, while the high threshold device will not, and the presence of a bit line current or voltage swing can be detected for a "1" or "0". In Fig. 20(a) and (b), the word line voltage must overcome the device threshold to turn it on, but there is no dc load on the word line, and the noise immunity is good. The circuit shown in Fig. 20(c) can operate with lower word line voltage because of the absence of a threshold between word and bit lines, but accordingly has no noise immunity and is susceptible to sneak paths as a result of the bidirectional source–drain coupling through on devices.

Since only a single device is required per bit of information and very efficient array layouts are possible, the density of bits on ROM chips is much higher than with read/write chips. 2K bit chips are routinely available, and at least two sources sell 4K bit chips. Because of the high bit densities possible, on-chip binary decoding is used. This is reflected in the chip performance; cycle times run 500 ns to about 1 μs.

The density of a ROM array can be approached in a read/write array with the MNOS device [43]–[49], a MOSFET structure in which the threshold may be set by tunneling of charge in or out of the gate insulator by a high voltage pulse applied to the gate. The threshold may be interrogated subsequently by a low voltage pulse, as in a ROM. Only a single device per bit is needed. The MNOS device is still in the laboratory stage of development at present.

VI. On-Chip Support Circuits

On-chip support circuits are used to decode the input address and select the corresponding information bit or bits. On-chip sense signal amplification can be provided; this is necessary where the MOSFET chip is designed to be completely compatible with a bipolar logic family, such as TTL. There are numerous methods of obtaining the necessary functions. The circuits shown in this section are examples of what can be done. ROM arrays have been shown for simplicity; the extension to read/write cells is straightforward.

A cross section of a decoded chip using simple dc coupled logic is shown in Fig. 21 [50]. Address inputs are inverted on chip. The output of a NOR block will be low if any input is high, and will be high only if all inputs are low. The selected word line drives the gates of all devices along it. One bit line is selected by the corresponding bit NOR block turning on the series transmission gate which connects the bit line to the dot–OR bus. The dot–OR bus is conditionally discharged to ground, depending upon the threshold voltage of the array device. The remainder of the output circuitry provides off-chip drive capability. It should be noted that the basic information storage array is word organized, and that selection is done at the periphery of the array similar to the $2\frac{1}{2}$ dimension organization used in magnetic film memories. This keeps the array simple and minimizes the decode circuitry.

The dc coupled circuitry has a number of drawbacks. It is slow because of gain ratio requirements between load and active devices, and the worst case path involves several load device source follower transients. Power dissipation is high because dc current flows in all unselected NOR blocks, and can flow in the inverters, the array, and the output drive circuits as well.

An example[12] of a decoded chip using pulsed support circuitry is shown in Fig. 22. The pulsed logic circuits are similar to the four phase nonoverlapping clock shift register circuits described previously. The sequence of operation is as follows: the word inverter and NOR block output nodes are precharged on ϕ_4 and ϕ_1, respectively. The inverter nodes are conditionally discharged on ϕ_1 and the NOR block nodes are conditionally discharged on ϕ_2. The bit inverter operation is similar to the word inverter, and uses ϕ_1 and ϕ_2. Thus as the end of ϕ_2, only one word line remains charged, and only one bit NOR block has all its inputs low. Also during ϕ_2 the bit line is unconditionally charged. On ϕ_3 all bit lines are conditionally discharged. All unselected bit lines have a discharge path through their respective bit NOR blocks; the selected bit line can be discharged only if the array device at the intersection of the selected bit and word lines has a thin gate oxide. Thus, at the end of ϕ_3, only the selected bit line can remain charged, and then only if the selected array device has a high threshold. ϕ_3 also serves to precharge the bit NOR block node. On ϕ_4 the bit NOR block output node is conditionally discharged. All unselected bit line inputs are low; therefore, the NOR block output is the inverse of the state of the selected bit line. The output may be sampled during either or both of the ϕ_1 and ϕ_2 periods.

No dc current flows during the operation of the chip, greatly reducing power dissipation. Dynamic power is dissipated due to the charging and discharging of the various information node capacitances on the chip, but this tends to

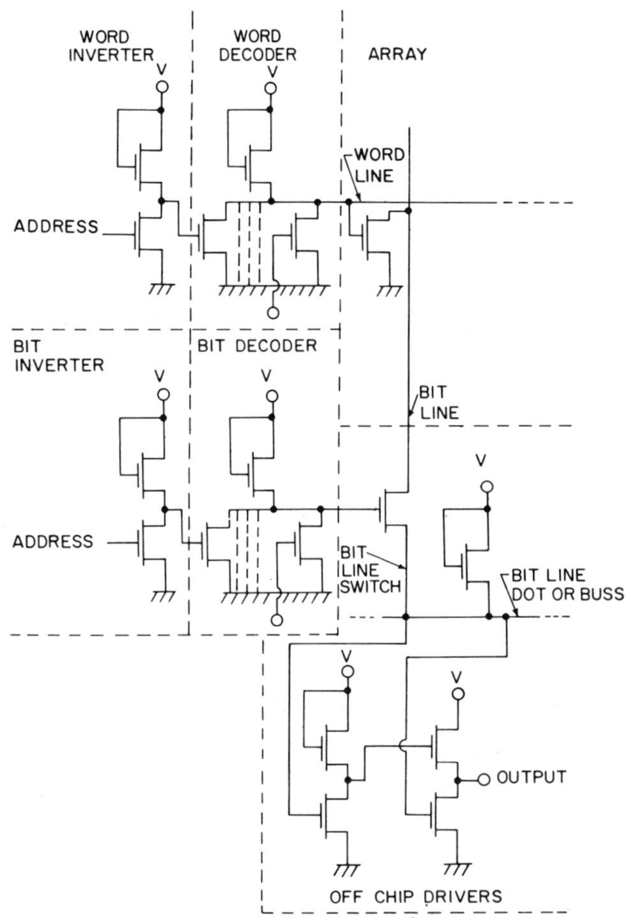

Fig. 21. Cross section of a fully decoded chip using dc logic.

[12] The circuits described here are similar to, but not identical with, the approach used in [51].

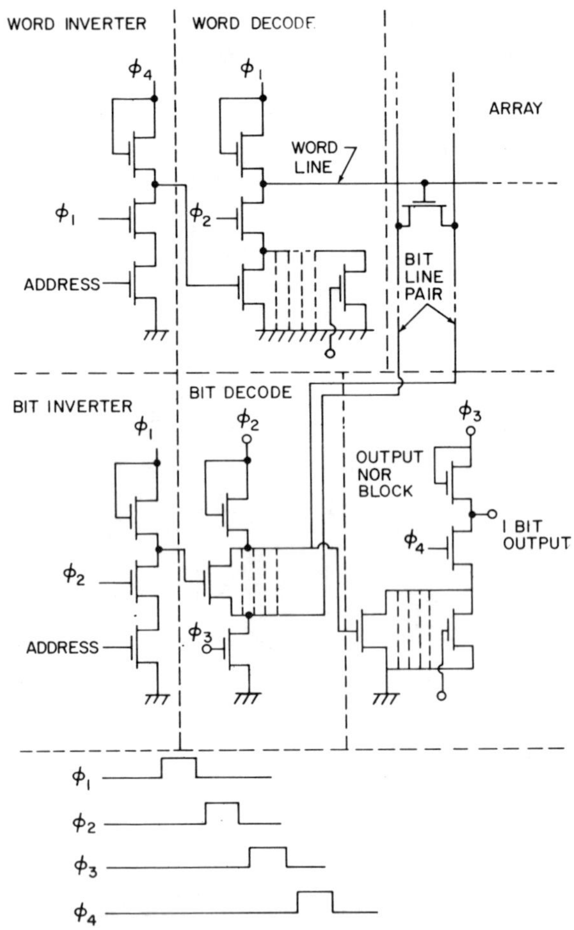

Fig. 22. Cross section of a fully decoded chip using pulsed logic.

be small in absolute value, using 2.5 mW at 1 MHz/100 pF for a 5 V swing. The pulsed logic eliminates gain ratio requirements with the attendant advantages. The result is a system which is faster and lower power than the dc coupled system, but which requires generation and timing of the clock pulses. On-chip generation of some clock pulses from others can reduce timing problems between pulses.

The circuits given above illustrate approaches to on-chip support circuits. There are, in fact, innumerable variations which are possible, each with particular advantages and disadvantages [22], [23], [51], [52].

A source–follower transient can be markedly sped up by use of the so-called "super one" technique [53], [22]. The circuit is shown in Fig. 23(a). Initially as the result of conventional pulsed logic operation, a voltage level V_g is stored on the gate of T_1. As long as ϕ_1 is low, the output remains low, and the feedback capacitance C_F has a voltage V_g across it. When ϕ_1 is pulsed, the load capacitance is charged through T_1. As the output rises, the voltage on the gate of T_1 is higher than the output voltage by the voltage across the feedback capacitor, and to the extent that C_F is much greater than the stray capacitance of the gate node, that voltage will remain close to the initial value of V_g V. Thus it is possible at the end of the transient to achieve a voltage on the gate of T_1 higher than any supply voltage

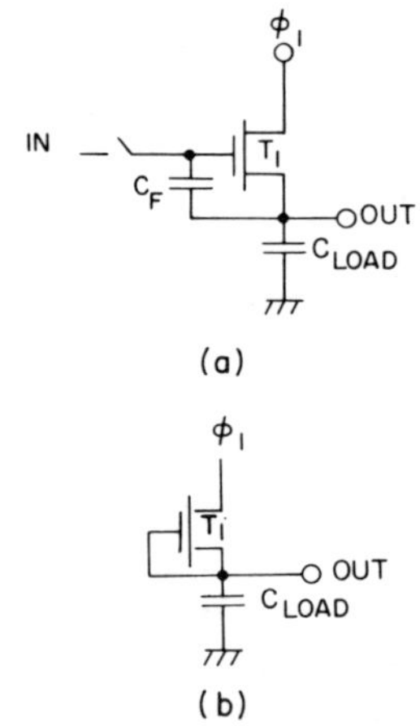

Fig. 23. (a) "Super one" circuit. (b) Depletion mode load device connection with similar operation.

present. In this manner the self-turning off action of the source–follower mode can be overcome, and in the limiting case then the voltage across the feedback capacitance does not change, the pull-up transient becomes identical with the common source transient.

Similar operation can be obtained using a depletion mode device connected as shown in Fig. 23(b) [54]. The built-in potential of the depletion mode threshold acts similarly to the feedback capacitor voltage of the "super one" circuit, except that the gate–source voltage is not reduced by charging the stray capacitance, and the gate node is not loaded by the feedback capacitance C_F.

VII. Epilogue

At present, the MOSFET device has applications in a wide variety of memory applications, due to unique characteristics of the devices and the technology. As overall semiconductors technology improves, MOSFET technology will share in the advances; in fact, MOSFET technology is newer and will probably undergo more rapid development. In addition, there are technology advances specifically applicable to MOSFET devices. These include n-channel technology, insulating substrates, thinner gate oxides, self-aligned and zero overlap gate structures, ion implantation, lower thresholds, charge free oxides, etc.; the result can be significant additional improvement in MOSFET devices, and it is not unreasonable that the performance gap with bipolars will narrow, and perhaps eventually vanish. Further, memory phenomena (as opposed to memory circuits) in MOSFET structures such as the MNOS and charge coupled devices [55], [56] have the potential of further reducing the area and complexity required to store an information bit.

References

[1] D. A. Hodges, "Large-capacity semiconductor memory," *Proc. IEEE*, vol. 56, July 1968, pp. 1148–1162.

[2] W. B. Sander, "Semiconductor memory circuits and technology," *AFIPS Fall Joint Computer Conf., Proc.*, Dec. 1968, pp. 1205–1211.

[3] "Semiconductor memory," *1970 Digest INTERMAG Conf.* (session 11), Apr. 1970, pp. 11.1–11.6, also *IEEE Trans. Magn.*, vol. MAG-6, Sept. 1970, pp. 584–589.

[4] J. W. Bremer, "A survey of mainframe computer memories," *Computer Design*, May 1970, pp. 63–71.
[5] R. Graham and M. Hoff, "Why semiconductor memories?" *Electron. Prod.*, Jan. 1970, pp. 28–34.
[6] R. H. Crawford, *MOSFET in Circuit Design*. N. Y.: McGraw-Hill, 1967, ch. 4, pp. 71–86.
[7] J. Karp and E. De Atley, "Use four-phase MOS IC logic," *Electron. Des.*, Apr. 1, 1967, pp. 62–66.
[8] B. G. Watkins, "A low-power multiphase circuit technique," *IEEE J. Solid-State Circuits*, vol. SC-2, Dec. 1967, pp. 213–220.
[9] L. Cohen, R. Rubenstin, and F. Wanlass, "MTOS four phase clock systems," *NEREM Rec.*, 1967, pp. 170–171.
[10] L. L. Boysel and J. P. Murphy, "Four phase LSI logic offers new approach to computer design," *Comput. Des.*, Apr. 1970, pp. 141–146.
[11] K. Goser, "Channel formation in an IGFET and its equivalent circuit for CAD," *Internat. Solid State Circuits Conf., Digest of Tech. Papers*, Feb. 1970, pp. 98–99.
[12] P. Pleshko and L. M. Terman, "An investigation of the potential of MOS transistor memories," *IEEE Trans. Electron. Comput.*, vol. EC-15, Aug. 1966, pp. 423–427.
[13] R. Igarashi, T. Kurosawa, and T. Yaito, "A 150 ns associative memory using integrated MOS transistors," *Internat. Solid State Circuits Conf., Digest of Tech. Papers*, Feb. 1966, pp. 104–105.
[14] F. S. Greene and W. B. Sander, "Address selection by combinatorial decoding of semiconductor memory arrays," *IEEE J. Solid-State Circuits* (Corresp.), vol. SC-4, Oct. 1969, pp. 295–296.
[15] T. W. Hartz, D. W. Hillis, J. Marley, R. C. Lutz, and C. R. Hoffman, "A main frame semiconductor memory for fourth generation computers," *AFIPS Fall Joint Computer Conf., Proc.*, Nov. 1969, pp. 479–488.
[16] Y. Tarui, Y. Hayashi, T. Koyanagi, H. Yamamoto, M. Shiraishi, and T. Kurosawa, "A 40 ns 144 bit n-channel MOS-LSI memory," *IEEE J. Solid-State Circuits*, vol. SC-4, Oct. 1969, pp. 271–279.
[17] W. M. Regitz and J. Karp, "A three transistor cell, 1024 bit, 500 ns MOS RAM," *Internat. Solid State Circuits Conf., Digest of Tech. Papers*, Feb. 1970, pp. 42–43.
[18] L. L. Vadasz, A. S. Grove, T. A. Rowe, and G. E. Moore, "Silicon-gate technology," *IEEE Spectrum*, vol. 6, Oct. 1969, pp. 28–35.
[19] J. D. Schmidt, "Integrated MOS random-access memory," *Solid-State Design*, Jan. 1965, pp. 21–25.
[20] D. E. Brewer, S. Nissim, and E. V. Podraza, "Low power computer memory system," *AFIPS Fall Joint Computer Conf., Proc.*, Nov. 1967, pp. 381–393.
[21] J. H. Friedrich, "A coincident-select MOS storage array," *Internat. Solid State Circuits Conf., Digest of Tech. Papers*, Feb. 1968, pp. 104–105.
[22] D. Lund, C. Allen, S. Anderson, and G. Tu, "Design of a megabit semiconductor memory system," *AFIPS Fall Joint Computer Conf., Proc.*, Dec. 1970, pp. 53–62.
[23] L. Boysel, W. Chan, and J. Faith, "Random access MOS memory packs more bits to the chip," *Electronics*, Feb. 16, 1970, pp. 109–115.
[24] R. H. Dennard, "Field-effect transistor memory," U.S. Patent 3 387 286, June 4, 1968.
[25] H. Z. Bogert, "Metal Oxide Silicon Integrated Circuits," *SCP Solid-State Technol.*, Mar. 1966, pp. 30–35.
[26] R. H. Crawford and B. Bazin, "Theory and design of MOS capacitor pull-up circuits," *IEEE J. Solid-State Circuits*, vol. SC-4, June 1969, pp. 145–158.
[27] F. L. J. Sangster, "Integrated MOS and bipolar analog delay lines using bucket-brigade capacitor store," *Internat. Solid State Circuits Conf., Digest of Tech. Papers*, Feb. 1970, pp. 74–75.
[28] General Instrument Corp. Tech. Spec., MEM 501, May 1965.
[29] M. N. Shen, "Two-way four phase MOSFET dynamic shift register," *IBM Tech. Disclosure Bull.*, vol. 12, no. 12, May 1970, p. 2149.
[30] F. M. Wanlass and C. T. Sah, "Nanowatt logic using field effect metal-oxide semiconductor triodes," *Internat. Solid State Circuits Conf., Digest of Tech. Papers*, Feb. 1963, pp. 32–33.
[31] J. R. Burns, "Switching response of complementary-symmetry MOS transistor logic circuits," *RCA Rev.*, Dec. 1964, pp. 627–661.
[32] E. J. Boleky, "The performance of complementary MOS transistors on insulating substrates," *RCA Rev.*, vol. 31, no. 2, June 1970, pp. 372–395.
[33] T. Klein, "Technology and performance of integrated complementary MOS circuits," *IEEE J. Solid-State Circuits*, vol. SC-4, June 1969, pp. 122–130.
[34] J. R. Burns and J. J. Gibson, "Complementary field effect transistor transmission gate," U.S. Patent 3 457 435, July 22, 1969.
[35] G. B. Herzog *et al.*, "Large scale integrated circuit arrays," RCA Interim Tech. Rep. 5, May 1967, pp. 46–47.
[36] J. F. Allison, F. P. Heiman, and J. R. Burns, "Silicon-on-sapphire complementary MOS memory cells," *IEEE J. Solid-State Circuits*, vol. SC-2, Dec. 1967, pp. 208–212.
[37] J. E. Meyer, J. R. Burns, and J. M. Scott, "High speed silicon-on-sapphire 50 stage shift register," *Internat. Solid State Circuits Conf., Digest of Tech. Papers*, Feb. 1970, pp. 200–201.
[38] G. Cheroff, D. L. Critchlow, R. H. Dennard, and L. M. Terman, "IGFET circuit performance, n-channel versus p-channel," *IEEE J. Solid-State Circuits*, vol. SC-4, Oct. 1969, pp. 267–271.
[39] J. R. Burns, NASA contract work, to be published.
[40] J. R. Burns and J. H. Scott, "Silicon-on-sapphire complementary MOS circuits for high speed associative memory," *AFIPS Fall Joint Computer Conf., Proc.*, Nov. 1969, pp. 469–477.
[41] J. A. Rajchman, "Computer memories and the impact of semiconductor technology," presented at Internat. Electron Devices Meeting, Oct. 1969.
[42] J. Wunner and R. Colino, "Applying the versatile MOS ROM," *Electron. Prod.*, Jan. 1970, pp. 35–40.
[43] H. A. R. Wegener, A. J. Lincoln, H. C. Pao, M. R. O'Connell, and R. E. Oleksiak, "The variable threshold transistor, a new electronically alterable, non-destructive read only storage device," presented at Internat. Electron Devices Meeting, Oct. 1967.
[44] J. T. Wallmark and J. H. Scott, "Switching and storage characteristics of MIS memory transistors," *RCA Rev.*, vol. 30, no. 2, June 1969, pp. 335–365.
[45] E. C. Ross and J. T. Wallmark, "Theory and switching behavior of MIS memory transistors," *RCA Rev.*, vol. 30, no. 2, June 1969, pp. 366–381.
[46] H. A. R. Wegener and F. A. Sewell, "Metal-insulator-semiconductor transistor for use as a nonvolatile digital storage element," Sperry Rand Tech. Rep. AFAL-TR-69-187, July 1969.
[47] D. Frohman-Bentchkowsky, "An integrated metal-nitride-oxide silicon (MNOS) memory," *Proc. IEEE* (Lett.), vol. 57, June 1969, pp. 1190–1192.
[48] ——, "The metal-nitride-oxide-silicon (MNOS) transistor—characteristics and applications," *Proc. IEEE*, vol. 58, Aug. 1970, pp. 1207–1219.
[49] F. W. Flad, C. J. Varker, and H. C. Lin, "The application of MNOS transistors in a preset counter with nonvolatile memory," *Internat. Solid State Circuits Conf., Digest of Tech. Papers*, Feb. 1969, pp. 46–47.
[50] L. Boysel, "Memory on a chip: a step toward large-scale integration," *Electronics*, Feb. 6, 1967, pp. 93–97.
[51] A. Varadi and R. Rubinstein, "2048 bit MTOS read only memory," *1968 Computer Group Conf. Digest*, June 1968.
[52] L. Boysel and J. Murphy, "Multiphase clocking achieves 100 ns MOS memory," *EDN*, June 10, 1968, pp. 50–55.
[53] A. K. Rapp, presented at the Internat. Solid State Circuits Conf., Session THE 7, Feb. 1969.
[54] J. MacDougall and K. Manchester, "Ion implantation offers a bagful of benefits for MOS," *Electronics*, June 22, 1970, pp. 86–90.
[55] W. Boyle and G. Smith, "Charge coupled semiconductor devices," *Bell Syst. Tech. J.*, vol. 49, no. 4, Apr. 1970, pp. 587–593.
[56] G. Amelio, M. Tompsett, and G. Smith, "Experimental verification of the charge coupled device concept," *Bell Syst. Tech. J.*, vol. 49, no. 4, Apr. 1970, pp. 593–600.

Three-Transistor-Cell 1024-Bit 500-ns MOS RAM

WILLIAM M. REGITZ, MEMBER, IEEE, AND JOEL A. KARP, MEMBER, IEEE

Abstract—**A semiconductor memory array has been designed, offering strong economical competition to magnetic technology for main memory usage. The array presented in this paper utilizes a new high-speed dynamic memory cell consisting of three minimum-geometry MOS transistors that require only four interconnect lines. The cell was integrated into a fully decoded 1024-bit array organized as 512 words by 2 bits. A performance of 500 ns for a READ or WRITE cycle and a 345-ns access time was demonstrated. The average power dissipation of 200 μW per bit under operating conditions, 30 μW per bit for standby power, and 5 μW per bit of battery power was also demonstrated.**

Introduction

OVER the past years there have been significant amounts of material published on the subject of semiconductor memories [1]–[4]. The majority of this work was done in the areas of buffers, scratch pads, ROMs, etc., with some efforts extending further into mainframe memories [5]. This paper encompasses the design of a semiconductor memory chip for mainframe applications, although it has other uses. A memory system was designed with this array as its main memory element to replace a core memory system on a one-to-one basis.

The primary objective within the development was to achieve a cost–performance product lower than available magnetic main memories. Certain cost–performance indexes were defined in order to meet this criterion. Conclusively, such a memory would necessitate the following features.

1) MOS circuitry [6] would give minimum cell size for any technology.

2) Dynamic circuitry [7], [8] that uses the MOS-device inherent storage capability yields a smaller number of devices per cell. Also minimum geometry devices could be used and fewer interconnect buses to each cell would be required. These factors all result in a minimization of total cell area.

3) Silicon-gate technology [9], [10], utilizing the self-aligning gate and the effective three levels of interconnect, yields a cell layout configuration with a smaller silicon area and lower capacitance on the given interconnect lines than other MOS technologies and, therefore, higher performance.

Some lower priority objectives were the following.

1) Low power—Since the array was to be designed as the basic storage element for mainframe memory and would require a battery backup power system, low power was an objective in the design.

2) Performance—All circuits were designed optimally for high performance within the guidelines of low power and minimum silicon area. However, a relatively high performance memory array resulted from these considerations.

Cell

Dynamic cells only require three transistors (see Fig. 1). Information is stored in the form of a charge on a capacitor C. This capacitance is the combination of the MOS

Manuscript received May 4, 1970; revised May 25, 1970.
W. M. Regitz is with Computer Control Division, Honeywell, Framingham, Mass. 01701.
J. Karp is with Intel Corporation, Mountain View, Calif. 94040.

Reprinted from *IEEE J. Solid-State Circuits,* vol. SC-5, pp. 181–186, Oct. 1970.

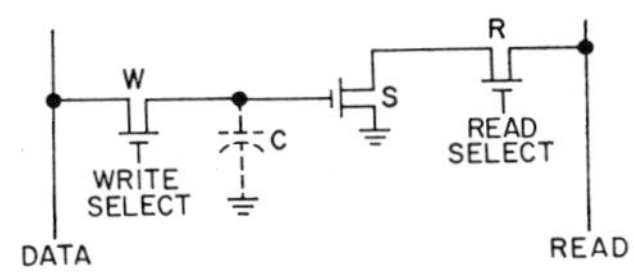

Fig. 1. Dynamic cell.

gate capacitance of transistor S and the capacitance of the drain diffusion region of transistor W. However, because this drain region is a reversed biased p-n junction, it has a finite reverse current that will cause the charge on the storage capacitance to leak off. The information must be refreshed at a frequency determined by the leakage current, total capacitance, and the voltage droop allowed on the storage node.

Refreshing memories of acoustical delay lines and cathode-ray tubes has been common for some time. Even ferrite cores once demanded a "post write disturb pulse" and still require data regeneration due to the destructive nature of the readout. Semiconductor memory cells of the three-transistor type, because of the cell simplicity, are the lowest in cost in main memory technology. Dynamic random-access memories are even less expensive than dynamic serial memories because of their circuit simplicity and greater production volume.

The three-transistor dynamic memory cell shown in Fig. 1 requires five interconnect lines, two select lines, one for reading (READ SELECT) and the other for writing (WRITE SELECT), a READ output line, a DATA input line, and one ground line. The READ and WRITE SELECT lines in the memory cell shown in Fig. 2 were combined into a common SELECT line, reducing the number of interconnect lines by one; all other lines remained the same. If two SELECT lines were used (one for reading and one for writing) the cell area (now less than 6 mils^2) would increase by 50 percent.

Combining the READ and WRITE SELECT lines necessitates the use of a trilevel signal to drive the SELECT line. The first level is off, the second, or intermediate level, is for reading, and the third level is for writing into the cell. The S transistor samples the state of the storage node, whereas the R transistor samples the state of the S transistor when the SELECT line is at the intermediate level. The W transistor is employed to refresh the stored information or to write into the cell. To explain the operation of the cell, let a charge on the storage capacitance C represent a ONE and let no charge represent a ZERO. Between memory cycles, both the READ and DATA lines are charged to $-V$. A cycle is initiated by taking the SELECT line to its intermediate level. If a ONE is stored, both the R and S transistors will be conductive and discharge the READ line. However, if a ZERO is stored the S transistor will be held off, leaving the READ line potential at $-V$. When the READ line potential is the complement of the stored information, the column amplifier is activated, transferring the complement of the information on the READ line to the DATA line, and at this same time the SELECT line goes to its high level to refresh the stored information or to write new information.

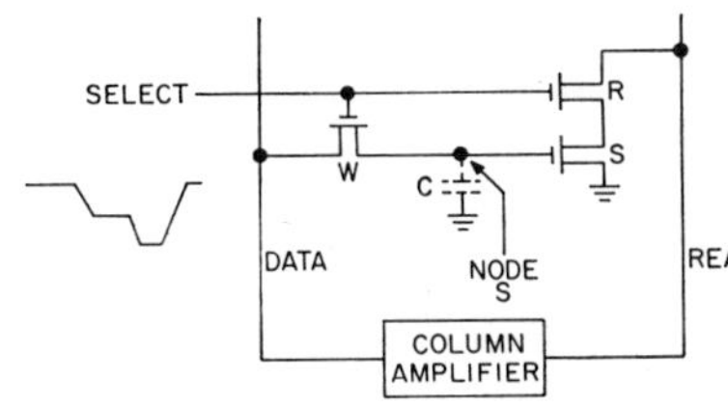

Fig. 2. Three-transistor cell.

With the aid of a MOS circuit analysis program [11], the equivalent circuit of the cell, obtained from the layout (Fig. 3), was characterized. Fig. 4 shows the results of the circuit analysis at typical threshold value (1.8 volts). These results were obtained by assuming a particular voltage on the READ line and by varying the intermediate level on the SELECT line. It is evident from Fig. 4 that if the intermediate level or the SELECT level is below 5.0 volts when a ZERO is stored in the cell, there is no discharge of the READ line. However, as the voltage increases, the READ line does begin to discharge. This discharge is caused by the storage node taking on a charge by way of the W transistor. At the same time, if these intermediate levels are applied when a ONE is stored (a charge on the storage node), the READ line discharges at a much faster rate. From this curve it can be seen that if the intermediate SELECT line voltage is held between 5 and 5.5 volts, a wide range of time is available between the discharge of the READ line associated with a stored ONE and that of a stored ZERO for proper cell operation. This analysis was continued over the full range of process parameters, temperature, and supply voltage variations with the results shown in Fig. 5.

Intermediate Voltage Generator

To have a complete working memory array, an analog circuit must be constructed to drive the SELECT line so that the output signal is a function of the process parameters, temperature, and supply voltage. Such a circuit is shown in Fig. 6. The circuit utilizes a negative feedback signal by way of $Q6$ and $Q7$ to determine the intermediate output level. Prior to the start of a cycle, XENBL is at ground, holding the intermediate voltage generator (IVG) off by way of the three-transistor inverter ($Q1$, $Q2$, and $Q3$) and by way of $Q5$. PRECH is at V_{DD} potential holding the node between $Q6$ and $Q7$ at ground potential and YENBL is at ground holding off $Q4$. The potential at the gate of $Q5$ is determined by the ratio of $Q9$ and $Q10$ and the potential at the gate of $Q9$. This voltage is a function of threshold voltage and is close to ground as threshold increases or is close to V_{DD} as threshold decreases. At the start of a cycle, PRECH goes to ground and XENBL goes to V_{DD} and $Q5$ turns on, moving IVG to a lower potential. As this voltage decreases, $Q6$ turns on, which turns on $Q7$ and turns off $Q5$, with the appropriate time delay and overdrive on $Q6$ and $Q7$ to obtain the desired output potential. This circuit was analyzed with the aid of the computer over

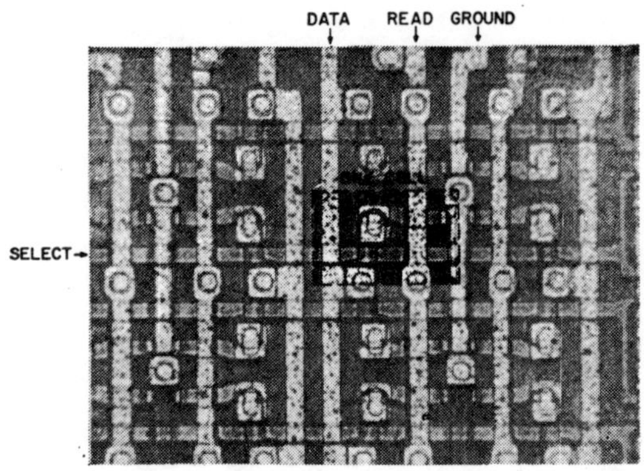

Fig. 3. Photomicrograph of cell.

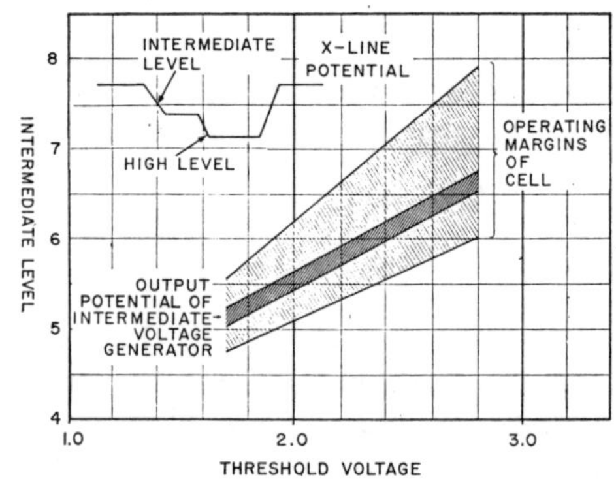

Fig. 5. Threshold voltage versus intermediate level.

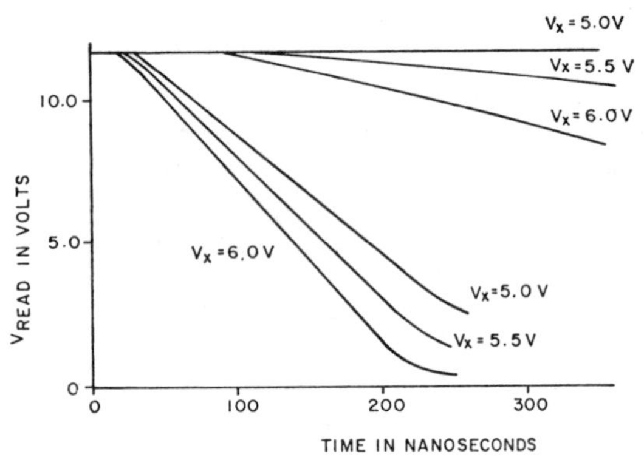

Fig. 4. READ voltage versus time.

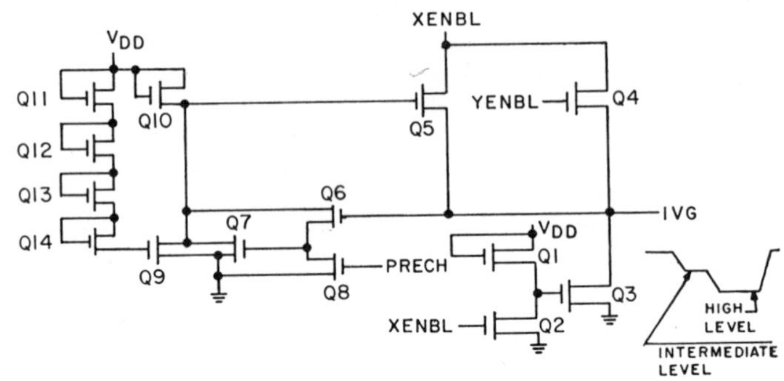

Fig. 6. Intermediate voltage generator.

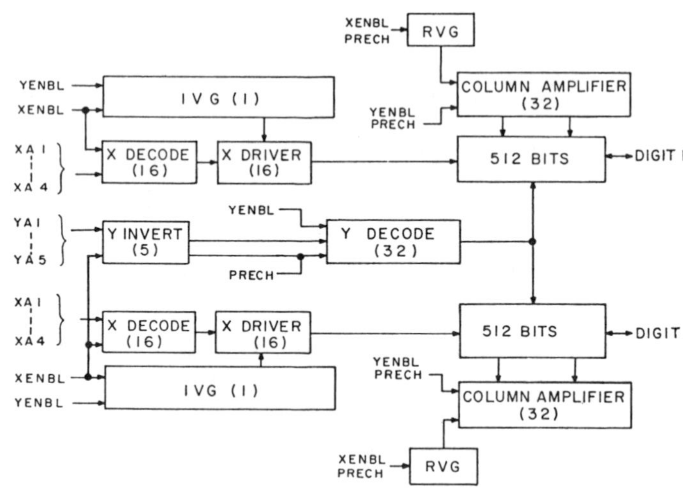

Fig. 7. Chip block diagram.

the range of temperature, process parameters, and voltage margins with the results shown in Fig. 5. As Fig. 5 illustrates, the operating margins of the cell far exceed the output variation of the IVG circuit.

Chip Block Diagram

The chip has a total storage capacity of 1024 bits (512 words by 2 bits, see Fig. 7). Each group of 512 words is selected by an x and y matrix of 16×32, respectively. (X is a row of cells and Y is a column of cells.) One of the 16 rows of cells or X lines is selected by a static decode circuit that does not require double-rail input signals for decoding and also does not require any standby power. The X-driver circuit is a simple noninverting buffer circuit.

One of the 32 Y lines or columns is selected by conventional methods. Double-rail signals are developed on the chip by the Y inverters (a dynamic inverter) and decoding is accomplished by a dynamic five-input NOR circuit. The output of the Y-decode circuit drives both groups of 512 bits in parallel.

The column amplifier circuit is an inverting buffer circuit shared by a column of cells. The circuit amplifies the low-level currents of the cell to a typical value of 400 μA and selects the desired column to the DIGIT line.

The READ voltage generator (RVG) is used to generate the voltage for the READ bus. This voltage tracks with the process parameters, supply voltage, and temperature in order to maximize the performance of the array.

Fig. 8 is a photomicrograph of the chip, which has the dimensions of 0.138 by 0.117 inch. The Y-decode circuits are located in the center of the chip. On each side of the Y-decode circuit is a group of 512 bits. Along both the top and bottom outside edges of the chip are the column amplifier circuits. The X-decode and drive circuits are at the left side of each group of 512 bits. The two IVG circuits are located at the upper left and lower right corners of the chip. The relationship of the input–output timing signals and requirements are shown in Fig. 9. The nine address inputs (four X and five Y address inputs) must be stable before a cycle begins and must remain stable throughout the cycle. The XENBL timing input, one of three timing input signals, is a 400-ns negative pulse.

Fig. 8. Photomicrograph of 1024-bit chip.

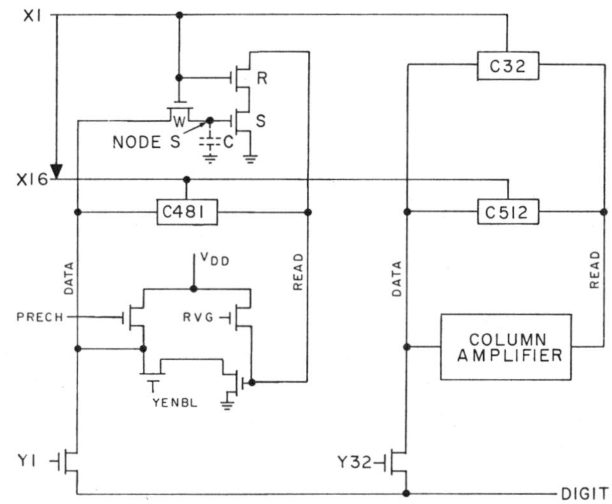

Fig. 10. Cell and column amplifier schematic.

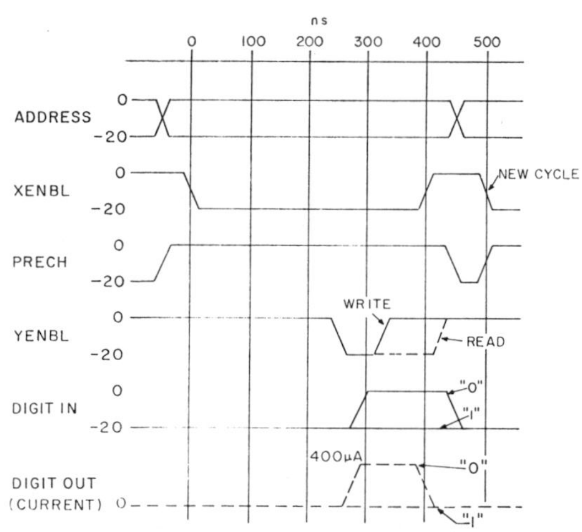

Fig. 9. Timing diagram.

Its function is to enable the X-decode circuits, invert the Y-address inputs, and enable the IVG circuit to go to its second or intermediate level. It should be pointed out that as long as XENBL is at ground potential, information in all the cells cannot be accessed, removed, or altered. PRECH timing input is active between cycles and preconditions all the dynamic nodes in the entire array. The YENBL timing input is at ground until 250 ns after XENBL. YENBL decodes the Y address, enables the column amplifier circuit, and moves the X line to the third or high level. During a WRITE cycle, YENBL is terminated early and disables the column amplifier, which enables the DIGIT line to write into the selected cell.

DIGIT is pulsed to ground during a WRITE cycle to write a ZERO into the selected cell and is held at -20 volts to write a ONE into the selected cell. The DIGIT-OUT signal is an output current occurring on the same I/O pin as DIGIT with an amplitude of 400 μA for a stored ZERO or less than 10 μA for a stored ONE.

The column amplifier circuit (see Fig. 10) is composed of four transistors. Each transistor, whose gates are connected to the PRECH input and RVG output, is used to precharge to a negative potential both the DATA and READ lines of each column of cells during the PRECH interval. The transistor whose gates are connected to the output of the Y-decode circuit ($Y1$ and $Y32$) is used to select one of the 32 DATA lines and connect to the DIGIT line.

The transistor whose gate is connected to the YENBL input is used to enable the column amplifier circuit at YENBL time and forces the READ and DATA lines to be complementary. During a WRITE cycle YENBL transfers control of the DATA line from the READ line to the external DIGIT line, allowing the DATA line to be at the same potential as the DIGIT line during a WRITE cycle.

At the start of either a READ or WRITE cycle the nine address inputs are stable, PRECH is turned off, and XENBL is turned on. While the selected X line is at the intermediate level, the R transistor samples the state of the storage node S by way of the S transistor. If a ONE is stored in the selected cell both the R and S transistors will be on and the capacitance associated with the READ line will discharge to ground. However, if a ZERO is stored in the selected cell, the S transistor will be off, leaving a charge on the capacitance associated with the READ line. At YENBL time the five Y-address bits are decoded, selecting one of the 32 DATA lines; the column amplifier circuit is enabled, creating the complement of the READ line on the DATA line and enabling the IVG circuit to go to its third or high level. During this interval, if a ONE was stored the READ-line potential would be near ground holding off the transistor whose gate is connected to the READ line. This condition is necessary to maintain the voltage on the DATA line. With one of the 32 Y-decode transistors selected, YENBL at -20 volts, and the DIGIT line terminated at -20 volts, a current path would be possible from ground through the READ transistor in the column amplifier circuit if a ZERO was stored in the selected cell. This ZERO charge at the storage node would hold off the S transistor and would not allow the READ-line potential to be discharged to ground. If a ONE was stored in the selected cell the READ-line capacitance would discharge, holding off the READ transistor and the output circuit.

With the output of the IVG circuit at its third or high level and the X line at V_{DD} potential, the W transistor is turned on and the potential that exists on the DATA line is transferred into the selected cell.

To write into a cell, the column amplifier is turned off, which transfers control of the DATA line from the READ line to the DIGIT line. Hence a ONE can be written into the cell if the DIGIT line is at V_{DD} potential or a ZERO can be written if the DIGIT line is at ground.

Refresh

As has been pointed out, information stored in the form of a charge on the MOS gate capacitance and the capacitance of a reversed-biased p-n junction must be refreshed. The 32 cell locations along an X line (Fig. 10) are refreshed when either a READ or WRITE operation is performed. Since one X line in each group of 512 bits (Fig. 9) is refreshed during a READ cycle, only 16 READ cycles, one at each X address in any sequence, are required to refresh a chip completely. In system application, these 16 READ cycles can be performed on all chips in parallel. Therefore, refreshing a complete memory system is independent of size and only requires 16 REFRESH (READ) cycles. Only 8 μs are required in a 500-ns memory system for REFRESH cycles. If these cycles were repeated at 2-ms intervals, the average time in the computer needed to perform an operation would be increased by 0.4 percent.

In order to obtain REFRESH rates of 2 ms or better over the temperature range of 0–70°C, the total leakage current at the storage node of the cell must be less than 50 pA. REFRESH measurements taken over the temperature range on the 1024-bit memory chip indicate that the leakage current is actually less than 10 pA.

Y Inverter and *Y* Decoder

The five Y-address inverters, which are three-transistor dynamic inverters, are shown in Fig. 11. The Y decoder uses a five-input NOR gate decoder, which is decoded by the YENBL input. When all of the particular address input signals are at ground, the output line remains near V_{DD} potential at YENBL time. If one or more address input signals are at V_{DD} potential, the output will switch to ground potential at YENBL time.

X Decoder

The X-decode circuit is of the ratio type but gated on at XENBL time and is shown in Fig. 12. Double-rail input signals to the decoder are not required. The double-rail signal is effectively obtained by placing the address input signal on a top transistor if a high state (V_{DD}) is required or on a bottom transistor if a ground state is required by this address signal for selection of that particular decoder.

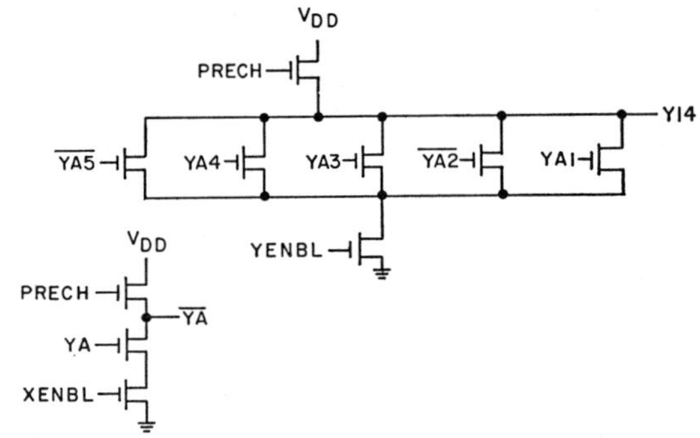

Fig. 11. *Y* decoder.

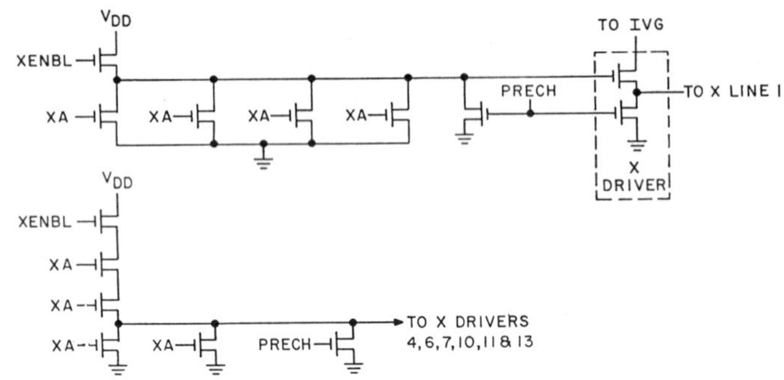

Fig. 12. *X* decoder.

These are five basic circuit configurations; two of these are shown in Fig. 12.

READ Voltage Generator

The READ voltage generator (RVG), shown in Fig. 13, is used to set the voltage potential on the READ line. The output potential of the RVG tracks the range of process parameters, temperature, and supply voltage. This tracking optimizes the time required to discharge the READ line and hence produces a faster access time.

Performance Characteristics

The important design characteristics are summarized in Table I.

Conclusion

The main objective was to design a basic memory array for high-volume mass-production semiconductor memory systems that have a lower cost than an equivalent magnetic memory system. The array presented here has a chip size of 0.117 by 0.138 inches designed with loose manufacturing tolerances of 10-micron metal lines and 10-micron separation between the metal lines. The design of the array was based on worst case conditions over the temperature range 0–70°C, process parameters, and input voltage margins (±5 percent). In addition, the array is in a dual in-line package. These factors should yield a cost lower than or equivalent to magnetic memory systems.

A 16-bit memory system has been constructed and successfully tested over the desired operating conditions

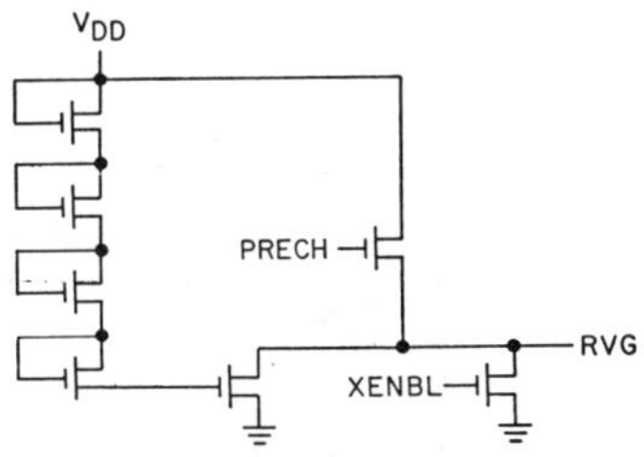

Fig. 13. READ voltage generator.

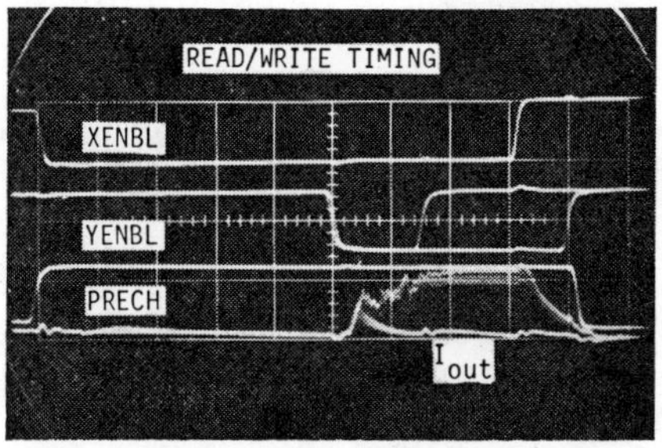

Fig. 14. Array timing.

TABLE I

Parameter	Array Performance	System Performance (8 × 16 kbit)
Access time	350 ns (max)	450 ns (max)
Cycle time	500 ns (min)	600 ns (min)
Operating power	200 μW/bit	300 μW/bit
Standby power	30 μW/bit	220 μW/bit
Battery power	5 μW/bit	40 μW/bit
REFRESH time	2 ms	2 ms
PRECH capacitance*	30 pF	
XENBL capacitance*	10 pF	
YENBL capacitance*	25 pF	
Address capacitance*	5 pF	
DIGIT capacitance*	3 pF	
Output current	333 μA	

* Input signal amplitude of 20 volts.

and, at present, the 1024-bit array (Fig. 8) is being assembled into an 8 × 16 kbit memory system. The array has shown to be completely operational as the timing signals and output current of Fig. 14 indicate (200 ns/bias).

References

[1] J. Schmidt, "Integrated MOS transistor random access memory," *Solid-State Des.*, vol. 6, p. 21, January 1965.
[2] D. E. Brewer, S. Nissim, and G. V. Podraza, "Low power computer memory system," *Fall Joint Computer Conf., AFIPS Proc.*, vol. 31. Washington, D. C.: Thompson, 1967, pp. 381–393.
[3] T. W. Hart, Jr., "Word-organized 64-bit MOS memory," *NEREM Rec.*, vol. 10, pp. 148–149, 1968.
[4] L. Boysel and J. Murphy, "Multiphase clocking achieves 100-ns MOS memory," *EDN*, vol. 13, pp. 50–55, June 1968.
[5] L. Boysel, W. Chan, and J. Faith, "Random-access MOS memory packs more bits to the chip," *Electronics*, vol. 43, pp. 109–115, February 16, 1970.
[6] L. Boysel, "Cutting system cost with MOS," *Electronics*, vol. 38, p. 105, January 20, 1965.
[7] J. Karp and E. deAtley, "Use four-phase MOS IC logic," *Electron. Des.*, vol. 15, pp. 62–66, April 1, 1967.
[8] L. Cohen, R. Rubenstein, and F. Wanlass, "MTOS four-phase clock systems," *NEREM Rec.*, vol. 9, pp. 170–171, 1967.
[9] L. L. Vadasz, A. S. Grove, T. A. Rowe, and G. E. Moore, "Silicon-gate technology," *IEEE Spectrum*, vol. 6, pp. 28–35, October 1969.
[10] F. Faggin and T. Klein, "A faster generation of MOS devices with low threshold is riding the crest of the new wave, silicon-gate ICs," *Electronics*, vol. 42, pp. 88–94, September 29, 1969.
[11] D. Frohman-Bentchkowsky and L. Vadasz, "Computer-aided design and characterization of digital MOS integrated circuits," *IEEE J. Solid-State Circuits*, vol. SC-4, pp. 57–64, April 1969.

INTEL CORP. 3065 Bowers Avenue, Santa Clara, California 95051 • (408) 246-7501

JULY 1971

FULLY DECODED RANDOM ACCESS 1024 BIT DYNAMIC MEMORY

PIN CONFIGURATION

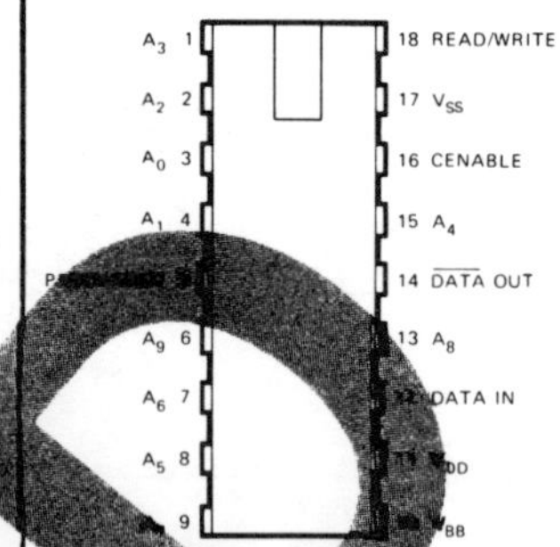

- **Low Power Dissipation — Dissipates Power Primarily on Selected Chips**
- **Access Time — 300 nsec**
- **Cycle Time — 580 nsec**
- **Refresh Period . . . 2 milliseconds for 0 — 70° C Ambient**
- **OR-Tie Capability**
- **Simple Memory Expansion — Chip Enable Input Lead**
- **Fully Decoded — on Chip Address Decode**
- **Inputs Protected — All Inputs Have Protection Against Static Charge**
- **Low Cost Packaging — 18 Pin Plastic Dual In-Line Configuration**

BLOCK DIAGRAM

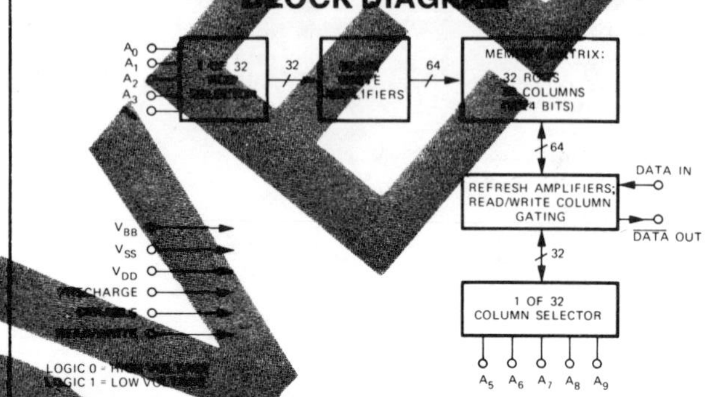

PACKAGE OUTLINE

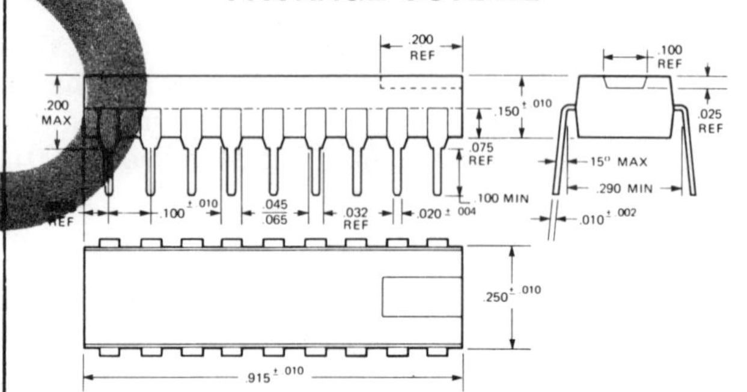

NOTE: Ceramic packaging will be available in 1971

The Intel 1103 is designed primarily for main memory applications where high performance, low cost, and large bit storage are important design objectives.

It is a 1024 word by 1 bit random access memory element using normally off *P*-channel MOS devices integrated on a monolithic array. It is fully decoded, permitting the use of an 18 pin dual in-line package. It uses dynamic circuitry and primarily dissipates power only during precharge.

Information stored in the memory is non-destructively read. Refreshing of all 1024 bits is accomplished in 32 read cycles and is required every two milliseconds.

A separate **cenable** (chip enable) lead allows easy selection of an individual package when outputs are OR-tied.

The Intel 1103 is fabricated with **silicon gate technology.** This **low threshold** technology allows the design and production of higher performance MOS circuits and provides a higher functional density on a monolithic chip than conventional MOS technologies.

Intel's silicon gate technology also provides excellent protection against contamination. This permits the use of low cost plastic packaging.

Maximum Guaranteed Ratings*

Temperature Under Bias	0°C to 70°C
Storage Temperature	−65°C to +150°C
All Input or Output Voltages with Respect to the Most Positive Supply Voltage, V_{BB}	−25V to 0.3V
Supply Voltages V_{DD} and V_{SS} with Respect to V_{BB}	−25V to 0.3V
Power Dissipation	1.0 W

*COMMENT:

Stresses above those listed under "Maximum Guaranteed Rating" may cause permanent damage to the device. This is a stress rating only and functional operation of the device at these or at any other condition above those indicated in the operational sections of this specification is not implied. Exposure to absolute maximum rating conditions for extended periods may affect device reliability.

D.C. and Operating Characteristics

T_A = 0°C to +70°C, $V_{SS}^{(1)}$ = 16V ± 5%, $(V_{BB} - V_{SS})^{(6)}$ = 3V to 4V, V_{DD} = 0V unless otherwise specified

SYMBOL	TEST	MIN.	TYP.	MAX.	UNIT	CONDITIONS
I_{LI}	**INPUT LOAD CURRENT** (ALL INPUT PINS)			**1**	**μA**	**V_{IN} = 0V**
I_{LO}	OUTPUT LEAKAGE CURRENT			1	μA	V_{OUT} = 0V
I_{BB}	**V_{BB} SUPPLY CURRENT**			**100**	**μA**	
$I_{DD1}^{(2)}$	SUPPLY CURRENT DURING T_{PC}		37	56	mA	ALL ADDRESSES = 0V PRECHARGE = 0V CENABLE = V_{SS}; T_A = 25°C
$I_{DD2}^{(2)}$	SUPPLY CURRENT DURING T_{OV}		38	59	mA	ALL ADDRESSES = 0V PRECHARGE = 0V CENABLE = 0V; T_A = 25°C
$I_{DD3}^{(2)}$	SUPPLY CURRENT DURING T_{POV}		5.5	11	mA	PRECHARGE = V_{SS} CENABLE = 0V; T_A = 25°C
$I_{DD4}^{(2)}$	SUPPLY CURRENT DURING T_{CP}		3	4	mA	PRECHARGE = V_{SS} CENABLE = V_{SS}; T_A = 25°C
$I_{DD\,AV}^{(5)}$	**AVERAGE SUPPLY CURRENT**		**17**	**25**	**mA**	**CYCLE TIME = 580 ns; PRECHARGE WIDTH = 190 ns; T_A = 25°C**
$V_{IL1}^{(7)}$	INPUT LOW VOLTAGE (ALL ADDRESS & DATA-IN LINES)	$V_{SS}-17$		$V_{SS}-14.2$	V	T_A = 0°C
$V_{IL2}^{(7)}$	INPUT LOW VOLTAGE (ALL ADDRESS & DATA-IN LINES)	$V_{SS}-17$		$V_{SS}-14.5$	V	T_A = 70°C
$V_{IL3}^{(7,8)}$	INPUT LOW VOLTAGE (PRECHARGE CENABLE & READ/WRITE INPUTS)	$V_{SS}-17$		$V_{SS}-14.7$	V	T_A = 0°C
$V_{IL4}^{(7,8)}$	INPUT LOW VOLTAGE (PRECHARGE CENABLE& READ/WRITE INPUTS)	$V_{SS}-17$		$V_{SS}-15.0$	V	T_A = 70°C
$V_{IH1}^{(7)}$	INPUT HIGH VOLTAGE (ALL INPUTS)	$V_{SS}-1$		$V_{SS}+1$	V	T_A = 0°C
$V_{IH2}^{(7)}$	INPUT HIGH VOLTAGE (ALL INPUTS)	$V_{SS}-0.7$		$V_{SS}+1$	V	T_A = 70°C
I_{OH1}	**OUTPUT HIGH CURRENT**	**600**	**900**	**4000**	**μA**	**T_A = 25°C** ; R_{LOAD} = 100 Ω [4]
I_{OH2}	**OUTPUT HIGH CURRENT**	**500**	**800**	**4000**	**μA**	**T_A = 70°C** ; R_{LOAD} = 100 Ω [4]
I_{OL}	OUTPUT LOW CURRENT		See Note 3			R_{LOAD} = 100 Ω [4]
V_{OH1}	OUTPUT HIGH VOLTAGE	60	90	400	mV	T_A = 25°C, R_{LOAD} = 100 Ω [4]
V_{OH2}	OUTPUT HIGH VOLTAGE	50	80	400	mV	T_A = 70°C, R_{LOAD} = 100 Ω [4]
V_{OL}	OUTPUT LOW VOLTAGE		See Note 3			

Note 1: The V_{SS} current drain is equal to $(I_{DD} + I_{OH})$ or $(I_{DD} + I_{OL})$.

Note 2: See Supply Current vs. Temperature (p. 3) for guaranteed current at the temperature extremes. These values are taken from a single pulse measurement.

Note 3: The output current when reading a low output is the leakage current of the 1103 plus external noise coupled into the output line from the clocks. V_{OL} equals I_{OL} across the load resistor.

Note 4: This value of load resistance is used for measurement purposes. In applications the resistance may range from 100 Ω to 1 kΩ.

Note 5: This parameter is periodically sampled and is not 100% tested.

Note 6: $(V_{BB} - V_{SS})$ supply should be applied at or before V_{SS}.

Note 7: The maximum values for V_{IL} and the minimum values for V_{IH} are linearly related to temperature between 0°C and 70°C. Thus any value in between 0°C and 70°C can be calculated by using a straight-line relationship.

Note 8: The maximum values for V_{IL} (for precharge, cenable & read/write) may be increased to $V_{SS}-14.2$ @ 0°C and $V_{SS}-14.5$ @ 70°C (same values as those specified for the address & data-in lines) with a 40 ns degradation (worst case) in t_{AC}, t_{PC}, t_{RC}, t_{WC}, t_{RWC}, t_{ACC1} and t_{ACC2}.

Supply Current vs Temperature

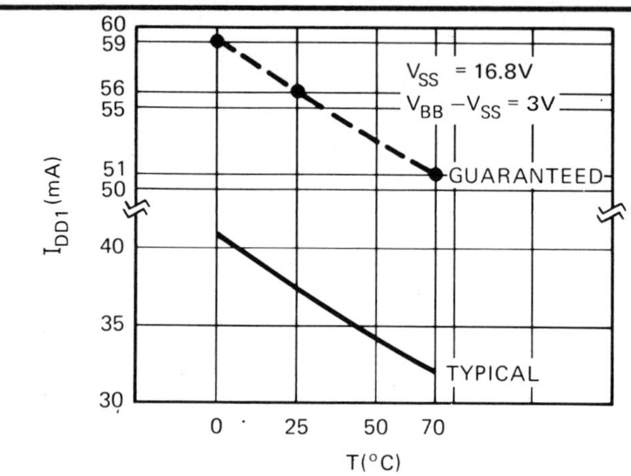

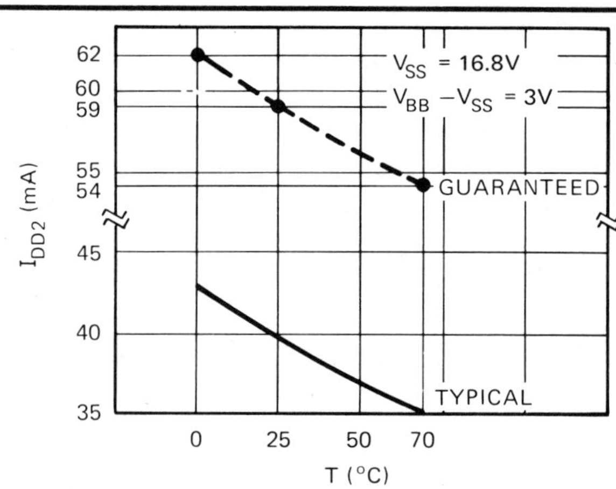

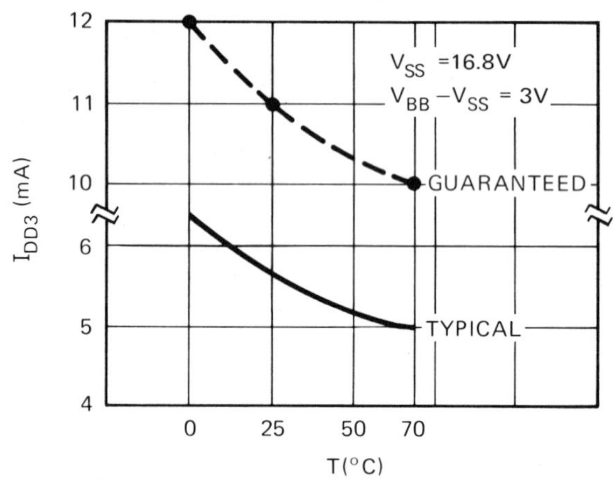

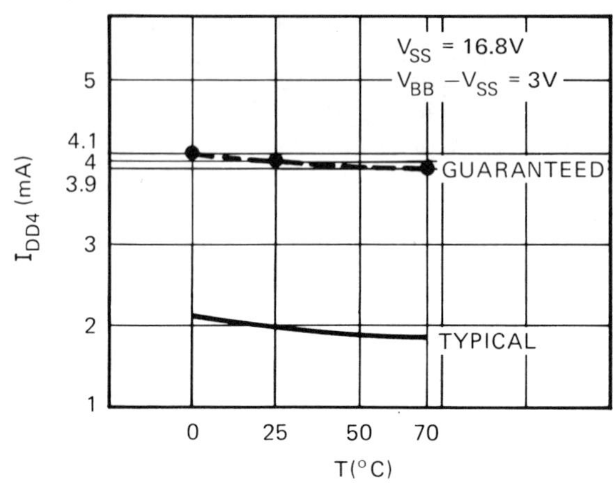

Typical Characteristics

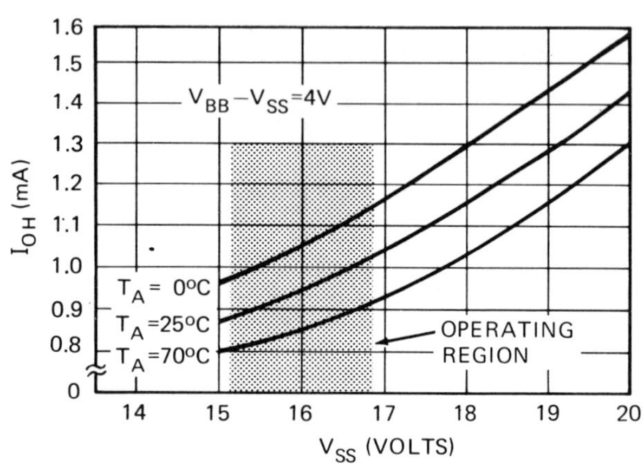

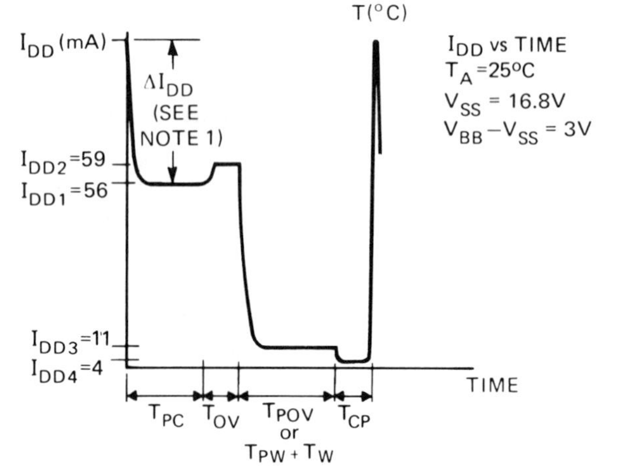

Note 1. ΔI_{DD} is due to charging of internal device node capacitance at precharge

Note 2. These values are taken from a single pulse measurement

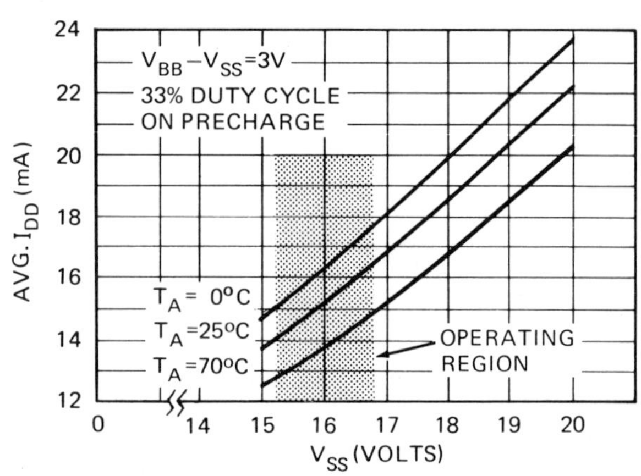

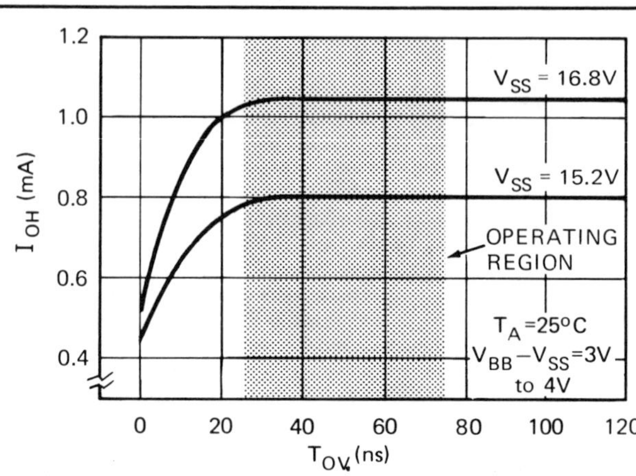

AC Characteristics $T_A = 0°C$ to $70°C$, $V_{SS} = 16 \pm 5\%$, $(V_{BB} - V_{SS}) = 3.0V$ to $4.0V$, $V_{DD} = 0V$

READ, WRITE, AND READ/WRITE CYCLE

SYMBOL	TEST	MIN.	TYP.	MAX.	UNIT	CONDITIONS
t_{REF}	**TIME BETWEEN REFRESH**			**2**	**ms**	
t_{AC} (1)	ADDRESS TO CENABLE SET UP TIME	115			ns	
t_{CA}	CENABLE TO ADDRESS HOLD TIME	20			ns	
t_{PC} (1)	PRECHARGE TO CENABLE DELAY	125			ns	
t_{OVL}	PRECHARGE & CENABLE OVERLAP, LOW	25		75	ns	
t_{CP}	CENABLE TO PRECHARGE DELAY	85			ns	
t_{OVH}	PRECHARGE & CENABLE OVERLAP, HIGH			140		

READ CYCLE

SYMBOL	TEST	MIN.	TYP.	MAX.	UNIT	CONDITIONS
t_{RC} (1)	**READ CYCLE**	**480**			**ns**	$t_T = 20$ ns, $C_{LOAD} = 100$ pF, $R_{LOAD} = 100\Omega$, $V_{REF} = 40$ mV (all rows)
t_{POV}	PRECHARGE TO END OF CENABLE	165		500	ns	
t_{PO}	END OF PRECHARGE TO OUTPUT DELAY			120	ns	
t_{ACC1} (1)	**ADDRESS TO OUTPUT ACCESS**	**300**			**ns**	$t_{ACmin} + t_{OVLmin} + t_{POmax} + 2\,t_T$
t_{ACC2} (1)	**PRECHARGE TO OUTPUT ACCESS**	**310**			**ns**	$t_{PCmin} + t_{OVLmin} + t_{POmax} + 2\,t_T$

WRITE OR READ/WRITE CYCLE

SYMBOL	TEST	MIN.	TYP.	MAX.	UNIT	CONDITIONS
t_{WC} (1)	**WRITE CYCLE**	**580**			**ns**	$t_T = 20$ ns
t_{RWC} (1)	**READ/WRITE CYCLE**	**580**			**ns**	$t_T = 20$ ns
t_{PW}	PRECHARGE TO READ/WRITE DELAY	165		500	ns	
t_{WP}	READ/WRITE PULSE WIDTH	80			ns	
t_W	READ/WRITE SET UP TIME	80			ns	
t_{DW}	DATA SET UP TIME	105			ns	
t_{DH}	DATA HOLD TIME	10			ns	
t_{PO}	END OF PRECHARGE TO OUTPUT DELAY			120	ns	$C_{LOAD} = 100$ pF, $R_{LOAD} = 100\Omega$, $V_{REF} = 40$ mV
t_P	TIME TO NEXT PRECHARGE	0				

Note 1: These times will degrade by 40 ns (worst case) if the maximum values for V_{IL} (for precharge, cenable and read/write inputs) go to $V_{SS}-14.2V$ @ 0°C and $V_{SS}-14.5V$ @ 70°C as defined on page 2.

*CAPACITANCE

SYMBOL	TEST	MIN.	TYP.	MAX.	UNIT	CONDITIONS
C_{AD}	ADDRESS CAPACITANCE		5	7	pF	$V_{IN} = V_{SS}$
C_{PR}	PRECHARGE CAPACITANCE		15	18	pF	$V_{IN} = V_{SS}$
C_{CE}	CENABLE CAPACITANCE		15	18	pF	$V_{IN} = V_{SS}$
C_{RW}	READ/WRITE CAPACITANCE		11	15	pF	$V_{IN} = V_{SS}$
C_{IN1}	DATA INPUT CAPACITANCE		4	5	pF	CENABLE = 0V, $V_{IN} = V_{SS}$
C_{IN2}	DATA INPUT CAPACITANCE		2	4	pF	CENABLE = V_{SS}, $V_{IN} = V_{SS}$
C_{OUT}	DATA OUTPUT CAPACITANCE		2	3	pF	$V_{OUT} = 0V$

Conditions for all capacitance rows: f = 1 MHz, All Unused Pins Are At A.C. Ground

*This parameter is periodically sampled and is not 100% tested. They are measured at worst case operating conditions. Capacitance are for plastic packages only.

WRITE CYCLE OR READ/WRITE CYCLE

Timing illustrated for minimum cycle.

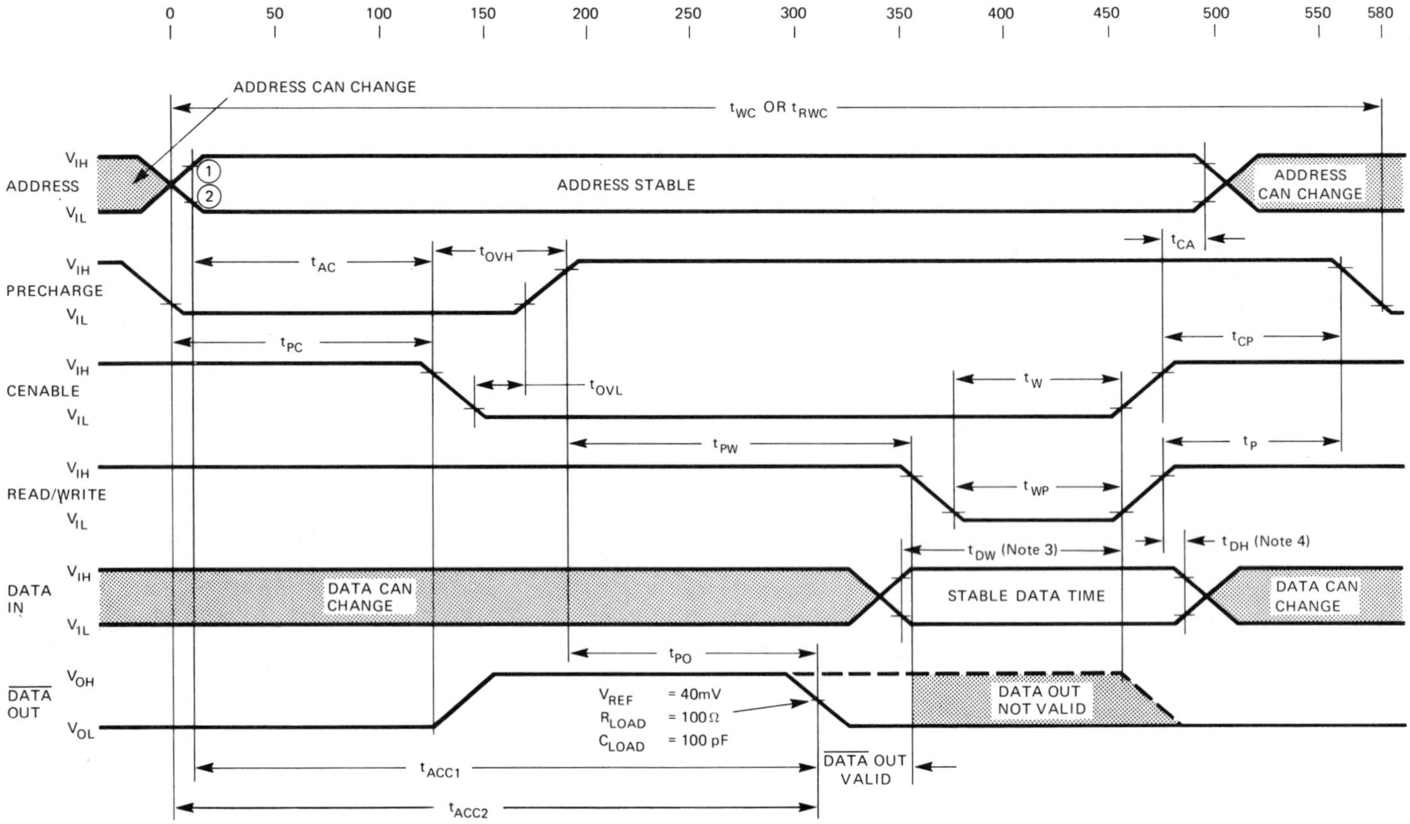

READ CYCLE

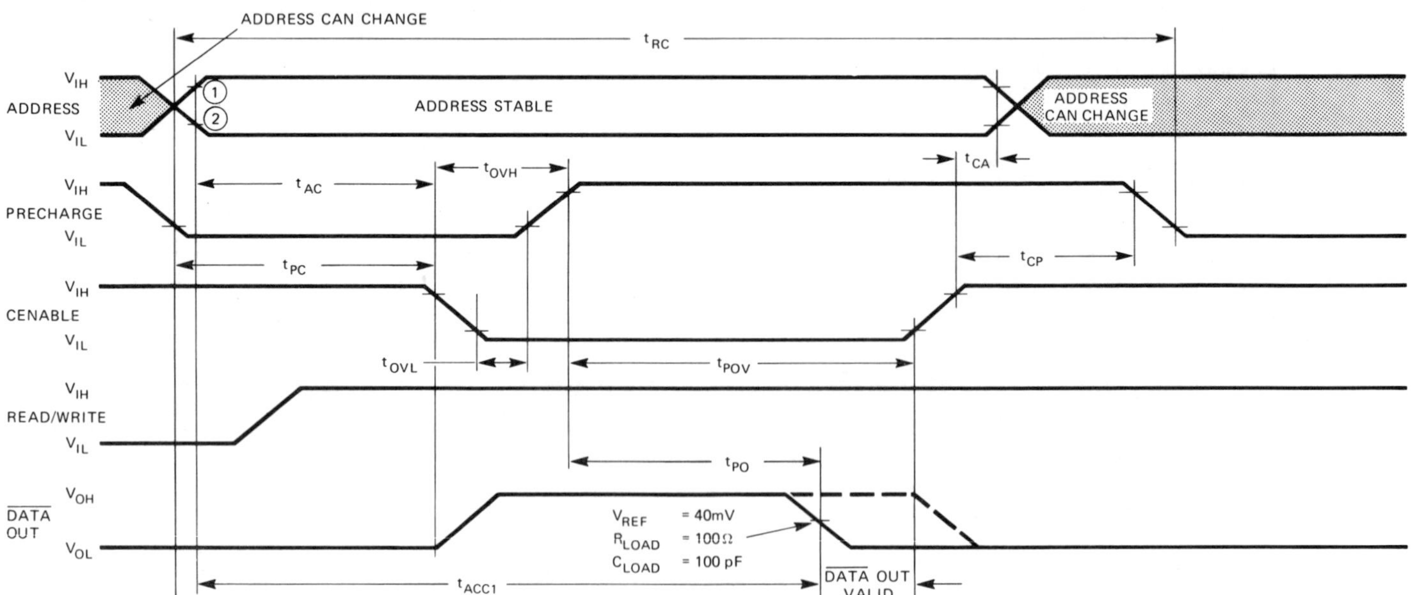

NOTE ① $V_{DD} + 2V$ } t_T IS DEFINED AS THE TRANSITIONS BETWEEN THESE TWO POINTS
NOTE ② $V_{SS} - 2V$

NOTE 3 t_{DW} IS REFERENCED TO POINT ② OF THE RISING EDGE OF CENABLE OR READ/WRITE WHICHEVER OCCURS FIRST

NOTE 4 t_{DH} IS REFERENCED TO POINT ① OF THE RISING EDGE OF CENABLE OR READ/WRITE WHICHEVER OCCURS FIRST

1103 Circuit Diagram

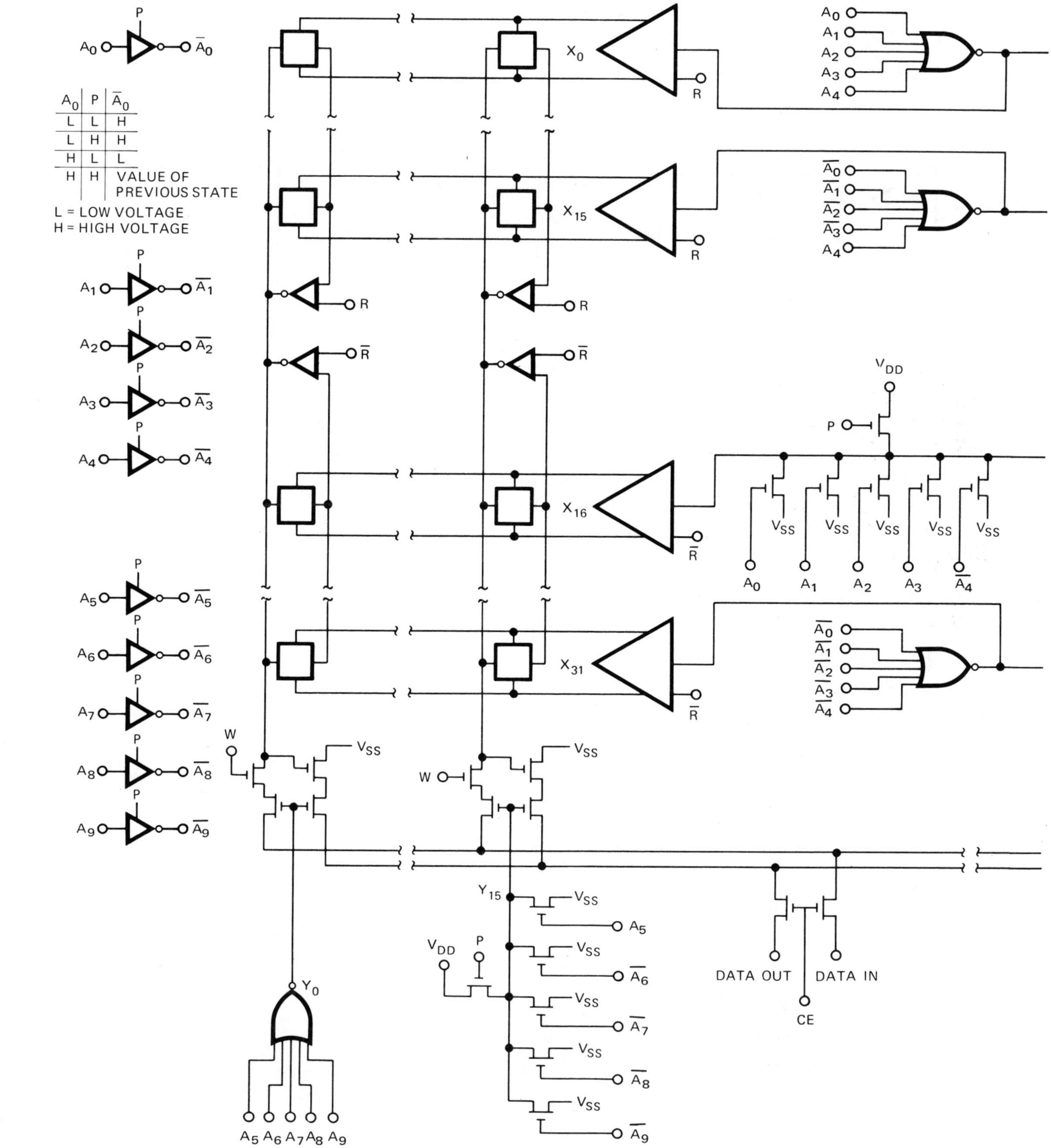

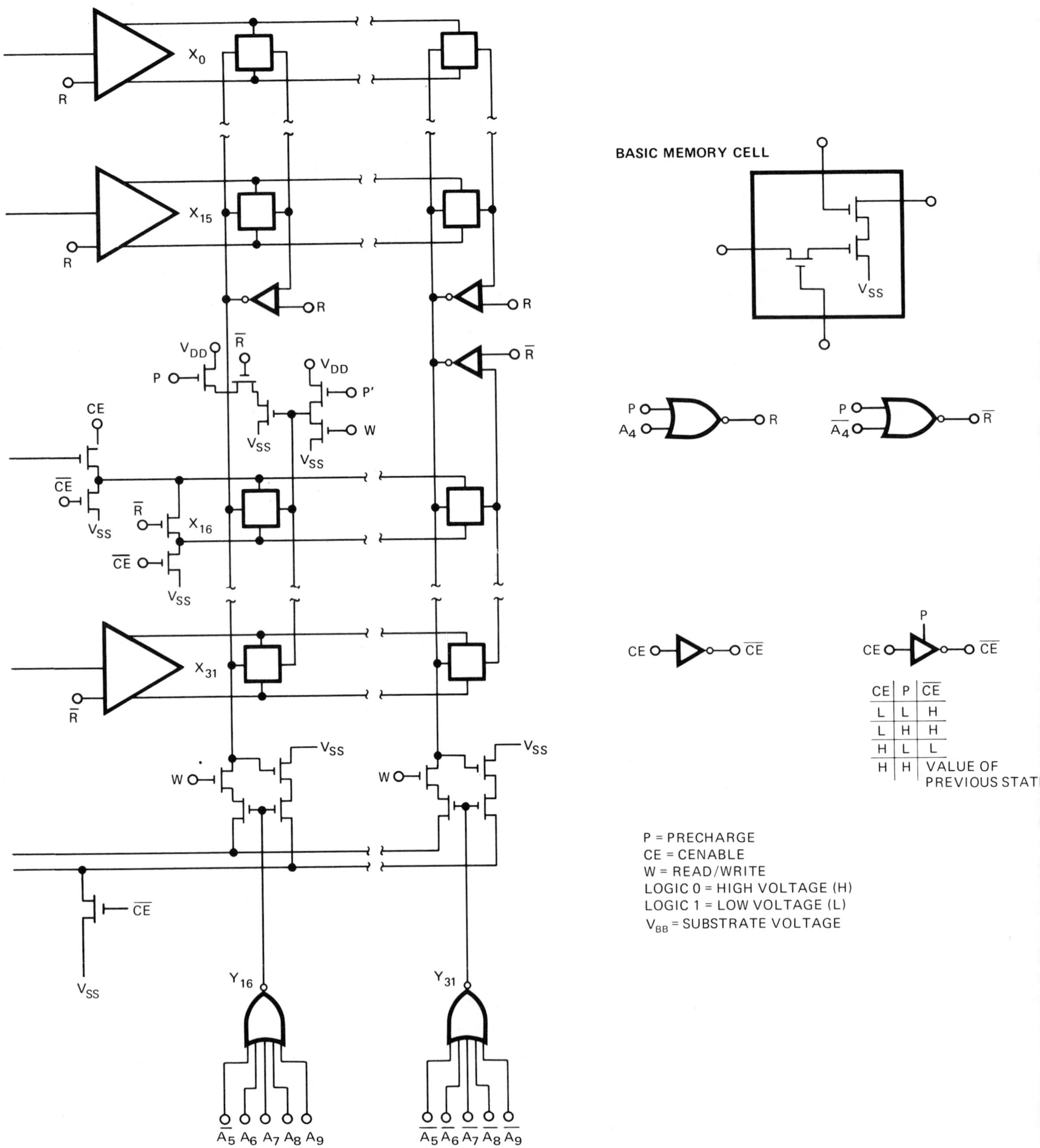

CE	P	$\overline{CE}$
L	L	H
L	H	H
H	L	L
H	H	VALUE OF PREVIOUS STATE

DEVICE OPERATION:

Device operation can be understood via the block diagram (Page 1), and the timing diagram (Page 5). Access begins t_{AC} before the negative transition of **cenable.** During this period, **precharge** is active, and the address becomes stable in row and column decoders. After the **cenable** transition the contents of the 32 cells along the selected row are written into the 32 on chip refresh amplifiers, one of which is required for each column in the array. At the positive transition of **precharge,** the contents of the refresh amplifiers are written back into their respective columns, and the output appears t_{PO} later. t_{PW} after the positive edge of **precharge,** new data on the data input lead may be written into the selected cell, using a **read/write** pulse of minimum duration t_{WP}.

The device exhibits four characteristics of considerable importance in system design:

- The memory cells are dynamic in nature, and thus require periodic data refreshing. This is accomplished by cycling through (Read Cycle) the 32 addresses of the A_0 through A_4 inputs. This operation must be performed at least every 2 milliseconds.
- The device dissipates maximum power while the **precharge** input is negative; as this input runs 40% duty cycle, during read cycle, dissipation is typically 320 mW. To reduce power dissipation in the memory system **precharge** may be held positive on all but the chips to which cenable is applied.
- Cell readout is nondestructive; the **read/write** cycle depicted in the timing diagram may be shortened for a **read only** or **refresh only** cycle, as the **read/write** pulse is only required for entry of new data. Thus a **read only** or **refresh only** cycle can be shorter than a **read/write,** or a **write only** cycle.
- Data written into the memory is inverted at the output, i.e., a high input (positive) is read as zero current out and a low written into the memory is read as .9 mA (typ.) out.

1103 SYSTEMS ORGANIZATION

Figure 2 shows a 4096 x 9 bit self contained asynchronous memory system. It communicates to the mainframe with three control signals.

Cycle Request is a signal which is generated by the mainframe demanding action by the memory.

Read/write is also generated by the mainframe in association with the request signal and indicates whether a **read** or **write** operation is requested.

Ready is generated by the memory system in response to the request signal issued by the mainframe. The memory system goes busy (not ready) at the receipt of the request. At the conclusion of the memory cycle, the ready line again becomes true, and additional requests may be accepted. The cycle length for a read cycle may be made shorter than for a write cycle.

The mainframe also supplies addresses and data for writing and accepts data read from the memory.

Application Information

To permit a memory with 1103's to interface with TTL logic and provide this simple mode of operation, certain peripheral circuits are required. These circuits are shown as blocks in Fig. 2.

Data input level-shift. Converts data supplied for writing at TTL levels to MOS levels. These circuits need drive only the light capacitive load represented by four (for this memory) data input leads.

Sense amplifiers and latches. These circuits convert the currents read from the devices to TTL levels and provide storage for subsequent use by the computer.

With a system such as shown in Fig. 2 a certain amount of noise may be introduced by capacitive coupling of **precharge**, and **cenable** signals to the sense lines. This noise may be reduced by running a dummy sense line in parallel and using a differential input sense amplifier. The transitions are then coupled equally into the sense line and the dummy line.

Address latches and clock drivers. These circuits accept and store the address. Clock drivers convert the TTL levels to MOS levels, providing sufficient capability to drive the capacitance represented by the full array of devices.

The address bits in excess of the first ten are decoded and drive **cenable** for each column. This decoded **cenable** signal acts as a column select within the memory array. Clock drivers are required for these signals also.

The multiplexer permits substitution of the refresh address for the address in the latches during refresh cycles. Only five bits are required. The **cenable** decoding is also overridden during refresh cycles.

Control and timing. These circuits generate all timing signals for the array. A detailed discussion of one such circuit is shown below.

Precharge and **read/write** drivers. These must be heavy duty drivers to drive the full array.

The organization shown in Fig. 2 uses common **precharge** and **read/write** drivers. By decoding **precharge** as well as **cenable,** the memory power consumption may be significantly reduced. For a 4096 word memory, full decoding of **precharge** can reduce DC power dissipation and clock driver losses by a factor of about 3. However, the extra level-shift circuits needed may offset some of these power savings. The decoding of **precharge** also reduces the noise coupled to the sense lines.

Dynamic MOS storage requires periodic refreshing of data. For the 1103, this refreshing may be accomplished by cycling through the 32 possible states of address lines A_0-A_4 in every 2 msec period.

To guarantee this condition for a random access memory, special refresh cycles may be executed by the controller. For a 600 nsec cycle the refresh operation requires less than one percent of available memory cycles.

Application Information

Fig. 3 shows how the devices in the memory array are connected. For a 4K x 9 memory, the number of rows N is 9, and the number of columns is 4. In each row, data input leads are connected in common. Similarly, data output leads in each row are connected. In each column, **cenable** leads are connected. Unless **precharge** and **read/write** decoding is used, the remaining terminals are bussed so that corresponding pins of all packages are connected.

Drive and Sense Circuits

Fig. 4 shows a line receiver used as a sense amplifier with strobing included. The 75108 has an open collector output and hence is ideal for large systems where several modules are tied to a common data bus.

The sense line and the optional dummy line run parallel through the array with only the sense line connected to the device. This organization is used to minimize noise due to capacitive coupling to the sense line. The voltage divider consisting of a 200Ω and an 18K resistor is used to establish the logic 1/logic 0 detection level threshold.

The circuits shown in Fig. 5 are examples of drivers which can be used when interfacing TTL to 1103 levels. The driver shown in Fig. 5A can be used to drive address and data lines. A more rugged driver is shown in Fig. 5B and can be used to drive address and data lines in very large arrays. An inexpensive driver for **precharge**, **cenable** and **read/write** is shown in Fig. 5C.

Control Circuits

The controller shown in Fig. 1 is used to generate the timing and refresh control signals for an 1103 memory. The upper flip-flop (normal-cycle control flip-flop) acknowledges requests for memory cycles by entering the busy state (Q output a logic 1) the lower flip-flop (refresh cycle control flip-flop) and single shot (multivibrator) perform refresh cycle control. Any cycle executed with the lower flip-flop on is considered a refresh cycle.

The multivibrator requests a refresh cycle every 60 μsec. Refresh cycles can only be initiated when the normal cycle control flip-flop is in the ready state (Q output a logic 1). Execution of either type of cycle is accomplished by sending a signal down the delay line. The cycle timing signals are then derived from taps along the delay line. To maintain suitable delay to rise-time ratios, two delay lines with a buffer stage in between them have been used. The signal from the buffer stage which is fed back to the driving stage of the first delay line establishes the minimum cycle length. The read/write pulse timing signal is used to clear the flip-flops and allow the start of another cycle.

The circuit provides immediate acknowledgement of a normal cycle request. However, if the request occurs during execution of a refresh cycle, the normal cycle flip-flop goes on and remains on, thereby requesting a normal cycle immediately upon completion of the refresh cycle. In this way refresh cycles are invisible to the memory requestor except for the effective increase in cycle time.

A five stage refresh address counter and a 5 bit wide multiplexer must be provided for refresh address generation and switching.

Application Information

Power Dissipation Considerations

To estimate power consumption of an 1103 memory, three sources of dissipation must be considered:

- **Memory DC power.** Memory DC power is approximately given by the product of the number of packages selected at one time, times the sum of the power dissipated during each portion of the cycle, added to the power dissipated by the unselected packages. For a 4K x 9 memory, with precharge decoding the DC memory, power is as follows:

$$9 \text{ devices} \times [\underbrace{.056 \times .3}_{\substack{\text{precharge} \\ (t_{PC})}} + \underbrace{.059 \times .095}_{\substack{\text{overlap} \\ (t_{OVL})}} + \underbrace{.011 \times .385}_{\substack{\text{cenable} \\ (t_{POV})}} + \underbrace{.004 \times .22}_{\substack{\text{cenable to next} \\ \text{precharge } (t_{CP})}}] \times \underset{V_{SS}}{16.8} + \underbrace{27 \times .004 \times 16.8}_{\substack{\text{unselected} \\ \text{devices}}} \approx 6.0\,\text{W}$$

- **Clock power.** Clock power is associated with current drawn from the power supplies to charge **precharge, cenable, read/write,** and **address** line capacitances. The worst case may be found by taking the product of memory frequency times total effective capacitance times the square of the clock drive voltage supply. For decoded precharge, R/W the total effective capacitance = sum of **address** capacitances x no. of devices in the array + sum of **precharge, read/write** and **cenable** capacitances times the number of devices driven.

  For a 4K x 9 memory with **precharge** decoding:

  $$C_{eff} = 10 \text{ address lines} \times 7\text{ pf} \times 36 \text{ devices} + {\sim}60\text{ pf} \times 9 \text{ devices} \cong 3000\text{ pf}$$

  $$\text{Worst case power} = \frac{10^6}{.600} \times 3000 \times 10^{-12} \times 16.8^2 \cong 1.5 \text{ watts.}$$

- **Level shift power.** Worst case is derived from the total number of level shifters times maximum dissipation per level shifter. For the level shifter shown with 470Ω load resistor, and 17 v supply, dissipation is approximately .6 watt. The 4K x 9 array requires 31 circuits. Worst case power would be about 18.5 watts.

These power dissipation figures may be higher than actually observed for the following reasons:

- Not all level shifters will operate at maximum dissipation. The 4 used for **precharge** will typically dissipate only 40–60% of the maximum value. The 4 used for **read/write** have very low duty cycles. **Address** and **data** level shifters dissipate the maximum values only for certain conditions.
- Capacitive charging dissipation may also be less than the computed values if few address bits change.
- When level shifters are added for **precharge** decoding, the duty cycle is reduced in proportion to the number of drivers added.

Precharge duty cycle should be a minimum to reduce the power dissipation.

Cenable Decode Circuits

Figure 6 shows a suggested circuit for controlling the **cenable** circuits. Address bits 10 and 11 are decoded by the leftmost gates. This decoding is overridden when refresh is true (REF is low). The CENT input supplies timing for the array.

Similar decoders may also be used for decoding **precharge** and **read/write.**

Application Information

Low Power Data Retention

Low power data retention mode is an operating mode for a dynamic MOS memory in which power consumption is reduced to the minimum value possible while still maintaining data in the memory. For most dynamic MOS memories the power consumption level in this mode is in the range of from 1 to 10 watts per million bits. These low power levels make it possible to maintain the data within the memory with small battery packs. Many of the problems associated with the normal volatility of semiconductor memory systems can be eliminated using this mode of operation.

In the low power data retention mode of an 1103, the memory is operated such that only refresh cycles are executed. Power dissipation in the memory array is then reduced to a minimum value. To insure low power dissipation elsewhere in the memory system, it is usually most desirable to execute the refresh cycles in bursts of 32 every 2 milliseconds, which eliminates the need for retaining information about which locations were refreshed and which were not. The entire memory is refreshed every 2 msec.

The 1103 has a potential current flow from V_{SS} to V_{DD} during the period between refresh bursts. However, this current can be minimized by insuring that address line A4 is low (V_{IL}) during the last cycle of the refresh burst, then returns high (V_{IH}) after the last cycle.

A typical implementation of low power data retention uses a time-out circuit which establishes when 2 milliseconds have elapsed. Power is then switched on to the TTL controller and the memory is forced into "burst" (i.e., burst of 32 refresh cycles) refresh mode. Upon completion of the burst of refresh cycles, controller power is shut off and the time-out circuit reset. This sequence is repeated until main system power is applied. Then upon completion of the burst of refresh cycles, the memory switches to the normal operating mode. Three power supplies must always be present: V_{BB}, V_{SS} and a supply for use by the TTL logic in the controller. The TTL logic supply is switched on to the control logic only during the execution of refresh bursts or during normal operation. The V_{BB} and V_{SS} supplies must be maintained at the memory array at all times. The memory array draws significant current from the V_{SS} supply only during the execution of memory cycles. To minimize power consumption in the system, level shifters should be designed such that with the V_{SS} supply maintained on the level shifters, and the TTL controller supply off, the level shifter output is at $+V_{SS}$ with no (or insignificant) power drawn from the V_{SS} or V_{BB} supplies.

The power supplies for a memory with low power data retention mode are usually backed up by some form of storage battery. For the best results, the batteries are located on the unregulated side of the power supply regulators. During normal operation of the system, the batteries are trickle charged. Whenever the unregulated voltage falls below the battery voltage, the batteries deliver current. The voltages to the memory experience no switching transients, and proper regulation of all memory voltages is maintained.

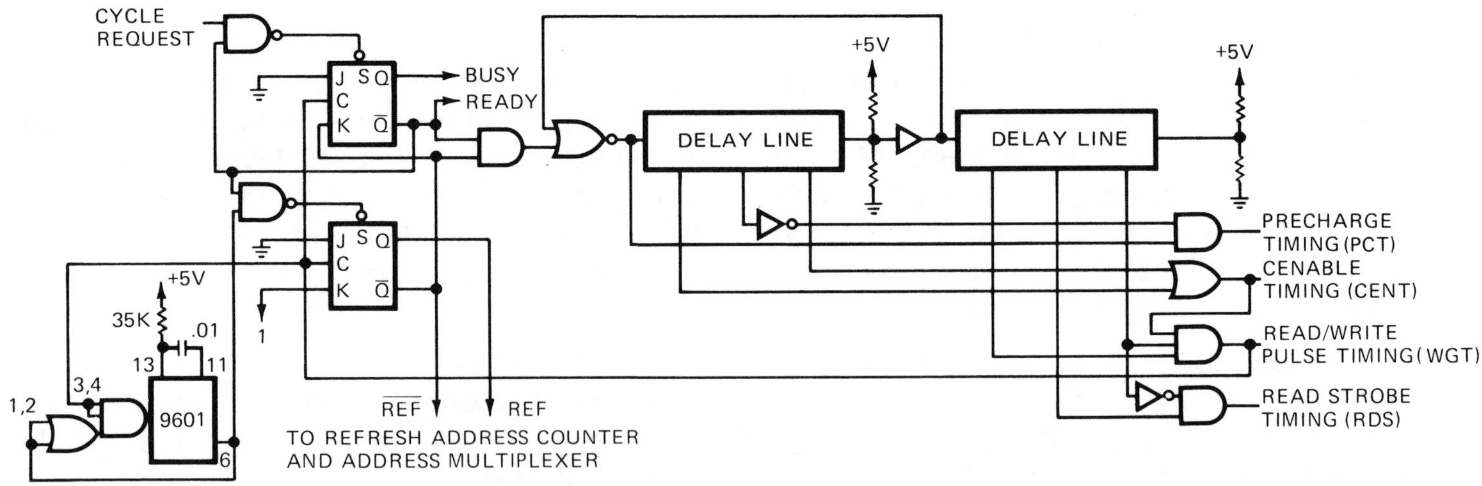

Fig. 1 Basic Timing Generator For 1103 Memory

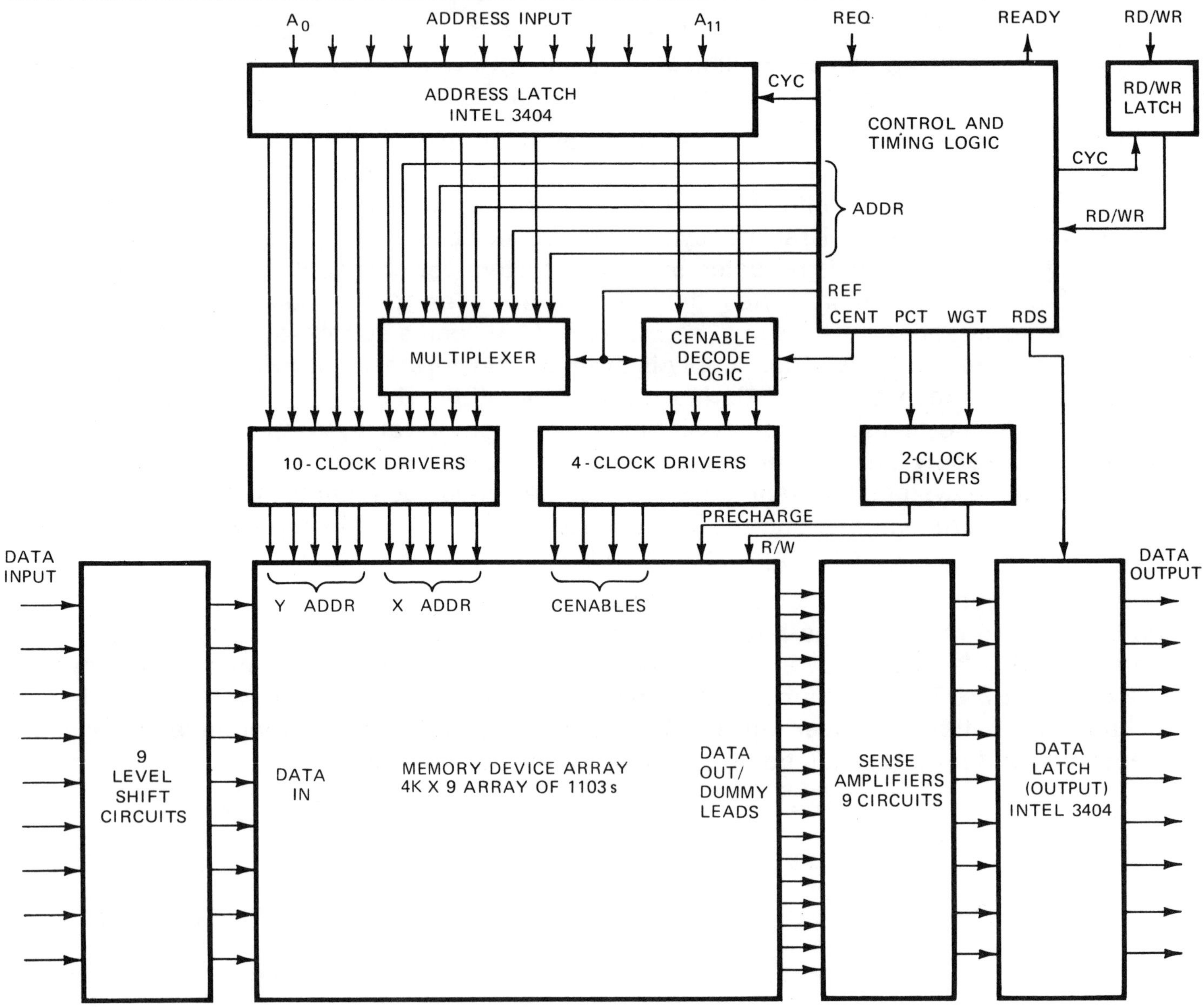

Fig. 2 1103 Memory System – 4096 9-Bit Bytes Block Diagram

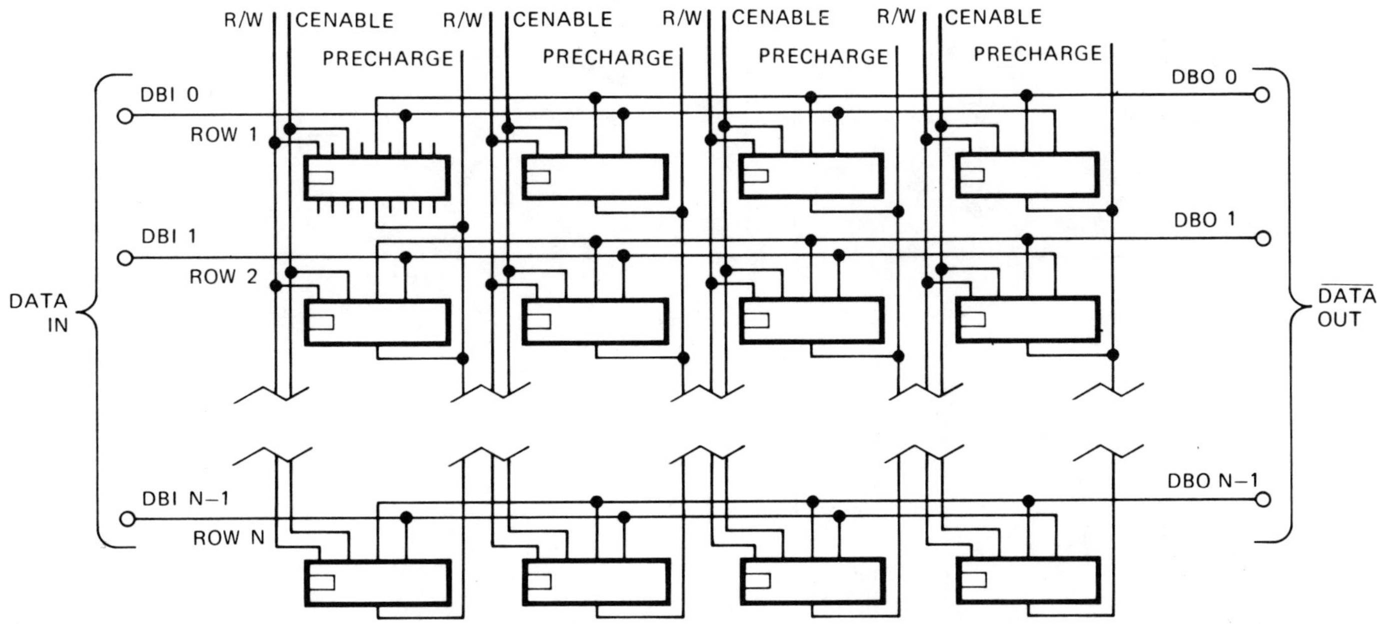

Fig. 3 1103 Memory Expansion

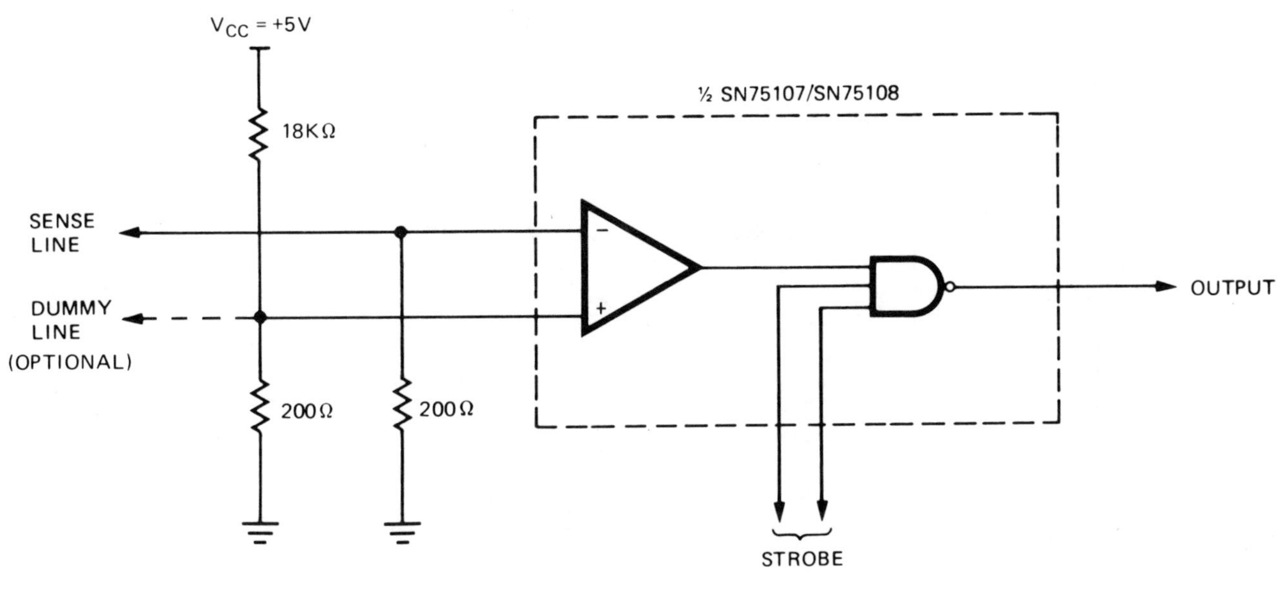

Fig. 4 Balanced Line Sensing

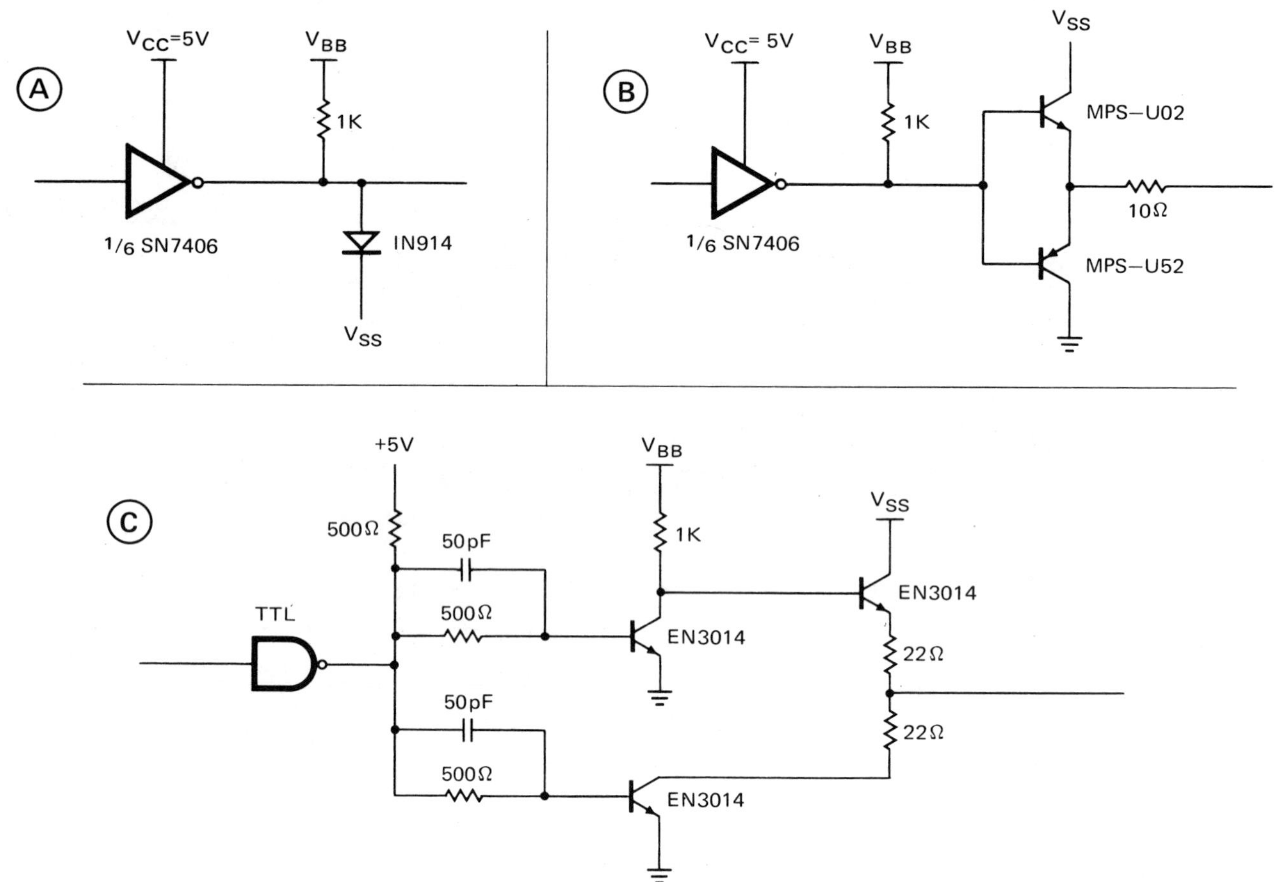

Fig. 5 TTL to 1103 Interface Circuits

INTERFACE CIRCUITS – TYPICAL OUTPUT CHARACTERISTICS (25°C)

	Driver (A)	Driver (B)	Driver (C)
I Low Voltage (V_{OL})			
No Load	0.25V	≈0V	0.04V
Sinking 3 mA	–	+0.97V	0.14V
II High Voltage (V_{OH})			
No Load	V_{SS} + 0.65V	V_{SS} ± 0.01V	V_{SS} ± 0.01V
Source - 3 mA	–	V_{SS} – 0.05V	V_{SS} – 0.12V
Sink x 0.5 mA	–	V_{SS} + 0.01 (1)	V_{SS} + 0.45V (1)
III Rise time (t_R) (2)			
10 pF	25 ns	25 ns	10 ns
50 pF	90 ns	25 ns	15 ns
100 pF	–	25 ns	20 ns
200 pF	–	25 ns	25 ns
470 pF	–	40 ns	50 ns
IV Fall time (t_F) (2)			
10 pF	12 ns	6 ns	6 ns
50 pF	20 ns	8 ns	9 ns
100 pF	–	10 ns	12 ns
200 pF	–	14 ns	20 ns
470 pF	–	24 ns	35 ns

1. This voltage level is a function of transistor reverse gain. A diode clamp to V_{SS} is recommended.
2. These values are measured between the 10% and 90% points.

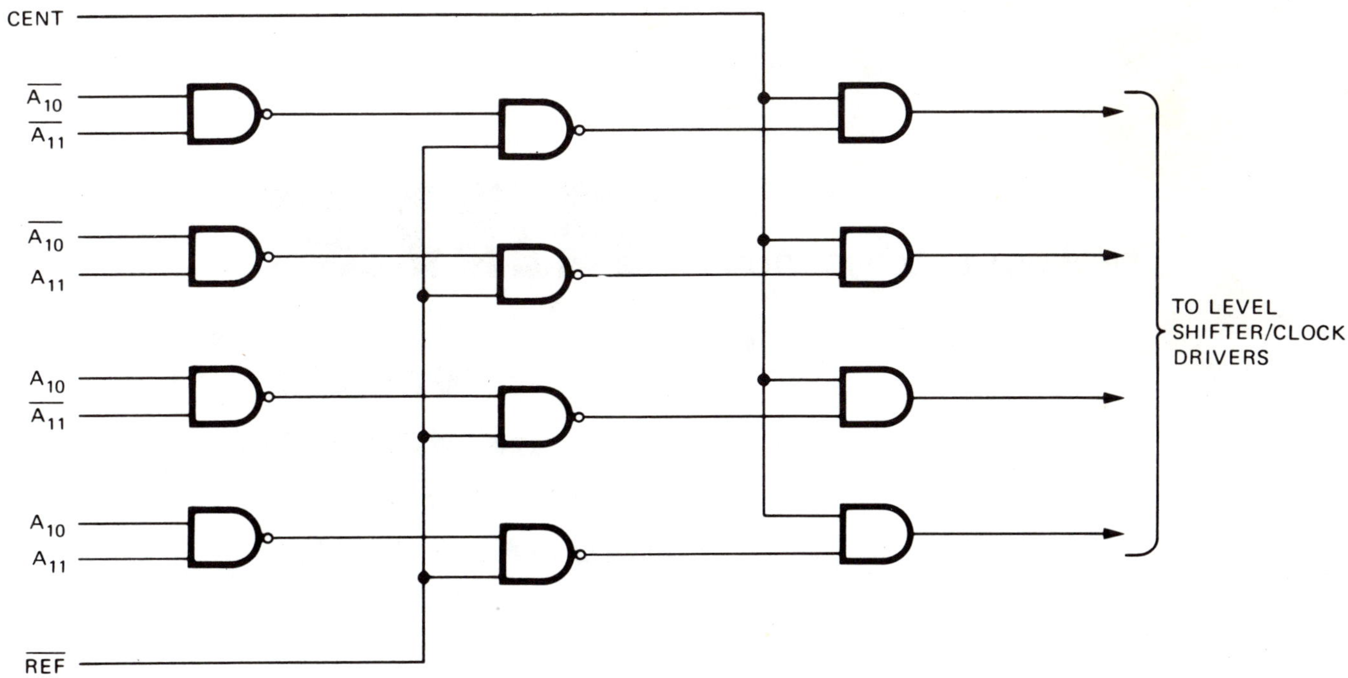

Fig.6 Cenable Decode Circuit

PRODUCT SPECIFICATIONS AMS

1024 BIT HIGH SPEED MOS LSI MEMORY RANDOM ACCESS AMS 6002

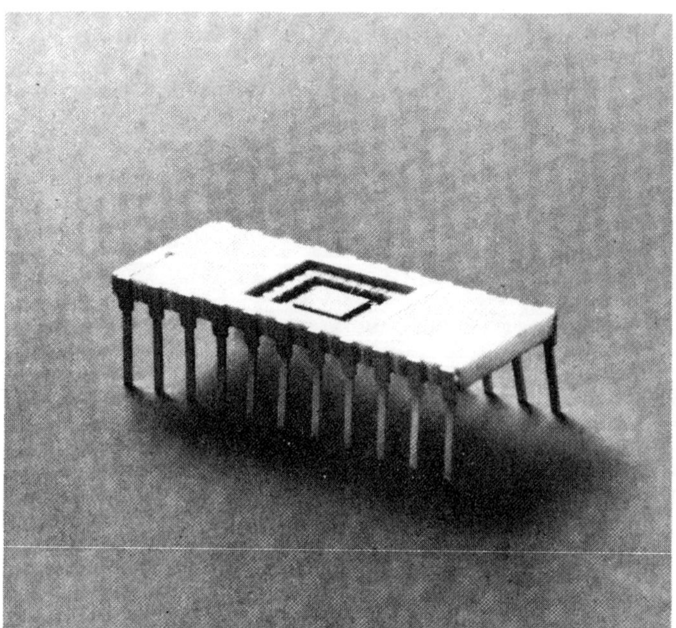

Reprinted with permission from *Product Specifications,* Oct. 1971.

AMS 6002 October, 1971
MOS LSI Random Access
Read/Write Memory Device
Order #91600211-6D

ADVANCED MEMORY SYSTEMS, INC.

FEATURES:

- 1024 Word x 1 Bit Organization
- Access – 150 nsec. max.
- Cycle – 250 nsec./290 nsec.
- Low Power Dissipation –
  180 mW, operating
  2 mW, standby
- Differential Output
- OR-tie Capability
- Chip Select – Simplifies Memory Expansion
- Inputs Static Charge Protected
- Fully Decoded
- Standard, Hermetic 22-pin DIP Package

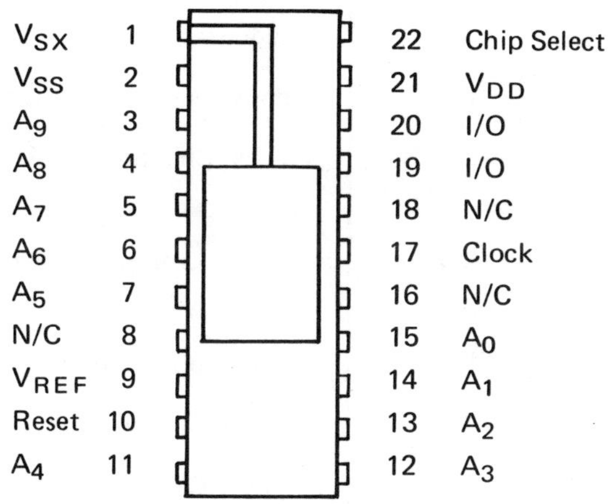

PIN CONFIGURATION

The AMS 6002 is a high speed, 1024 word by 1 bit, dynamic random-access memory device fabricated with industry proven, P-channel metal gate MOS processing. It is designed for use in low cost, high performance memory applications.

The 6002 achieves its high performance at a low power dissipation with a four-transistor storage cell and unique support circuitry. The low capacitance inputs minimize driver circuit power requirements, simplifies TTL to MOS conversion, and reduces overall system costs.

The memory is fully decoded and its differential outputs can be OR-tied. The Chip Select input allows the selection of individual components in arrays with greater than 1K words.

The information stored in the memory is non-destructively read, but since it is dynamic it must be refreshed periodically. Refreshing of the total memory requires 32 cycles every 2 milliseconds. The chip need not be selected during refresh.

The 6002 is housed in a ceramic, hermetically sealed, 22-pin dual-in-line package and operation is specified over a 0 – 70°C temperature range.

MAXIMUM RATINGS

Supply Voltage:

V_{DD} & V_{REF}	with respect to V_{SS}	-27v to 0.5v
V_{DD} & V_{REF}	with respect to V_{SX}	-30v to 0.5v
V_{SX}	with respect to V_{SS}	-0.5v to 10v

Input Voltages:

All inputs	with respect to V_{SS}	-30v to 0.5v

Storage Temperature: -55°C to +125°C

FUNCTIONAL CHARACTERISTICS:

T_A = 0°C to 70°C, V_{DD} = 0v, V_{REF} = +7v ± 5%, V_{SS} = +20v ± 5%, $V_{SX} - V_{SS}$[1] = 2.5v ± 5%

Symbol	Characteristics	Min.	Typ.	Max.	Unit	Condition
I_{LIA}	Input Leakage Current Addresses			1	μA	$V_{IN} = V_{DD}$
I_{LI}	Input Leakage Current Reset, Clock, Chip Select			10	μA	$V_{IN} = V_{DD}$
I_{LO}	Output Leakage Current			2	μA	$V_{IN} = V_{REF}$
I_{SX}	V_{SX} Supply Current		1		μA	
I_{SS_1}	Supply Current			100	μA	All inputs at V_{SS}
I_{SS_2}	Supply Current		9	15	mA	R at V_{DD}; C, CS at V_{SS} T_A = 25°C
I_{SS_3}	Supply Current		18	25	mA	All addresses at V_{DD}; R, C, CS at V_{SS} T_A = 25°C
V_{IH}	Input High Voltage	V_{SS} - 2.0		V_{SS}	V	All inputs
V_{IL}	Input Low Voltage	V_{DD} - 2.0	V_{DD}	V_{DD} + 2.0	V	All inputs except I/O & R
V_{ILR}	Input Low Voltage,Reset	V_{DD} - 7.0	V_{DD}	V_{DD} + 2.0	V	See Note 2
$V_{I/O}$	Input Low Voltage, I/O	V_{REF} - 1.0	V_{REF}	V_{REF} + 1.0	V	
I_{DO}	Differential Output	100			μA	See Note 3

Note 1. ($V_{SX} - V_{SS}$) must be applied prior to V_{SS}.

Note 2. See A.C. Characteristics, taking Reset Input Low Voltage to V_{DD} - 5v improves cycle time.

Note 3. Differential Output Current 90 nsec. after clock and chip select go negative.

A. C. CHARACTERISTICS:*

T_A = 0°C to 70°C, V_{DD} = 0v, V_{REF} = +7.0v ± 5%, V_{SS} = 20v ± 5%, V_{SX} - V_{SS} = 2.5v ± 5%

Read/Refresh/Write Cycle

Symbol	Characteristics		Min.	Typ.	Max.	Unit	Condition
T_{REF}	Time between Refresh		2			ms	70°C
***T_{RP}	Reset Pulse Width		90			ns	V_{ILR} = V_{DD} - 5.0v
T_{RP}	Reset Pulse Width		130			ns	V_{ILR} = V_{DD}
T_{DD}	Decode Delay		60			ns	
T_{CP}	Clock Pulse Width		100			ns	
**T_{CSP}	Chip Select Pulse Width		100			ns	
T_{CRD}	Clock to Reset Delay		0			ns	
T_{CSRD}	Chip Select to Reset Delay		0			ns	

Read/Refresh Cycle

Symbol	Characteristics		Min.	Typ.	Max.	Unit	Condition
***T_{RC}	Read Cycle		250			ns	V_{ILR} = V_{DD} - 5.0v
T_{RC}	Read Cycle		290			ns	V_{ILR} = V_{DD}
***T_{REFC}	Refresh Cycle		250			ns	V_{ILR} = V_{DD} - 5.0v
T_{REFC}	Refresh Cycle		290			ns	V_{ILR} = V_{DD}
T_{CD}	Select to Output Delay				90	ns	I_{DO} = 100μA R_L = 400Ω C_L = 10pF
T_{ACC}	Address Valid to Output				150	ns	

Write Cycle

Symbol	Characteristics		Min.	Typ.	Max.	Unit	Conditions
***T_{WC}	Write Cycle		250			ns	V_{ILR} = V_{DD} - 5.0v
T_{WC}	Write Cycle		290			ns	V_{ILR} = V_{DD}
T_{DS}	Data Setup Time		0			ns	
T_{DH}	Data Hold Time		0			ns	

*All time measurements at V_{SS} - 3.0v unless otherwise specified, T_R, T_F $\leqslant$20ns.
**Chip Select not used during Refresh Cycle
***Minimum cycle time only (i.e. V_{ILR} = V_{DD} - 5.0v).

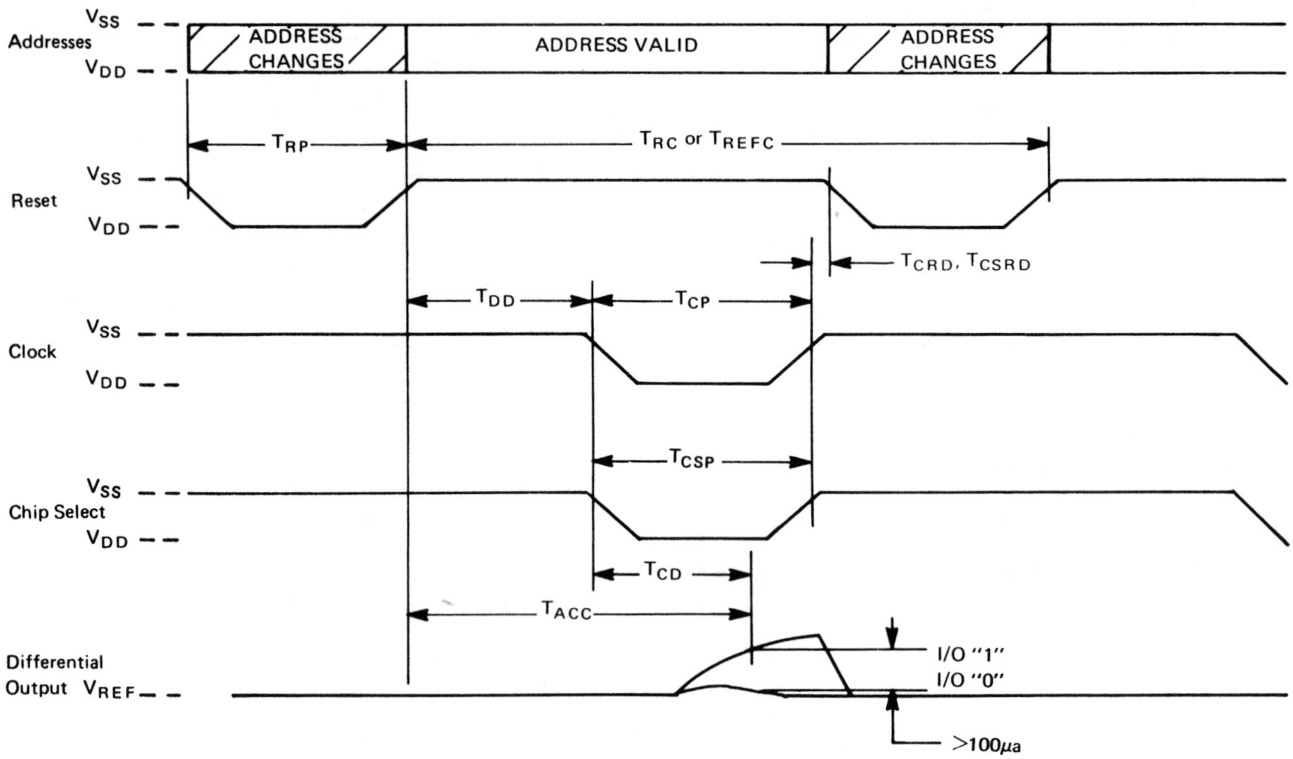

READ or REFRESH CYCLE

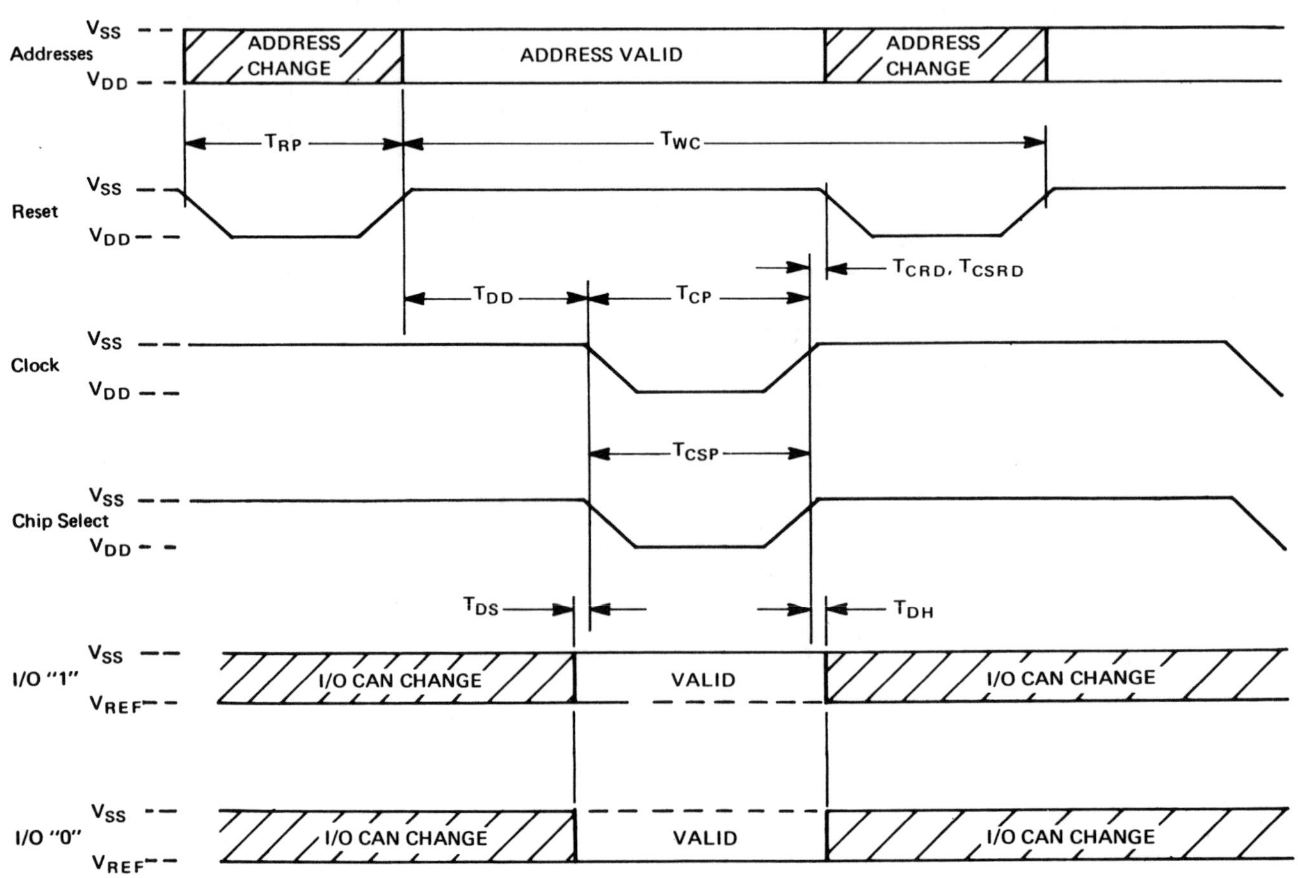

WRITE CYCLE

TYPICAL CHARACTERISTICS

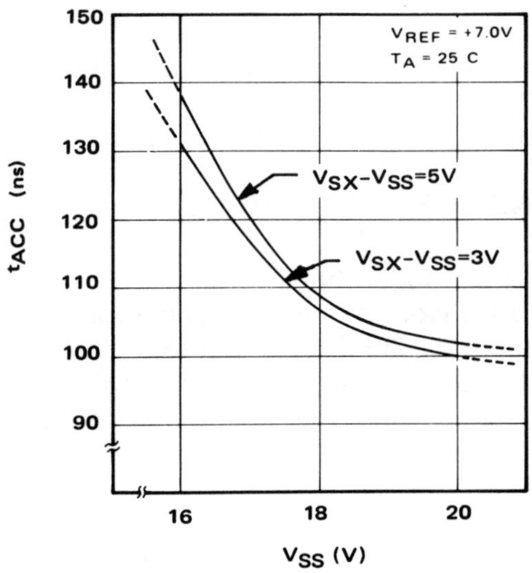

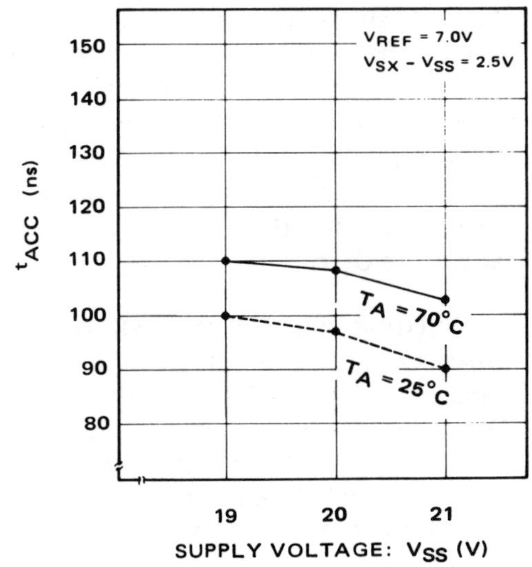

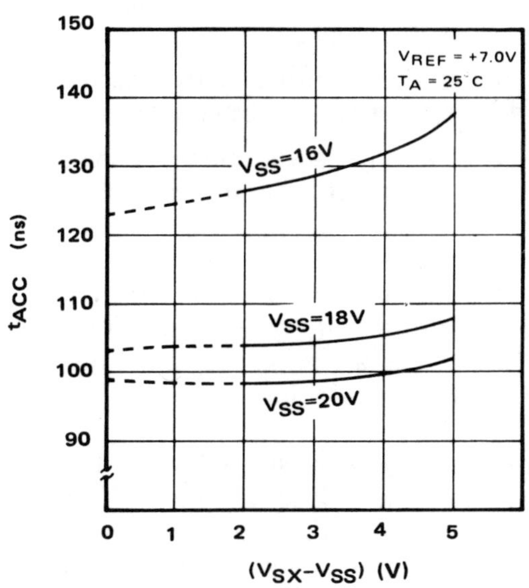

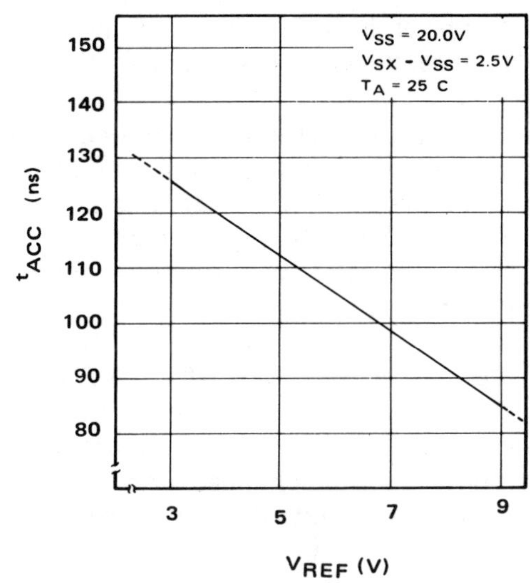

CAPACITANCES:

Symbol	Characteristics	Typ.	Max.	Unit	Condition	
C_{AD}	Input Capacitance, Address	2.5	3.5	pf	$V_{IN} = V_{SS}$	f = 1 MHz
C_R	Input Capacitance, Reset	30.0	40.0	pf	$V_{IN} = V_{SS}$	
C_C	Input Capacitance, Clock	15.0	18.0	pf	$V_{IN} = V_{SS}$	
C_{CS}	Input Capacitance, Chip Select	15.0	18.0	pf	$V_{IN} = V_{SS}$	
$C_{I/O}$	Capacitance, I/O (not selected)	2.5	3.5	pf	$V_{IN} = V_{SS}$	

DEVICE OPERATION:

The storage element in the AMS 6002 is the four-transistor cell shown in Figure 1. (The transistors connected at A and B are not part of the cell.) The actual operation of the cell is straight forward. The cell is selected by taking the clock input to V_{DD} and is a D.C. stable flip-flop at that time. When the clock is at V_{SS} the cell does not conduct and information is stored on one of the parasitic capacitors C_1 or C_2.

Writing data in the cell is simply executed with a pair of switches such as the transistors attached at A and B. If the write "1" transistor is turned on it will force A to close to V_{SS}. This shuts off any conduction which may have been occuring in Q_2. At the same time the capacitor, C_1, will be charged and Q_1 will begin to conduct current. When the cell is selected for a subsequent Read operation, current will exist in the Q_1, Q_4 and R_{S1} side of the cell, and a differential amplifier between A and B would sense the voltage difference and detect the presence of a "1".

To understand how the cell is integrated into a 1024 array, consider the block diagram in Figure 2 and the timing information on Page 5. The Reset input when in the LOW state performs two functions, which are: (1) precharge the device and (2) turn off all the address inverters. On the 6002 the address inputs are the drain of an MOS device, and because they are off during the reset period very low input capacitances are seen. Address inputs can be changed prior to Reset going LOW, provided that the previous cycle is complete, but the input capacitances are greater at that time.

After the Reset input is taken HIGH, there is a period required for the row and column decoders to function. This is the T_{DD} time in the timing diagrams. The clock and chip select periods occur next, with the clock input gating the row decoder while the chip select input enables the column decoders. Information from the selected cell is then transferred to the I/O lines which are also gated by the chip select input. After the output data has been sensed the clock and chip select are returned to the positive state and the device is ready for the next cycle.

In the 6002, the row decode output is tied to the clock input of the cell in Figure 1. Thus any time a row is selected all 32 cells are in a stable state. This selection will refresh those 32 cells, regardless of the state of the chip select input. In order to refresh the 1024 bits all 32 rows must be selected. This is accomplished by exercising all 32 combinations of the A_0 through A_4 addresses.

TIMING AND CONTROL CIRCUITS:

The control logic shown in Figure 3 can be used for a 4K word by 18 bit asynchronous memory using the AMS 6002. All logic is standard SN7400 series unless otherwise noted. Slight improvement in access and cycle times is possible if series 74S is used in selected places.

There are three basic functions which are performed by this logic. They are:

1. Generate the proper timing relationships of the clocks;
2. Provide the memory with internal addresses for the refresh cycles, and
3. Accept and generate control signals for communication with processor.

In addition, this logic does the chip select decoding, and the multiplexing of the external addresses and the internally generated refresh addresses. Figure 4 is a timing diagram for a system using this control logic.

The memory cycle is started with the application of a Memory Request signal. This positive going signal has a duration of from 30 to 200 nsec. The minimum cycle time between any two Memory Requests or Memory Request & Refresh cycle signals is 300 nsec (see timing diagram).

The address lines can change no later than the leading edge of a Memory Request signal. However, the address lines may change when the memory is not busy, i.e., Memory Busy line is "High." Upon the application of a Memory Request signal, the Input latch (SN74H74) is cleared and Reset drivers (SN75452) are driven "High." It is necessary to drive the Resets "High" due to the decode delay required by AMS 6002 between Reset and Chip Select inputs. With Reset drivers going "High," the single shots in series (SN74123) are activated. The outputs of the single shots provide the timing for the clock and Chip Select Inputs to the memory array

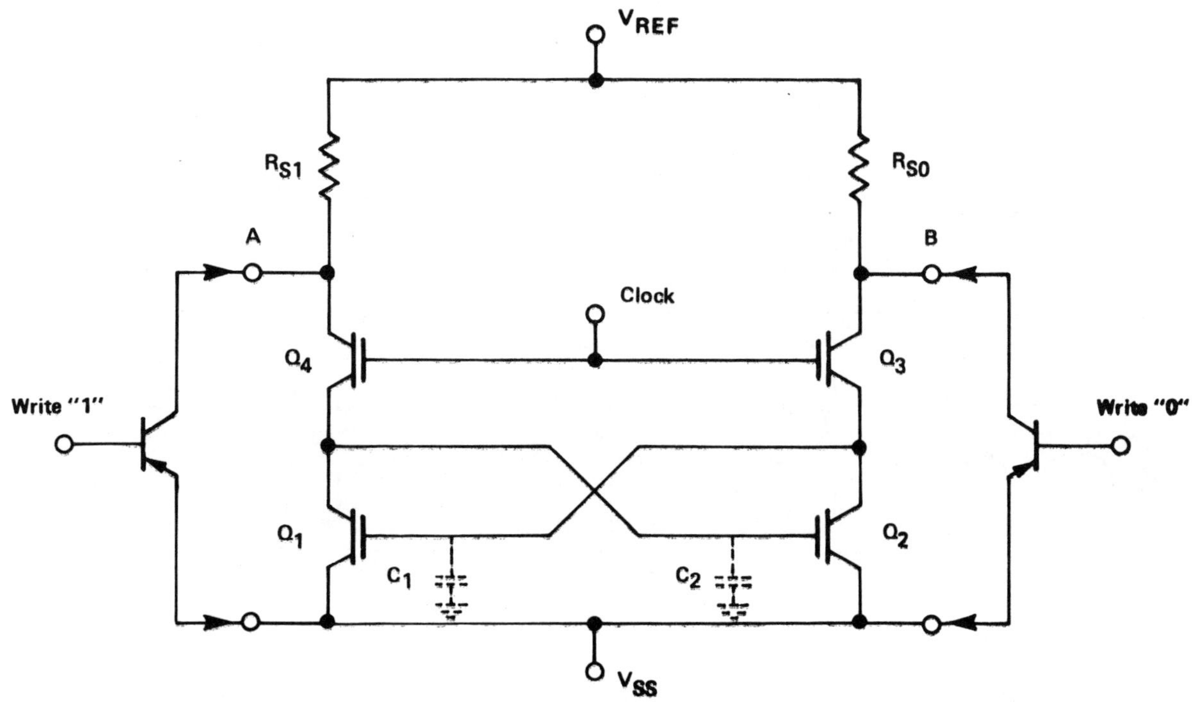

Figure 1 MEMORY CELL

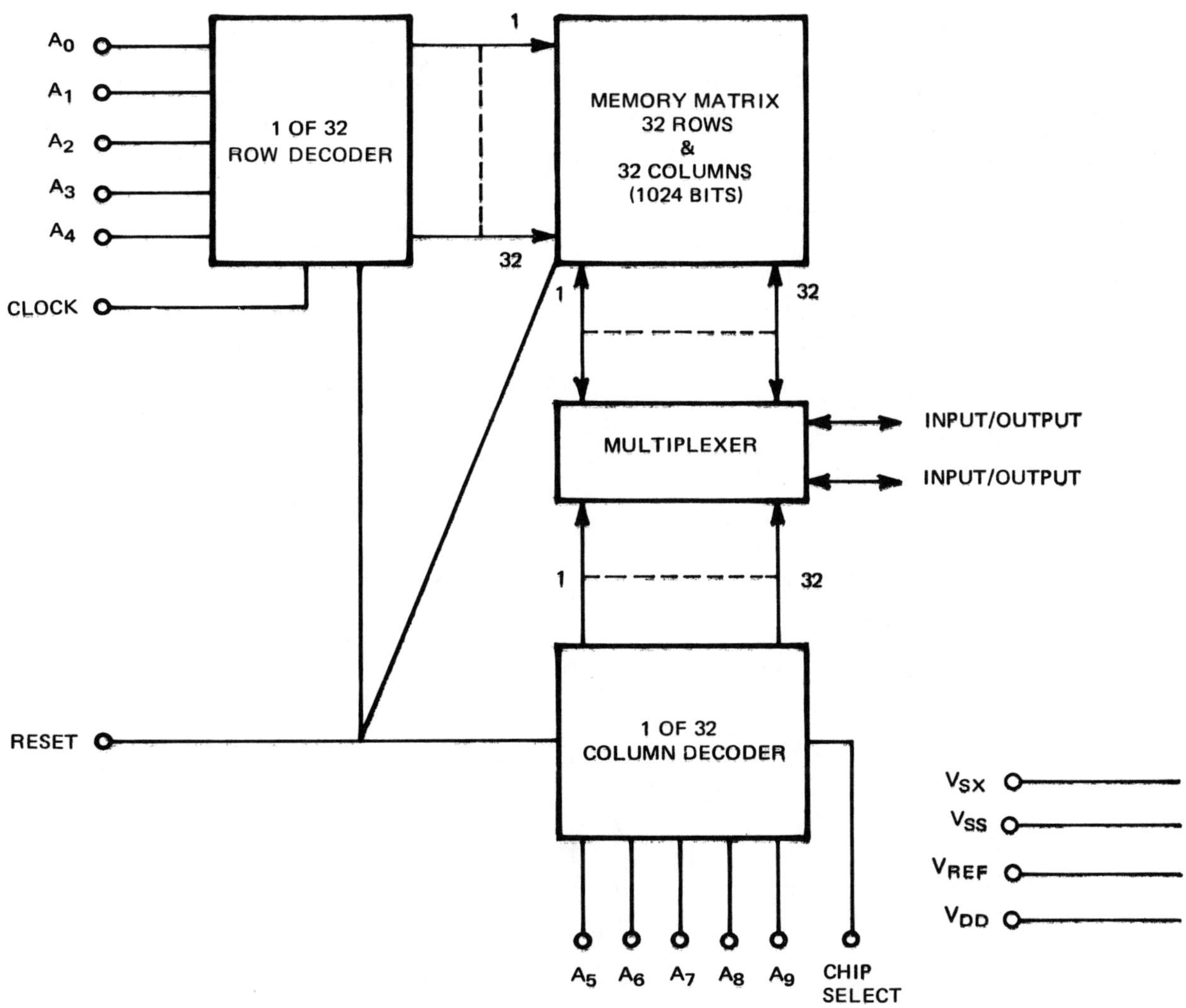

Figure 2 AMS 6002 BLOCK DIAGRAM

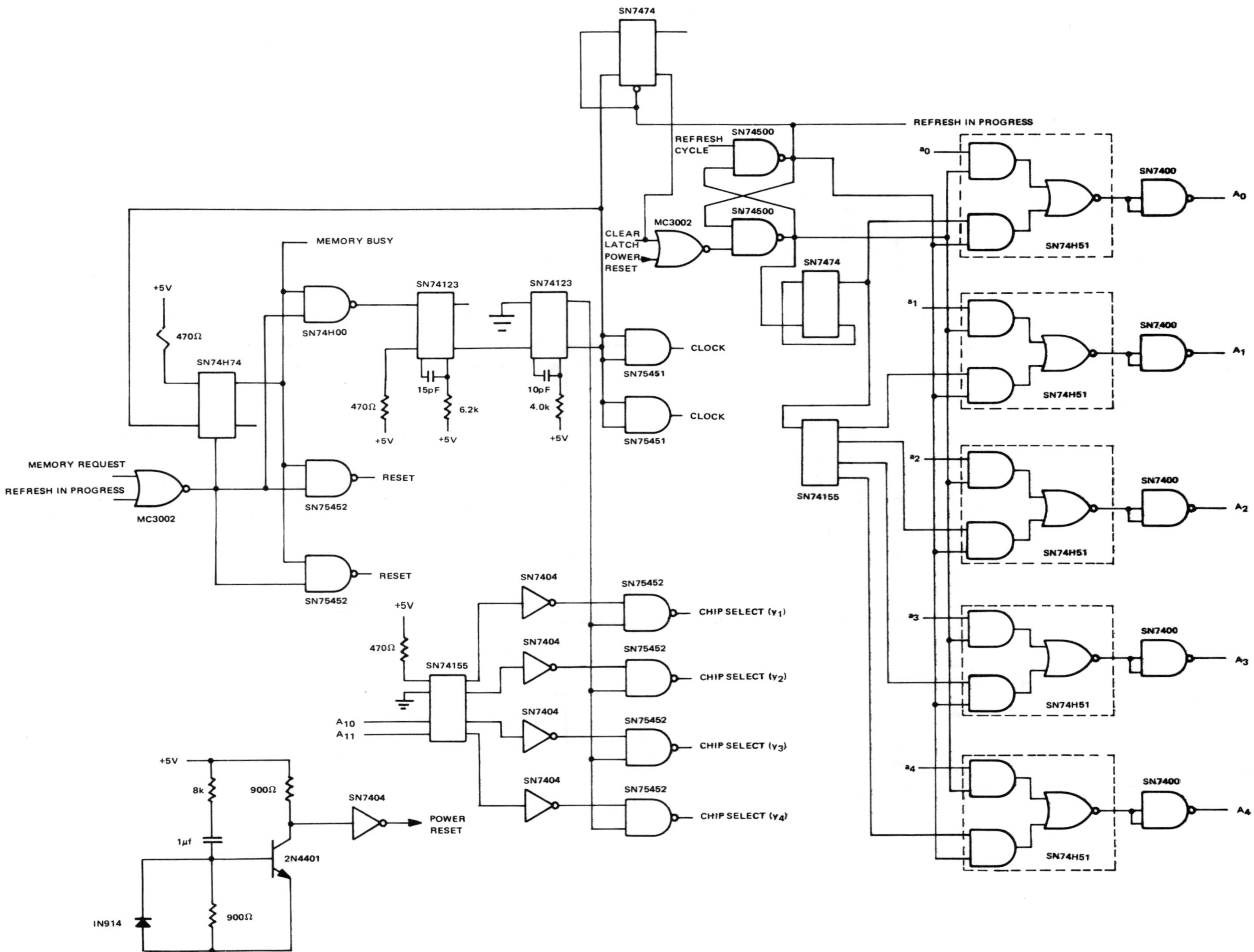

Figure 3 TIMING AND CONTROL LOGIC DIAGRAM

through buffers. The positive going edge of the Chip Select Input forces the Resets "Low." Reset lines remain "Low" until the next Memory Request or Refresh signal is applied. The cycle time of the memory is determined by the RC networks connected to the single shots.

As shown in the timing diagram, a Read cycle is performed only when both Input/Output lines are "Low." This is accomplished by bringing the Write line "Low" during a Read operation. Because of recovery time considerations of the bit driver/sense amplifier circuitry, it is advisable to always have the Write line in the "Low" level except during the period shown in the timing diagram.

The Write operation consists of supplying a positive going pulse at the time shown in the timing diagram. Data In must be valid at least 20 nsec before and after the Write pulse.

One of the features of the control logic shown in Figure 4 is that the Refresh operation is externally controlled. This feature allows the user the option to perform a Refresh cycle. In applications such as in the CRT Refresh, where the memory is sequentially addressed, no memory Refresh cycle is required, provided all addresses are accessed every 2 milliseconds. A Refresh cycle will start when a negative going edge is applied to the Refresh Control. The Refresh Control pulse may vary between 30 nsec and 200 nsec. The addresses (A_0-A_4), necessary for Refresh operation, are generated by a 5-bit counter (SN74155) and a D-type flip-flop, SN7474, and are gated into the memory by the Refresh Control signal.

The additional control lines are Data Strobe and Refresh in Progress. Data Strobe (externally provided) is used for strobing the Data Output Registers. Refresh in Progress (internally provided) is used for indicating to the user that the memory is being refreshed; no Read or Write operation may be introduced when memory is being refreshed.

With the control logic described in this section, memory sizes beyond the 4K x 16/18 configuration can be implemented simply by designing the memory storage cards with card select inputs and open collector outputs. Thus a 16K x 16/18 can be assembled with four storage cards and one control card.

LEVEL CONVERSION AND DRIVING CIRCUITS:

The circuits shown in Figures 5 and 6 are examples of level shifters and drivers which can be used to interface the AMS 6002 with 7400 Series TTL circuits.

The driver in Figure 5 can drive capacitive loads of 1000pf, the equivalent of 330 address inputs on the 6002 with rise and fall times of less than 40nsec. An additional booster stage in parallel, consisting of the complementary emitter followers can be added to the same level shifter to drive additional loads or to improve the transition times. The 10Ω resistor in series with the outputs is for current limiting and the diode is added to speed up the risetime.

As noted earlier in the specification, the AMS 6002 has a faster cycle time if the Reset input has a larger signal swing (typically 5v) than the remaining inputs. One means of implementing this in a system without adding an additional supply is to operate the memory between +25v and +5v with V_{SS} = +25v and V_{DD} = V_{CC} = +5v. The Reset input would then be driven between Ground and +25v while the addresses, clock, and chip select would be driven with a +5v to +25v signal. The driver shown in Figure 6 can be used in such a memory system as it has an output swing of +4v to +25v. The down level of the driver can be adjusted to other values by adjusting the R_2/R_1 resistor ratio.

BIT DRIVER/SENSE AMPLIFIER

The data from the AMS 6002 is in the form of a differential current. To interface to TTL logic, a commercially available differential amplifier and a conversion circuit can be used.

A differential amplifier of the MC1711L type, a zener diode such as IN5230B in series with 200Ω resistor for level conversion is all that is required to sense and convert the differential current output to TTL logic levels. Such a sensing and conversion network is shown in Figure 7.

The output current from the AMS 6002 approximates the characteristics of a ramp function rather than a step function. This ramp characteristic causes the MC1711L type operational amplifier to respond within 10 nsec as compared to a typical response delay of 40 nsec for a step input.

To protect the sense amplifier from the large voltage difference on the Input/Output line during Write operations, a voltage divider with series-resistors is used. These resistors being in series with a current source (50–400 μA) have negligible effect on the access time; however they affect the Input/Output line recovery time by increasing the RC time constant.

The Write operation can be simply executed with a pair of PNP or NPN transistors acting as switches. The bit driver transistor, whether a PNP inverter or an NPN emitter follower, will raise one of the two bit/sense lines to (or close to) V_{SS}, which in turn shorts one side of the four-transistor cell to V_{SS}, shutting off current in the other side. For detailed discussion, please consult AMS Application Note 28. Figure 7 and 8 show two possible bit driver/sense amplifier circuits for the AMS 6002. An integrated bit driver/sense amplifier will be available in early 1972.

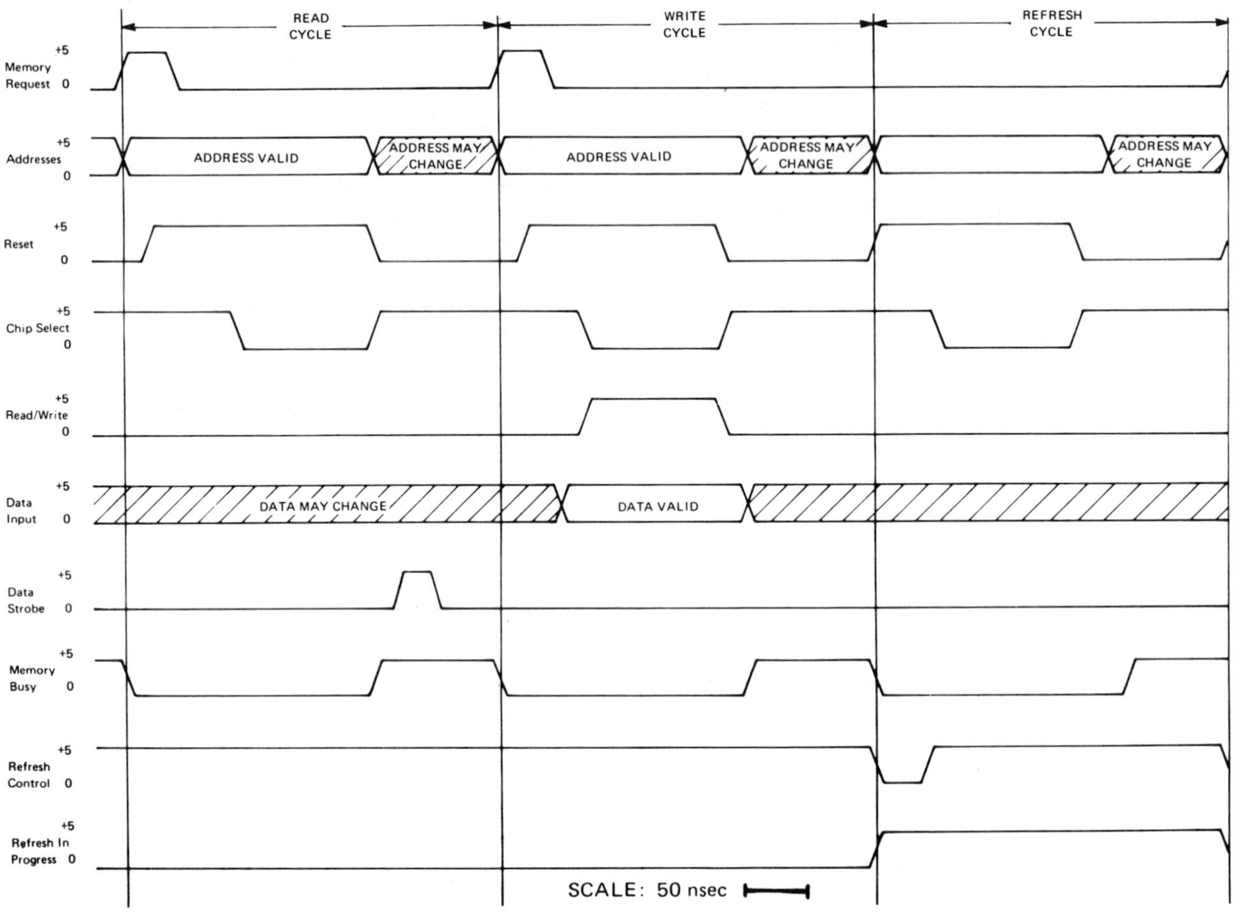

Figure 4 SYSTEM TIMING DIAGRAM

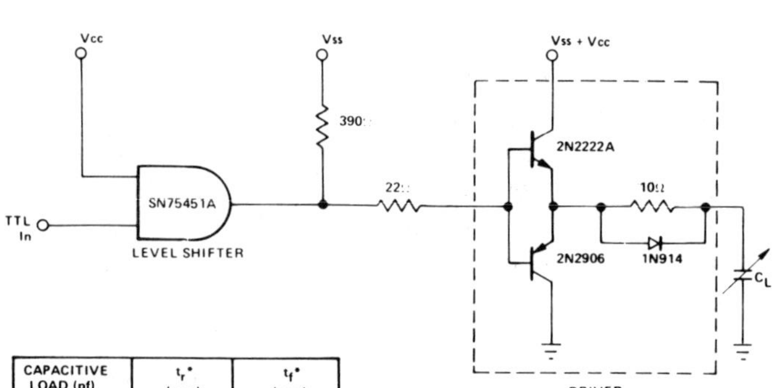

CAPACITIVE LOAD (pf)	t_r* (nsec)	t_f* (nsec)
0	11	9
250	18	13
500	26	19
1000	38	28

Figure 5

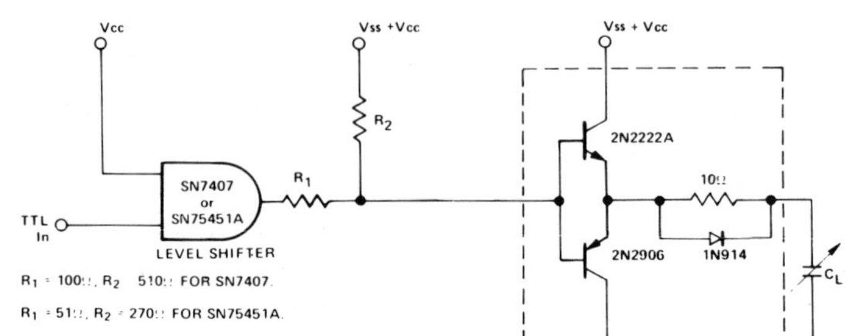

CAPACITIVE LOAD (pf)	t_r* (nsec)	t_f* (nsec)
0	5	5
250	8	8
500	13	12
1000	22	21
1300	30	26

Figure 6

*Measurements made from 10% to 90% points, values are for SN75451A and are typical.

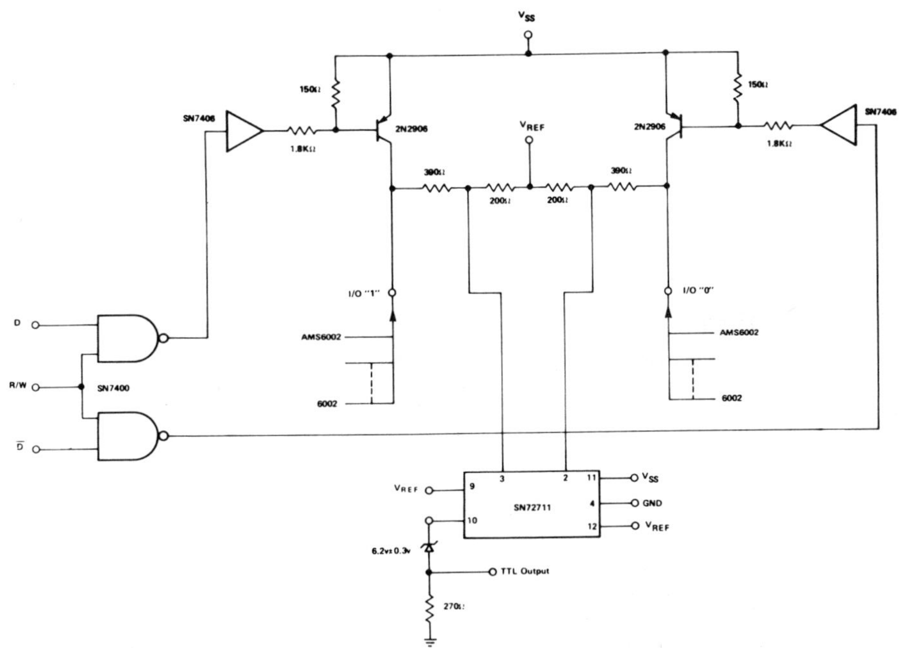

Figure 7 AMS 6002 BIT DRIVER/SENSE AMPLIFIER WITH PNP DRIVER

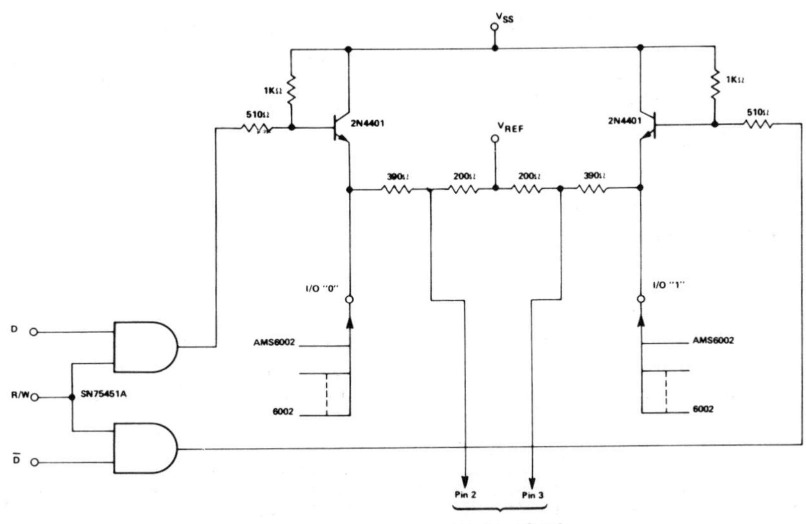

Figure 8 AMS 6002 BIT DRIVER, EMITTER FOLLOWER TYPE

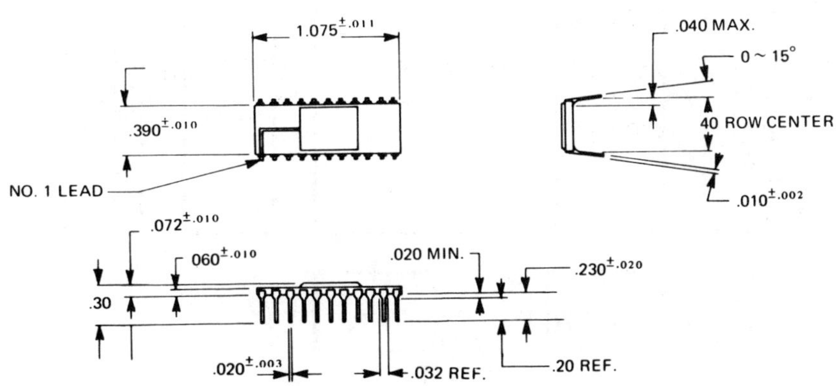

40 ROW CENTER 22 LEAD DUAL IN-LINE CERAMIC PAK
ORDER No. 91600211–6D

A 4096-Bit Dynamic MOS RAM

J. A. Karp, W. M. Regitz and S. Chou

Intel Corp.

Santa Clara, Cal.

AN MOS SEMICONDUCTOR MEMORY ARRAY, designed for large main memory, will be described. High density has been achieved with a unique cell configuration, in combination with N-channel silicon-gate technology, while still maintaining manufacturing tolerances consistent with high volume and low cost production.

A block diagram of the memory array is shown in Figure 1. The 12-bit address buffer register accepts TTL inputs; 2.4-V minimum. The address is then decoded into 1 of 64 rows and 1 of 64 columns, thus selecting one of the 4096 cells contained in a 64 x 64 matrix. The decoders use 6 input dynamic NOR gates which accomplish high speed address decoding at minimum power dissipation. The address buffers, which also utilize dynamic circuitry, not only convert the TTL levels to MOS levels, but also serve as a register, thus requiring stable address for only 100 ns; Figure 2.

The 4096-bit memory uses a single high-level clock (CE) from which all internal timing signals are triggered. The precharge clock input, normally used in dynamic RAMs to precondition all internal nodes, has been eliminated. The function of this clock is performed by the $\overline{CE}$ driver; thus all dynamic nodes are preconditioned automatically between active memory cycles. The additional internal timing pulses in Figure 2 (R, $\overline{R}$ and Row Voltage) are representative of the types of signals that have been generated on the chip.

The cell has three minimum geometry transistors and occupies an area of 1.8 sq. mils. This small area per bit is the

Memory array characteristics
4096 bits: 4096 words by 1 bit
TTL compatible inputs; except clock
Single clock input; high level
Cell size < 2 mil^2; 3 transistors
Access time < 300 ns
Cycle time < 500 ns
Active power < 100 μW/bit
Power supplies: +12 V, +5 V, -5 V
Standby power < 1 μW/bit
Standard 22-pin package

Memory array characteristics.

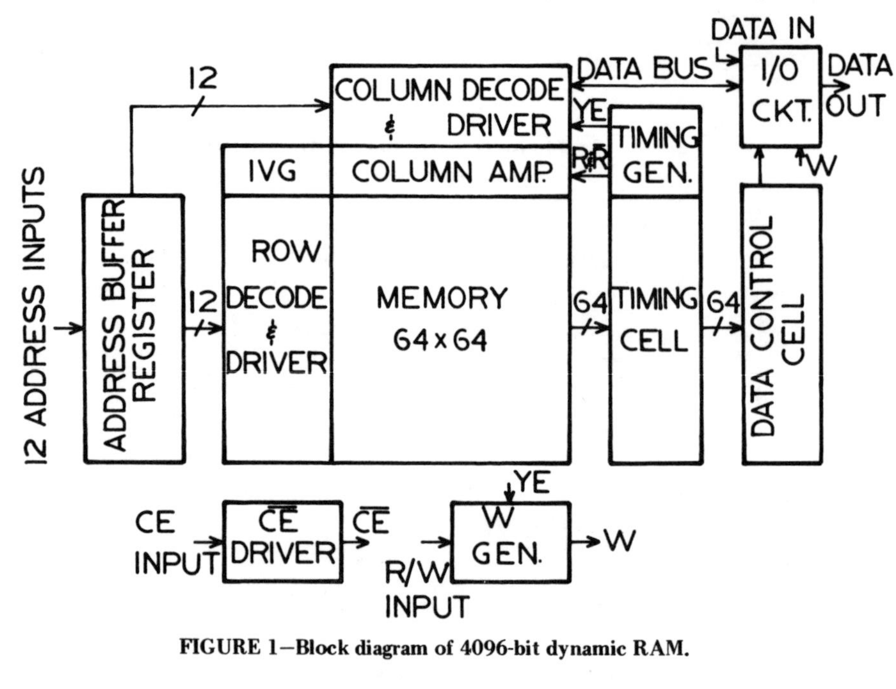

FIGURE 1—Block diagram of 4096-bit dynamic RAM.

Reprinted from *1972 ISSCC Digest*, pp. 10–11.

result of the cell configuration (Figure 3) which needs only 2-1/2 interconnect lines per cell as follows:

A)–A single selection line [1] activates a row of cells for both reading and writing. The Row Voltage is a tri-state signal gated on to the Row Select line by the Row Decoder. Reading of the cell is accomplished during the intermediate level of this signal and writing or refreshing the cell is accomplished during the high level of this signal. The high level is terminated as soon as proper information is gated to the cell in order to conserve power.

B)–A single column line [2] allows transfer of information into or out of the cell. This line is connected to the column amplifier which is used to sense the information in the cell, or write new information into the cell.

C)–A ground line is shared between two adjacent columns of cells, resulting in 1/2 connection per cell.

Since the information is stored dynamically in the cell, it must be refreshed periodically. To refresh the 4096-bit array, it is only necessary to perform memory cycles at each of the 64 row addresses.

The memory can perform read, write or read-modify-write cycles controlled by a single TTL level input signal.

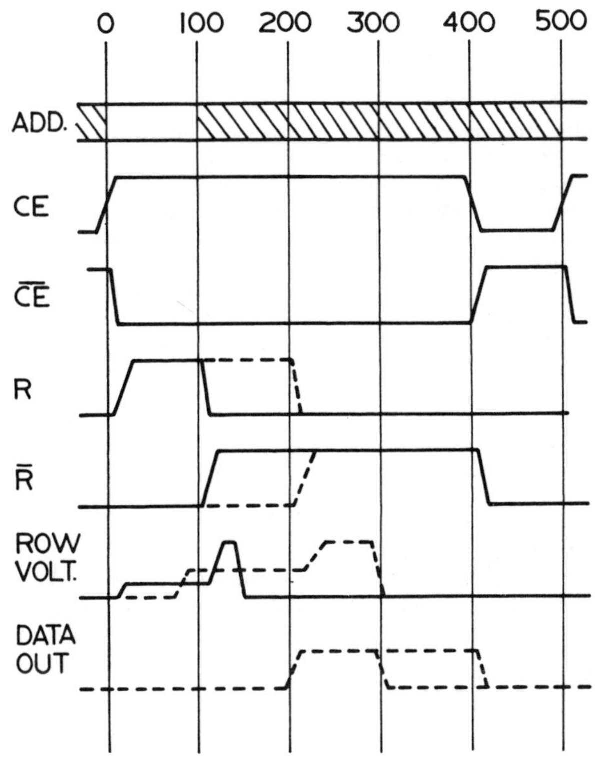

FIGURE 2–Timing diagram showing inputs (ADD, CE) and internally-generated signals ($\overline{CE}$, R, $\overline{R}$, ROW VOLTAGE). Dashed lined shows worst case as predicted by computer analysis.

[1] Regitz, W. M., Karp, J. A., "Three-Transistor Cell 1024-Bit 500-ns MOS RAM", *IEEE Journal of Solid-State Circuits,* Vol. SC-5, No. 5; Oct., 1970.

[2] "Field-Effect Transistor Capacitor Storage Cell," United States Patent 3,585,613.

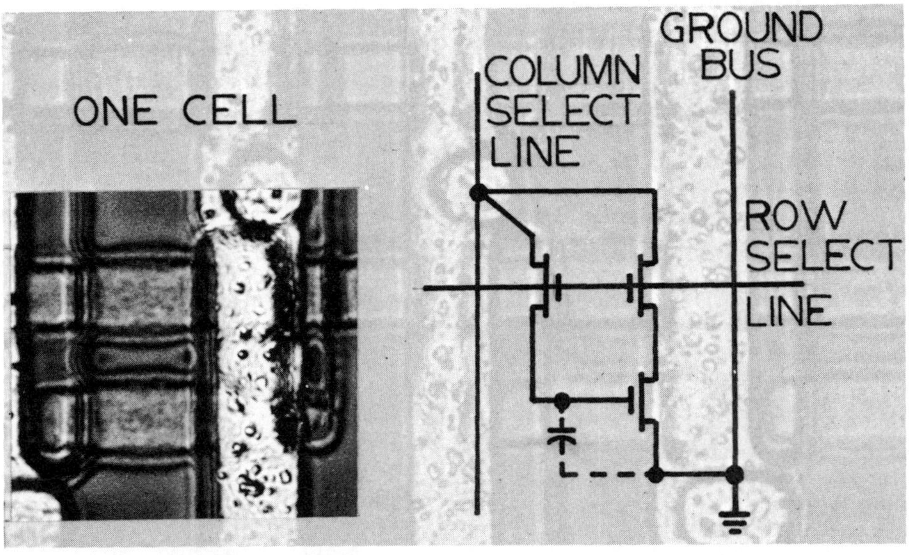

FIGURE 3–Microphotograph and schematic of 3-transistor cell.

A Fully Decoded 2048-Bit Electrically Programmable FAMOS Read-Only Memory

DOV FROHMAN-BENTCHKOWSKY, MEMBER, IEEE

***Abstract*—This paper describes a fully decoded 2048-bit electrically programmable read-only memory implemented with a novel floating-gate avalanche-injection MOS (FAMOS) charge-storage device as the basic nonvolatile memory element. The memory is organized as 256 words of 8 bits, it is fully TTL compatible, and can be operated in both the static or dynamic mode. The memory array was successfully fabricated with silicon gate MOS technology yielding functional devices with access times of 800 ns in the static mode and 500 ns in the dynamic mode of operation. The memory chip is assembled in a 24-lead dual-in-line package.**

SEMICONDUCTOR read-only memories (ROM) are presently implemented in a variety of digital system and computer applications. Most available semiconductor ROMs are programmed permanently at the integrated-circuit fabrication stage by a custom mask that defines the desired information pattern. As a result, program changes in microprogramming applications as well as pattern changes during the debugging phase of digital systems involve the generation of a new mechanical mask for every modified ROM pattern. In addition to being an expensive step, it also limits the flexibility of ROM applications because of the delay involved in the production process. These limitations of mask programmable ROMs have led to a growing interest in electrically programmable semiconductor ROMs in which the permanent information pattern is recorded by application of an electrical signal. This allows programming changes to be affected without the expense and time delay involved in generation of a custom mask.

The different proposed electrically programmable ROMs can be divided into two main categories: 1) ROMs in which a permanent (irreversible) change in the memory metal interconnection pattern is affected by an electrical pulse; and 2) alterable ROMs in which a reversible change in active memory device characteristics is induced electrically.

The first category includes fusible-type ROMs, which are mainly bipolar memories with capacities of up to 512 memory bits. Their main disadvantage is that they cannot be reprogrammed, which prevents complete functional testing before shipment, as well as pattern modification in case of programming error or needed change.

The search for alterable (charge storage) ROMs stems from the need for low-cost fully tested field-programmable ROMs as well as an attempt to provide a substitute for the nonvolatile storage capability (retention of stored information without an external power source) of magnetic memories. Most proposed alterable ROMs rely on charge storage in a dielectric that forms part of the gate of an insulated-gate field-effect transistor. Feasibility has been demonstrated for a metal-nitride-oxide-silicon (MNOS) memory [1], [2] a metal-aluminum-oxide-silicon (MAS) memory [3], and a dual-gate MNOS memory [4]. Difficulties in controlling the electrical characteristics of the storage dielectrics and additional fabrication steps required to achieve on-the-chip decoding have limited the realization of these approaches to undecoded memory arrays of up to 256 memory bits.

Recently, feasibility of the ovonic amorphous semiconductor memory device has been demonstrated by fabrication of an undecoded 256-bit memory array [5].

The introduction of a novel MOS memory element—the floating-gate avalanche-injection MOS (FAMOS) charge-storage device—has led to the fabrication of a fully decoded 2048-bit electrically programmable ROM. The memory is organized as 256 words of 8 bits, it is fully TTL compatible and can be operated in both the static or dynamic decoding and sensing mode. The monolithic memory array was successfully fabricated with silicon-gate MOS technology, yielding functional devices with access times below 800 ns in the static mode and less than 500 ns in the dynamic mode of operation. The memory chip is assembled in a 24-lead dual-in-line package.

I. Device Structure and Operation

A cross section of the FAMOS structure is shown in Fig. 1 with its suggested electrical symbol. It is essentially a p-channel silicon-gate MOS field-effect transistor [6] in which no electrical contact is made to the silicon gate. The floating polysilicon gate is isolated from the silicon substrate by a SiO_2 layer of approximately 1000 Å and from the top surface by 1.0 μ of vapor-deposited oxide. Operation of the FAMOS memory structure depends on charge transport to the floating gate by avalanche injection of electrons from either the source or drain p-n junctions. A junction voltage in excess of -30 V applied to a p-channel FAMOS device (Fig. 2) with 1000 Å of oxide and 5–8-$\Omega \cdot$cm substrate resistivity will result in the onset of injection of high-energy electrons from the p-n junction surface avalanche region to the floating silicon gate. The gate charging current is of the

Manuscript received March 16, 1971. This paper was presented at the ISSCC, Philadelphia, Pa., February 1971.

The author is with the Intel Corporation, Mountain View, Calif.

Reprinted from *IEEE J. Solid-State Circuits,* vol. SC-6, pp. 301–306, Oct. 1971.

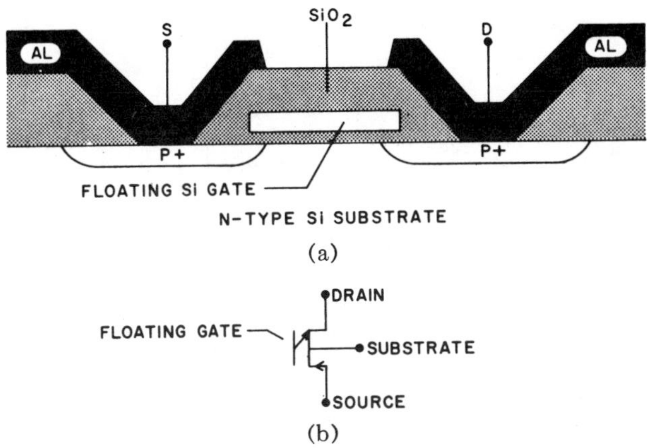

Fig. 1. (a) Cross section of FAMOS structure. (b) Suggested electrical symbol.

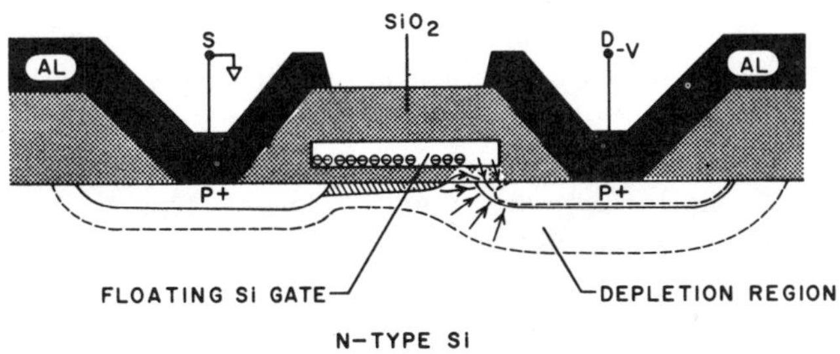

Fig. 2. Cross section of FAMOS device under bias.

order of 10^{-7} A/cm². Since the silicon gate is floating the electron current through the oxide results in the accumulation of a negative charge on the gate. For a p-channel FAMOS transistor this negative charge will induce a conductive inversion layer connecting source and drain. The amount of charge transferred to the floating gate as a function of the amplitude and duration of the applied junction voltage is shown in Fig. 3. The presence or absence of charge can be sensed by measuring the conductance between the source and drain regions.

Once the applied junction voltage is removed, no discharge path is available for the accumulated electrons since the gate is surrounded by thermal oxide, which is a very low conductivity dielectric. The electric field in the structure after the removal of junction voltage is due only to the accumulated electron charge and is not sufficient to cause charge transport across the polysilicon-thermal-oxide energy barrier. The maximum stored charge of 4×10^{12} electrons/cm² ($V_W = 50$ V, Fig. 3) results in an electric field of approximately 2×10^6 V/cm across the thermal oxide. Assuming current transport by Fowler–Nordheim emission from the polysilicon gate into the oxide, the estimated discharge current (for a polysilicon–SiO_2 energy barrier of at least 3.2 eV) is of the order of 10^{-40} A/cm² at 300°K.

Charge decay plots as a function of time at 125° and 300°C are shown in Fig. 4. The rapid initial decay saturating with time cannot be explained by electron transport across the oxide. Its temperature and electric-field dependence correspond to that of positive-charge buildup at the SiO_2–Si interface that occurs at high electric fields

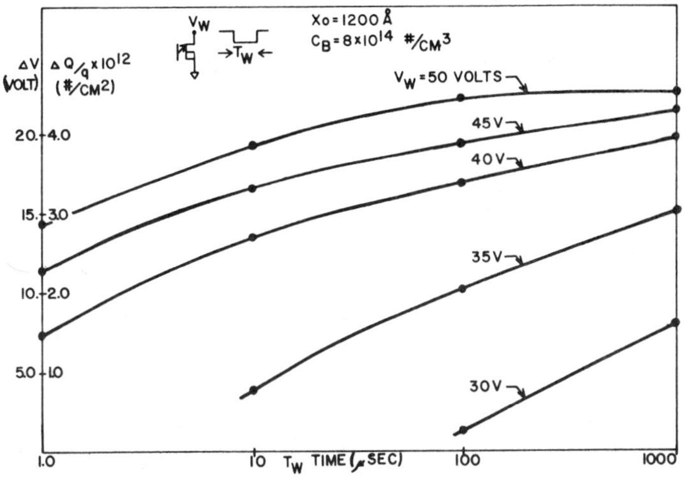

Fig. 3. Charge accumulation on the floating gate as a function of charging pulsewidth for different values of pulse amplitude.

under negative gate bias conditions [7], [8]. The activation energy of charge decay was measured to be 1.0 eV. An extrapolation of the 300°C charge decay results indicates that 70 percent of the initial induced charge can be retained for as long as 10 years at 125°C.

Since the gate electrode is not electrically accessible, the charge cannot be removed by an electrical pulse. However, the initial condition of no electronic charge on the gate can be restored by two nonelectrical methods. Illumination of the unpackaged device with ultraviolet light will result in the flow of a photocurrent from the floating gate back to the silicon substrate thereby discharging the gate to its initial condition. This erase method allows complete testing of a complex memory array before the package goes through final seal. Once the package is sealed, information can be erased by exposure to X-ray radiation. The radiation dose required is in excess of 5×10^4 rad, which is many orders of magnitude higher than the average yearly atmospheric radiation dose and easily attainable with available commercial X-ray generators.

II. Memory Cell

The FAMOS charge-storage device described above can be used as the basic storage element in a large memory array. A circuit schematic of the memory cell and its associated decoding circuitry is shown in Fig. 5. The decode and sense circuits shown are common to both the PROGRAM and READ modes. To select a bit to be programed the address-decode inputs as well as the V_{DD} and V_{GG} lines are energized to −50 V. Programming of a memory bit is accomplished by coincidence selection of the X and Y select lines. The applied programming pulse is transferred to the selected FAMOS device that turns normally on due to the electron charge transferred to its floating gate. All other memory bits are not programmed due to either the lack of a pulse on the Y select line or the absence of a transfer pulse on the X select line. The programming signal V_P is a −50-V 5.0-ms pulse. The amount of charge stored in a memory cell in re-

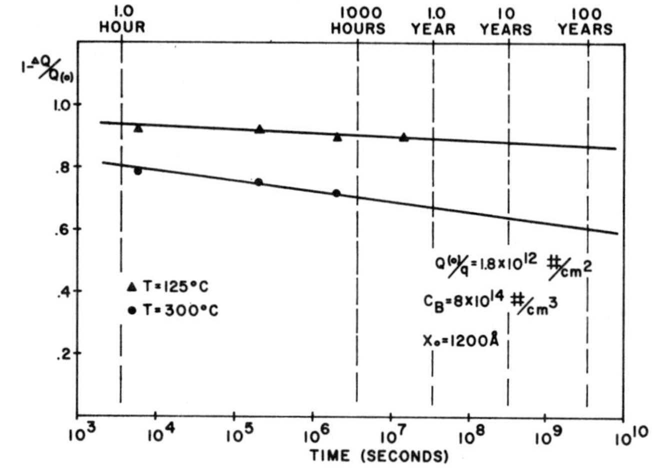

Fig. 4. Charge decay in a FAMOS device as a function of time at 125° and 300°C.

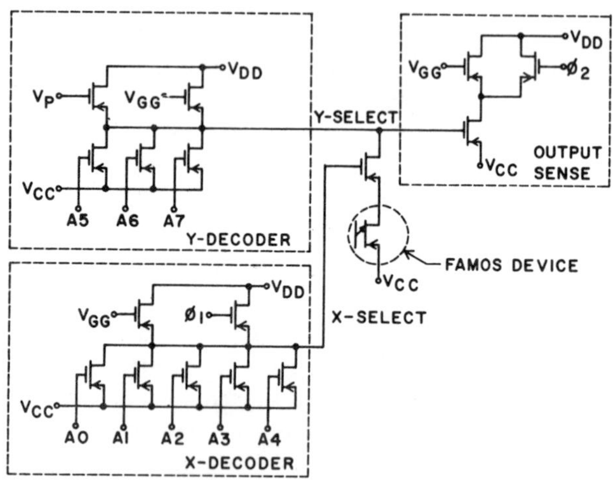

Fig. 5. Memory cell with its associated decode and sense circuits.

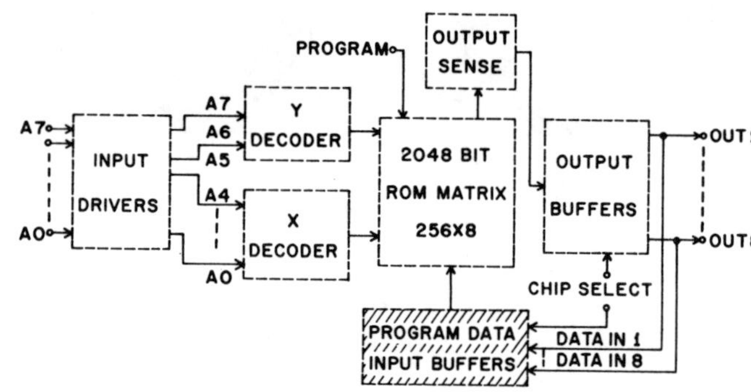

Fig. 6. Block diagram of memory organization.

sponse to a programming pulse is typically 3.0 × 10^{-7} C/cm², which is equivalent to 10 V on the gate of a conventional MOS transistor. The load current required to program a memory bit is approximately 5.0 mA. The increase in programming pulse amplitude and duration to charge the memory cell compared to the data presented in Fig. 3 is due to the additional voltage drop across the decode circuits. These high applied voltages in the PROGRAM mode could give rise to parasitic programming paths due to field inversions over the thick oxide regions on the chip. This potential problem was accounted for in both the circuit design and chip layout by providing a highly conductive programming path as compared to the parasitic path as well as careful routing of high-voltage lines on the chip to avoid sensitive circuit nodes.

Memory-cell operation in the READ mode is similar to that of other ROMs. A memory bit is selected by a coincidence of signals on the X and Y select lines. However, compared to the programming mode the READ mode voltages are substantially lower. Since the programming threshold is −30.0 V, the maximum READ voltage of −15 V guarantees a wide margin to avoid disturbing the information during the READ mode. Information in the selected memory cell is sampled by the output sense circuit. If a "0" is stored in the cell (charge on the floating gate) the FAMOS device is ON and the level at the input of the sense circuit is close to V_{CC}. A "1" corresponding to no charge stored in the cell is reflected by a more negative level (close to V_{DD}) at the input of the sense circuit. Information can be decoded and sensed in either the static (no clocks) or dynamic mode (2 clocks). The static mode of operation eliminates the need for clocks at the expense of increased power dissipation and reduced speed, while the dynamic mode offers advantages in both performance categories. The option of the two modes of operation is achieved by parallel load transistors in the decode and sense circuitry, one connected to the clock lines and the other to V_{GG} as shown in Fig. 5. The load device connected to the V_P terminal is connected to ϕ_1 in the dynamic READ mode. Mode selection is done by activating the clocks with V_{GG} connected to V_{CC} in the dynamic mode. V_{GG} is activated in the static mode with the clocks connected to V_{CC}.

III. Memory Organization and Operation

The electrically programmable memory chip consists of a monolithic array of 2048 FAMOS devices organized as 256 words of 8 bits. A block diagram of memory organization is shown in Fig. 6. All circuit blocks are common to both the PROGRAM and READ modes with the exception of the PROGRAM data input buffers. In the PROGRAM mode the eight output terminals are used as data inputs to determine the information pattern in the eight bits of each word, while word address selection is performed by the X and Y decoders through the input drivers. The PROGRAM data input circuitry for one of eight outputs is shown in Fig. 7. To inhibit the programming of a bit, a negative voltage is applied to the data input terminal. This voltage level is transferred to the inhibit transistor (chip select is enabled), which turns on and overrides the Y select signal. To allow programming of a selected bit, the data input terminal is kept at ground. Initially all 2048 bits are in the "1" state corresponding to normally OFF FAMOS devices (no charge on the floating gate). Information is introduced by selectively programming "0" in the proper bit locations through charging the FAMOS devices from the PROGRAM terminal. The supply (V_{DD}, V_{GG}, V_P), address ($A_0 - A_8$) and data input (D_{IN1}

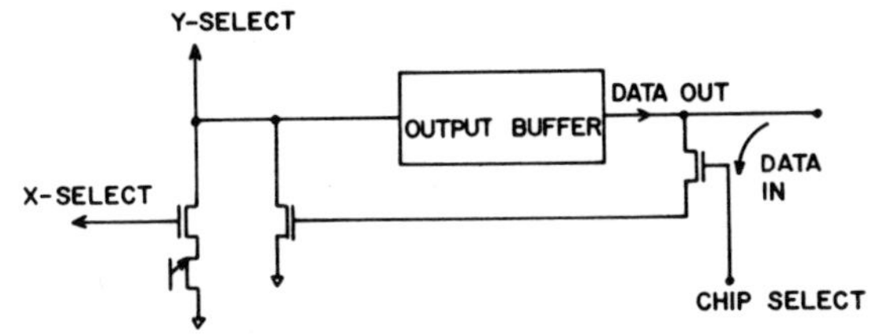

Fig. 7. Circuit description of PROGRAM data input buffers.

APPLIED VOLTAGE (VOLTS) / MODE	V_{DD}	V_{GG}	V_P	ϕ_1	ϕ_2	V_{CC}	ADDRESSES / DATA IN
PROGRAM	-50.0	-50.0	-50.0	V_{CC}	V_{CC}	0.0	-40.0
STATIC READ	-9.0	-9.0	V_{CC}	V_{CC}	V_{CC}	+5.0	TTL
DYNAMIC READ	-9.0	V_{CC}	-9.0	-9.0	-9.0	+5.0	TTL

Fig. 8. Applied voltage level to the memory in the PROGRAM and READ modes of operation.

$- D_{IN8}$) voltages in the PROGRAM mode are detailed in Fig. 8. A timing diagram for the applied voltage sequence in the PROGRAM mode is shown in Fig. 9. In the READ mode the PROGRAM data input buffers are inhibited by the chip select signal to cut off the feedback path from the output to the memory array established in the PROGRAM mode. Memory operation in the READ mode is the same as in conventional mask programmable ROMs. The input and output buffers provide for full TTL compatibility and addressing is accomplished by the X and Y decoders operating at the voltage levels detailed in Fig. 8. A selected memory cell with a charged FAMOS device will be reflected by a low "0" TTL level at the output, while a memory bit that is not charged will result in a high "1" TTL level. Both the static and dynamic modes of decoding and sensing are available in a single package through use of parallel load transistors as described in Section II. Mode selection is done by activation of either the clock lines (ϕ_1, ϕ_2) or the V_{GG} terminal in the dynamic and static modes, respectively. Typical access times are 400 ns in the dynamic mode and 700 ns in the static mode. A photomicrograph of the chip with the designation of the different circuit blocks is shown in Fig. 10.

As can be seen from the above description, electrical programming of the memory is conceptually the same as operation in the READ mode with the exception of the voltage levels (Fig. 8). Hence, the memory can be easily programmed from punched paper tape or other data input devices through an electrical programming terminal. Once programmed, the information pattern in the memory can be erased (restored to the all "1" state) by ultraviolet light before packaging or by exposure to X-ray radiation as described in Section I. The ability to erase before packaging allows for complete testing of the memory chip prior to shipment, which is a distinct advantage over once-programmable ROMs. Erasure of information in packaged devices by placing them in a commercially available X-ray generator allows for correction of programming errors as well as unpredictable future pattern modifications.

IV. Reliability

One of the most important performance factors in an electrically programmable ROM (as in all other products) is its long-term reliability. Reliability in this case incorporates both the long-term functionality of the standard MOS transistors and the long-term information retention of the FAMOS devices. Reliability data on silicon-gate MOS memory products have been accumulated over a period of two years, resulting in many device hours without failure. This record is directly applicable to the programmable ROM since it is fabricated with the same silicon-gate MOS technology. As to long-term information retention, the charge-storage programmable ROM has a predictable time-temperature dependence for stored information.

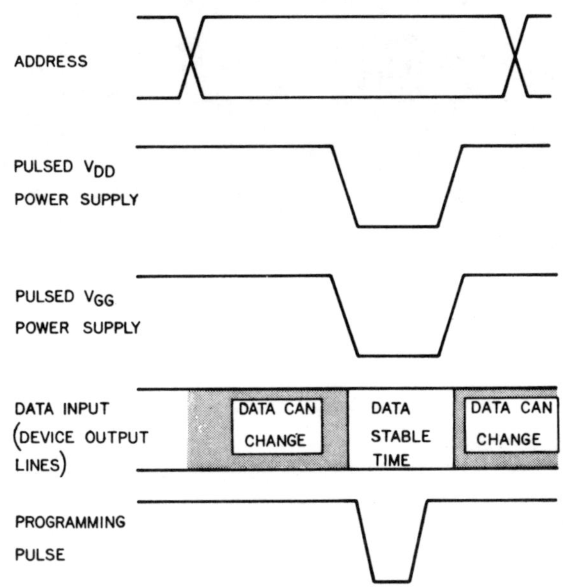

Fig. 9. Timing diagram for the PROGRAM mode.

The long-term decay (Fig. 3) has a logarithmic dependence on time with a slope of approximately 1.0 V per 5 decades of storage time at 125°C. This rate of charge decay extrapolates to storage retention times greater than 10 years at 125°C. The predictable time-temperature behavior of stored charge corresponding to an activation energy of 1.0 eV, allows for accelerated life testing of the programmable ROM. To guarantee 10-years retention at 125°C, the device has to be subjected to 300°C storage for only 10 h. This procedure is similar to existing high-temperature stress testing in MOS products to detect potential contamination in the dielectric.

A summary of reliability results on the 2048-bit electrically programmable ROM is shown in Fig. 11. The extrapolation of time from 125° and 200°C to the operating temperature of 85°C is based on an activation energy for charge decay of 1.0 eV. The bias configuration in the temperature/bias test corresponds to worst case operating voltages in the READ mode. No failures have been

Fig. 10. Photomicrograph of the memory chip with designation of the different circuit blocks.

1601 RELIABILITY RESULTS

TEST	T_A	UNITS	HOURS	FAILURES	EQ. TIME AT 85°C
TEMP/BIAS	125°C	15	1500	0	5 YEARS
STORAGE	125°C	10	1500	0	5 YEARS
STORAGE	200°C	10	250	0	100 YEARS

1.0 HOUR AT 125°C = 35 HOURS AT 85°C

Fig. 11. Reliability results for storage retention of the programmable ROM.

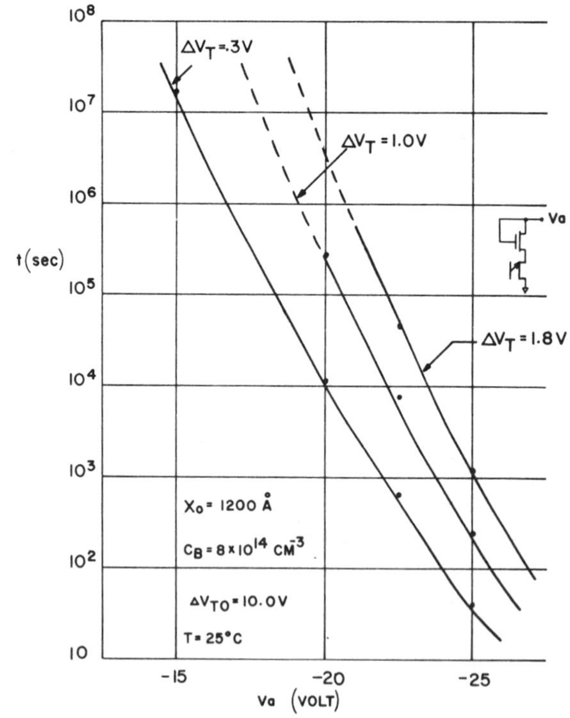

Fig. 12. Retention of the uncharged memory state as a function of applied voltage to the memory cell.

observed in any of the long-term information-retention tests.

An additional reliability factor is the retention of the uncharged "1" memory state. It was pointed out in Section III that operation in the READ mode differs from the PROGRAM mode only in the lowering of the operating voltages from -50 to -14 V. This brings up the question of whether an uncharged memory bit can be slowly charged by repeated selection in the READ mode. Retention of the uncharged memory state as measured on a typical memory cell is shown in Fig. 12. The time for a shift in FAMOS device turn-on voltage as a function of applied voltage to the memory cell is plotted for different values of turnon voltage shift. For an operating voltage in the READ mode of -14 V, a shift of $V_T = 1.0$ V is expected to take place after more than 100 years. A shift of at least 5.0 V is required before the uncharged memory state "1" is interpreted as a "0" by the sense circuitry. Another important aspect of reliability is the extent of preshipment testing. As pointed out in previous sections, because of its reprogrammable feature, the memory array can be completely tested before shipment for functionality and programming, as well as an accelerated high-temperature test to guarantee long-term retention. After the tests are performed, the information pattern is erased and the programmable ROM is shipped with a pattern of all "1" ready for programming.

V. Applications

Despite the attempts of semiconductor manufacturers to lower mask charges and turn-around time for mask-programmable ROMs, the economics and design delay involved have limited the versatility of semiconductor ROM applications. The availability of a low-cost electrically field-programmable ROM opens new application areas, which heretofore have not been economically feasible, mainly those in which relatively small quantities of different ROM patterns are required. For example, the implementation of computer lookup tables that provide a hard-wired means of performing certain routine computer calculations. The variety of possible patterns makes the use of mask-programmable ROMs costly and impractical, while ideally suited for field-programmable ROMs. Other application areas that fall into the same category are binary-sequence generators, microprogramming control of central processors in computer terminals, and programmed logic arrays.

On the other hand, in areas such as code converters, character generators, and other applications in which large quantities of a few patterns are required the field programmable ROM augments the capability of mask programmable ROMs. In these applications, the initial custom mask cost is insignificant compared to the total cost of the high-volume standard ROM pattern. However, in order to finalize a decision on a given high-volume standard pattern a method of debugging is needed to avoid the costly phase of correcting potential errors. In this initial pattern definition phase the electrically programmable ROM is most economical and flexible. Since the flexibility of field-programmable ROMs is generally achieved at the expense of chip area, once the pattern has been defined, the high-volume order is placed for mask-programmable ROMs, which are lower cost in high quantities. Hence the combination of the two programming methods offers both the flexibility and cost incentive required to proliferate the use of semiconductor read-only memories in high-volume applications.

VI. Summary

The introduction of a novel semiconductor memory element—the FAMOS charge storage device—and its implementation in a fully decoded 2048-bit electrically programmable read-only memory constitutes a significant advance in the state of the art of semiconductor memories. It is the first available large-capacity programmable ROM in which the information pattern is recorded electrically by way of a reversible change in memory-device characteristics. This was achieved by a combination of a new charge-storage structure and circuit techniques that allow its implementation in large-scale memory arrays with existing MOS processing techniques.

Acknowledgment

The author wishes to thank L. L. Vadasz, A. S. Grove, and G. E. Moore for many helpful discussions, G. Pasco for fabrication of the memory devices, and G. Greenwood for his help in testing and instrumentation.

References

[1] H. A. Wegener, "MNOS memories," *Digest Intermag. Conf.*, Apr. 1970.
[2] D. Frohman-Bentchkowsky, "An integrated metal-nitride-oxide-silicon (MNOS) memory," *Proc. IEEE* (Lett.), vol. 57, June 1969, pp. 1190–1192.
[3] S. Nakanuma *et al.*, "A read-only memory using MAS transistors," *ISSCC Digest Tech. Papers*, Feb. 1970, pp. 68–69.
[4] H. G. Dill and T. N. Toombs, "A new MNOS charge storage effect," *Solid-State Electron.* vol. 12, 1969, pp. 981–987.
[5] R. G. Neale, D. L. Nelson, and G. E. Moore, "Amorphous semiconductors (Part I)," *Electronics*, Sept. 1970.
[6] L. L. Vadasz, A. S. Grove, T. A. Rowe, and G. E. Moore, "Silicon-gate technology," *IEEE Spectrum*, vol. 6, 1969, pp. 28–35.
[7] B. E. Deal, M. Sklar, A. S. Grove, and E. H. Snow, *J. Electrochem. Soc., Solid-State Sci.*, vol. 114, 1967, p. 266.
[8] S. R. Hofstein, *Solid-State Electron.*, vol. 10, 1967, p. 657.

Part IV
Interconnections, Packaging, and Economics

Editor's Introductory Comments

Integrated circuit packaging and interconnection techniques have a major impact on the cost and reliability of the resulting components and systems. Part IV, 1 and 2 provide an excellent discussion of the relative merits of alternative packaging and interconnection methods. Very likely the dual inline package will be widely used for some time to come, despite its drawbacks. Beam lead and flip chip (solder reflow) techniques are not being adopted rapidly outside of Bell Laboratories and IBM due to the major capital and engineering expenses required of both makers and users of components using these techniques.

Specific information on manufacturing costs and yields is understandably hard to obtain from integrated circuit manufacturers. The Editor's estimates of current values for these parameters are provided in Part IV,3. These figures have not been supplied or endorsed by any manufacturer and thus must be viewed with caution. Even if they are substantially in error, however, they show that the major part of the cost of new memory components must be engineering and marketing costs.

A Critique of Chip-Joining Techniques

L. F. Miller
IBM Components Division
East Fishkill Facility
Hopewell Junction, N.Y.

Six particularly interesting chip joining methods are intensively analyzed and compared to show their special advantages and disadvantages. The methods were chosen both to represent the leading industrial contenders and to provide insight into still other techniques, and include: wire bonding, flip-chips with molten and nonmolten joints, spider bonds and lead frames, chips back-joined to the substrate and connected by vacuum metallizing, and beam leads. The major topics discussed in detail are: the criteria for judgment, key characteristics, main advantages and disadvantages of each method, comparison of reliability, comparison of process details and design flexibility.

Introduction

This critique is a companion to a previous survey paper which discussed a wide variety of chip-joining methods described in the literature.[1] This article continues by providing critical analyses of some of those methods.

Just as a survey has generic limitations and is subject to dispute, this comparison of methods also has inherent shortcomings and pitfalls. Controlled experiments (which are difficult to design) are normally required to make valid and meaningful comparisons. Yet, here we find that intuitive leaps and combinations of inductive and deductive reasoning must be used, due to the lack of such controlled and organized data. Each manufacturer or experimenter typically emphasizes a single format or process; it is quite rare that statistically valid comparisons are made of several methods in order to make a choice and if such have been done the literature apparently has not been informed. This lack of direct comparison is also due to the complexity of making statistically valid comparisons on processes as complex as those being discussed here; each minor perturbation may be quite meaningful with respect to cost, reliability, practicality, etc. Furthermore, although certain specifications do exist for aspects of stress testing and reliability, there is truly no universal standard to which electronic structures must be compared. Thus, the criteria used here must be somewhat arbitrary and subjective. However, as with the survey, even with these barriers to logical purity and statistical validity it is nevertheless considered worthwhile to present such a critical analysis—if for no other reason than to excite discussion and prod the proper experiments into existence, which will prove or disprove individual aspects of contention.

The previous paper stressed several points which are still considered valid: the constraints of other module fabrication parameters often dictate which chip-joining scheme is optimum for a particular construction; no single chip-joining scheme can be considered optimum for every set of parameters; furthermore, even with a given module type, such factors as the existence of certain kinds of tooling or quantities of units involved are pertinent in the choice of a chip-joining method for any particular organization. Care must be taken to recognize which postulations are not based on data, or which existing data do not really pertain to a slightly different system.

It is recognized that other factors beyond technical superiority are sometimes used by manufacturers in choosing production schemes. Government contracts, for example, will sometimes insist on a particular scheme, and once the tooling has been established for a contract and the experience attained, there is a tendency for that procedure to be continued, due to inertia and cost factors. Device or component availability is often pertinent in today's market; the smaller manufacturer generally adapts his production techniques to the components he can buy. This probably has made experimentation with flip-chips particularly difficult, since devices with pads have been difficult to obtain over the last several years. There are also psychological factors involved, whereby a solid tech-

Reprinted with permission from *Solid State Technology*, pp. 50–62, Apr. 1970.

nical image must be presented to customers, and admissions of non-optimum technical directions are anathema. This can have the effect of elongating the usage of obsoleted technologies.

Because of the enormous number of potentially usable chip-joining schemes, this discussion is restricted to a group of six methods which were chosen for two main reasons: they have been used in commercial production operations and are in the public eye, and they represent in a microcosm a wide gamut of processing, formats, and materials. It is believed that these six schemes provide a sound tutorial platform and permit this discussion to stand as a yardstick for other schemes not described. Where the literature is deficient. such designations as "probably", "usually", "should", etc. will be used to distinguish such conclusions from those backed by data or direct knowledge.

The six main schemes are shown in Fig. 1. They include two back-joined configurations (wire-bonding[2] and imbedded devices[3]), three flip-chip variations (controlled-collapse,[4] non-molten pads,[5] and spider-bonding,[6]), and beam-leads.[7] Although it is an oversimplification, non-molten pads represent any flip-chip configuration which involves a relatively rigid terminal pad, such as thermo-compression bonds or ultrasonic bonds. The flipped beam-lead gold configuration is chosen because it seems to be the leading contender at this point. Spider-bonding is chosen to represent the many possibilities of lead frame uses. This tutorial article will be most clear with a thorough understanding of references 1–7, and the reader is encouraged to refer to them.

The definitions and descriptions of the criteria which will be used for assessing these six typical types of chip-joining schemes are discussed. The schemes are compared, mainly by comparative charts, supplemented by discussion wherever necessary.

Table I—Some Typical Criteria.

Guide Rules

Multi-chip fabrication requires inexpensive land repair technique
Spring contacts are not adequate
Keep stresses and masses small (thin leads)
Match thermal expansions of materials as much as possible
No entrapped contaminants
Metal resistances >5 Micro-ohm-cm probably acceptable
Au:Al joints not acceptable if high temperatures are involved

Necessities

Meets process temperature limit defined by device
Parts are not fragile, can be handled
For multilayer ceramic, replaceable chips; second or third placement has same metallurgy and reliability—surrounding devices not affected
Automatable, low tool cost
No predictable metallurgical failure mechanisms
At least one flexible joint to device (unless surrounded or mechanically supported)
Minimum pad stresses

Desirable

Simplest device alignment—avoid buildup of alignment tolerances
Cheapest lead type (screened > punched > etched > vacuum?)
At least 40°C temperature difference between processes
Demonstrated high reliability
Low cost
Applicable to chip carriers and multichip constructions
Minimum connections, but not at sacrifice of reliability
Fewer parts—but not if few parts are extremely expensive
Permits backside joint to thermal conductor

Minor

Permits prior in-situ device test before committment to multichip surface
Can permit minor design changes
Cost does not directly increase with number or size of leads (yield not included)
Permits 0.002″ lands

Criteria

The criteria for comparing methods can be grossly considered in three main categories: Cost, Reliability, and Throughput-Yield-Practicality. As these main categories are subdivided, it becomes apparent that some criteria are more important than others, and that this degree of importance may vary according to the nature of the product. For example, some products can tolerate a failure every now and again, while other highly critical devices can afford to pay almost any price for guaranteed non-interrupted service. The trade-off of reliability *vs.* cost hinges on the ultimate cost and difficulty of making a repair. It is an understatement that there is considerable difference in designing for a space program compared to designing a television set. In order to have some basis for discussion, however, we will assume here that a system is desired which has reliability at least in the order of 0.001% failures per thousand hours per joint at a high confidence level, in normal human ambients.

The main topics for comparisons of methods include the following:

ECONOMICS,
DESIGN FLEXIBILITY,
RELIABILITY,
REPAIRABILITY, and
PROCESS and CONTROLS.

A typical list of criteria which can be subdivided into those which are absolutely necessary, those which are desirable, and those which would be nice but which are really minor considerations is presented in Table I.

Economic Criteria

Although exact costing is a complex undertaking, dependent on many elements other than direct costs, certain factors are virtually universal. For example, it is generally preferred that the cost of chip-joining should not change very much as the number of joints varies. This of course is an impossibility, since more joints often necessitate lower yields, but it is also apparent that some methods do not require changes in tooling for changes in formats, while others do.

Perhaps the main factor in integrated circuit cost is the efficiency of the usage of the semiconductor surface. Since devices are usually processed in wafer form, the cost per wafer is fixed, and the number of good devices which can

	Typical Example	Metallurgy: Device Metals	Metallurgy: Chip Pads	Metallurgy: Module Lands	Joining Process
1 Beam Lead	Bell Labs-Western Electric Beam Lead Sealed Junction	Si, PtSi, Ti, Pt, Au	Au	Cr or (Ti) Au	Thermocompression of Au Leads to Au Lands: a) Individually b) Wobble Table c) Heated Ram d) Compliant Bond
2 Molten Pad	IBM Corp Controlled Collapse Reflow	Si Al Cr-Cu- Au	Pb/Sn 95/5	Ag/Pd 80/20, Pb/Sn 90/10	Reflow solder pad chip to fluxed tinned module lands.
3 Non-Molten Pad	Honeywell Co. - Al ultrasonic Westinghouse Au T.C.	Si Al Si, Al	Al Tu or Cr or Mo, Au	Al Cr, Au	Ultrasonically bond Al pads to Al lands or Thermocompression bond Au pads to Au lands
4 Spider	Motorola Corp. Spider	Si Al	Al	Stamped Al	Ultrasonically Bond Device to spider, Spider to Lead Frame
5 Imbedded Chip	General Electric Co. S.T.D.	Si	Many Types	Cr Au, Cu, Au	Push chip into plastic, guided by plated Cu studs. Evaporate Au over surface, connecting chip to studs
6 Wire Bond		Si Au or Al	None	Many Types	Thermocompression Au or Al wires individually to Au or Al chip lands and to module lands or pins.

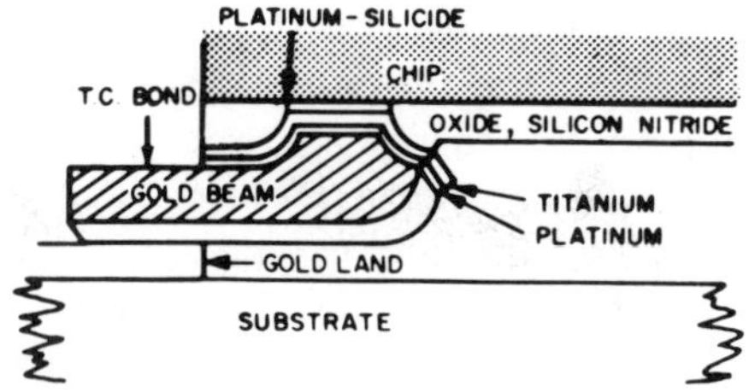

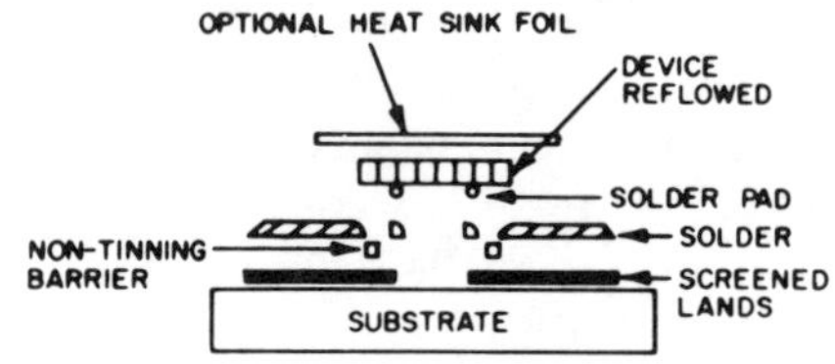

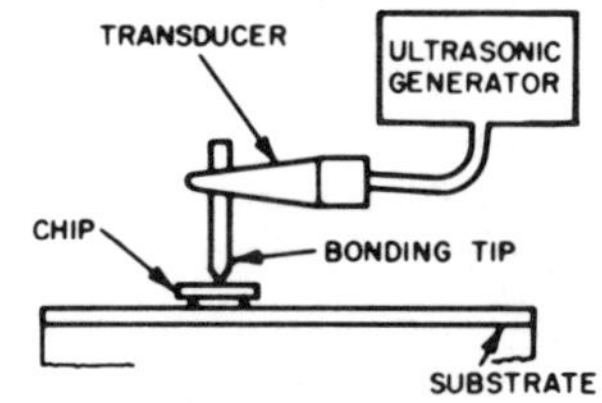

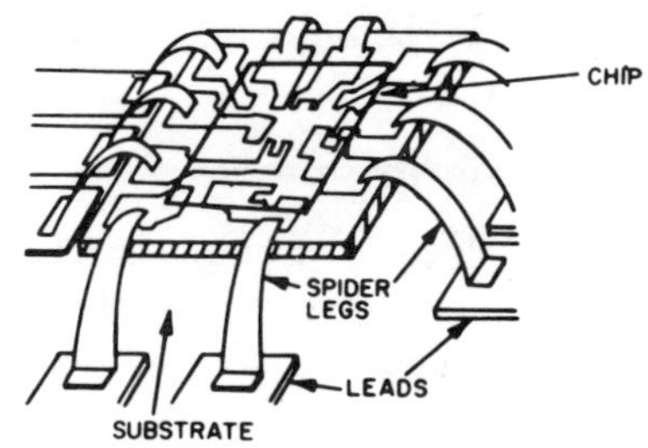

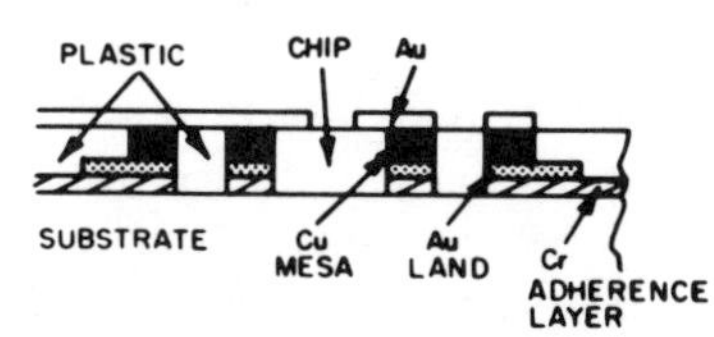

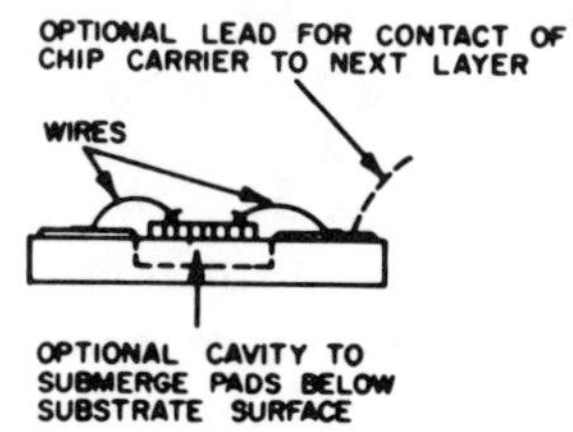

Fig. 1—Description of six chip joining techniques.

be attained from that wafer dominate the cost. It is astounding for normal sized wafers in the order of 1-2 inches in diameter how drastically a few mils change in the chip size affects the total number of good chips that can be obtained per wafer.

Yield is obviously an important consideration, but it is an extremely difficult aspect to assess without specific experience. Yield is dependent on such factors as competence, incentive, environment, and capitalization as well as the technology itself; it is sometimes difficult to attain equivalent yields from two separate facilities. Thus, yield will not be used as one of the criteria, although it is certainly important. Time spent at the chip-joining station is predictable, however. The chip-joining station is often a rate-controlling factor in the manufacturing process, and therefore, time there should be minimized. Another controllable factor is the expense of the materials used in the fabrication although the type of material is often less important than the time and effort spent in fabrication. Thus, although gold or platinum are considerably more expensive than aluminum or silver, there can be little difference in the cost of a single module based on using either the expensive or cheaper materials. Of course, for large-scale production, the sum of such small differences makes a vast difference in the total cost of a program, but it is the fabrication cost of the material that is most pertinent. A specially etched lead frame of cheap material may be far more costly than a screened deposit of precious metal. Vacuum metallized inexpensive refractory metal may take more time and thus be more costly than metallizing with a more precious metal which evaporates faster.

Some of the other obvious economic factors include the number of actual steps and the number of parts which must be fabricated, the ability for the tooling to be utilized as fully as possible (or at least readily adaptable to other uses), the degree of equipment maintenance if the equipment must be kept sharply tuned, etc.

Design Flexibility Criteria

This involves the amount of freedom each method has in fitting the many requirements of varied product lines—including geometrical changes—as well as electrical functionality and adaptability to various environments. The upper size limit for chips, for example, depends on the chip-joining scheme; joints which are either flexible or ductile can take up some of the stresses induced in thermal cycling, which rigidly fastened joints cannot stand. The amount of heat any device can dissipate is dependent on the number and size of the thermal paths leading from the heat producing areas.[8] Thus, thin or long leads are not much help, while thick sturdy joints are obviously much better for thermal dissipation. Some methods are restricted either to thin-film or thick-film techniques, while others can readily be adapted to either; this option is sometimes needed to permit changes as technologies advance. Some methods are more adaptable to a wide range of pad sizes and shapes and the ability to put few or many devices on a single module surface. Some of the methods are also more capable of providing pads in the center portions of the device, while others are restricted to peripheral leads. Because some joints are stronger, it is feasible to use a heat sink which adds weight and mass to the chip.

Electrical characteristics are important in the consideration of design flexibility. Device speed is dependent upon such factors as conductance, capacitance, spacings, inductance, and cross talk. Difficulty can be experienced with long parallel leads which slow down the circuit. Very often, the shortest electrical path with the highest conductivity and surrounding regions of low dielectric constant are preferred. Sometimes, therefore, the ability to make connections on very small centers becomes valuable.

There are differences in the adaptability for various uses and ambients since some of these methods permit the use of higher or lower temperatures than others. Certain packages are less susceptible to environmental factors such as humidity, radiation, etc. (Quite often of course, the rest of the package is just as important in these respects.)

Reliability Criteria

These criteria are described later in considerable detail, but a few comments are pertinent here. A proper assessment of reliability should relate not only to the theoretical properties of the materials and processes and geometries, but also to the realities of reproducibility and probable contaminations. It is also an aspect on which the least amount of comparative data is available. There are so many ways of evaluating reliability, so many statistical methods for interpreting the evaluation, and so little industry-wide agreement as to which of these are actually valid and predictive of field life, that it is particularly difficult at this time to view the issue clearly. Furthermore, reliability is dependent on intended usage, such as the difference between an electronic component in a space satellite, a unit in a Viet-Nam jungle, and a computer in an air conditioned building.

Discussions of reliability generally fall into three classifications:

1. Extensive field data
2. Statistical analyses of limited tests, and
3. Postulations based on analogy and theoretical reasoning—which may or may not have relationship to the real world.

The first type has been presented on SLT modules,[9] where more than 30 billion module hours in the field have been reported. The controlled collapse method reliability is extrapolated from this data by laboratory stress tests and limited field input.[10] Other extensive programs, such as Minuteman, have also been reported.[11] The second type (statistical analyses of limited tests) has been made on ultrasonic and thermocompression bonds.[12] Unfortunately, the third type of discussion (postulations based on analogy and theoretical reasoning) provides the bulk of the literature.

It should be apparent, although not universally accepted, that despite metallurgical simplicity, process variability can produce reliability variations. Furthermore, the degree of control can be a prime factor, leading to the interesting possibility that managerial psychology and philosophy can outweigh technology.

Other aspects relating to reliability criteria will be discussed in the section dealing with the Reliability Comparison Chart.

Repairability Criteria

Not all constructions require repairability, but as modules become more complex, due to pressures of speed and miniaturization, this need will become greater. In any event, consideration must be made as to the economic feasibility of any repair scheme, since semiconductor parts are reaching the point where they are almost cheap enough to be thrown away. When many integrated circuits are put on a single surface however, the economic justification becomes stronger. We do not address here the repair of broken conductors on the modules, but only the techniques for removing a defective chip and replacing it with another. It is important that such replacement not disturb adjacent chips or the module itself, and that the substitute chip either has exactly the same characteristics as its neighbors, or, if the joint is slightly different metallurgically, that the reliability aspects are sufficiently understood and reproducible. This is not an insignificant task, since multiple operations might be expected to make minor changes either in intermetallic compound concentration or residues, due to the multiple exposures to either heat or pressure or both.

The type of repair often limits the circuit density; some schemes require merely pulling the chip from the module surface, while others require a sideways plowing motion which precludes closely placed adjacent chips. Replacement schemes also vary considerably in cost and complexity, those requiring dressing of individual pad sites may be more costly than schemes where no dressing or bulk dressing can be performed.

Process and Control Criteria

As previously noted, reliability is a function of the process and control, and quality control and assurance procedures not only sharpen up a production line, but can also weed out defective products. Even beyond this obvious observation, it should be understood that some processes require tighter controls than others to be reproducible. Not only is it important to understand the range of tolerable process variations, but also to ascertain whether normal industrially available equipment is reasonably capable of maintaining such tolerances without excessive maintenance. The canny semiconductor manufacturer would prefer not to hire technicians with the finesse of jewelers, but rather, whenever possible, high school drop-outs. A single furnace, as an example, with careful handling might be expected to maintain a narrow temperature range; it is a far different thing to control a battery of such furnaces with the same range. Many â beautiful laboratory scheme has created hardship in a manufacturing context for this reason.

It is wise to extrapolate what will happen if the controls do go slightly askew. If the pressure in a process increases somewhat, are the chips damaged? If the temperature rises a few degrees, are sufficient brittle intermetallics formed to completely change the reliability ground rules? If one device is slightly different in size from another, does this preclude automating the process?

Now that we have briefly tasted the types of criteria which can be used for making our judgments, we proceed to the actual critique of these six methods.

Discussion

This section discusses the six prototype chip joining methods by means of tables and accompanying comments. These tables relate to the major criteria just discussed, with the first one summarizing the most important characteristics. The values are comparative. Strong or weak designations imply either a comparison with the other methods, or a particularly noticeable characteristic.

Key Characteristics

Table II presents data for general relationships and comparison of key characteristics as discussed below.

1. Chip Handling

This pertains to the structural strength of the device and its ability to be handled relatively roughly with impunity. The controlled-collapse and non-molten chips, for example, can be readily probed and misaligned, and then re-oriented in a simple vibrating bowl. Similarly, devices without pads can certainly be handled, but it is considerably more difficult to align them mechanically. Generally, optical alignment is required since there are no protrusions to sense mechanically. It is commonly acceded that beam-leads should be kept in wafer orientation and the good devices picked out by computer control in order to avoid nightmarish difficulty with mis-oriented devices, since the beams might entangle each other.[13] Beams and leads are considered more fragile than device pads.[14]

In the spider method, once the devices are fastened to the spider, none of the leads can be drastically bent without fear of damaging the chip-to-lead joint. However, since this method must be highly automated to achieve reasonable cost, the control of the processing equipment may preclude such damage. Certainly, only production experience can illuminate this factor.

2. Automation

This relates both to the potential for automating the chip-joining process, and to the actual degree of demonstrated automation. For example, both the spider and controlled-collapse techniques have been automated to a high degree, and this degree of automation helps make these methods particularly attractive. Indeed, beam-lead handling has also been automated, but at a more sophisticated level, where each chip is picked out by computer control. In this case, automation may not be a cost improvement, but perhaps even a liability. It is more difficult to automate the other processes, since the imbedded devices and wire-bonded devices (as previously mentioned) require optical sensing rather than mechanical alignment, and the non-molten pads do not have the self-alignment features of the controlled-collapse flip-chip. The non-molten pads must be put down exactly in registration with the lands to which they are joined, while several mils of leeway are available with the controlled-collapse methods.

Since wire-bonds are commonly performed by individual operators, this is not considered to be a truly automated process, even though certain aspects of chip handling may be done by automatic machinery.

Table II—Key Characteristics

	Beam Lead	Controlled Collapse	Non-Molten Pad	Spider	Imbedded Device	Wires
1 Chip Handling	*	S	S	*	A	S
2 Automation	A	S	A	S	A	W
3 Rapid Turnover	A	A	A	W	W	S
4 Reliability	*	S	*	*	*	A-W
5 Chip Cost	W	A	A	A	S	S
6 Joining Cost	A	S	A	A	A	W
7 Replaceability	A	A	A	W	W	W
8 Thermal Dissipation	W	A	A	A-W	A	S
9 Smallest Leads	S	W	W	W	S	A
10 Inside Connections	W	S	U	A	A	S

S—Strong W—Weak A—Average *—Requires Comment U—Unknown

3. Rapid Turnover

This addresses the ease with which a few parts of each type can be made. Some methods require extensive changes in equipment and processing to adapt to different sized chips. For example, new tooling would be required to punch out the frames for the spider if the chip size or number of pads were changed, unless the leads are made by etching techniques (which are probably more expensive). Similarly, new evaporation masks would have to be made for the imbedded devices, even if the same chip were only relocated on the surface of the module. However, wire-bonding is particularly adaptable since each joint is made individually. There seems to be little major differentiation among the other methods since each requires a change in the conductor pattern, which would require new masks. However, if the compliant-bonding technique is used with beam-leads, the compliant member would have to be changed for each format, just as with the spider-lead frame.[15]

4. Reliability

Data and extrapolations show and predict good joint reliability for controlled collapse devices in the order of about a hundred mils square—approximately 10^{-7}% joint failures per thousand hours due to thermal stressing.[10] Alternatively, field returns from wire bonds have been rather erratic—sometimes showing excellent results, and sometimes showing failures—probably due to the variability of operators and equipment.[16]

With the lack of reliability data from production facilities, it is difficult to provide a precise figure for beam-lead potential reliability. This will probably depend on the reproduceability of grain size, which of the several available bonding methods are used (compliant bonding,[15] heated ram,[17] wobble table,[18] individual finger bonding, welding, etc.), the amount of lift-up (or "bugging"), the degree of surface contamination, etc. However, to permit decision making, some extrapolations must be made, and it is expected that beam-leads will be more reliable than individually bonded wires. Due to thermal expansion problems, the author expects that spider-bonds and imbedded devices will not show as high reliability as either beam-leads or controlled collapse devices. Similarly, the non-molten pads, due to their relative low ductility, are also expected to be more subject to stress cracking. Further detailed comparisons on reliability will be found in Table III.

5. Chip Cost

Imbedded and wire-bonded devices are inherently cheaper than any of the others which require special pads or joining members on the device. Beam-lead devices have been described as being particularly expensive, due to the relatively long etching and plating times involved in the fabrication.[14] It should be recognized, however, that large scale production of any of these methods will probably minimize chip cost differences. If test sites are desired in the discarded kerf sections between chips, the interdigitated beams preclude their presence. In such cases, test sites might be put elsewhere on the wafer, with the loss of some potentially usable silicon; in certain cases, this can cause an appreciable loss of yield.

6. Joining Cost

The following factors are particularly pertinent: yield, time at the chip-joining station, and complexity and expense of equipment. Wire-bonds are particularly expensive due to the long time involved in making each joint, but yields are high since each joint is replaceable. On the other hand, the controlled-collapse technique takes a fraction of a second for the device to be positioned. All the other methods require that the device be acted upon at the chip-joining station, which can create a bottleneck, which is particularly pertinent for large scale production. All thermocompression or ultrasonic bonds have a slight time delay built into the chip-joining station, generally in the order of several seconds, although this can be reduced with the proper ultrasonic bonding equipment.

7. Replaceability

This refers to the removal of a complete device and replacing it with a new one resulting in substantially the

Table III—Reliability Comparison

Characteristic	Beam	Molten	Non-Molten	Spider	Imbedded Chip	Wires
1 Bond Strength	A	S	S-A	S	*	W
2 Inspectability	A*	W*	W*	A*	W*	A*
3 Process Dependence	W-A	S	A	A	U	W
4 Minimum Connections	A	A	A	W	A	W
5 High Usable Temperature (>200°C)	S	W	S	S	U	S
6 Subject to Damage During Processing	A	S	A	U-W*	A	W
7 Usable in Normal Ambients	A	A	A	A	A*	A
8 Flexible Joint	S	A	W	U	*	S
9 Simple Metallurgy	A	W	S	S	S	S
10 Probable Relative Overall Reliability	S	S	A	A	A	W

S—Strong W—Weak A—Average *—Requires comment U—Unknown

same metallurgy and structure. The imbedded devices and wire-bonded devices will be more difficult to replace than the flip-chip or beam-lead styles. Residues of gold/silicon or other eutectic back joints are generally left. With the wire-bonding format, however, this may not indeed be a reliability exposure and the wires could certainly be individually replaced. The spider devices will probably be replaceable, but since each lead will have to be lifted, it might be more complex to do this than with a flip-chip which can be either remelted or sheared off. Similarly, although beam-leads may leave a residue on the lands where the chip is removed, this residue may not interfere with subsequent chip replacement.[18]

Thermocompression or ultrasonically-bonded flip-chips have been removed by shearing them off, but this requires that the joint be of some intermediate strength or else the lands may be ripped from the module during the removal action.[19] The easiest replacement seems to be with the controlled-collapse or beam-lead systems, and the most difficult seems to be with the imbedded devices.

8. Thermal Dissipation

In general, back-bonded devices provide the best thermal dissipation format for intermediate power levels with air cooling. The imbedded devices are surrounded by insulating material on the edges, and may be slightly less effective than wire-bonded devices. (These considerations are with no heat-sinks on the devices.) In such a format, spider-bonded devices can be variable with respect to thermal dissipation, depending upon whether or not they are firmly fastened to a thermally conducting substrate. A similar phenomenon is involved with beam-leads, where if the devices are lifted above the substrate, thermal dissipation will probably be among the worst of this group. For low power densities, all of these methods are acceptable; for intermediate levels with air cooling, the back-joined devices are best; for very high power levels, where immersion cooling is required, again they may all be adequate, with some advantage for the flip-chip configuration.

9. Smallest Leads

This refers to the minimum lead spacing realistically achievable by the methods. Beam-leads have been demonstrated with one mil lines, and this is the smallest spacing available today. On the other hand, the flip-chip configurations are usually limited to 3 mil pads on 6 mil centers, although they can be squeezed a little beyond that. Similarly, it is difficult to fabricate lead frames with extremely small dimensions. Since imbedded devices are metallized by either vacuum deposition or etching, they may approach the miniscule dimensions of the smallest beam leads. With wire-bonding tight spacings are hindered by operator and equipment finesse. An important consideration here is whether such extremely fine dimensions are really necessary for most formats, since extremely fine lines sometimes are not capable of carrying either the heat or current density required of many circuits.

10. Inside Connections

This is associated with the ability to make connections to the chip inside the periphery, (sometimes called an area array). It will tend to be more important in the future, as monolithic memories and other sophisticated MSI and LSI configurations evolve. It has been demonstrated by the controlled-collapse technique, and certainly by wire-bonding. Imbedded chips can probably have lines evaporated to make contacts to the center of the device, but sometimes these configurations require multi-level metallizing, which may be a bit difficult or expensive. It is uncertain whether non-molten pads can make such internal connections, due to their strong need for planarity. This may not be true if the pads are tapered to permit the tallest to be crushed during the joining operation. Beam-leads will probably have difficulty making such internal connections, if used in their present format. Spider leads might be extended to the center of the device, but again the ability of multi-level structures to permit crossovers might be more difficult with this method than by joining a controlled-collapse chip to a multi-level substrate.

No method is universally superior in all respects as each has its particular strong and weak points.[20]

Reliability Comparison (Table III)

1. Bond Strength

Bond strength, of course, varies according to geometry, so that the diameter of the contact to the chip or to the module lands and the total volume of material in the joint may dictate the ultimate strength as well as the processing and material details. However, since such geometries are defined (within limits) by the need for efficient use of the silicon, some reasonable estimations can be made. Typically, it is expected that the thermocompression and reflow flip-chip joints will be among the highest in the group, averaging somewhere between 30 and 50 grams per joint. Since some compromise is commonly made with the strength of ultrasonic joints (which can certainly be as strong as the others) in order to shear the chip from the lands for replacement purposes without also pulling off the lands, typical strengths of 20 to 30 grams per joint are reported.[21] Spider-bonds are in the order of 20–25 grams when the process is under control.[6] Wire-bonds are typically the weakest of the group, showing strengths in the order of 7 to 10 grams for a typical 1 mil gold wire. However, this is often a significant variable and is one of the reasons some sections of the industry are trying to replace wire-bonding. Beam-leads will probably be somewhat stronger than wire thermocompression bonds for two main reasons: the flatter lead geometry, and the availability of compliant bonding which does not excessively cut into the beam. Unlike the other five methods, it is virtually impossible to measure the bond strength of imbedded chips, since the contact leads are vacuum metallized across the chip contacts. However, thin-film technology experience indicates that the initial joint can indeed be reliable and strong if the surfaces are clean. Bond strengths of the others are commonly measured destructively and with air-blast techniques.[22]

2. Inspectability

We have previously noted that although some joints can be optically viewed more easily than others, there is no way to verify a good joint by visual inspection. The joints which are most readily viewable include wires, beams, and—depending on the format—spider-bonds. Flip-chip configurations can be viewed from the side, but this is somewhat difficult. Since silicon is transparent to certain infrared wavelengths, such joints can also be viewed through the module and chip by infrared examination. The imbedded-chip joints are under the vacuum metallizing, and are uninspectable.

3. Process Dependence

All methods will show some dependence of reliability on variations in the process, but the molten joints appear to be among the least susceptible. Next are the non-molten and spider joints, since the main variable would be pressure and ultrasonic frequency, or the pressure and wide range of tolerable temperature controls required for thermocompression bonding. Since the surfaces of the imbedded chip must be scrupulously clean, any contamination could be disastrous, and the ability to consistently achieve this degree of cleanliness is unknown. Both wire-bonds and beam-leads are potentially susceptible to excessive crushing and unclean surfaces, but they may be less critical in this respect if a technique such as compliant bonding is used, which precludes the notching caused by the common wedge tool. Also, see item 11 in Table III.

4. Minimum Connections

All methods except spider-bonding and wire-bonding require only one joint from the chip to the module lands. The latter two require a separate bond at the chip and at the land, thus potentially increasing the reliability exposure. This is a rather strong factor in consideration of reliability, and may be a significant detriment of these two methods.

5. High Usable Temperature

To avoid metallurgical reactions, molten joints are usually restricted as to temperature usage, considerably below the liquidus. The remaining methods seem indistinguishable in this respect when considered metallurgically. However, there may be other limitations to high temperature usage which involve thermal expansion stresses, and these are not addressed here. It is suspected that the more rigid flip-chip joints—such as thermocompression or ultrasonic bonds—will not be usable at high temperatures for this reason, but it is unfair to downgrade them without confirmation. Similarly, no data is available on thermal cycling for the imbedded chips. Intuitively, it seems that this would be an area of difficulty, but a private communication[23] indicates that the fluorocarbon material surrounding the chip can yield and flow slightly during thermal cycling to overcome this potential defect. Certainly, more information is needed here.

6. Processing Damage

Since wire-bonding is usually a manual operation, there are reports of scratching the chips, which creates real damage. Although no direct information is available, one may suspect the fragility of the spider and the long extension of the joint down the metal leads may require delicacy during handling. This will depend on the processing details and equipment. Initial reports indicate that handling may not be a significant problem.[6] Beam-leads can undoubtedly be probed outside the chip periphery, and if bent may be capable of being straightened. Non-molten pads are more subject to probing deformations or stresses which are unrepairable as compared to molten joints (which can reform during the reflow operation). The strongest member of this group seems to be the molten joint, which can be crushed out of shape without severe detriment.

7. Usable in Normal Ambients

Certainly none of these methods would be in production if they were not considered reliable under normal ambient conditions. As previously mentioned, more information is required about imbedded chips.

8. Flexible Joints

Obviously, the strongest members of this group are long beam-leads and wires, which are inherently flexible. It has been noted that short beams may transmit thermal expansion stresses to the chip.[24] Next come the molten joints, where solder was chosen specifically for purposes of ductility. Aluminum non-molten joints are not considered particularly ductile, but a slight improvement might be inherent in non-molten thermocompression gold joints. It is not known whether the thickness of the lead frames required for proper assembly will be too great for good flexibility, and this should be investigated carefully. There is a possibility that the fluorocarbon around the embedded chips can provide some flexibility to take up thermal stressing, and this should also be observed.

9. Simple Metallurgy

The metallurgies vary considerably in complexity; the most complex is the metallurgy of the molten joint; this is followed by the gold beam-lead metallurgy. Even simple aluminum-aluminum bonds are aggravating since oxide contaminations can interfere with the joints.

10. Probable Overall Relative Reliability

It has been noted that total reliability is dependent upon a myriad of factors which include geometries, matching of materials, and even the sophistication and size of the production operation.

Based on all of these factors, the author expects that the most reliable of the methods will be beam-leads and solder reflow-molten joints, with wire-bonds representing the lowest overall reliability—due mainly to variabilities in the production operation. With the information now at hand, it is not realistic to differentiate among the remaining three methods although initial data infers that the spiders may be slightly more reliable than wires. Of course, this estimation is worthless if one changes the ground rules, such as requiring usage at temperature extremes or environments particularly deleterious to a specific metallurgical combination.

Process Comparison (Table IV)

1, 2. Heat and Pressure

To establish bonds, some of the methods require heat, some pressure, and some both. If two quantities must be controlled carefully instead of only one, the potential for some variation in the process becomes greater. From this standpoint, it may be that the imbedded chip metallizing is least critical, while the wire-bonds are the most critical. Thermocompression heat control is probably less critical than with reflowed joints, and the beam-lead chips may actually encounter less heat than reflowed chips if the heat is supplied by the bonding tool only to the leads.

3. Critical Time

Some of the processes must be controlled within a fraction of a second, or detrimental results may occur; for example, ultrasonic bonds—if vibrated for too long a time—may become weakened. However, if the pressure and temperature in a thermocompression bond is adequately controlled, the time of pressure application is probably not too important, as would be the case with the heating of reflowed devices. In the latter case, of course, the time should be minimized to prevent intermetallic formation and possible spalling; but here the time can be controlled in parts of minutes and not of seconds.

4. Obviously Automatable

The only techniques which may not readily meet this requirement are the imbedded chips and wire-bonds, and even the former, with a great deal of effort, might be made automatable. The solder reflow method has received the most automation to date, followed closely by beam-leads and spiders. However, beam-leads seem to require a great deal of automation for any reasonable quantity to be joined economically. Computers are used to select good chips for placement,[25] and this is indeed more expensive than being able to test devices and separate them into groups with similar properties which are reoriented automatically (such as can be done with molten or non-molten devices quite readily). Mechanical handling of these devices is facilitated by the presence of bumps on the chips, but if devices are not made symmetrical, they probably can be handled without the pads. There is still some conjecture as to whether beam-leads can be handled by vibrating bowls due to the intermeshing of the beams and other mechanical difficulties.[18]

5. Critical Planarity

It is generally accepted that thermocompression or ultrasonically bonded flip-chips need excellent control of the levelness and evenness of all surfaces, often requiring more expensive polished substrates.[26] Certainly some, but probably not a very wide,[21] tolerance is allowable if the pads are long and tapered and can be crushed. However, the flexible beam-leads can be extended to non-planar surfaces, and solder pads permit irregularities as great as several mils. Spider-bonds do not require planar surfaces, since the leads of the frame can be individually depressed against the chip surface. However, the imbedded chips do require planar surfaces to prevent the vacuum metallizing from bending over corners and thus creating weak spots or electrical shorts.

6. Critical Alignment

Certainly all chips are difficult to align because of their microscopic size, but the controlled-collapse chips have a self-compensating factor of registration within several mils, since surface tension pulls the chip into the desired position. It is certainly easier to see the beam-leads, but this does not necessarily make alignment easier. As previously noted, infrared microscopes can be used for aligning devices with opaque bumps, even in flip-chip configurations. The S.T.D. imbedded chips are aligned against copper plated studs.[27] It is important to keep in mind that mechanical alignment equipment does not require optical viewing, and thus all non-self-aligning methods become more equal.

Table IV—Process Comparison

Characteristic	Beam	Molten	Non-Molten	Spider	Imbedded Chip	Wires
1 Requires Heat Control	See Discussion					
2 Requires Pressure	See Discussion					
3 Critical Time (Sec.)	See Discussion					
4 Obviously Automatable	X*	X	X	X		
5 Critical Planarity			X		X	
6 Critical Alignment	X		X	X	X	X
7 Useful for Small Quantities	A	A	A-W	W	W	S
8 Many Dicing Options		X	X	X		X
9 Simply Reoriented		X	X	X		
10 Relative Process Criticality	S	S	A	A	A	W
11 Joining Station is Not a Delay Point	A	S	A	A	A	W
12 Probe Pressures to Unjoined Chip Probably not Harmful	A	S	W	W	W	W
13 Relative Handleability	W	S	S	W	W	A

X—Applies S—Strong W—Weak A—Average

7. Useful for Small Quantities

This is probably wire-bonding's strongest point, where it inherently has tremendous flexibility. On the other side of the scale, the spider configurations and imbedded chips are less flexible since any format change requires extensive retooling and remasking. It is somewhat difficult to really differentiate among the remainder, except that the non-molten configurations may require setting up new process parameters as the chip size, pad sizes, shapes, or quantities are varied; only the placement equipment settings are required for controlled-collapse chips.

8. Many Dicing Options

The beam-leads and imbedded chips may require exact edge dimensional control, and thus may have to be etched. This can be expensive, and in the case of beam-leads may require prior device thinning.

Flip-chips are capable of using slurry saw or scribe-and-break techniques, although at least one manufacturer also etches to orient his chips from the edge.[28]

9. Simply Reoriented

This was previously mentioned in parameter 4, but is inserted here because it is considered important. Only chips with pads seem to be reoriented easily by vibrating bowl techniques. As previously mentioned, the need for reorienting can be bypassed by keeping the devices in orientation throughout the entire processing operation, but this is considered by many technologists to be more expensive and cumbersome.

10. Relative Process Criticality

This sums up parameters 1, 2, and 3 and is indicative of the expected uniformity of reliability when normal process variables are taken into consideration. This is related to line (3) of Table III, but the emphasis is on process and not on design and materials. It is anticipated that beam-leads with compliant bonding and the controlled-collapse methods will be the least critical with respect to process. Wire-bonding has already been shown to be critical in this respect, and actually, there is not enough information to evaluate the remainder. Some ranges for ultrasonic bonding have been published and indicate a potentially broad useful range of processing variables.[29]

11. Delay at Joining Station

The strongest member of this group is the controlled-collapse method where chips are positioned rapidly onto the flux covered lands. The remaining processes all require a joining operation at the station creating some delay; wire-bonds are particularly notorious for long delays.

12. Pressures to Unjoined Chip Probably not Harmful

The strongest members of this group are the molten pads since they can reform and heal any compressed pad regions after electrical testing and probing. There may be some scratching or slight crushing of gold beams, but this will not tend to be a serious problem in most circumstances. The gold beams can be probed without any contact with the chip itself, whereas with the other methods, some damage can occur due to probe slipping and scratching of the metallization. Of course, glass passivated chips are less susceptible to such damage. On occasion, ultrasonic or thermocompression tools on the chip have been observed to cause damage.

13. Relative Handleability

Generally, flip-chips can be relatively roughly handled or mixed up, while this will probably not be true with beam-leads, as previously mentioned. The strongest members of this group seem to be the flip-chip configurations and pad-less devices.

Design Flexibility Comparison (Table V)

1. Power Dissipation—Air Cooled

It is difficult to precisely define low, medium, or high power characteristics, since the heat sinking characteristics, module arrangements, and manner in which air circulates past the system have as much pertinence as the power input per volume. However, it is important that heat be removed from the sections where it is being created as rapidly as possible to avoid device damage. Isothermal chip surfaces are the most desirable. None of these systems is completely acceptable when air cooled, with no heat-sinks at high power densities, (in the order of over 1 watt or so per chip); conversely, all are acceptable for low power dissipation. At the intermediate power levels some differentiation can be seen, and in this case the back-joined configurations tend to be superior because heat is pulled directly from a large area of the chip. Next in order of acceptability are the devices with large pads, such as the molten or non-molten joints. The weakest of this group are beam-leads and spiders which possess thin leads and present a high thermal resistance. As previously mentioned, if heat sinks or filled plastics are placed against the chip surfaces, these ground rules are no longer valid—but this involves extra cost. Furthermore, the designer must be extremely careful of excessive stresses when devices are surrounded with materials that expand at substantially different rates than the substrate or chip.

2. High Power, Liquid Cooled

For special circumstances requiring the modules to be immersed in a coolant, it will probably be found that there is little differentiation among any of these methods. Boiling fluid will be most effective for pulling heat from the device when the total contact area is maximized. Thus, some advantages might be inherent in flip-chip configurations, but data are not yet available.

3. Perimeter Pads

Obviously, any of these techniques is capable of interconnecting pads around the periphery of the chip.

4. Internal Pads

With some methods, it is considerably more difficult to connect pads not directly accessible from the edge. For example, although an argument can be made that air isolation techniques permit such internal contacts with beam-leads, the author believes this to be an excessively difficult technique. Similarly, although some of the leads of the spider can be extended into the device center, it may be difficult to achieve certain configurations because of the inherent need to criss-cross some of the leads; this could set up problems with electrical shorting. As previously mentioned, the main problem area with non-molten pads is the planarity requirement, and it was noted that an ultrasonically bonded chip may swivel under the anvil if some of the pads are too high. With proper controls and greater expense, the planarity might be controlled enough to permit this to be feasible. Imbedded chips are capable of having multiple layers metallized over them to establish electrical contact, but this may provide a distinct yield loss at the worst possible time, that is, when the entire module is committed with chips. The strongest technique for internal pad capability appears to be the molten-pads, which are lenient with respect to planarity, and wires, which can be bonded anywhere. In the latter case, some configurations require great care to prevent the wires from touching.

5. Direct Heat Sink Attachment

It is difficult to bond a metallic heat sink to a chip which has already been bonded on its back side, such as the imbedded chips or wire-bonded devices. Of course, a case could be made that such devices are already joined to a heat sink, but it is realistic to note that ceramic materials other than beryllia are at least an order of magnitude less thermally conductive than metallic heat sinks. Undoubtedly, heat sinks can be attached to beam-lead devices, but there is conjecture whether the thin beams are capable

Table V—Design Flexibility

Characteristic		Beam	Molten	Non-Molten	Spider	Imbedded Chip	Wires
1 Power Dissipation Air Cooled	Low	A	A	A	A	A	A
	Med.	W	A	A	W	S	S
	High	W	W	W	W	W	W
2 High Power, Liquid Cooled		A	A	A	A	A	A
3 Perimeter Pads		A	A	A	A	A	A
4 Internal Pads		W	S	A	W	A	S
5 Can Permit Direct Heat Sink		W*	A	A	A*	W	W
6 Short Electrical Path		A	A	A	W	A	W
7 Back Joint or Flip-Chip		A	W	W	W	W	W
8 Thin or Thick Films		A	A	W	A	W	A
9 Narrow I.O. Spacings		S	A	A	A	S	A
10 Requires Thin Wafer		W	A	A	A	A	A
11 Amount of Semiconductor Real Estate		A	A	A	A	A	A
12 Air Isolation		S	W	W	W	W	W
13 Many Formats for a Module		A	S	A	W	W	S
14 Possible Inductance Problems		A*	A	A	W	A	W

S—Strong W—Weak A—Average *—Requires Comment

of supporting the extra mass, particularly when shock or vibration tests are applied. The other members of the group are acceptable in this respect, and heat sinks more than thirty or forty times heavier than the chip can be applied to the controlled-collapse devices without appreciable change in the chip-to-substrate height.[31]

6. Short Electrical Path

Since speed is a function of path length, the strongest members of this group have the shortest paths and it is difficult to really separate beam-leads, molten joints, non-molten joints, or imbedded chips in this respect. However, it would seem that both the spider-leads and wires tend to be substantially longer than the others. Of course, this may not be the case, depending on the electrical path from the joints out to the next connection, and this second conductor pattern may be just as critical. Presently, this is not considered to be a particularly important aspect.

7. Back-Joint or Flip-Chip

Only the beam-leads seem to be readily adaptable to any configuration, but the leads may have to be elongated appreciably if joined back-down on a flat substrate, causing a possible severe sacrifice of silicon real estate. The remainder are considerably more constrained, and require more engineering to accomplish such flexibility. For example, solder-pad devices can be joined back-down and interconnected by means of decal, overlays, or straps; however, some of the simplicity of the original reflow process will have been sacrificed. It is difficult to conceive how wire-bonded chips can be joined face down.

8. Thin- or Thick-Films

Only two techniques, the non-molten pads (again due to planarity constraints) and the imbedded chips (requiring vacuum metallizing for interconnections), are restricted to only thin films; the remainder have considerable flexibility in this respect. The spiders, for example, can be attached to thick- or thin-films on the modules, as can wires.

9. Narrow I/O Spacings

The beam-leads and imbedded chips are probably capable of the smallest spacings of the group, but this differentiation is in the order of several mils. Although for certain special circuits this may be pertinent, the author does not believe that it is important in the majority of applications.

10. Thin Wafers

Any of the processes requiring etching for separation from the wafer will normally require a thin wafer, such as beam-leads and the Hughes flip-chip. The other methods are more flexible. Thin wafers may be advantageous for heat dissipation but tend to warp with multilevel metallization; thus they can be advantageous or not, depending on the format. In any case, they may be somewhat more costly due to the extra processing and handling problems.

11. Semiconductor Real Estate

Contrary to some opinions, the author believes that there is no great difference among any of these methods. Before conclusions can be drawn with respect to pad contact area, amount of kerf required for dicing, etc., the specific details of the remainder of the device must be known. Thus, in individual cases, one method might be slightly advantageous over the others, but it is difficult to predict. However, the interdigitated beam-leads do not readily permit the location of test sites or markings in the kerfs; these can be useful and their sacrifice is unfortunate, particularly as devices get more complex. If test sites are essential, beam-lead configurations may use up valuable space to contain them.

12. Air Isolation

Only the beam-leads are capable of air isolation, but although this was described as a significant advantage, the industry does not seem to have responded particularly, perhaps because of the large silicon real estate loss.

13. Many Formats For a Module

The strongest technique for this type of flexibility appears to be the molten joints; there is no change in the process for chips of different size or shape or intermixes—all that needs to be done is for the chip replacement machine (or operator) to position each chip in the proper place. Of course, the land pattern on the module must be changed—but this is a universal requirement—except perhaps for wire-bonding, which is the next strongest of this group. The beam-leads and non-molten joints may require some change in the bonding parameters to adjust for different numbers of pads or chip sizes, with the exception of compliant bonding, which requires a new throw-away compliant member. It would seem that the spider and imbedded chip methods are the weakest of this group; the former requires a new lead frame, while the latter requires different placement of the studs on the substrate as well as the vacuum metallized interconnection pattern.

14. Possible Inductance Problems

There are two edges to this sword: some electrical designers worry about the inductance potentially inherent in an extended lead, while others enjoy the ability to be able to doctor the inductance by changing the lead length. Thus, beam-leads are either a joy or a nightmare depending on the point of view and requirements. There does not seem to be a major problem with bulky pads, but the spiders and wires do run the risk of long lead length which could cause a problem.

In summation of Table V, it is evident that each method has its own advantages and disadvantages with respect to flexibility, and it is erroneous to approach a technology merely because it is extendable in a particular direction (unless, of course, it can be established that it is the exact direction required by a particular manufacturer). With a constantly changing integrated circuit technology, it requires considerable astuteness to predict the path of flexibility which would be most pertinent to future generations of devices.

Conclusion

The series of comparative charts represent one person's view of a complex technology. Some of these views are obviously open to dispute, and it is expected that with so many decisions, subsequent data will show some of them to be erroneous. Furthermore, it is virtually impossible in any limited article to include the many possible criteria or innuendos which might be pertinent, and this was not attempted. It is hoped that such comparisons will provide a basis for future judgments and other comparisons.

References

1. L. F. Miller, "A Survey of Chip-Joining Techniques," *Proc. IEEE Comp. Conf.,* April 30–May 2, 1969, pp. 60–76.
2. E. F. Koshinz, "Semiconductor: Wire Bonding and Face Bonding Considerations and Comparisons", *Proc. SAE Microelectronic Packaging Conf.,* Nov. 1968.
3. Article—*Electronic News,* February 19, 1968.
4. L. F. Miller, "Controlled Collapse Reflow Chip Joining," *IBM J. Res. Develop.* **13,** 239–250, May 1969.
5. T. J. Matcovich, "Direct Bonding Processes and Hybrid Circuit Applications," *Elec. Pack. & Prod.,* Nov. 1967, p. 101.
6. R. Helda, "Spider Bonding Technique with IC's", *Proc. Int. Electronic Circuit Pack. Symp. Wescon,* Aug. 20–21, 1969, 4/3.
7. Special Issue on Beam-Leads, Western Electric: *The Engineer,* December 1967.
8. A. P. Mandel and P. K. Vahey, "Thermal Design Criteria for Hybrid Microelectronic Modules", *Proc. Hybrid Microelectronics Symp.,* Oct. 28–30, 1968, p. 31–42.
9. E. F. Platz, "Reliability of Hybrid Microelectronics", 1968 WESCON Symposium, Designing with Hybrid Microelectronic Circuits.
10. K. C. Norris and A. H. Landzberg, "Reliability of Controlled Chip Collapse Interconnections", *IBM J. Res. Develop.,* vol. 13, May 1969, p. 266–271.
11. See, for example, D. R. Fewer, *Proc. IEEE, Electronic Comp. Conf.,* 1966; T. R. Myers, "Flip-Chip Microcircuit Bonding Systems", *Proc. IEEE Comp. Conf.,* April 30–May 2, 1969, pp. 131–144.
12. J. E. McCormick, "On the Reliability of Microconnections," *Elec. Pack. & Prod.,* June 1968, p. 187–190.
13. M. P. Eleftherion, *et al,* "Handling and Bonding of Beam Lead Sealed-Junction Integrated Circuits", *Proc. IEEE Comp. Conf.,* 1968, p. 149.
14. L. Curran, "In Search of a Lasting Bond", *Electronics,* Nov. 25, 1968, p. 72–80.
15. A. Coucoulas and B. H. Cranston, "Compliant Bonding—A New Technique for Forming Microelectronic Components," *IEEE Trans. Electronic Dev.,* **ED-15,** no. 9, Sept. 1968, p. 664–674.
16. I.C.E., Inc., "Integrated Circuit Engineering—Basic Techniques," Section 7.15, 1966 Ed.
17. R. H. Cushman, "Mechanical Thermal Pulse Metal Joining," *Symposium Record, 7th International Electronics Circuit Packaging Symposium,* 1966, p. 4–12.
18. J. E. Clark, "Wobble Table for Thermocompression Bonding Beam-Lead Silicon Integrated Circuits", *Proc. 1968, Int. Electronic Circuit Pack. Symp.* Aug. 19–20, 1968, p. 211.
19. P. Scharf *et al.,* "Flip-Component Technology", *Proc. 1967 IEEE Components Conference,* p. 269–275.
20. L. F. Miller, slides used in 1969 IEEE Elect. Comp. Conf. presentation, (Ref. 1).
21. J. S. Cubert *et al.,* "Face Down Bonding of Monolithic Integrated Circuit Logic Arrays," *Proc. 1966 IEEE Components Conference,* p. 156–166.
22. K. Clark *et al.,* USP 3,413,839.
23. Mr. R. J. Clark, G. E. Co.—private communication.
24. D. Boswell, "Mechanical Design of Chip Components for 'Flip' and Short Beam-Lead Mounting," *Proc. Hybrid Microelectronics Symp.,* Sept. 29, 30, 1969, pp. 5–11.
25. B. Dale, "The Application of Beam-Lead Devices to Hybrid Microcircuits," *Proc. 1968 IEEE Elec. Comp. Conf.,* p. 84–92.
26. M. B. Shamash, "New Device Bonding Techniques," Presentation *Nepcon,* Feb. 12, 1969.
27. R. K. Field, "The New World of 'Leaded Chips'," *Electronic Engineer,* August 1968, p. 100–107.
28. G. E. Gorgenyi, "The Hughes Dimensionally Controlled Glass Ambient Flip Chip", presented at *Nepcon,* Feb. 12, 1969.
29. R. P. Moore, "An Evaluation Study of Ultrasonic Face Bonding of Integrated Circuit Chips to Evaporated Aluminum Conductors", *Proc. IEEE Components Conference,* 1968, p. 427–435.
30. W. Workman, "Failure Modes of Integrated Circuits and their Relationship to Reliability", *Microelectronics and Rel.* **7,** 1968, p. 257–264.
31. P. Lin *et al,* "Design Considerations of a Flip-Chip Joining Technique," *Proc. Hybrid Microelectronics Symp.,* Sept. 29, 30, 1969, p. 313–322.

Microcircuit Packaging and Assembly—State of the Art

Gerald K. Fehr
Intel Corporation
Mountain View, California

Microcircuit packages (primarily hermetic) available today are reviewed and the ability to accomplish their functions is discussed. Assembly techniques and their associated costs are presented. Particular packages discussed include TO-5's, flat packs, dual-in-line packages, Sentinel packages, and various multi-chip packages. It is concluded that there is not yet a totally satisfactory low cost packaging and assembly technique, particularly for a small company which must limit its automation expenditure.

PACKAGING HAS BECOME a major consideration in the manufacture of electronic devices. Commercial markets require easy-to-handle, low cost circuits with high electrical and environmental reliability. In the past, the bulk of the microcircuit engineering effort was placed upon the semiconductor element. However, the point has been reached where the packaging costs can, and many times do, exceed the chip costs, and the majority of device failures are either assembly or package oriented. Efforts are now being directed to the assembly and packaging area. The cost element in particular brought about a rash of assembly and packaging developments, such as flip-chip, beam-lead, spider-bond, etc.

The purpose of this article is to discuss packages (primarily hermetic) available today and how well they accomplish their functions (with emphasis on cost). To fully consider the total packaging cost, it also becomes necessary to discuss the assembly technique and its associated costs.

Package Functions

The package functions include the following:

1. Chip Protection (Environment Isolation)

The construction of the package must be such that it protects the electronic device from the surrounding environment. Common tests to insure adequate protection are:

a. Hermeticity
b. Thermal shock and cycling
c. Mechanical shock and centrifuge
d. Humidity tests.

2. Compatibility with System Requirements

The basic properties of the materials used in the package manufacture and the design utilized can greatly affect the circuit performance. Of particular interest in this area are such factors as:

a. Heat transfer capability
b. Radiation hardening
c. Electrical properties, such as the lead resistance and capacitance.

3. Mechanical Configuration

This category contains the requirements of size, weight, shape, number of leads, type of leads, and lead location. Particular points to consider are:

a. Effect on size of finished equipment
b. Usable device space per size of package
c. Ease of device assembly and test.

4. Interface Between the Die and the Electronic System

The package supports the chip and provides the electrical connection to the outside world in a convenient and reliable manner. Points to consider are:

a. Minimum voltage drop within connection elements
b. Isolation between the circuits
c. Ease of assembly of packages into equipment.

5. Cost Objectives

The cost of the package is usually a major part of the device cost. Not only must the manufacturing cost of the package be considerd, but also the cost of handling the package during assembly and test of the device, and assembly of the final equipment. The most economical de-

Reprinted with permission from *Solid State Technology*, pp. 41–47, Aug. 1970.

Fig. 1—(Left to Right) TO-18, Coined TO-5, eight lead TO-5, ten lead TO-5.

sign may not use the cheapest package. The ability to use automatic handling during the processing may well justify a package several times the cost of a cheaper package.

Standard Assembly Techniques

The standard assembly technique will be defined for purposes of this article to be the process using eutectic die attachment of the back of the die to a substrate or lead frame pad, and wire bonding between the chip bonding pads and the lead frame or metalization pads on the package. It includes both ultrasonic bonding of aluminum wire and thermal compression bonding of gold wire.

The standard die attach of relatively large dice (approximately .100″ square) ranges in cost from one cent to five cents each, depending on whether a semi-automatic die attach machine (500/hour) or die attach by hand (150/hour) is used. The wire bonding costs of these dice (16 lead) is approximately seven and one-half cents each (75/hour). The wire bonding is the single most expensive operation in the standard assembly process.

Chip replacement can be performed by heating the package to the eutectic temperature and removing the die as the eutectic mix melts. If there is gold remaining on the die attach area, another die can be attached. Use of a eutectic braze preform will enhance the die attach. However, this is the simple part of the repair work. The most difficult part is to open the seal in such a way that the package can be resealed after the repair. Due to the added cost of the repair and the low probability of getting a hermetic reseal after the repair, chip replacement is very seldom done on single chip packages. It just has not been economical, even when technically feasible.

Packages for Standard Assembly (Single Chip)

1. Radial Lead Type

The most common radial lead packages are the TO-5 and the TO-18 (Fig. 1). They are most often made using a matched seal technology (the thermal expansions of the glass and the metal are very closely matched). The matched seal TO-5's and TO-18's have proven to be of outstanding reliability. The TO-5 package family is made with leads numbering from two to twelve in a 0.200″ or .230″ diameter pin circle. The leads range from 0.016″ to 0.026″ diameter with lengths between 0.5 to 1.5 inches. The TO-18 is simply a smaller version (0.100″ diameter pin circle) of the TO-5. The round shape provides an even distribution of the stresses in the seal. The long seal path (0.070″ to 0.100″) insures reliable hermeticity giving lead pull strengths of 10 to 20 pounds. The final enclosure technique of welding results in a very high seal yield (99% + to 1 x 10^{-8} cc He/sec). After welding the lid on the package, it is virtually impossible to reopen this type of package to do any repair work. However, due to the low cost of the package and its normal use as a single chip package, repair would most likely be uneconomical even if it were feasible.

The package cost ranges from \$50 per thousand for a three lead TO-5 package (no gold) to \$125 per thousand for eight to ten lead TO-5 packages (no gold) in volumes of 25 to 50 thousand. Utilizing the coined TO-5 with a compression seal will reduce the cost and increase the heat transfer of the package. Compression seals result when the material choice is such that during the cooling process after manufacture of the seal, all seals are placed in compression. A typical package would use 1010 steel eyelets, Corning 9010 glass, and 52 alloy leads. The lead seal is essentially a matched seal, while the eyelet seal is in compression.

It is necessary to have enough metal strength in the package to insure that the metal is not over stressed when the compression is reduced upon cooling. The seal must be round or nearly round to insure uniform stresses. The glass meniscus on these packages must be restricted to a minimum height so that stress moments from the compression seal do not crack the glass above the steel platform. The package has good thermal characteristics.

Disadvantages of this package are:

a. The ratio of allowable die size per package outline (.100″ square)
b. Very inconvenient to use—leads are easily bent and require the use of carriers for automatic handling
c. Not particularly light or small
d. Limited number of leads (twelve).

2. Flat-Packs

The flat-packs are produced in both round and rectangular shapes (Fig. 2). The common sizes range from ¼″ x ⅛″ to 1″ x 2″. The ¼″ x ⅛″ and ¼″ x ¼″ are the larger volume units at this time. Most of the packages available utilize a matched seal technology, with either a metal or ceramic base and ring. The leads are kept thin (.004″) and flexible to reduce weight and to prevent stresses from being transmitted to the seal. The main advantages are:

a. Light weight and small size
b. Relatively large die area per package size.

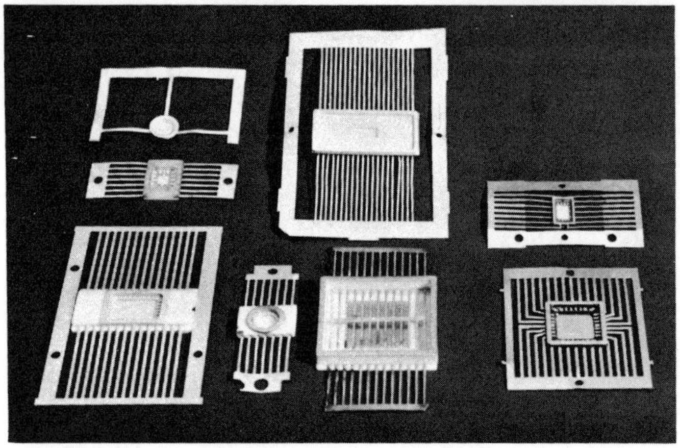

Fig. 2—Flat-packs

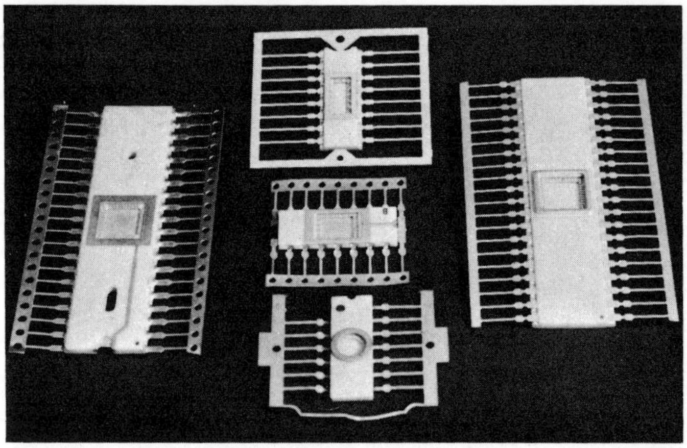

Fig. 3—Ceramic multilayer DIP's.

This package type lends itself to expansion to large sizes to accommodate ceramic based hybrid circuits containing several chips. Depending on its construction, it can either be welded, brazed, or solder-glass sealed. Its major disadvantages are a relatively high cost (about 25¢ each for the small package in large volume [100K], up to \$5 to \$6 each for small quantities, such as 1,000, of the large packages). It is not a particularly easy package to handle and requires a carrier for automatic handling. The smaller packages with the short seal patch (.030″) have had a general history of hermeticity problems when mishandled, or when the packages weren't made under very controlled conditions.

3. Dual-In-Line Packages (DIP's)

There are many types of construction used for DIP's; each has its advantages and disadvantages.

The multilayer ceramic-to-metal seal package is shown in Fig. 3. Since the seal is formed between pieces of identical ceramics, this package will withstand temperature shocks better than any other construction. It also can utilize reliably a very short seal patch (25 to 35 mils), resulting in a large cavity to package size ratio. Packages with .170 inch wide cavities can be bent in .300 inch lead centers. To gain the width, the leads are brazed on the package bottom, causing some handling problems. The top brazed package results in smaller cavity width per package width, which is particularly important in the 16 lead package with lead rows on .300 inch centers. One drawback of any of these packages presently available is their relatively low lead strength; the leads bend very easily. Another critical point is to insure, particularly in the wide cavity bottom braze packages, that there is a braze fillet between the top of the lead and the side metalization connection. Without the fillet, the lead cracks away from the side relatively easily. The package has a tendency to bow and warp during manufacture, causing some sealing problems. This package is probably the most expensive DIP in common usage. A 16-pin package costs from 75 cents to one dollar each in quantities of 10 to 25 thousand. It is too thin to be handled by some automatic insertion equipment.

The side braze package shown in Fig. 4 is slightly lower in cost (65 to 70 cents each in like quantities) than that of the ceramic-to-metal. It utilizes a glass between the two ceramic pieces which closely matches the expansion of the ceramic. Its temperature cycle capabilities, while not as good as the ceramic-to-metal package, are adequate. Its cavity width is limited to .150 inches for a .300 inch center-to-center lead package, which is adequate for most LSI (large scale integration) products. It has a relatively clean bottom (very little metal on the bottom) and fairly rigid leads. Its disadvantage is the cost, plus the fact that it is different from most other packages; however, it can be used in most automatic insertion equipment.

The glass-to-metal sealed DIP (Fig. 5) is priced from 50 to 75 cents for equivalent volumes. In large volumes, the price can drop to 25 to 35 cents. Its seal path should be at least .050 to .060 inch. It must have .015 to .020 inches from the edge of the package to the lead bend, so that it can be properly clamped to avoid cracking the glass when bending the leads. This, plus the seal length required, limits the cavity width for a .300 lead center-to-center package to approximately .100 to .120 inches. This is a very serious limitation for LSI manufacturers. The thermal shock capability is lower

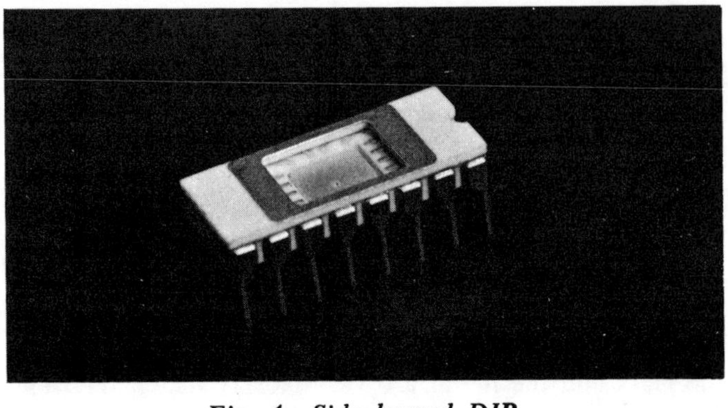

Fig. 4—Side brazed DIP.

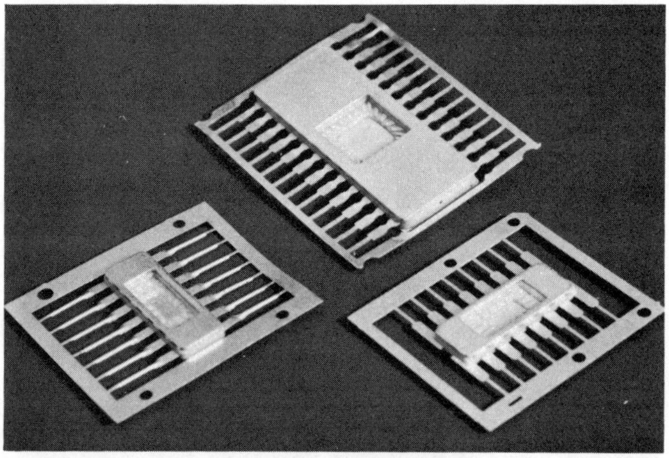

Fig. 5—Glass-to-Metal DIP's.

than the other two packages. The main selling point for this package is cost.

The C-DIP package shown in Fig. 6 is assembled on the component assembly line. The header oriented costs are in the 10 to 15 cent range. The technology required for successful sealing is critical and needs to be well controlled. Problems can result if the glass is either over or under devitrified. This package has about the same limitations as the glass-to-metal DIP previously discussed. While it has a lower cost than the previous package, its quality is more sensitive to the manufacturing conditions and requires a well controlled line.

The plastic injection molded DIP is the least expensive DIP (Fig. 7). The package is solid, and hence has little problem with mechanical shock or centrifuge. To stay on a .300 inch center-to-center line limits the die bond width to about .140 to .150 inches. Disadvantages are its nonhermeticity (by definition, as it can't be checked), lower resistance to moisture, high thermal resistance, temperature cycle failures, and temperature limitations. However, when used in a controlled environment, it offers adequate protection to the die at a greatly reduced cost. The plastic package cost varies from 3 to 12 cents, depending on the cost of the lead frame.

4. Miscellaneous Packages

This category includes the packages of unique design due to particular usage, or because an individual company likes them. They are used in much lower volume than either the dual-in-line or the flat-pack. Examples are the high power packages and the IBM package outline (Fig. 8).

Note the use of pins in two of the packages to make a plug-in package. The IBM package (4 rows of 4 leads) is produced by placing Kovar pins in metallized holes in the ceramic and then brazing to make the seal. This package is quite rugged and passes stringent requirements.

The second package with pins in Fig. 8 is produced by placing Kovar pins through a glass preform and the pin and flange seal made at the same time by fusing the glass (much the same as producing a TO-5). A metallized ceramic substrate is then placed over the pin (as shown in Fig. 8). Solder connections are made between the pin and the metallized ceramic. The package hermeticity can be checked before committing the chip. Also, the electrical characteristics of the device on the ceramic substrate can be tested before committing the package to assembly.

Face-Down Bonding

Face-down bonding combines the chip attach and bonding operations. The most common face-down bonding methods are beam-lead, flip-chip, and reflow solder techniques. In order to make contact only at the pad locations, the contacts on either the chip or header must be raised. The choice is usually made on a cost basis and the availability of chips with bumps. Since the die can be processed in wafer form while the packages are individually treated, it is usually cheaper to place the raised contacts on the chip.

The main advantage of this technique is the elimination of the die attach and the cost reduction of the bonding operation. The bond operation cost ranges from one-half cent to two and one-half cents, depending on the amount of automation. This technique requires that there be a different package metalization pattern

Fig. 6—C-DIP.

Fig. 7—Plastic DIP.

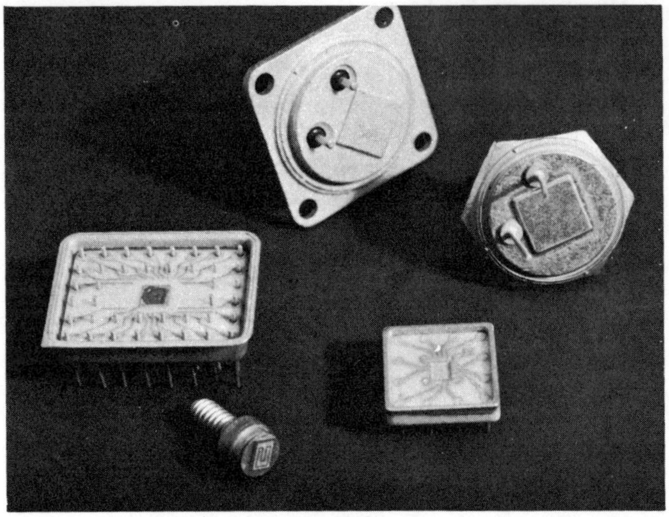

Fig. 8—Miscellaneous packages.

Fig. 9—Beam-lead packages.

or lead frame for each device with different pad locations. This added cost, along with the cost of automating many of the techniques, makes this process just not feasible for single chip packages unless the volume is very large. Face-down bonding is feasible for multi-chip and hybrid packages in much smaller volumes (e.g., <100K/ month).

Face-Down Bonded Packages (Single Chip)

No matter which type of raised contacts are used, certain requirements are placed upon the package. These are:

1. The package lead pattern must have a tight resolution. Lead widths and lead spacings are commonly .004″ x .004″; however, .003″ x .003″ are used. This is relatively easily attained through photographic techniques; however, it is crowding the state of the art for high production silk screening or lead punching.
2. An electrically non-conductive base must be used. It may be either a metallized ceramic or a lead frame which can be built into a plastic package.
3. Since the raised contacts are usually between .0005 and .002″ in thickness, the lead pattern flatness must be held very tightly. Solder reflow techniques (.005″ contact thickness) allow greater pattern flatness variation; however, the lead pattern usually requires .006″ to .008″ lead separation.
4. Thermal expansion problems due to the expansion difference between the substrate and silicon may show up during repeated thermal cycling. This can be solved by trade-offs between die size, lead flexibility, substrate expansion, and temperature cycling limitations on the device.
5. Since contact to the chip is made only through the raised contacts and any material placed around the chip, there is a higher thermal resistance path than when a chip is die attached to a heat sink. Depending on the operating power of the device, it may or may not be a problem.

The beam-lead technology was given a large boost by the Sentinel Program (now Safeguard). Figure 9 shows a Sentinel package. This package has an etched pattern and utilizes thermal compression bonding of the leads.

The Fair-Pak is shown in Fig. 10. Due to the package cost of about 25 to 30 cents, the assembly cost of the device was above the cost of standard plastic encapsulation and hence it died an early death at one of the West Coast firms. It has since been given new life by a beam-lead user.

A new technique which has been a success is the spider-bond, where the chip is attached to a lead frame (Fig. 11). The illustration shows a one piece lead frame. The center has been made thinner to allow tight lead spacing. More common is the use of a thin (.002″ to .004″) aluminum inner lead frame and a .010″ thick Kovar outer lead frame. The chip is attached to the thin lead frame and then the combination (chip and lead

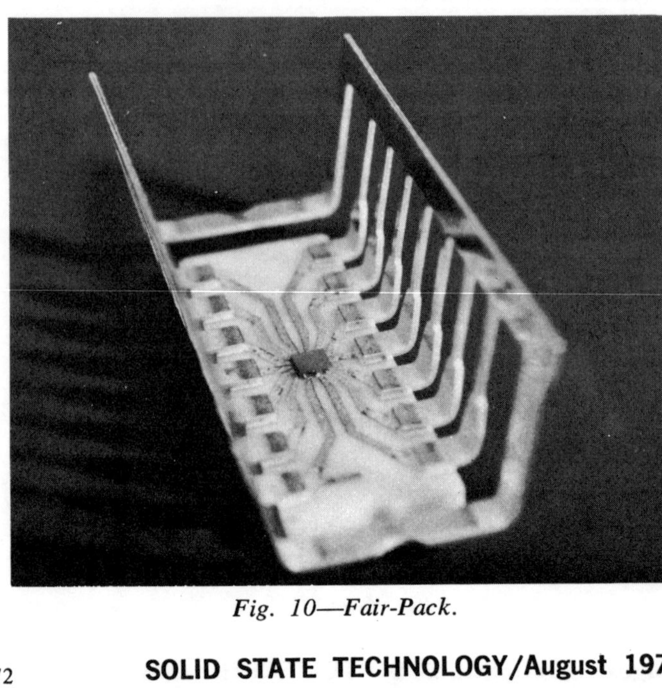

Fig. 10—Fair-Pack.

Fig. 11—Spider-bond lead frame.

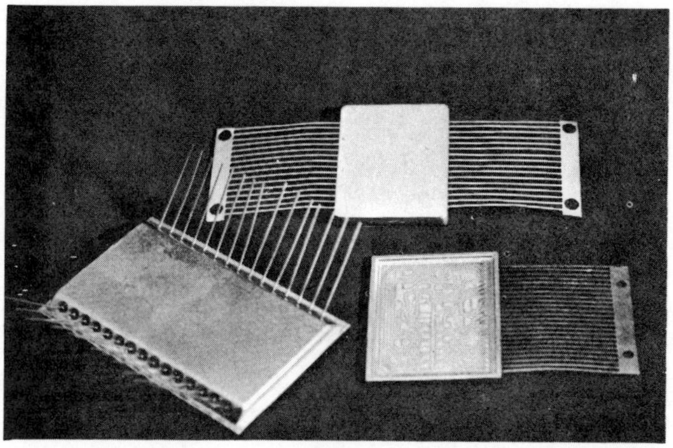

Fig. 12—Multichip packages.

frame) attached to the outer lead frame. In a highly automated system this is a low cost assembly technique. However, the high cost of equipment, the small number of each particular device with its own pad locations, small number of frames purchased, etc., again rules this technique out for all but the large users who can make about one million devices of one type per month.

Multi-Chip Packages

One way of reducing the package cost is to increase the amount of electronics per package. The complexity in the chip can be such that its selling price can afford a high package cost, or several chips can be grouped in one package. It is in this area that the little package development done in the past few years has appeared. Multi-chip packages utilizing standard assemblies are shown in Fig. 12.

The ceramic insert can be fully tested before the hermetic package is committed to the product. The concept above was to put just enough electronics into a package to be able to afford an expensive package. However, the face-down bonding technique will greatly reduce costs because it reduces assembly labor and can afford a large package (Fig. 13). The Sentinel package (Fig. 14) utilizes a multi-chip approach to build the various devices. Multi-level packages again play a big part in face-down bonding, as they allow easy interconnection of the devices inside the package.

The use of many chips in a package, their high cost if very complex, and the high cost of large packages, leads one to again think of repair techniques. It has been well established that almost all of the assembly techniques have good repair possibilities. The most difficult part is opening the package in such a way that it can be resealed. A welded package is almost impossible to reopen without destroying the seal surface.

Glass seals are also very difficult to open and reseal. Due to the roughness of the material left upon opening, the chances of resealing hermetically are small.

The use of plastic packages offers no chance of repair. It is difficult to do failure analysis work. The use of an epoxy to glue a ceramic or metal cap on the package does have a very good chance of repair and reseal. This is often done on commercial products, but is not acceptable to the military.

The only hermetic sealing technique which works well for repair and reseal is the braze seal, particularly low temperature Pb-Sn solder seals. It is relatively easy to heat up the seal, remove the lid, complete your repair, and resolder the lid onto the package. Because this material wets the metal surface easily the repair can be done many times. The package and lid should be designed such that a solder build-up at the seal will not affect the device.

Fig. 13—Multi-flip-chip substrate.

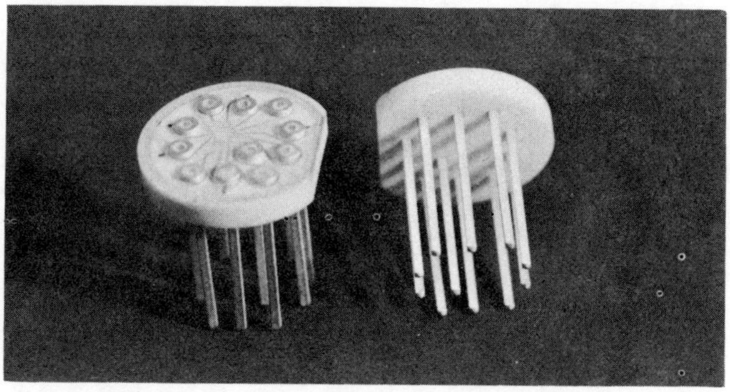

Fig. 14—Beam-lead multichip package.

Table 1—Comparison of Assembly Techniques

Assembly Process	16 Lead Package Cost	Die Attach Cost	Bonding Cost	.125″ Sq. Die Die Cost Ratio	Equipment Investment	Hermetic	Thermal Resistance	Good Multi Chip Tech.	Remarks
Standard D/A & Wire Bond									
-Hermetic	40-50¢	1-5¢	7.5¢	1	$ 50K	Yes	Good	Yes	
-Plastic	10¢	1-5¢	7.5¢	1	$100K	No	Poor	No	
Beam-Lead (Bell Lab Type)	25-50¢	—	2.5¢	1-½	$ 50K	Yes	Poor	Yes	Requires Au metallization system
Flip-Chip	25-50¢	—	2.5¢	1-½	$ 50K	Yes	Poor	Yes	Can be—All Al system —Solder ball
Spider-Bond	12.5¢	—	.5¢	1	$500K	No	Poor	No	—Highly automated —Requires a million devices per month with same pad locations to be feasible (lead frame design)
Spider-Bond Plastic Film	12.5¢	—	2.5¢	1	$150K	No	Poor	No	Use of H-film to produce a beam lead device—reasonably low priced inner lead frame —Some automation

Summary

Table I shows a comparison of various assembly techniques. Certain critical items are shown; in particular, how they would affect a small company producing complex devices. Guidelines for the chart are as follows:

—large die (>.100″ square)
—complex chips (>100 bits)
—relatively small production of an individual device type (< 100K per month)
—use of 16 lead dual-in-line package.

The cost figures show that there is not yet a good technique for low cost packaging and assembly for a small company with large complex dice. However, the use of many dice in a package or a very complex die will reduce the packaging cost per function. Plastic will most certainly continue as a low cost package.

Chip Yield and Manufacturing Costs

DAVID A. HODGES, MEMBER, IEEE

TABLE 4-A

	SSI	MSI	LSI
Active Area (mils)	30 X 30	60 X 60	120 X 120
Chip Size (mils)	50 X 50	80 X 80	140 X 140
Probe Yield (percent)	80	50	15
Chips/2 in diameter slice	1200	400	150
Good Chips	960	200	22
Cost/Chip	2¢	10¢	90¢

Assumed: \$20 slice fabrication cost; 200 defects/in^2.

Table 4-A. Approximate 1971 integrated circuit chip manufacturing costs.

TABLE 4-B

	SSI	MSI	LSI
Tested Chip	\$0.02	\$0.10	\$0.90
Package (DIP)	0.03	0.04	1.00
Assembly	0.05	0.05	0.50
Final Test•	0.02	0.02	0.20
	0.12	0.21	2.60
70% Yield at Final Test	X1.4	X1.4	X1.4
Factory Cost	\$0.17	\$0.30	\$3.60

Table 4-B. Approximate 1971 integrated circuit package and assembly costs.

Table 4-A shows the writer's estimate of 1972 integrated circuit manufacturing capability. The figures apply to proven designs that have been in successful manufacture for six months or more. Small-scale integration (SSI) includes items such as dual gates or single flip-flops. Medium-scale integration (MSI) includes items of typical complexity 20–60 gates, such as a four-bit binary counter. Large-scale integration (LSI) includes both MOS and bipolar memory components of complexity 256–1024 bits.

Active area refers to all the chip area occupied by circuit components; it excludes borders and bonding pad area. Chip size shows the overall silicon chip dimensions. Probe yield refers to the yield measured by dc tests before the chips have been separated. The \$20 slice fabrication cost applies to either bipolar or silicon gate MOS processing. This figure is probably valid for normal operations by the more efficient manufacturers. It includes direct labor and factory overhead, but not any engineering costs, marketing costs, or profit.

The defect density (200/in^2) is an empirical constant in a mathematical model for chip yield as a function of chip active area. A number of papers on this subject have been published proposing similar models [1]-[5]. The model proposed by Dingwall [3] is most rigorously documented and is also simple to use. It states chip probe yield Y in terms of defect density D_0 and active area A, as follows:

$$Y = \left(1 + \frac{D_0 A}{3}\right)^{-3}$$

The value 200/in^2 is quite typical for normal operations by efficient producers of both bipolar and silicon gate devices. This defect density is a function of design rules; the 200/in^2 figure is about right for 0.4-mil (10-μ) minimum metal line widths, with other dimensions comparable. This figure is not improving rapidly. Smaller geometries usually result in sharply increased defect density.

The model suggests that doubling the shown LSI chip active area would reduce probe yield from about 15 to about 3 percent. Since there are only about 70 chips of this size on a 2-in diameter wafer, a chip cost of \$10 is indicated. Further calculations lead to the determination of optimum chip size under any specified conditions [6].

Once a figure for chip cost is available, total factory cost of a packaged integrated circuit can be estimated using the data on packaging costs provided in IV,2. A summary is given in Table 4-B. SSI and MSI components are assumed to be packaged in plastic or silicone, with assembly outside the U.S.A. LSI is assumed to be packaged in a solder-sealed ceramic package with assembly in California. Cheaper packaging and assembly methods are, of course, applicable to LSI.

It has been emphasized that these cost figures do not include engineering, marketing, or profit. Engineering and marketing costs are probably the dominant factors in the overall cost of LSI today; this will continue to be true as long as the market life of products continues to be short due to rapid innovation.

References

[1] B. T. Murphy, "Cost-size optima of monolithic integrated circuits," *Proc. IEEE*, vol. 52, pp. 1537–1545, Dec. 1964.

[2] R. B. Seeds, "Yield and cost analysis of bipolar LSI," presented at the 1967 Int. Electron Devices Meeting, Washington, D.C.

[3] A. G. F. Dingwall, "High-yield processed bipolar arrays," presented at the 1968 Int. Electron Devices Meeting, Washington, D.C.

[4] R. N. Noyce, "Making integrated electronics technology work," *IEEE Spectrum*, vol. 5, pp. 63–66. May 1968.

[5] G. E. Moore, "What level of LSI is best for you?" *Electronics*, pp. 126–130, Feb. 16, 1970.

[6] J. H. Wuorinen and D. A. Hodges, "Component technologies for future systems," in *Information Processing 68*. Amsterdam: North-Holland, 1969.

Part V
Reliability, Maintainability, and Testing

Editor's Introductory Comments

Reliability assessment and improvement are major problem areas for semiconductor memory, particularly when large systems are considered. This is clear from Fig. 5-1, which shows memory system mean time between failures (MTBF) as a function of memory storage capacity. The assumed chip failure rate in Fig. 5-1 is one to two orders of magnitude better than median current field service results for commercial grade integrated circuits in dual inline packages used in ground-based systems. In other words, MTBF of 10^3 - 10^4 h would be expected today for a memory of 10^6 bits storage capacity, employing 1000 bit chips. This is not adequate in many applications.

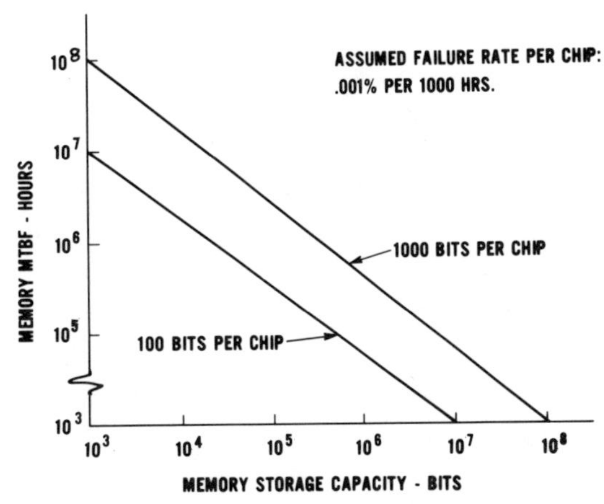

Fig. 5-1. Memory system mean time between failures as a function of storage capacity.

This failure rate range of 0.01 to 0.1 percent per 1000 h for a single chip in a dual inline package does not tell the whole story. Reliability may vary by orders of magnitude from one lot to another of nominally identical integrated circuits. A range from 0.0007 percent per 1000 h to 20 percent per 1000 h, for large lots of similar integrated circuits, is tabulated in a recent publication of the Reliability Analysis Center [1]. High failure rates can usually be traced to poor supervision of labor or poor process controls at some stage of the manufacturing process.

One answer to these problems involves a high degree of automation in the manufacturing process, very thorough testing of the product, and failure analysis of every failed part. IBM has pursued this path with real success, as indicated in V,2. A failure rate of 0.001 percent per 1000 h is documented from field data for small-scale integrated circuits employing IBM's SLT packaging technology. No failure rate data have been reported yet for the semiconductor main memory in the System 370 Model 145, but it is not unreasonable to expect the failure rate to be in this approximate range per chip; perhaps it will be better, since continuing improvements are being made in SLT.

Returning to commercial devices in dual inline packages, a number of interesting observations may be made. Available data show that most failures are related to interconnections. Failure rates are not systematically different for hermetic and nonhermetic packages, except in very high humidity conditions. Failure rates are similar for devices with a similar number of interconnections, regardless of chip complexity. Finally, there is no systematic difference in the failure rates of bipolar and MOS components, although, of course, there are much less data for MOS.

In summary, a conservative first-order model for the reliability of a memory or other large ground-based system employing commercial grade integrated circuits might assume a typical failure rate of 0.1 percent per 1000 h per integrated circuit. Occasional manufactured lots showing a failure rate two or three orders of magnitude worse than this should be expected. Possibly well-planned screening techniques could be effective in isolating bad lots.

Reference

[1] "Microcircuit failure rates—December 1971," Reliability Analysis Center, IIT Res. Inst., Chicago, Ill.

Some Reliability Considerations Pertaining to LSI Technology

MICHAEL M. SCHLACTER, MEMBER, IEEE, RALPH S. KEEN, JR., AND GEORGE L. SCHNABLE, SENIOR MEMBER, IEEE

Abstract—**Large-scale integration (LSI) has been attractive from the systems point of view because of its promise of high reliability as well as lower costs. New trends that have developed in LSI technology, however, introduce new failure modes that potentially can offset any anticipated gains in reliability. This paper discusses the new trends that have developed in LSI, and the requirements for refinements in process technology, closer process control, and improvements in design that are considered to be essential for the full realization of high LSI reliability. Particular attention is paid to advanced multilevel-structure processing and to the use of specially designed test vehicles for purposes of process improvement and process control.**

Introduction

THE general incentive for the development of large-scale integration (LSI) has been the prospect of lowering the cost of large electronic systems that are based on integrated circuits while simultaneously increasing system reliability. This is accomplished by the judicious partitioning of systems into component arrays such that the cost per function accomplished in the total systems is reduced to a minimum. The minimum is determined by a tradeoff between high assembly and packaging costs per function that occur when the number of functions per chip is low, and low manufacturing yield that generally accompanies the use of extremely large chips [1]–[4].

All other conditions remaining the same, higher reliability would appear to be a companion advantage of the lower costs accomplished by the LSI approach [5]. It would be attained by the use of fewer wire bonds and external interconnections and by a reduction in the number of packages and of circuit components such as drivers. Since increased circuit complexity results in lower total packaging costs, then for some applications it will be economically possible to use hermetically packaged LSI circuits rather than plastic encapsulated conventional components.

Available data on the failure rates of integrated circuits under various levels of stress indicate that failures of wire bonds are one of the principal limitations of integrated-circuit reliability [6]–[11]. Data on other methods of making connections to the integrated-circuit chip also indicate reliability limitations [12]–[15]. Thus the degree to which reliability improvement can be obtained by use of LSI depends, to a large extent, on the degree of success in partitioning the system to obtain high gate-to-pin ratios. The most favorable improvement in gate-to-pin ratios can be obtained with serial shift registers that require the same number of pins regardless of register length. The resulting improvement in gate-to-pin ratio is directly proportional to the increase in circuit complexity. Another advantage of LSI is that the number of output buffers required per gate falls off with added complexity. Increases in array complexity also reduce the number of packages required, and therefore reduce the probability of failures in a system due to package failure [16]–[18].

However, new trends have developed in LSI fabrication, some of which can introduce new device failure modes or aggravate existing ones [3], [5]. These trends are multilevel metalization, deposited dielectric, smaller geometry, higher functional packaging, larger chip size, newer packages, advanced circuitry, higher power density, less complete functional testing, and lower unit volume production. Although the concept of LSI suggests a very high potential reliability, it is a reliability that can be realized only if improvements in circuit design, process techniques, and procedures to control processing are introduced that ensure that the possible adverse effects of the new trends in LSI will not seriously affect integrated-circuit reliability. To accomplish this, progressively greater use is being made of sophisticated tools of analysis that are now available. In addition, new techniques are being developed for the evaluation and assessment of LSI processing sequences and device performance. For example, much of the data for this study were obtained through the use of scanning electron microscopy [19], [20], interference contrast microscopy [21], and specially designed test vehicles [5], [22]–[25].

Test Vehicles

Test vehicles are integrated-circuit chips containing test structures specifically designed to provide information about the effects of integrated-circuit processing on fundamental material and device parameters. Test vehicles are quite useful for process improvement engineer-

Manuscript received August 24, 1970; revised February 23, 1971. This paper was presented at the ISSCC, Philadelphia, Pa., Feb. 1970.

M. M. Schlacter was with the Microelectronics Division, Philco-Ford Corporation, Blue Bell, Pa. 19422. He is now with the Aeronutronic Division, Philco-Ford Corporation, Newport Beach, Calif. 92664.

R. S. Keen, Jr., was with the Microelectronics Division, Philco-Ford Corporation, Blue Bell, Pa. 19422. He is now with the Materials Division, Inselek Company, Princeton, N. J. 08540.

G. L. Schnable was with the Microelectronics Division, Philco-Ford Corporation, Blue Bell, Pa. 19422. He is now with RCA, David Sarnoff Research Center, Princeton, N. J. 08540.

Reprinted from *IEEE J. Solid-State Circuits*, vol. SC-6, pp. 327–334, Oct. 1971.

ing and manufacturing process control, as well as wafer screening and reliability assessment. In some cases test vehicles may be used on every chip per wafer, and in other cases test vehicles may be used on only select chip locations per wafer. In some processes entire lots of wafers with test vehicles may be processed, and in other cases only an occasional wafer containing test vehicles may be processed. The information obtainable from test vehicles includes information about bulk effects (that is, effects in the silicon), information about surface effects (phenomena at the Si–SiO_2 interface), dielectric effects (effects in SiO_2 or in other insulators), and metalization effects (phenomena in aluminum or other metalization systems).

There are two general types of information obtainable from test vehicles: 1) the average properties of the semiconductor, oxide and metal regions on the wafers; and 2) information about properties dependent on the incidence of localized defects. Average properties, most of which can be obtained from relatively small area test vehicles, include doping density N_D, sheet resistivity of diffused regions ρ_s, oxide charge Q_{ss}, oxide charge stability Q_0, surface recombination velocity s, and discrete device properties I_R, V_{BR}, β, V_{GST}. Test vehicles can also provide information about oxide breakdown voltage, metalization sheet resistivity, via resistance and related parameters.

The yield and reliability of complex integrated circuits are also determined or limited, to a large extent, by the presence of localized defects. Larger area test vehicles can provide information about oxide defect density, localized variations in Q_{ss}, localized avalanche or high generation regions, or localized areas of open or constricted metal at steps. Figs. 1 and 2 illustrate two examples of test vehicles that were designed for evaluation of MOS processing. For the test vehicle in Fig. 1, the tests were designed to provide information about material properties such as oxide breakdown strength, conductor and diffusion resistivity, contact resistance, and film thickness. For the test vehicle in Fig. 2, the tests were designed to provide information about surface phenomena such as mobile and immobile charge density, fast state density and surface ions, and the effect that those phenomena have on device parameters such as threshold voltage and surface breakdown voltage.

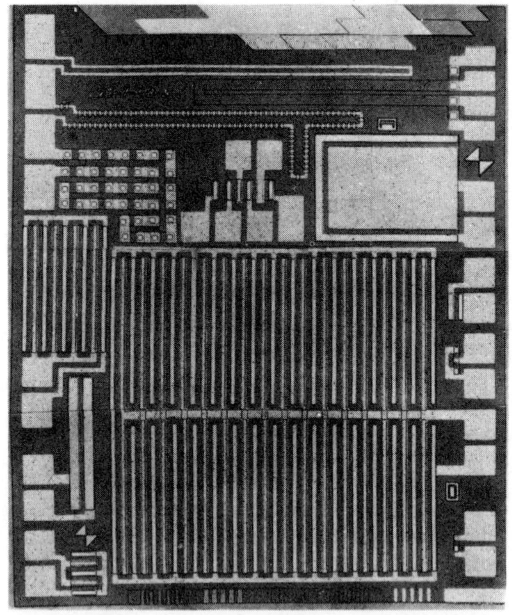

Fig. 1. MOS material parameters test vehicle. 47 mils × 59 mils.

Fig. 2. MOS surface phenomena test vehicle. 47 mils × 59 mils.

Failure Mechanisms of Conventional Integrated Circuits

While some data on LSI reliability exist, the amount available is not extensive. However, knowledge of the failure mechanisms prevalent in conventional integrated circuits permits the prediction of some of the failure mechanisms that will seriously limit the reliability of LSI circuits. These failure mechanisms include ion migration, [26], [27], microcracks in metalization at oxide steps [28]–[30], electromigration [31]–[33], shorts through oxide layers [5], [34]–[36], and open wire bonds [6]–[11].

In digital bipolar integrated circuits, failures are principally because of open bonds, open metal at steps, and electromigration. MOS devices contain thin oxide in certain regions. One of the most common failure modes in MOS devices, therefore, is shorts or leakage paths through localized regions in the thin oxide [10]. Also, MOS devices, in contrast to bipolar devices, are fabricated on higher resistivity material and operate at higher voltages than bipolar devices; thus they are more sensitive to surface-related failure mechanisms [22] than to failure mechanisms that involve continuity of metalization or high current density in metals and contacts.

Surface-related failure mechanisms are caused by changes that occur in the electrical properties of the Si–SiO_2 interface as a result of motion of mobile charge in or on the SiO_2, trapping of charge in the SiO_2, or existence of fast states at the Si–SiO_2 interface. Instability in the fundamental parameters of that interface can cause channeling either under or beyond metal, can change the voltage at which punchthrough occurs, and can increase diode reverse current. Circuits can also degrade due to parasitic lateral bipolar transistor action, or due to changes in the threshold voltage, or field inversion voltage of MOS transistors.

Localized defects in the silicon and in the thermally grown SiO_2 are a factor in integrated-circuit reliability and are a major factor affecting the yield when they occur in active device areas. Thus, yield falls off severely as circuit area increases and the probability of occurrence of a localized defect that can cause reliability problems increases with increasing chip area. However, the yield and reliability of very large chips decreases less rapidly than the exponential decrease that is predicted for purely random, distinguishable defects. Some have attributed this observation to the fact that localized defects do not occur completely randomly [1], [37]; others have attributed it to the fact that the defects should be treated as indistinguishable rather than distinguishable events [38]. Nevertheless, due to the large active areas used in LSI, localized defects will be an important factor in yield and reliability and will create a need for process refinements that reduce the incidence of material defects.

New Failure Mechanisms of LSI

Information is being accumulated by integrated-circuit manufacturers on the failure modes that are directly related to the trends in LSI. In many cases, process revisions aimed at improving yield and reliability have already been made. Several categories of failure, which are the result of using multilevel metal, can occur in integrated circuits. These categories are metalization discontinuities [39], shorts through deposited dielectric [40], effects of the deposited dielectric on silicon surface potential [22], and effects of the deposited dielectric on the properties of the metalization [31], [41]. Other failure modes result from the use of smaller geometry, larger packages, and higher power density.

Multilevel Metalization Discontinuities

The principal locations at which either total or partial metalization discontinuities occur are at the metal–metal interface where metalizations of different levels make contact (via contact), at the edge of the etched vias in deposited dielectric, which permit the metal–metal contact, and along the line in the top view at which an upper-level metalization crosses the edge of an underlying one.

Techniques for obtaining reliable low-resistance via contacts include the use of suitable heat treatments [42] or of RF sputter cleaning [43], [44] prior to deposition of the second-level metal. Metal continuity at via contacts can effectively be evaluated by means of test patterns that provide a large number of via contacts in series [5]. Comparison of the resistance of such patterns with the sheet resistivity of first- and second-level metal permits calculation of via resistance. These structures, when subjected to stresses such as temperature cycling or high current density can provide additional information about the reliability of the multilevel metalized pattern.

Fig. 3. Vapor-plated PSG delineated with tapered edges.

Techniques for ensuring continuous metal over delineated dielectric steps at via contacts strive for control of the angle at the edge of the delineated dielectric. One technique is to use two via masks with slightly different cut sizes so that the step occurs as two smaller steps rather than a single large step. Another technique is to use two different types of deposited dielectric that differ in etch rate, causing a less steep step to be etched in the contact window.

Processes have been developed that permit almost any desired edge angle to be obtained. For example, Fig. 3 illustrates an island of chemically vapor-deposited phosphosilicate glass (PSG) [45], which has been delineated by a special technique. The PSG is 1 μ thick and the taper of the delineated edge, which is approximately linear as evidenced by the interference fringe spacings, extends over a lateral distance of 10 μ. This corresponds to an angle of less than 6° with the plane of the substrate. Such a structure would be expected to eliminate all possibilities of a reliability hazard.

Another possible cause of failure in multilevel arrays arises because of the possibility of open second-level metal lines where they cross delineated first-level metal. These metal lines may be open or greatly reduced in cross-sectional area as a result of notches due to photoresist (Fig. 4), microcracks at deep steps (Fig. 5), the formation of tunnels under the metal (Fig. 6), filamentation of the metal (Fig. 7), shadowing due to evaporation from a point source at a low angle relative to the substrate, or to a number of other effects. Figs 4–7, which illustrate some of these phenomena, are scanning electron micrographs of structures with two levels of Al, each 1 μ thick and 0.6 mil wide, separated by a chemically vapor-deposited layer of PSG also 1 μ thick.

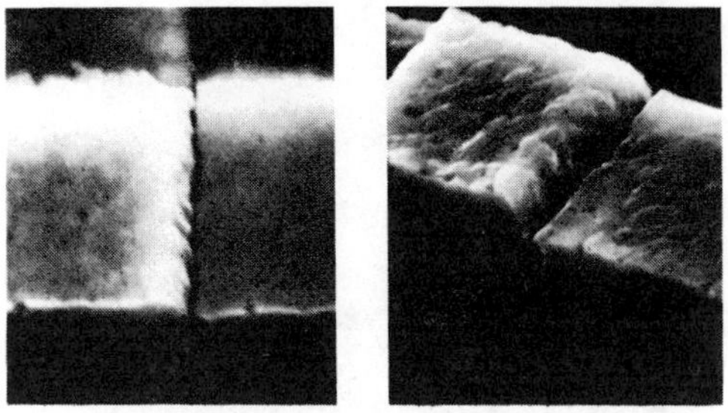

(a) (b)

Fig. 4. Photolithographic notching. (a) Top view. (b) Oblique view.

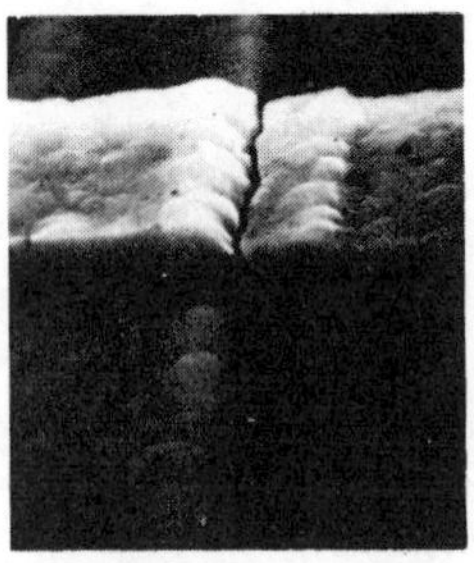

Fig. 5. Microcrack at a deep step.

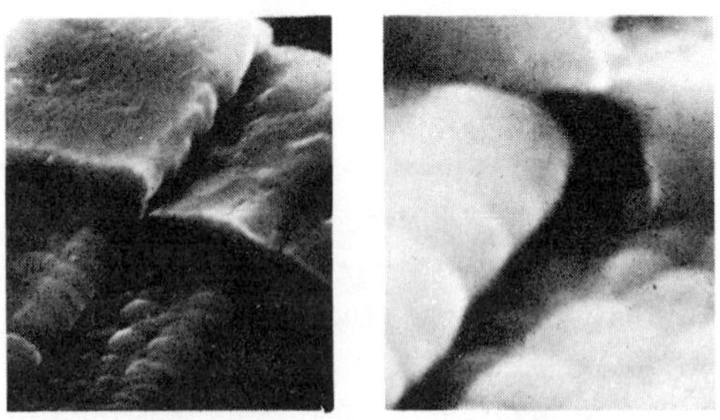

(a) (b)

Fig. 6. Tunnels under metal.

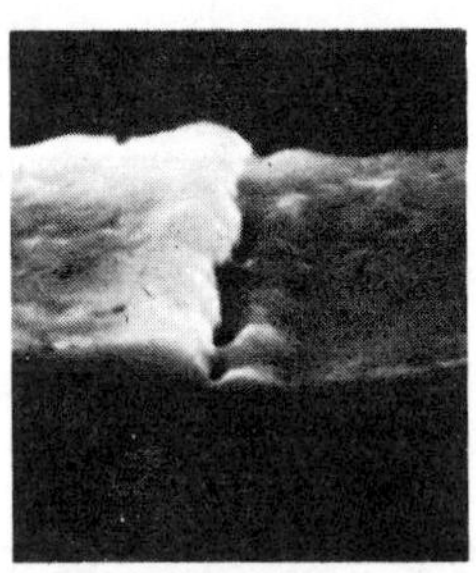

Fig. 7. Filamentation of metal.

Fig. 8. Al pattern delineated with tapered edges.

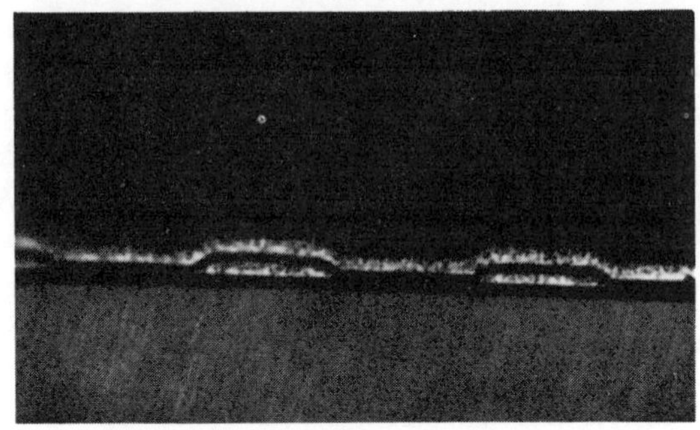

Fig. 9. Cross section of a two-level metal structure in which the first-level metal pattern has tapered edges.

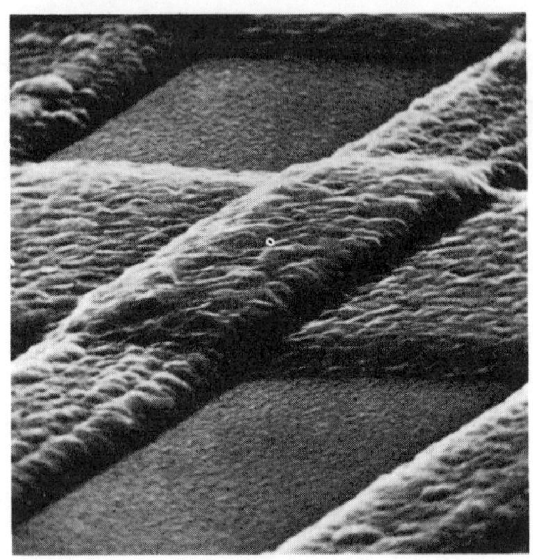

Fig. 10. Two-level metal structure in which both levels of metal have tapered edges. 1.5-mil-wide line crossing over 0.5-mil-wide line.

One method that overcomes the problem of metalization discontinuities is the use of process techniques similar to those used in etching tapered contact vias, which result in a gradual taper of the delineated edge of the first-level metalization. Fig. 8 shows a tapered delineated Al pattern in which Al stripes 1 μ thick and 0.5 mil wide, have edge angles of approximately 20° relative to the plane of the surface. This type of taper is very desirable in first-level metal to ensure that second-level metal crossovers at such delineated steps are not a reliability problem. This fact is illustrated in Figs. 9 and 10. Fig. 9 is a cross section showing tapered Al (1 μ thick and 0.5 mil wide) being crossed by second-level Al. Fig. 10 is a scanning electron micrograph of tapered first-level Al lines being crossed by a second-level Al line, which is also tapered. Both levels of Al and the vapor-plated PSG film, which separates them, are 1 μ thick. Fig. 10 illustrates that the tapering process eliminates microcracks, tunnels, notches, and filamentation.

The problem of discontinuities in metalizations at crossovers depends strongly on the angle of the delineated edge of the bottom-level metal. What ever edge angle exists in the bottom-level metal, the process of dielectric deposition, in certain cases, increases that angle and thus makes it more difficult to cover with continuous metal in the next level. It has been shown that this is very likely to occur under conditions where the bottom-level metal is thick and the angle at its edge is near vertical. An example of the possible effects of vapor plating on increasing the angle of the delineated first-level metal is shown in Fig. 11(a) and (b). In Fig. 11(a) the chemically vapor-plated SiO_2 is 12 000 Å thick and the step over which it was deposited in 6000 Å high. In this case, the step in the SiO_2 is no more difficult to cross with continuous metal than the original step over which the SiO_2 was deposited. In Fig. 11(b) the thickness of the chemically vapor-plated SiO_2 and the step height are both 18 000 Å. The greater height of the underlying step has caused the SiO_2 to become knobby. The knobby overhanging growth of SiO_2 would cause shadowing during metal evaporation, preventing the metal from being continuous over the step. Although this is a factor if the angle in the bottom-level metal is excessively steep, it is not a problem with metal that has been delineated with sides that have a taper of 45° or less.

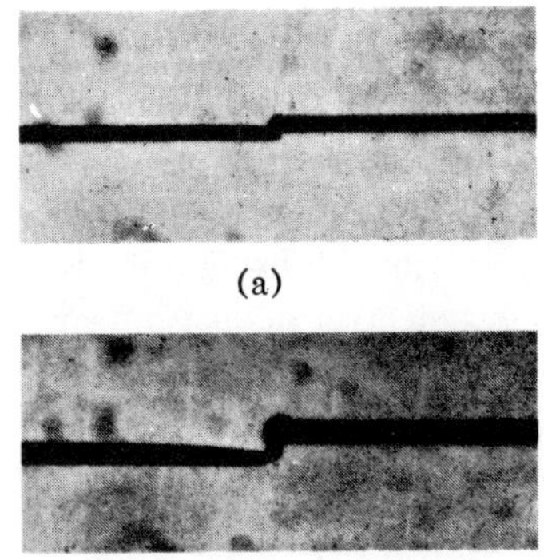

(a)

(b)

Fig. 11. Effect of deposited dielectric on the surface topology above an underlying step.

Shorts Through Deposited Dielectric

For high yields and for reliable multilevel metalized arrays the coverage of dielectric over delineated first-level metal must be continuous and uniform. Densities of pinholes and weak spots must be very low in areas where dielectric covers the edge of delineated first-level metal and in areas where it covers metal that is essentially planar. The extent to which pinholes and weak spots occur are determined by the deposited material and its stress, the conditions of deposition, the cleanliness of the process areas, the possibility of metal–dielectric interaction, and the topography of the underlying substrate [45], [46].

Test vehicles are quite useful in evaluating the densities of planar and edge pinholes or thin spots. Fig. 12, for example, shows a plot of cumulative pinhole density for a dielectric between first- and second-level metal. This plot was obtained using a vehicle for which the bottom electrode was a 1000-mil-long Al resistor, 1 mil wide and 5000 Å thick, the deposited dielectric was vapor-plated SiO_2, 7000 Å thick, and the top electrode was Al, 10 000 Å thick. The measurement conditions were such that localized dielectric breakdown resulted in self-healing discharges rather than shorts [5]. The self-healing effect has been described by Klein[34]. Self-healing occurs with the metalization and dielectric thicknesses typically used in multilevel metalized arrays. The type of plot shown in Fig. 12 provides information about the density of pinholes and of thin spots that can be expected

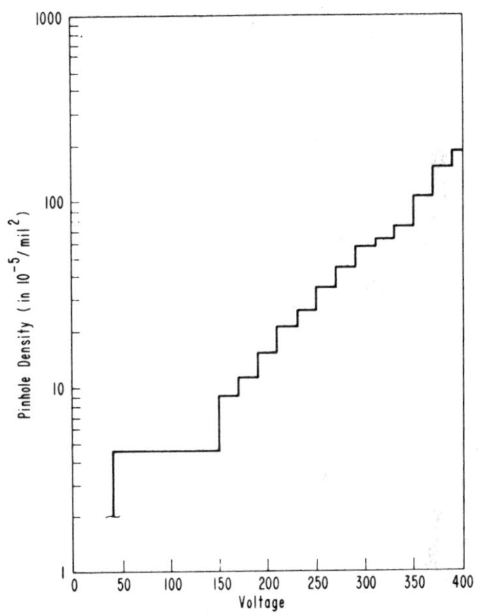

Fig. 12. Cumulative pinhole density in deposited dielectric versus applied voltage.

with a given process at various levels of applied voltage. It also permits an extrapolation to lower voltages to make predictions about yield and reliability of very high quality dielectric films under conditions that would otherwise require measurements to be made on a very large number of samples.

Inspection of conventionally delineated test vehicles after pinhole testing indicates that when the first-level metal line is 1 mil wide, approximately twice as many pinholes occur over the edge of the bottom-level metal as over the top of it. A significant improvement can be made by delineating the first-level metal with tapered sides. The relative number of pinholes that occur over the edge of the underlying metal stripe can be greatly reduced by an amount that depends on the angle to which the sides of the metalization have been tapered. Thus, tapering of the lower levels of metalization is a process refinement that promotes high LSI reliability by offsetting two distinct failure modes: opens in upper levels of metalization and shorting through deposited dielectric.

Effects of Deposited Dielectric on Silicon Surface Potential

The deposition of a dielectric either for passivation purposes or as a dielectric between metal layers in multilevel metalized LSI can have a number of surface-related reliability implications. The materials and temperatures involved in the dielectric deposition process can result in changes in the immobile charge in the oxide, changes in the stability of the charge in the oxide under bias at elevated temperatures, or changes in the surface recombination velocity. Each of these properties of the Si–SiO_2 interface can influence the reliability as well as the performance of the integrated circuit, since the level and the stability of these parameters is an important determinant of reliability.

Table I indicates the effects that two different chemically deposited dielectrics, namely SiO_2 and PSG, have on Si surface potential [45]. The data were determined by MOS C–V measurements and the shift in C–V curves as a result of applying bias at 300°C. It indicates that both dielectrics increase the mobile and immobile charge densities compared to the original thermally grown oxide. However, the amounts of the increases are considerably less with vapor-plated PSG, indicating that it is the superior choice for LSI applications.

TABLE I
EFFECT OF DEPOSITED DIELECTRIC ON MOBILE AND IMMOBILE CHARGE DENSITIES

Multilevel Structure		Charge Density	
First Layer	Second Layer	Mobile (300°C) $10^{11}/cm^2$	Immobile $10^{11}/cm^2$
2000-Å SiO_2[a]	none	<0.1	1.4
2000-Å SiO_2[a]	4500-Å SiO_2[b]	>8.0	4.4
2000-Å SiO_2[a]	2000-Å PSG[b]	0.6	2.4

[a] Thermal oxide.
[b] Vapor-plated dielectric.

Effects of Deposited Dielectric on Properties of Metalization

An important effect that the deposited dielectric has, which affects LSI reliability, is an effect on the susceptibility of the underlying metal to fail as a result of electromigration. Several investigators, for example, have reported that deposited SiO_2 reduces the rate of electromigration in underlying metal [31], [47], whereas the converse has been reported by other investigators [45]. From these conflicting reports one concludes that the process used for dielectric deposition is an important factor in the reliability of the metalization in LSI.

Effects of Geometry

Small geometry is accomplished in integrated circuits by a relaxation of design rules that permits the use of smaller linewidths and smaller spacings between lines [48]. The use of smaller geometry in LSI is a trend that can be associated with the aggravation of certain yield and reliability problems that already exist to a limited degree in medium-scale and conventional-scale integrated circuits. A decrease in the geometry of an integrated-circuit pattern tends to increase its susceptibility to registration and resolution problems, alignment effects, bridgeing by conductive particles, corrosion effects, and photolithographic problems. However, the purpose of using small geometry is to increase function density to permit given circuits to be fabricated on chips of smaller size. The reason is that the probability per chip of encountering a localized material defect depends on the amount of the active device area of the chip.

The dependence of defect occurrence on active device area can be determined by making use of specially designed test patterns with a series of different active areas and perimeters. Fig. 13 shows data on oxide breakdown voltage obtained from four sizes of test structures on the test vehicle show in Fig. 1. The structures were MOST structures in which the gate oxide was thermally grown SiO_2 approximately 1500 Å thick. The channel length for each of the test structures was 0.4 mil, as drawn, and the channel widths were 0.5, 5.0, 50, and 500 mils. The smallest MOST represented the approximate size of a single conventional transistor in an MOS array, whereas the largest represented the aggregate size of all the MOSTs in an array approximately the size of a 250-bit shift register. For the smallest test vehicle, the breakdown typically occurred at 120 V, which is characteristic of a defect-free structure. Breakdown generally occurred at the corners where the highest electric fields exist. For sucessively larger test vehicles there is an increasing incidence of breakdown at lower voltages. For test structures with a channel width of 500 mils, most breakdowns occur at an applied potential of approximately 50 V. This indicates that with the large areas that occur in complex chips there is a very high probability of each circuit containing at least one defect of the type that reduces the effective dielectric strength of the oxide to less that half that of defect-free structures.

Fig. 14 shows the effect that increasing the area and periphery of diffused p^+-n junction diodes has on the incidence of devices with excessive leakage currents. Test structures contained on the test vehicle shown in Fig. 1, modified so that gate metal did not cover the p-n junction at the surface, were used. The junction areas and perimeters for the different sizes of test structures that were used were, respectively, 1.0 mils^2 and 7.0 mils, 9 mils^2 and 20 mils, 80 mils^2 and 140 mils, and 700 mils^2 and 1000 mils. The smallest diodes conducted 10 nA at approximately 90 V; the mechanism for this conduction being avalanche at the surface. As the area of the junction is increased, a larger number of devices which pass 10 nA at less than 90 V are encountered because of the increasing probability that the junction depletion layer occurs in an area in which a defect exists. Some of the large area junctions

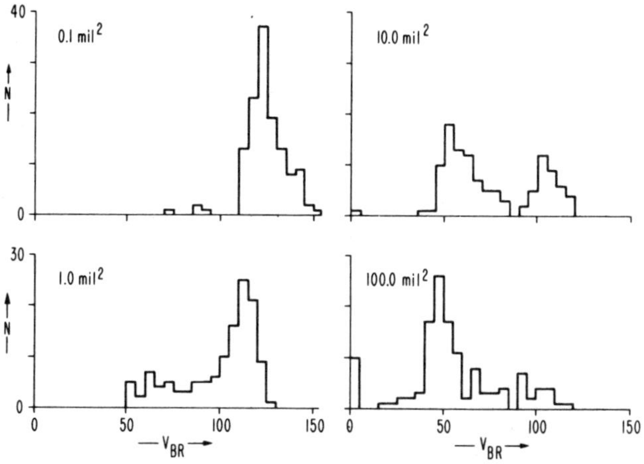

Fig. 13. Effect of area on oxide breakdown voltage.

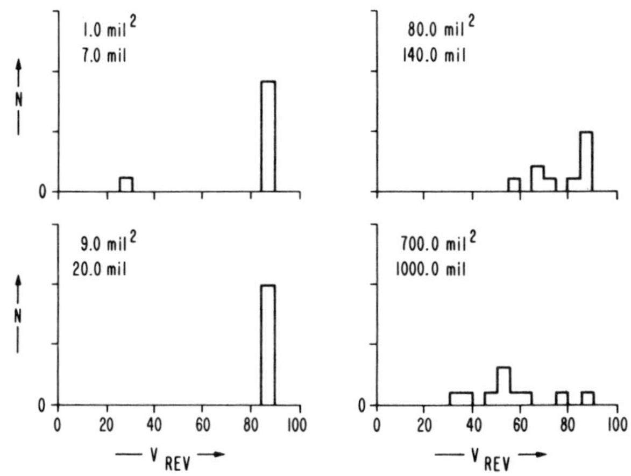

Fig. 14. Effect of area and perimeter on diode voltage at 10-nA reverse current.

continue to have low leakage, indicating that bulk lifetime and surface recombination velocity were not the limiting factors in the leakage of large area junctions.

Strategically designed test vehicles can be constructed to yield information on fundamental material and device parameters. Use of the test vehicles will establish what the norms for those parameters should be under given sets of conditions. Thus, properly designed test vehicles can be a valuable tool for assessing new manufacturing processes and for optimizing the process conditions to obtain the highest yield and to obtain structures that are least likely to contain process-induced defects that may subsequently have an adverse effect on reliability.

Conclusion

Digital circuits based on bipolar transistors are particularly sensitive to metalization-related increases in resistance, whereas circuits based on MOS transistors are particularly sensitive to surface effects and to shorts in thin oxide. Available data indicate that the principal additional modes of failure in multilevel metalized bipolar arrays include open circuits at vias or oxide steps, opens in metal lines as a result of electromigration, and shorts between metal layers at imperfections in the deposited dielectric. In multilevel metalized arrays, particular attention must be paid to the angle of the edge of photolithographically delineated regions. Special techniques have already been developed to ensure that the angle is not excessively steep.

To fully attain the improved reliability capabilities of LSI, it is concluded that a number of integrated-circuit process refinements and additional process controls must be employed. For bipolar circuits the factors that influence metal coverage over oxide steps and the factors that determine susceptibility to electromigration must be closely controlled. Multilevel metalized arrays require control of the contour of the edges of delineated metal and dielectric. In MOS circuits, close control of surface properties and stability must be maintained and more effective input protection against static electricity effects is also desirable. All large-area circuits require improved processing that yields lower densities of localized materials defects.

Although LSI devices may be somewhat less reliable per packaged device than conventional integrated circuits, it is concluded that they will be substantially more reliable per function accomplished, and that complete systems consisting of a given number of functions will also be more reliable when based on LSI rather than conventional arrays.

Acknowledgment

The authors wish to think E. Eichen of the Ford Scientific Laboratory for the scanning electron microscopy and the electron beam X-ray microprobe analysis, and D. Diehl and D. Moyer of the Microelectronics Division, Philco-Ford Corporation for assistance with much of the experimental work.

References

[1] R. B. Seeds, "Yield and cost analysis of bipolar LSI," presented at the Int. Electron Devices Meeting, Washington, D. C., Oct. 1967, Paper 1.1.

[2] ——, "Yield, economic, and logistic models for complex digital arrays," *Digest 1967 IEEE Int. Conv.*, New York, N. Y., Mar. 1967, pp. 498–499.

[3] M. Schlacter *et al.*, "Large scale integrated circuit arrays," Microelectron. Div., Philco-Ford Corp., Blue Bell, Pa., Contract AF33(615)-3620, for Air Force Avionics Lab., Wright-Patterson AFB, Ohio, AFAL-TR-69-23, June 1969.

[4] G. E. Moore, "What level of LSI is best for you?," *Electronics*, vol. 43, Feb. 16, 1970, pp. 126–130.

[5] G. Schnable *et al.*, Procedural guidelines for the reliability assessment of large-scale integrated circuits," Microelectron. Div., Philco-Ford Corp., Blue Bell, Pa., Final Tech. Rep., Contract F30602-68-C-0133, for Rome Air Develop. Ctr., Griffiss AFB, N. Y., RADC-TR-69-220, Aug. 1969.

[6] P. J. Besser, A. B. Menefee, and P. H. Eisenberg, "Failure analysis of minuteman integrated circuit failures," Autonetics Div., North American Rockwell Corp., for Rome Air Develop. Ctr., Air Force Systems Command, Griffiss AFB, N. Y., Contract F30602-67-C-0184, Final Rep., Oct. 1968.

[7] J. Brauer, "Poor QC yields bad IC's," *Electron. Eng.*, vol. 25, Aug. 1966, pp. 78–82.

[8] ——, "Microcircuit testing: Matching the value with the cost," *Microelectronics*, vol. 2, June 1969, pp. 17–23, and July 1969, pp. 23–31.

[9] P. Gott and R. Soltau, "Quality defects in integrated circuits," Microelectron. Div., Philco-Ford Corp., Blue Bell, Pa., for Rome Air Develop. Ctr., Griffiss AFB, N. Y., Contract F30602-67-C-0153, Final Tech. Rep., Dec. 1968.

[10] L. C. Hamiter, Jr., "How reliable are MOS IC's? As good as

bipolars, says NASA," *Electronics,* vol. 42, June 23, 1969, pp. 106–110.
[11] J. E. McCormick, "On the reliability of microconnections," *Electron. Packag. Prod.,* vol. 8, June 1968, pp. 187–189.
[12] K. C. Norris and A. H. Landzberg, "Reliability of controlled collapse interconnections," *IBM J. Res. Dev.,* vol. 13, May 1969, pp. 266–271.
[13] H. T. Hochman, "The art of building LSIs," *IEEE Spectrum,* vol. 6, Sept. 1969, pp. 29–36.
[14] F. L. Howland and J. E. Clark, "Bonding of beam lead integrated circuits," presented at the Int. Electron Devices Meeting, Washington, D. C., Oct. 1967, Paper 12.1.
[15] R. J. Clark, "IC chip interconnection: Replacing the flying lead," *Electron. Packag. Prod.,* vol. 9, Dec. 1969, pp. 54–62.
[16] J. A. Narud, C. D. Phillips, and W. C. Seelbach, "Complex monolithic arrays: Some aspects of design and fabrication," *Microelectronics,* vol. 2, July 1969, pp. 12–17.
[17] R. L. Petritz, "Current status of large scale integration technology," *IEEE J. Solid-State Circuits,* vol. SC-2, Dec. 1967, pp. 130–147.
[18] R. W. Keyes, "Physical problems and limits in computer logic," *IEEE Spectrum,* vol. 6, May 1969, pp. 36–45.
[19] P. R. Thornton, *Scanning Electron Microscopy.* London: Chapman and Hall, 1968.
[20] S. Kimoto and J. C. Russ, "The characteristics and applications of the electron microscope," *Amer. Sci.,* vol. 57, 1969, pp. 112–133.
[21] F. Herzog, "Polarization interferometric methods in incident-light microscopy, with special reference to material testing," *Ind. Anz.,* no. 60, July 27, 1962.
[22] E. S. Schlegel and G. L. Schnable, "Application of test structures for the study of surface effects in LSI circuitry," *IEEE Trans. Electron Devices,* vol. ED-16, Apr. 1969, pp. 386–393.
[23] J. L. Farley, "A simplified procedure for reliability evaluation of MOS LSI circuits," *EEE,* vol. 15, Aug. 1967, pp. 86–88.
[24] E. A. Herr, D. W. Baker, and A. Fox, "The TEG—A test element for the control of quality and reliability of integrated circuits," *Proc. 1968 Annu. Symp. Reliability,* vol. 1, 1968, pp. 201–216.
[25] E. S. Schlegel, "Study of failure modes of multilevel large scale integrated circuits," Microelectron. Div., Philco-Ford Corp., Blue Bell, Pa., Interim Sci. Rep. 4, Contract NAS12-544, for NASA/ERC, Cambridge, Mass., May 1969.
[26] E. H. Snow and B. E. Deal, "Polarization effects in insulating films on silicon—A review," *Trans. Met. Soc. AIME,* vol. 242, Mar. 1968, pp. 512–523.
[27] E. S. Schlegel, G. L. Schnable, R. F. Schwarz, and J. P. Spratt, "Behavior of surface ions on semiconductor devices," *IEEE Trans. Electron Devices,* vol. ED-15, Dec. 1968, pp. 973–979.
[28] R. J. Anstead and S. R. Floyd, "Thermal effects on the integrity of aluminum to silicon contacts in silicon integrated circuits," *IEEE Trans. Electron Devices,* vol. ED-16, Apr. 1969, pp. 381–386.
[29] J. F. Campbell, Jr., and W. H. Shepherd, "The aluminum microcrack," presented at the Electrochemical Society Meeting, Detroit, Mich., Oct. 1969, Paper 198.
[30] T. Yanagawa and I. Takekoshi, "Failure analysis of evaporated metal interconnections at contact windows," *IEEE Trans. Electron Devices,* vol. ED-17, Nov. 1970, pp. 964–970.
[31] J. R. Black, "Electromigration failure modes in aluminum metallization for semiconductor devices," *Proc. IEEE,* vol. 57, Sept. 1969, pp. 1594–1598.
[32] M. J. Attardo and R. Rosenberg, "Electromigration damage in aluminum film conductors," *J. Appl. Phys.,* vol. 41, May 1970, pp. 2381–2386.
[33] I. A. Blech and H. Sello, "The failure of thin aluminum current carrying stripes on oxidized silicon," *Phys. Failure Electron.,* vol. 5, 1967, pp. 496–505.
[34] N. Klein, "The mechanism of self-healing electrical breakdown in MOS structures," *IEEE Trans. Electron Devices,* vol. ED-13, Nov. 1966, pp. 788–805.
[35] C. Fritzsche, "Der dielektrische Durchschlag in SiO_2—Schichten auf Silizium," *Z. Angew. Phys.* vol. 24, 1967, pp. 48–52.
[36] N. J. Chou and J. M. Eldridge, "Effects of material and processing parameters on the dielectric strength of thermally grown SiO_2 films," *J. Electrochem. Soc.,* vol. 117, Oct. 1970, pp. 1287–1293.
[37] A. G. Dingwall, "High-yield-processed bipolar LSI arrays," presented at the Int. Electron Devices Meeting, Washington, D. C., Oct. 1968, Paper 14.1.
[38] J. E. Price, "A new look at yield of integrated circuits," *Proc. IEEE* (Lett.), vol. 58, Aug. 1970, pp. 1290–1291.
[39] R. H. Cox, H. Hentzschel, and T. E. Hartman, "Characterization of multilevel interconnection defects," presented at the Electrochemical Society Meeting, Detroit, Mich., Oct. 1969, Paper 230.
[40] G. T. Mitchell, R. S. Keen, J. B. Kons, and R. C. Dunbar, "A reliable multilevel digital switching matrix," presented at the Int. Electron Devices Meeting, Washington, D. C., Oct. 1969, Paper 5.5.
[41] S. M. Spitzer and S. Schwartz, "The effects of dielectric overcoating on electromigration in aluminum interconnections," *IEEE Trans. Electron Devices,* vol. ED-16, Apr. 1969, pp. 348–350.
[42] G. L. Schnable and R. S. Keen, Jr., "Aluminum metallization—Advantages and limitations for integrated circuit applications," *Proc. IEEE,* vol. 57, Sept. 1969, pp. 1570–1580.
[43] ——, "Failure mechanisms in large-scale integrated circuits," *IEEE Trans. Electron Devices,* vol. ED-16, Apr. 1969, pp. 322–332.
[44] A. W. Horsley, "Glow-discharge unit makes multilayer deposition easy," *Electronics,* vol. 42, Jan. 1969, pp. 84–86.
[45] M. M. Schlacter, E. S. Schlegel, R. S. Keen, Jr., R. Lathlaen, and G. L. Schnable, "Advantages of vapor-plated phosphosilicate films in large-scale integrated circuit arrays," *IEEE Trans. Electron Devices,* vol. ED-17, Dec. 1970, pp. 1077–1083.
[46] W. R. McMahon and T. H. Ramsey, "Analysis of stresses in multilevel metal-insulator structures," presented at the Electrochemical Society Meeting, New York, N. Y., May 1969, Paper 53.
[47] J. Black, "Metallization failures in integrated circuits," Motorola Inc., Final Rep. RADC-TR-68-243, for Rome Air Develop. Ctr., Griffiss AFB, N. Y., Contract AF30(602)-67, Oct. 1968.
[48] R. Luce, "Research and development of high speed processor arrays," Philco-Ford Corp., Interim Summary Rep. 4, for M.I.T. Lincoln Lab., Cambridge, Mass., Contract AF19(628)-5167, Dec. 1969.

SOLID LOGIC TECHNOLOGY COMPUTER CIRCUITS — BILLION HOUR RELIABILITY DATA

E. F. PLATZ

IBM Components Division, East Fishkill Facility

Hopewell Junction, N.Y. 12533

IBM's Solid Logic Technology (SLT) was developed to fill a need for high performance, low cost computer circuits. These microminiature hybrid integrated circuits are the logic building blocks of IBM's System/360. This paper will describe the reliability of SLT and the Failure Analysis Information Retrieval program (FAIR) being conducted on System/360. Although actual module malfunctions have been even fewer than anticipated, those that have occurred have fallen into significant categories which the FAIR program has pinpointed.

Before getting into the mechanics of the FAIR program and the results, it would be well to describe IBM's Solid Logic Technology module. Figure 1 shows the SLT module with its aluminum protective cap in place, as well as the module with the cap removed. You will note that there is no attempt on the part of the aluminum capping operation to provide a hermetic seal. Protection against humidity or moisture is brought about by the introduction of the protective silicon gel coating over the screened resistor and land interconnexions as well as over the active compounds themselves. Again, there is no attempt at a hermetic seal here, but merely a protective barrier to deter corrosion of the metallic materials. Figure 2 shows an enlargement of the semiconductor chips. You will note that the transistor chip consists of a planar silicon device with oxide protection with the overall surface and aluminum metallurgy protected further by means of IBM's patented glass passivation process as reported by Riseman and Perri in *Electronics*, October 3, 1966. Connexion to the outside world (the next level of packaging) is made by reflow soldered connexions involving the use of the three 5 mil "balls" observed on the top of the chip.

The diode chip is actually two diodes on one chip. It also has three balls for interconnexions providing a common terminal for the anode with individual connexions for the cathodes.

IBM's SLT circuits cover a wide range of speeds and functions, but for the purposes of this paper we are considering only the 30 nsec family—the logic workhorse in the Models 20, 30, 40, 44 and 50 computers in the 360 line. In the 30 nanosec family, it has been determined, by noting the distribution of actual component usage, that the average module consists of one transistor chip, two dual diode chips and three resistors.

At the outset of SLT production while System/360 was still being developed, it was necessary to project the reliability of the product. Up to that time, little use had been made in IBM systems of silicon planar devices. Until then, most of the components in IBM's semiconductor logic circuitry were germanium mesa transistors. Projecting reliability rates for the new circuits involved design analysis, as well as making use of the sparse failure rate and failure mode data available on prior technologies in IBM systems. These extrapolations were performed prior to seeing more than sketches of the proposed product. Nevertheless, we projected reliability rates and drew specifications for the use of SLT in IBM systems. As often happens, when making such projections, we subsequently found ourselves to have been pessimistic in the projections of some failure mechanisms and optimistic in others. All in all, our current field

Reprinted with permission from *Microelectronics and Reliability*, pp. 55–59, 1969.

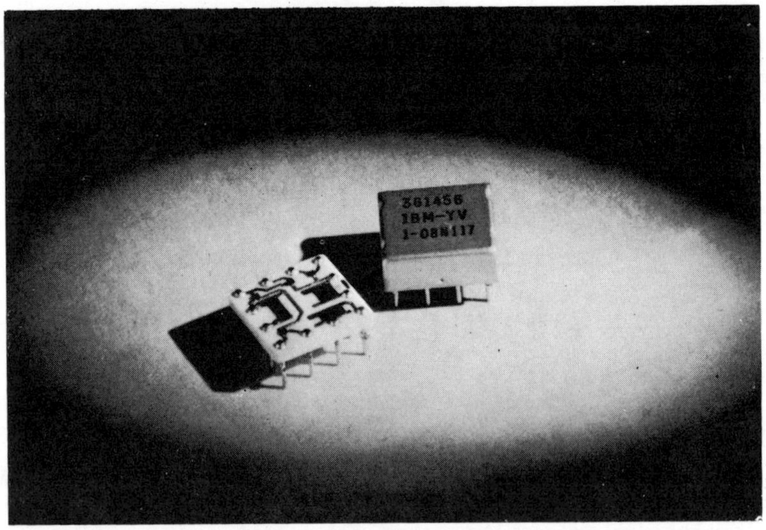

FIG. 1. SLT module.

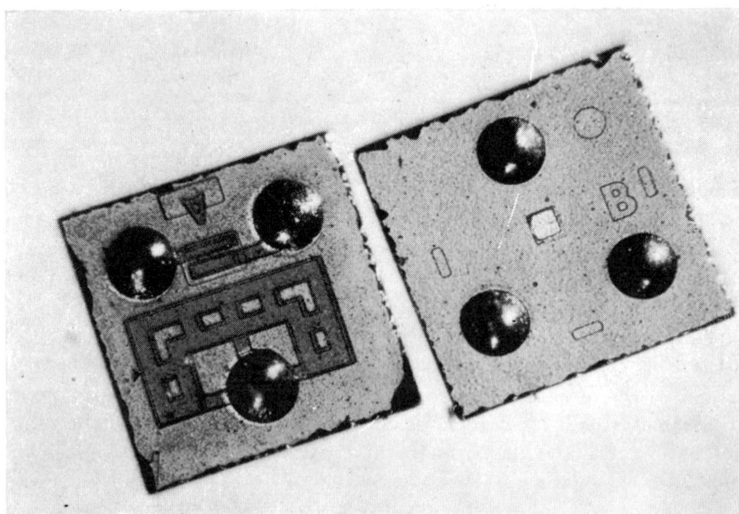

FIG. 2. Semiconductor chips (enlarged).

measurements indicate that our overall estimates were pessimistic.

At the time we made these projections, we recognized that attempts at measuring failure rates in the actual environment of the magnitude projected would require extremely large sample sizes and/or extremely long periods of actual use time. Moreover, the decision on the part of IBM to introduce this new technology across the board in all System/360 at once made it mandatory that a very closely monitored program be implemented to assess the reliability of the product. This need led to the development of the FAIR program. It has subsequently yielded invaluable information to IBM. Field return programs of one type or another cover all IBM systems, but the main thrust of measurement and observation is on the System/360 Model 40 and larger machines. These machines provide the large sample sizes required to make accurate assessments, as well as the rapid accumulation of power-on-hours.

To summarize then, the FAIR program was implemented to perform essentially three functions:

(1) Measure the field performance of the SLT componentry.
(2) Identify any failure modes and mechanisms.
(3) Provide a sounder basis for the projection of reliability of future technologies.

The remainder of this paper is aimed at demonstrating the ability of the FAIR program to meet these requirements.

To measure accurately field performance of components, our reliability people need three critical pieces of information:

(1) the number of failures;
(2) the number of hours of operation;
(3) the total sample population.

Obtaining such information in detail is a relatively simple task when experiments are performed in the laboratory, but when the "laboratory" consists of hundreds of IBM computers of the size of

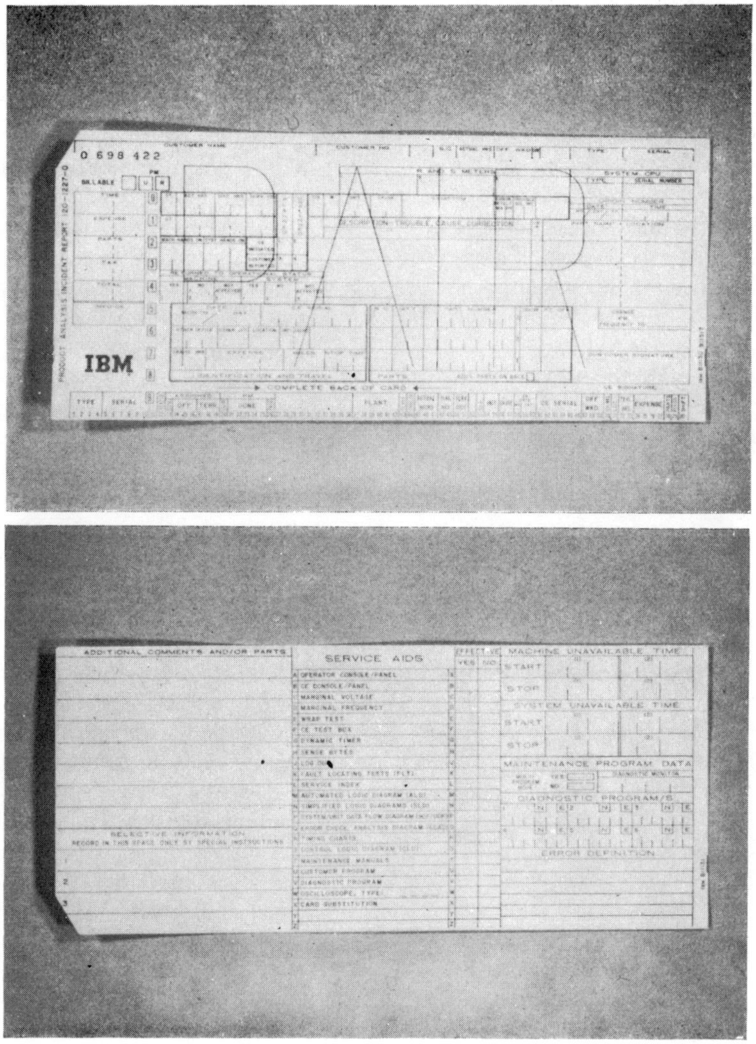

FIG. 3. PAIR card. Front (top); back (bottom).

Model 40s and 50s distributed all over the United States, the retrieval of such information becomes difficult. Obtaining this information required the use of several continuously updated documentation systems as well as the close co-operation of five major divisions of the IBM Corporation. A large part of the burden of documentation fell upon the customer engineer whose real responsibility is the diagnosis and repairing of IBM systems. Upon completing his repair, he is required to complete a "Product Analysis Incident Report" (PAIR) card as well as a systems log sheet, and return the failed component in a pre-addressed mailer carton provided by the Components Division. The subsequent analysis of these documents and the component yields the necessary information regarding the number of failed components in the study group. It should be noted that not only was the incident reported, but also the component itself was returned. By comparing the information return and the component return systems, we established that the great majority of all removed components have been returned for analysis. Armed with the ability of determining just how many components have not been returned, we can accurately assess the true failure rate by pro-rating the observations of the analyzed removals to the known quantity of total removals.

Let us take a look at the documents which are used to retrieve information regarding the incident. The first mentioned is the PAIR card. This card is an expanded version of the standard incident report form used on all IBM systems. It is used to report the installation and maintenance activity on new products and products requiring special analysis. The PAIR provides a high-speed reporting media that communicates additional data on product use, availability, performance and quality. The report is directed to engineering, manufacturing product testing and the field engineering technical groups at the manufacturing site. Both sides of a PAIR card are shown in Fig. 3. Although quite a bit of information is included on such a card for other purposes, the following items are of interest to us. First, the identification of the system from which the removal occurred. This identification includes the type and serial numbers of the system as well as the machine involved. Second, a written description of the trouble caused and the correction involved in repairing the system. Third, the quantity and part numbers of the actual parts removed. Fourth, an indication of the number of hours taken to diagnose and effect the cure.

The second document referred to was the System Log. A System Log is maintained for all audited systems. The Log identifies related maintenance activities and reflects the relationship between the performance of system components (as reported on PAIRS) and the performance of the system itself. The System Log and its related PAIR cards are the bases of system use, availability and performance analysis. Of specific interest to the FAIR program is the entry on the Log providing a further description of the incident involving the removal, particularly aimed at identifying the time of the removal. It is from these Logs that the power-on-hour accumulation of the system is calculated, as well as the actual power-on-time of any given component removed.

The third form referred to is the pre-addressed mailer carton. In addition to being a mechanically sound container in which to return the component, the carton also serves as a source of information. Again, the identification of the part number, as well as the system, and machine serial numbers are included to assure cross-identification of components. Further information included on the carton involves specific comments regarding the component removed. Also, a coded identification of the electrical location of the removed component is given.

A word or two about just what constitutes a failure might well be introduced at this point. The definition of failure used in the FAIR program is a true, practical reliability analysis in terms of the measure of impact on IBM systems. A failed component is defined first as one which was removed as a result of repairing a system having suffered a malfunction. In other words, a removed component must have contributed to a system malfunction. Any components removed because of engineering changes or other system up-dating techniques cannot be classed as failures. Secondly, electrical testing of the component must indicate that it fails to meet its specifications. The so-called NDF (No Defect Found), therefore, is not classed as a failure. I might note that the actual occurrence of NDFs in the returns from IBM systems is not out of line with the expected number of such occurrences based upon the number of components removed per call. The third requirement upon determining a failure is that the component should give no evidence of having been applied outside the ratings for that component. Any evidence of burn-out or electrical overload which can be shown to have been caused by excessive application of currents or voltages would automatically exclude the device from being grouped as a failure. It should be noted that 100 per cent failure analysis is necessary in order to determine adequately the true extent of the performance in the field.

Having established the number of failures occurring in any given observation period, it remains to determine the population and the power-on-hour occurrence during that same interval. Population count is determined by scanning a continuously up-dated configuration tape for all systems involved. Although for our purposes it would be most advantageous if the sample being monitored were stable, this is not the case. Machine-user requirements vary from time to time, and engineering changes are introduced occasionally to improve operational characteristics of systems. Thus, from month to month, the actual component population of a given machine configuration can change considerably. Only by maintaining continuous monitoring of all changes made to all audit systems can an accurate accounting be made of the actual component count on test. By merging the latest up-to-date system configuration with the bills of material for the various machines making up the system, the actual component count by part number can be established for each installation.

The determination of power-on-hours is made by a hand analysis of the systems log maintained at each site. This way, each interruption of a system resulting in a component removal is identified insofar as the actual time of removal. Accurate knowledge is therefore known of the actual number of power-on-hours accumulated by the component up to its time of removal. We assume that the number of power-on-hours of the particular machine from which the component was removed also represents the true number of power-on-hours for the component. Analyses of the accuracy of this assumption have been made, and it can be concluded that an insignificantly small error is introduced in this manner.

By combining these pieces of information—the number of failures, population and the power-on-hours accumulated—we calculated the reliability rate of IBM's SLT. This information-gathering system has been in operation for three years and has resulted in an accumulation of more than 10 billion module hours of reliability data for the SLT family under discussion. I would like to stress that this is actual use information involving a wide

Go2d0

CORRECTED VERSION

range of product vintage as well as a wide range of applications. The advantage of such a retrieval system is obvious, in that no assumptions of acceleration factors need be made, nor are we required to extrapolate failure rates from stress tests beyond the range of the data. Measured in the program just described, SLT has demonstrated a module failure rate of less than 0·003 per cent per 1000 hr for product manufactured as early as 1964 and as late as our most recent reports. In addition to the vintage range of the product, it should be recognized that systems power-on-hours ranging from less than 2000 to more than 12,000 are included. As indicated earlier, the product is more reliable than originally projected.

Thus, even with the mix of product early on the learning curve for SLT, as well as some short-term hours of operation product, a mean time to failure of the order of 33 million hours has been demonstrated. Based on analyses of sub-sets of the data including only later vintage product (1966), indications are that the mean time to failure approximates 100 million hours for SLT. These are actual observations, not extrapolations from stress conditions.

The second major purpose of the FAIR program is to identify principal failure modes and mechanisms. With the return of all removed components along with the information surrounding the occurrence of the removal, it becomes a simple task indeed to seek out and identify major failure mechanisms. In the case of SLT, the original projections anticipated resistor drift problems, occasional corrosion of the substrate metallurgy, problems due to inversion of the semiconductor surfaces and damaged aluminum metallurgy on the semiconductors themselves. Failure due to mechanical opens of the ball-joint contacts was of course anticipated, and a portion of the original projection was attributed to this mode. As it turned out, however, component failure due to corrosion of metallurgy, inversion of semiconductor surfaces, or problems with the resistors has been insignificantly small. Some problem attributable to scratched aluminum lands has been observed, but the only significant problem with SLT has been mechanical opening of the reflow solder ball joint. Fortunately this problem was detected very early in the program (in fact it was detected in life tests prior to the introduction of any system to customers' offices), and improvements were made in the interconnexion technique. Nevertheless, it continues to be a factor. It should be noted that even with this somewhat large (68 per cent) contributor to the overall failure rate of SLT, the product still far surpasses our expectations in overall reliability. Even though something in excess of 30 billion resistor hours of observations is included in the data, there has not yet been one component failure attributable to the resistor. Such an observation indicates that the failure rate of the SLT resistor is less than 0·000008 per cent per 1000 hr at the 90 per cent confidence level. Again, in terms of mean time to failure, this indicates that the SLT resistor has a mean time to failure in excess of 12 billion hours. In order of decreasing severity, identifiable failure modes of SLT are:

Malfunction modes

Ball joint	68
Lands, mechanical damage	16
Surface defects	7
Miscellaneous	9
	100%

The third major purpose of the FAIR program is to provide the basis for future reliability projections. It can easily be seen that a closely monitored program, such as FAIR, is providing us with the necessary information to project to future technologies. The step from IBM's SLT, a hybrid technology, to complete integration in a monolithic structure is considerably smaller than the step indicated earlier, that is, from germanium mesa devices to a silicon planar technology. Monolithic and hybrid packaging techniques are almost identical. It is expected, therefore, that reliability projections to such monolithic technologies will be more easily made, and will be made with higher degress of accuracy than were possible in the past. Of even more value, obviously, is the guidance provided by the assessment and identification of the major failure modes and mechanisms in SLT. Indications are that increased reliability lies in the direction of decreased number of interconnexions per logic function. This consideration would seem to point toward large-scale integration as a way to go for increased reliability.

This factor, in addition to yield and cost factors, is being weighed by IBM and is, of course, being reflected in the direction to be taken in the new technologies and development. One aspect in achieving high levels of reliability in component manufacturing is the quantity of product manufactured. There are indications that extremely high volume (as in SLT) tends toward greater reproducibility and greater learning from a reliability viewpoint. As we progress to large-scale integration, the smaller numbers of specialized functions produced may impact the experience factor otherwise to be gained with large production.

The FAIR program is continuing to show us that improved reliability is to be found in large-scale integration. Whether the other factors involved will outweigh the information learned from FAIR remains to be seen, but again, it seems that only through actual manufacturing and use experience can accurate answers be obtained.

Corrected page 191 for Semiconductor Memories, edited by David A. Hodges.

MOS Integrated Circuit Reliability

GEORGE L. SCHNABLE, Senior Member, IEEE, HENRY J. EWALD, and EARL S. SCHLEGEL, Member, IEEE

Abstract—**This paper presents information on the reliability of MOS integrated circuits based on p-channel enhancement-mode transistors, and describes their failure modes and mechanisms. The principal failure mechanisms were ion migration at the surface and oxide shorting. The results of experimental studies of the effects of variations in construction, processing, and levels of stress are presented, and are compared with other available information on MOS integrated circuit reliability. The failure rate for commercially available complex MOS arrays is on the order of 0.001 to 0.01 per 1000 h of operating life at 125°C for arrays containing approximately 600 p-channel transistors. This corresponds to a failure rate on the order of 5×10^{-6} to 5×10^{-5} per equivalent gate per 1000 h. The effects of device complexity, operating temperature, and other factors are discussed. A reliability prediction equation for MOS integrated circuits is derived from available information. An overall activation energy for functional failure mechanisms of approximately 5 kcal/mole (≈0.2 eV/molecule) is considered applicable to typical MOS integrated circuits. Thus, the failure rate of MOS devices operated at 50°C ambient temperature can be predicted to be on the order of 10^{-6} to 10^{-5} per equivalent gate per 1000 h.**

Reader Aids:
Purpose: Report of experimental work
Special math needed for explanations: None
Special math needed for results: None
Results useful to: Device engineers, reliability physicists

Introduction

The MOS integrated circuit market is growing very rapidly and it is therefore very important that a broad understanding be developed of the failure rates and failure mechanisms of these circuits.

MOS and bipolar integrated circuits have a number of similar failure mechanisms; in addition, MOS integrated circuits have a number of failure mechanisms that are unique to MOS devices or are relatively unimportant in digital bipolar integrated circuits [1], [2]. On the other hand, certain failure mechanisms that are important in bipolar integrated circuits are relatively uncommon in MOS arrays. These differences are not unexpected in view of the significant differences in design, fabrication, and operating principles of these two types of planar silicon integrated circuits [3]. Since most MOS integrated circuits represent a complexity in excess of 100 equivalent gates per chip, MOS integrated circuits offer an excellent vehicle for studies of the effects of chip complexity on reliability. The understanding of the effects of chip complexity on reliability, and of the failure rates of packages and wire bonds, can be used to make predictions about the effects of system partitioning on equipment reliability.

Manuscript received May 14, 1971; revised August 5, 1971. This work was supported in part by the U.S. Air Force, Rome Air Development Center, Rome, N.Y., under Contract F30602-69-C-0252.

G.L. Schnable was with the Microelectronics Division of the Philco-Ford Corporation, Blue Bell, Pa. He is now with the RCA Laboratories, Princeton, N.J.

H.J. Ewald was with the Microelectronics Division of the Philco-Ford Corporation, Blue Bell, Pa. He is now with the Ford Motor Company, Dearborn, Mich.

E.S. Schlegel was with the Microelectronics Division of the Philco-Ford Corporation, Blue Bell, Pa. He is now with the Westinghouse Research Laboratories, Pittsburgh, Pa.

Experimental Results

Life Tests

During the past several years, life tests were run on a total of 5295 devices manufactured by Philco-Ford's Lansdale, Pa., operation for a cumulative total of 16.8 million device-hours. The device types included a dual 50-bit ratio-type dynamic serial shift register, a 256-bit ratioless-type dynamic serial shift register, a 1024-bit static READ-ONLY memory, and other complex digital arrays. The devices were fabricated by several types of processes involving variations in design, passivation, and packaging.

All devices were fabricated using 5-Ω·cm n-type silicon of (111) orientation. Source and drain electrodes were planar diffused, and the gate dielectric was thermally grown silicon dioxide. The metallization was aluminum, single level, and deposited by electron-beam evaporation. The gate threshold voltage was approximately 5 V.

Phosphosilicate glass (PSG) 1-μm thick was used as a passivation layer on all MOS devices, and was chemical vapor deposited at 400°C using SiH_4, PH_3, and 0_2. The PSG, which contains 3 percent phosphorus by weight [4], was deposited under conditions that result in a low stress film, and is thus free of cracks even when subjected to subsequent processing temperatures as high as 550°C.

The overall failure rate on operating life for recently manufactured dual 50-bit ratio-type dynamic serial shift registers was less than 0.004/1000 h at 125°C, with approximately one-half of the failures due to ion migration effects. This type of failure is recoverable, in that baking for a few hours at 200°C without applied bias restores the device to its original characteristics. Most of the other failures were due to shorts through the gate (thin) oxide between the silicon and metal.

The failure rate was not constant, but decreased with time, particularly during the first several hundred hours of operation at 125°C. The average failure rate during the first 168 h of life was approximately three times that for the following 2000 h.

Table I shows the relative distribution of failure mechanisms in all types of MOS devices manufactured in the Lansdale, Pa., production facilitiy during the last several years. The stress

Reprinted from *IEEE Trans. Rel.*, vol. R-21, pp. 12–19, Feb. 1972.

TABLE I

Relative Distribution of Failure Mechanisms in MOS Integrated Circuits

	Percentage of Failures
Ion migration	64
Shorts through oxide	31
Hermeticity or package defects	3
Open wire bonds	1
Particulate shorts	1
	100

TABLE II

Relative Failure Rates of MOS LSI Circuits on Extended (≥1000 h) Life Tests for Various Types of Stresses

Type of Stress	Relative Failure Rate
Ring counter operation at 125°C	100
Ring counter operation at 150°C	150
Ring counter operation at 75°C	50
Ring counter operation at 125°C with periodic cycles of the ambient temperature down to −55°C	230
Storage at 125°C with dc reverse bias applied	190
Storage at 150°C with zero bias applied	15
Storage at 75°C with zero bias applied	10

TABLE III

Effect of Various Process, Construction, and Design Features on Functional Failure Rates Relative to Standard Construction MOS Circuits Operated at 125°C

Type of Construction	Relative Failure Rate
Standard construction, 1970, including 1-μm-thick PSG passivation, hermetic sealing (ten-lead TO-100)	100
Standard construction, 1970, after screening test and 168-h burn-in	40
With 2-μm-thick PSG layer	100
Without passivating glass layer	120
Circuits manufactured before 1970	200

tests were principally 125°C operating life and 125°C reverse bias life. These statistics are based on the sample of 5295 devices, and included a total of several hundred failed devices.

Table II gives relative failure rates for different variations in the conditions for functional stressing, storage, and ring counter [5] operation.

In Table III the effects of various process, construction, and design features are related to the functional failure rates of Philco-Ford MOS LSI circuits of standard construction operated at 125°C for 1000 h. The standard construction included wafer processing techniques designed to minimize the incidence of crystallographic defects or oxide defects [6], techniques to insure against mobile ions in gate or field oxides [6], [7], a 1-μm-thick phosphosilicate passivation layer over the entire chip surface [4], and hermetic sealing.

Glass passivation caused the failure rate due to degradation (i.e., the circuit was functional but leakage or other parameters were not within specifications) to decrease by a factor of two, and improved the overall failure rate, including both functional and degradation failures, by approximately 20 percent [4].

Microcircuit life test data of PSG passivated devices show that the stability in ceramic dual in-line packages (cerdip) and in ten-lead TO-100 packages is approximately equal. The cerdip packaging operation involves a bake at over 500°C to complete the hermetic seal. Test structure data taken on MOS capacitors with 0.2-mil (0.05-mm) line width [7] covered with PSG similarly show that the stability of devices packaged in cerdip is comparable to that in TO-5 packages (less than 5×10^{10} alkali ions/cm^2).

Failure Mechanisms

Several types of surface ion migration effects were observed, with one commonly occurring type attributed to a negative ion-type migration phenomenon in the field (thick) oxide [8], [9]. Negative ion-type instability occurred in the field oxide under metal crossover lines, and also in unmetallized field oxide areas, where the mechanism involved the combined effects of surface-ion migration plus negative-ion migration. In some cases, failures were due to negative-ion migration which resulted in surface depletion of the surface n-type region, and this increased minority carrier generation. In other cases, actual field inversion occurred.

Oxide shorts were found principally through the thin gate oxide in the areas where the gate metal overlaps p-type source and drain regions. Those shorts attributable to voltage overstress (static electricity discharge) were more commonly found under metallic areas that were connected to lower capacitance inputs. Those shorts that occurred during operating life, at voltages within device ratings, have been attributed to localized defects in the thin thermally grown oxide.

Field Returns and Failures

Field returns provided additional information on MOS integrated circuit reliability. Field returns were electrically tested and the degraded and inoperable devices have been subjected to failure analyses. The analyses indicated that a large group was damaged by electrical stress in excess of maximum ratings. Of those field failures that showed no evidence that the ratings had been exceeded, shorts through gate oxides and ion migration in or on the oxide surface were the major mechanisms of failure.

Life Tests of Devices of Other Manufacturers

Life tests conducted at Philco-Ford on commercially available MOS devices from five different manufacturers for over 412 000 device-hours of high-temperature operating life showed that the failure rate per equivalent gate averaged 0.00018 per 1000 h (75°C and 125°C combined). The principal failure mechanisms were oxide shorts and ion migration at the surface.

TABLE IV

Relative Distribution of Locations at which Inputs were Shorted as a Result of Static Electricity Discharges During Device Handling

Device Terminal	Relative Distribution of Shorted Devices (percent)	Length of Edge of Avalanche Diode Under Metal (mil)	Approximate Input Capacitance of Packaged Device (pF)
Input 1	32	40	2.5
Input 2	52	18	2.5
Clock 1	8	60	35
Clock 2	8	56	35

TABLE V

Applied Voltage that Caused Gate Oxide Breakdown for Two Types of Input Protection Devices

Construction of Input Protection Device	Median Capacitor Voltage at which Oxide Breakdown Occurred: 10-kΩ Series Resistor (volts)	500-ΩSeries Resistor (volts)
p^+-n field-enhanced avalanche diode	260	210
p-n-p transistor designed for simultaneous field-enhanced avalanche and punchthrough	>600	260

Static Discharge Effects

An analysis of ratio-type dynamic serial shift registers that were inadvertently subjected to static electricity discharges during device handling is given in Table IV. The clock lines had several orders of magnitude more area of metal over thin oxide than the logic inputs, and thus had higher capacitance to the substrate.

Effectiveness of Input Protection Devices

Oxide shorts may occur because of electrical breakdown of the thin gate oxide during testing or operation within maximum voltage ratings, or as a result of transients or static electricity that raises the voltage applied to certain device pins above maximum rated voltage. Because MOS devices have a high input impedance, they are particularly susceptible to damage due to static electricity discharges. Most commercially available MOS integrated circuits contain some type of input protection to reduce susceptibility to damage due to static electricity. However, input protection devices vary widely in effectiveness.

Experiments in which input protection devices were evaluated by discharging a 1.2-μF capacitor through several values of resistance have been conducted to compare empirically the effectiveness of various types of input protection devices. The results of these tests are given in Table V for input protection devices on 5-$\Omega \cdot$cm substrates (n-type).

The field-enhanced avalanche diode consisted of a diffused p^+-n diode, a portion of which was covered with thin (gate) oxide and had overlying metal that was at substrate potential. The avalanche voltage was approximately 40 V. The lateral p-n-p transistor structure, which consisted of two adjacent p-type regions with the n-type region between them covered with thin (gate) oxide and overlying metal, is designed to undergo simultaneously both field-enhanced surface avalanche and punchthrough at approximately 40 V. Both the overlying metal and one p-type region are grounded to the substrate.

Discussion of Experimental Results

Ion migration may occur both in the thermally grown silicon dioxide layer or laterally along the dielectric-ambient interface.

The negative ion-type instability in the field oxide of MOS devices has recently been shown to occur in devices fabricated by a number of manufacturers [9]-[11]. The instability, which is attributed to the presence of immobile negative ions plus mobile positive charge (presumably sodium ions), has an activation energy of 32 kcal/mole (≈1.3 eV/molecule) [9].

In typical MOS integrated circuits, the area of metal over thin oxide over n-type regions is considerably greater than that of metal over thin oxide over p^+ regions. However, shorts occur principally over p^+ regions because the electrical field with negative potential on the metal relative to the substrate is higher in the oxide over p^+ regions than over n-type regions.

The greater incidence of shorts in logic inputs than in clock leads is attributed to the lower input capacitance. This is because less charge is required to exceed the dielectric strength of the 0.16-μm gate oxide (approximately 120 V). The difference between logic inputs 1 and 2 is attributed to the lower on-resistance of a larger protection device on input 1. The improved characteristics of the p-n-p protection device, which is designed to undergo both punchthrough and avalanche, is attributed to its having a lower on-resistance than the other structure. Moreover, it provides better protection since it cannot "walk-out" (drift to higher avalanche breakdown voltages) during successive discharges. Input protection devices containing two adjacent p-type regions have been described in the literature [12], [13].

Since the principal failure mechanisms of MOS arrays are surface ion migration effects and shorts through gate oxide, it is not surprising that operating or dc bias aging tests cause a higher incidence of failures than storage or environmental tests.

Of the failure mechanisms observed, alkali ion migration in silicon dioxide has an activation energy of 32 kcal/mole (≈1.3 eV/molecule) [14]. By contrast, surface-ion migration [15] and oxide shorting [16] are not particularly temperature sensitive. The effective activation energy for functional failure would thus be expected to be significantly lower than 32 kcal/mole (≈1.3 eV/molecule).

Glass passivation reduced the possibility of surface-ion migration effects by increasing the distance between surface charge and the underlying n-type substrate [15]. Also, the glass passivation process resulted in lowering of the surface recombination velocity in unmetallized oxide regions, thus lowering leakage currents [4]. Phosphosilicate glass provides the additional benefit of immobilizing any alkali ion contamination that might be present in the chemical vapor deposited layer [4].

Particulate shorts, caused by conductive particles that bridge adjacent metallization stripes, are effectively eliminated by use of a passivation glass [17].

In the case of serial shift registers, we have considered each bit to be equivalent to two gates. MOS shift registers have also been characterized this way by Khambata [18]. On this basis, the observed failure rate of less than 0.004/1000 h for a dual 50-bit serial shift register at 125°C corresponds to a failure rate of less than 2×10^{-5}/1000 h per equivalent gate.

Since the failure rate is not constant, but decreases with increasing time, published data based on a small group of devices tested for many thousands of hours cannot be compared directly with data on a large group of devices tested for a thousand hours. A decreasing failure rate is characteristic of life testing which detects individual devices embodying deficiencies resulting from manufacturing inadequacies or errors rather than a wearout mechanism. The higher initial failure rate indicates that screening can be used to improve produce reliability.

In general, field return data on recently manufactured devices are less complete than data for devices delivered some time ago. Newer devices, embodying improvements in design and processing, have substantially lower failure rates in accelerated stress tests. It is expected that recently manufactured devices will have even lower failure rates than those of the newest devices for which life-test data are available.

Reliability Information from Other Sources

Vendor Survey

A vendor survey of MOS integrated circuit reliability and failure mechanisms was made as part of Contract F30602-69-C-0252 with Rome Air Development Center, Rome, N.Y. The consensus of replies was that oxide shorts and surface effects constitute the principal mechanisms of failure in MOS arrays [19]. Failure rates ranging from 0.0001 to 0.05/1000 h at 125°C were reported. Input protection and screening by voltage stressing were cited as available techniques for reducing the incidence of oxide shorts. Burn-in was reported to be effective in increasing reliability. Information on the effect of temperature on failure rates indicated an activation energy on the order of 5 kcal/mole (≈0.2 eV/molecule).

Published Information on MOS Reliability and on Failure Mechanisms

Studies of MOS surface instabilities have been extensively reported in the literature and have been surveyed in a number of publications [10], [20]-[25]. Specific information about the reliability of MOS integrated circuits has, however, been less readily available, although a number of reports and papers on MOS integrated circuit reliability have appeared in the last several years [5], [26]-[32], particularly for p-channel enhancement-mode devices containing SiO_2 dielectric. Some information is also available on potential instabilities in metal-nitride-oxide-silicon devices [33], metal-Al_2O_3-SiO_2-Si devices [34], and silicon-gate devices [35]. The failure rates given in the references cited above range from 5×10^{-5}/1000 h at 0.90 confidence level to 0.05/1000 h.

The effect of complexity on reliability of integrated circuits, whether MOS or bipolar, is somewhat controversial. Some authors have stated that MSI or LSI failure rates are, or will be, independent of the functional complexity of the device [36], [37]. Others have implied that chip reliability will be unchanged with an increase in complexity, but that the number of wire bonds will be the determining factor in reliability. It has been suggested that although the reliability per single packaged chip will be lower, the reliability per function will be higher [38], [39]. In a number of cases, users of MOS LSI devices have reported improved system reliability [40].

Effects of MOS Integrated Circuit Construction on Reliability

MOS integrated circuits are fabricated by more than 30 manufacturers using a variety of materials, processes, and designs [40], [41]. The starting substrate may be of (111) or (100) orientation, and the substrate donor density can range from 5×10^{14} to 5×10^{15} atoms/cm^3. Dielectrics being using include thermally grown SiO_2, and bilayer dielectrics consisting of SiO_2 covered with PSG, silicon nitride, or aluminum oxide. Metallization is almost always aluminum. In some cases, polycrystalline silicon is used for gates and for some interconnections.

There are also a considerable number of variations in processing sequences, design rules, and circuit types. For example, both thick oxide and thin oxide processes are used. Diffused n^+ channel stoppers, or n-type field diffusions in all areas except gates, may be used to insure against field inversion. Ion implantation is used by several manufacturers.

In addition to arrays based on p-channel MOS transistors, complementary circuits and circuits based on n-channel devices are available.

A deposited glass passivation layer is widely, but not universally, used in MOS arrays. Devices are typically enclosed in hermetically sealed packages, although plastic encapsulated devices are available from a number of manufacturers.

Many more recently introduced process modifications, such as the use of a diffused n^+ channel-stopper in low threshold voltage devices, the use of silicon nitride (Si_3N_4) under the gate metal, and the use of polycrystalline silicon gates, have added complexity to the MOS process. While the additional steps required to achieve these design or material modifications do not necessarily adversely affect reliability, additional process controls are necessary to insure that the new steps do not introduce any new failure mechanisms or aggravate any existing reliability limitations. The use of chemical-vapor-deposited Si_3N_4 over SiO_2 in gate oxides requires, for example, that the thickness, dielectric constant, and specific electrical conductivity of the Si_3N_4 layer be controlled. If the specific conductivity of the Si_3N_4 is too high, charge buildup at the SiO_2-Si_3N_4 interface can cause reliability problems [33].

In the case of silicon gate devices, the fabrication techniques must produce polycrystalline silicon interconnection lines that cross steps in the underlying SiO_2 without reliability problems due to cycling to low temperatures. While the polycrystalline silicon has the same thermal coefficient of expansion as the silicon substrate, there is a mismatch between polycrystalline silicon and SiO_2. This results in placing the polycrystalline silicon which crosses steep oxide steps in tension at low temperatures. Crossover continuity problems can be avoided by using a fabrication technique that produces a tapered (beveled) oxide step [3], [6] rather than a steep step. Silicon gate technology also requires making a reliable ohmic contact between aluminum metal and polycrystalline silicon. Excessive solid-state dissolution of polycrystalline silicon during the contact alloying step must be avoided to insure reliable low-resistance contacts.

Beam-lead techniques with platinum silicide ohmic contacts require a processing sequence that eliminates any native oxide on the silicon in contact cuts, whereas aluminum is self-fluxing, and is thus able to form reliable ohmic contacts to silicon when a native oxide film is present in the contact areas at the time of aluminum evaporation [42].

The use of n^+ channel stoppers between (but not contacting) p^+ regions in low-threshold-voltage MOS circuits has been used to prevent field inversion problems. Generally, narrow n^+ lines are used. If breaks in the n^+ channel stoppers occur under metal crossovers, the device will not function at all. This is because the crossovers generally are biased so as to invert the surface if an n^+ channel is not there. If breaks in the n^+ channel stoppers occur elsewhere, a temporarily functional device results, but it will slowly degrade due to surface ion migration.

Reliability Prediction

Reliability prediction requires extensive knowledge of the particular integrated circuits and their intended usage. In the case of MOS integrated circuits, the number of possible variations is rather large and thus reliability prediction can be difficult and controversial [43], [44].

Important Factors in MOS Integrated Circuit Reliability

In considering the reliability implications of the various materials, processes, and designs, it is obviously not possible to make generalizations that apply unequivocably to all available MOS types. It is possible, however, to indicate which factors in design and manufacture must be closely controlled to insure good product reliability. For high reliability, one should do the following.

1) Minimize the density of localized defects in the thin (gate) oxide.

2) Minimize gate oxide electrical instabilities under bias at elevated temperature.

3) Minimize effects of field oxide electrical instabilities under bias at elevated temperature.

4) Employ designs that do not require critical control of electrical parameters. Such designs provide tolerance to slight shifts in parameters.

5) Employ techniques and controls to insure against field inversion or field-enhanced punchthrough.

6) Employ designs and inspections to avoid partially missing gate metal.

7) Employ a deposited glass passivation layer over all of the chip (except the central area of bonding pads).

8) Use adequate controls on bonding.

9) Employ input protection devices.

10) Use hermetically sealed packages and 100 percent leak testing.

11) Employ 100 percent mechanical and thermal environmental screens to insure against failures due to mechanical failure mechanisms.

12) Employ 100 percent electrical burn-in at an elevated temperature to remove devices with time-bias-temperature related failure mechanisms.

13) Employ 100 percent voltage stressing with dynamic operation to apply voltage stresses to internal circuit nodes.

Reliability Prediction Equation

In considering the foregoing experimental results, plus available published information and results of a vendor survey under RADC sponsorship, a reliability prediction equation has been established [19], and tentative values for its parameters have been proposed. As additional information on MOS integrated circuit reliability becomes available, modifications in the values of these constants may be appropriate. In most cases, the symbols and definitions are similar to those contained in [45] or [46].

A mathematical model for predicting the failure rate of silicon planar monolithic MOS microcircuits is given by the equation

$$\lambda_M = \lambda_B \pi_T \pi_P \pi_E \pi_Q \pi_F \pi_C + \lambda_W \pi_T \tag{1}$$

where the following hold. λ is a failure rate; subscripts refer to: M, the microcircuit; B, basic failure rate for a particular type, or small family group of types, of circuits built with a given process, type of construction, set of design rules, etc.; W, wire bonds. π is an adjustment factor; subscripts refer to: T, the operating temperature of the circuit; P, package type, size, materials, sealing conditions, and the number of lead-throughs; E, type of environment; Q, stringency of screening; F, fabrication conditions; C, circuit size, and complexity as determined by the number of elements, the total critical area, the total critical periphery, the number of crossovers, etc.

Equation (1) was prepared for predicting the reliability of circuits that meet the following set of conditions.

1) The circuits are commercially available MOS circuits manufactured with commonly used materials, processes, and techniques.

2) Early failures and gross defectives have been removed by quality screening.

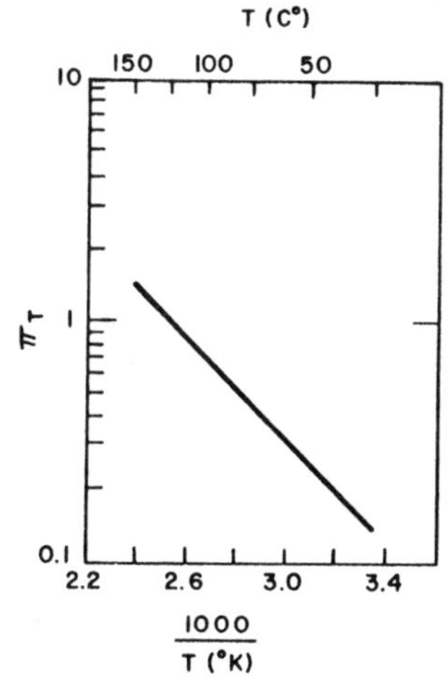

Fig. 1. Adjustment factor for temperature, π_P.

3) There is evidence from qualification or other tests that the devices are typical of their type.

4) No extreme or unique conditions exist in assembly or in subsequent handling or application that might significantly degrade the reliability.

5) The directly applied or field-induced voltages and currents are held within the maximum ratings specified by the manufacturer.

6) No voltage surges, transients, or spikes are allowed to reach the devices.

The values of each of the factors in (1) follow. Within reasonable limits, the user may interpolate between or extrapolate beyond the values given in these tables.

Failure Rates:

Basic, λ_B: Unless more specifically defined by other available information, the basic failure rate λ_B for MOS microcircuits can be assumed to 0.01/1000 h for operation in a 125°C ambient.

Adjustment Factor for the Operating Temperature, π_T: Based on an assumption that the activation energy for the average degradation on operating life is 5 kcal/mole, Fig. 1 gives the adjustment factor for differences in the actual operating temperature.

Wire Bonds, λ_W: Current industry wire bonding methods vary in the type of equipment used to place a bond and in the type of bond. There are variations in the metal combinations used in the wire, package land, and chip land. An adjustment factor calculated using (2) is for a system using ultrasonic bonding on the chip land, and wedge bonding on the package land, aluminum wire, gold plated packages, and aluminum metallization on the chip. The adjustment factor may require modification for other bonding systems.

$$\lambda_W = (2 \times 10^{-6}/1000\ \text{h})W \qquad (2)$$

where W is the number of wires connecting different points within the package, as well as those to package leads.

Adjustment Factors for Package Type, π_p: The following values of π_p are for: 1) a chip glassed with a material that does not introduce an instability, 2) a gold-silicon eutectic chip-to-package bond, and 3) a gold wire thermal compression or Al-1 percent silicon wire with ultrasonic bonds. For the values of π_p with a hermetic package:

$$\pi_p = 1 + 0.05L \qquad (3)$$

where L is the number of leads in excess of ten. Without glassing, add 3 to π_p as defined in (3), and with plastic package, add 5 to π_p as defined in (3).

Environment and Quality of Screening π_E and π_Q: The values for π_E and π_Q can be taken from Tables VI and VII which are from [45]. The values for π_E range from 1.0 for a laboratory environment to 10 for a missile environment. The values of π_Q range from 1 for an optimum screen to 30 for a lower grade screen.

TABLE VI

Screening Adjustment Factor, π_Q

Quality Grade	π_Q
Optimum screen includes a) Vendor, line and product qualifications; b) Line discipline on an interference basis; c) Failure feedback (tight loop) with continuous corrective; d) Screens and burn-in; e) Traceability of test data.	1
Upper grade includes a) Screens and burn-in comparable to RADC Spec. 2867 [47] based on limit testing (sample subjected to destructive tests to establish absolute limits of stresses which devices can withstand) to identify major failure modes and mechanisms to which screens are tailored; b) Feedback from screening results only.	2
Average grade includes a) Normal production grade and lot acceptance testing on a sampling basis; b) No 100 percent screening beyond routine vendor procedure for electrical parameters and hermeticity.	15
Lower grade includes routine vendor procedures that are applied to all production devices on a 100 percent basis.	30

TABLE VII

Environmental Adjustment Factor, π_E

Environment	π_E
Laboratory	1.0
Satellite, orbit	1.5
Ground, fixed	2.0
Ground, portable	5.0
Ground, mobile	7.0
Airborne, inhabited	5.0
Airborne, uninhabited	7.0
Satellite, launch	8.0
Missile	10.0

TABLE VIII

Adjustment Factor for Differences in Fabrication Conditions, π_F

Circuit that has been in production for at least one year on a given line with line discipline on an interference basis, with failure feedback with continuous corrective action and with firm process controls to prevent instabilities in both the gate and field oxides: $\pi_F = 1$
Circuit in production less than one year on a given line, add 10 to π_F.
Without firm process controls to prevent oxide instabilities, add 20 to π_F.
Without input protection, add 10 to π_F.

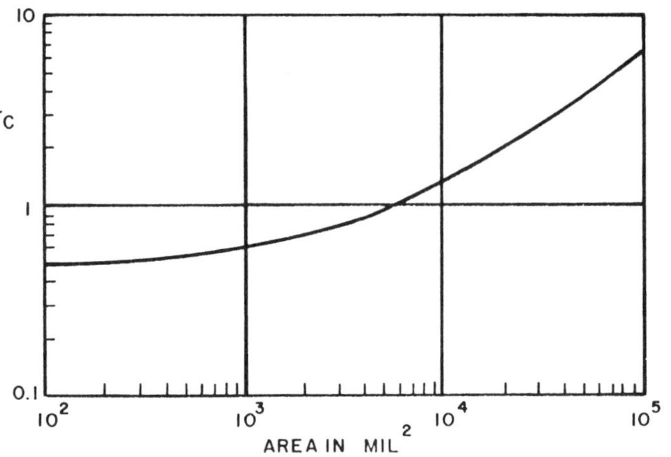

Fig. 2. Adjustment factor for circuit size and complexity, π_C.

Fabrication Conditions, π_F: The values for π_F are given in Table VIII.

Circuit Size and Complexity, π_C: π_C is given in Fig. 2 and was calculated by (4).

$$\pi_C = 0.5 + 0.5\left(\frac{A}{5000}\right)^{0.8} \tag{4}$$

where A is the active area of the chip in square mils (1 sq mil $\cong$ 645 sq μm). The active area of the chip includes everything except the border regions (scribe lines, contact lands, and test devices) of the chip.

We have chosen to define π_C in terms of area rather than in terms of gates or bits because of the difficulties involved in defining gates and bits for different types of MOS circuits. (Typically, the active area is 50 mil^2/bit for a shift register, 40 mil^2/bit for a static RAM, and 25 mil^2/NOR gate in a logic circuit.)

Equation (4) was derived with the following assumptions.

1) π_C should be unity for an area of 5000 mil^2.

2) Circuits having a chip area of 5000 mil^2 are assumed to have a failure rate that is half due to area-independent effects and half due to area-dependent effects.

3) The effects of area on the failure rate are assumed to be less than if the number of defects were linearly proportional to area. It is likely that a lower average defect density exists in larger area chips in order to achieve a good yield. Also, it is believed that many localized defects are clustered and nonrandomly distributed over the area of a wafer.

If the circuit includes nodes in which the developed voltage is higher than any of the applied voltages, multiply the value obtained from Fig. 2 by three.

References

[1] G. L. Schnable *et al.*, "Procedural guidelines for the reliability assessment of large-scale integrated circuits," Philco-Ford Corp., Blue Bell, Pa., Final Tech. Rep. prepared for RADC, RADC-TR-69-220, Aug. 1969.

[2] G.L. Schnable, E.S. Schlegel, and R.S. Keen, "Failure mechanisms and analyses in large-scale integrated circuits," in *Proc. 2nd Annu. Seminar on Failure Anal., Philadelphia Sect., IEEE*, pp. 37-40, May 1969.

[3] G. L. Schnable and R.S. Keen, "On failure mechanisms in large-scale integrated circuits," in *Advances in Electronics and Electron Physics*, vol. 30, L. Marton, Ed. New York: Academic, 1971, pp. 79-138.

[4] M.M. Schlacter, E.S. Schlegel, R.S. Keen, R. Lathlaen, and G.L. Schnable, "Advantages of vapor-plated phosphosilicate films in large-scale integrated circuit arrays," *IEEE Trans. Electron Devices*, vol. ED-17, pp. 1077-1083, Dec. 1970.

[5] E. Pfiffner and J.W. Pfeiffer, "MOS reliability," Philco-Ford Corp., Blue Bell, Pa., Nov. 1967.

[6] G.L. Schnable, R.S. Keen, and M.M. Schlacter, "The impact of reliability requirements on LSI technology," *IEEE Int. Solid-State Circuits Conf. Dig. Tech. Papers*, vol. XIII, 1970, pp. 146-147.

[7] E.S. Schlegel and G.L. Schnable, "The application of test structures for the study of surface effects in LSI circuitry," *IEEE Trans. Electron Devices*, vol. ED-16, pp. 386-393, Apr. 1969.

[8] "Study of failure modes of multilevel large-scale integrated circuits," Philco-Ford Corp., Blue Bell, Pa., Rep. 1-5, prepared for NASA ERC under Contract NAS 12-544, Nov. 1967-Nov. 1969.

[9] E.S. Schlegel and G.L. Schnable, "A negative-ion type instability in MOS devices," *Electrochem. Soc. Extended Abstr.*, pp. 69-71, May 1971.

[10] B.E. Deal, "Measurement and control of dielectric film properties during semiconductor device processing, in *Silicon Device Processing*, NBS Spec. Pub. 337, Nov. 1970, pp. 36-50.

[11] F.H. Reynolds, R.W. Parrott, and D. Braithwaite, "Use of tests at elevated temperatures to accelerate the life of an M.O.S. integrated circuit," *Proc. Inst. Elec. Eng.*, vol. 118, pp. 475-485, 1971.

[12] D.E. Farina and D.R. Borror, "Protective circuit for insulated gate metal oxide semiconductor field-effect device," U.S. Patent 3 395 290, issued July 30, 1968.

[13] R.E. Pace, J.D. Trotter, and W.C. Wheeler, "Overvoltage protective circuit for insulated gate field effect transistor," U.S. Patent 3 403 270, issued Sept. 24, 1968.

[14] E.H. Snow, A.S. Grove, B.E. Deal, and C.T. Sah, "Ion transport phenomena in insulating films," *J. Appl. Phys.*, vol. 36, pp. 1664-1673, 1965.

[15] E.S. Schlegel, R.S. Keen, and G.L. Schnable, "The effects of insulator surface-ion migration on MOS and bipolar integrated circuits," in *Proc. 8th Annu. IEEE Rel. Phys.*, 1971, pp. 9-16.

[16] N.J. Chou and J.M. Eldridge, "Effects of material and processing parameters on the dielectric strength of thermally grown SiO_2 films," *J. Electrochem. Soc.*, vol. 117, pp. 1287-1293, 1970.

[17] G.L. Schnable, "Glass passivation of integrated circuits by chemical vapor deposition of oxide films," in *IEEE 71 Int. Conv. Dig.*, 1971, pp. 586-587.

[18] A.J. Khambata, *Introduction to Large-Scale Integration*. New York: Wiley-Interscience, 1969.

[19] H. Ewald, "Procedural guidelines for the reliability assessment of MOS microcircuits." Philco-Ford Corp., Final Rep. prepared for RADC, RADC-TR-71-178, Sept. 1971.

[20] E.H. Snow and B.E. Deal, "Polarization effects in insulating films on silicon-A review," *Trans. Met. Soc. AIME*, vol. 242, pp. 512-523, 1968.

[21] S.R. Hofstein, "Proton and sodium transport in SiO_2 films,"

IEEE Trans. Electron Devices, vol. ED-14, pp. 749-759, Nov. 1967.

[22] A. Goetzberger and S.M. Sze, "Metal-insulator-semiconductor (MIS) physics," in *Applied Solid State Science*, R. Wolfe, Ed. New York: Academic, 1969, pp. 154-238.

[23] P.V. Gray, "The silicon-silicon dioxide system," *Proc. IEEE*, vol. 57, pp. 1543-1551, Sept. 1969.

[24] J.R. Szedon and R.M. Handy, "Characterization, control and use of dielectric charge effects in silicon technology," *J. Vac. Sci. Technol.*, vol. 6, pp. 1-12, 1969.

[25] D.R. Kerr, "A review of instability mechanisms in passivation films," in *Proc. 8th Annu. IEEE Rel. Phys.*, 1971, pp. 1-8.

[26] L.C. Hamiter, Jr., "How reliable are MOS IC's? As good as bipolars, says NASA," *Electronics,* vol. 42, no. 13, pp. 106-110, 1969.

[27] "Reliability report, MOS integrated circuits," National Semiconductor Corp., Santa Clara, Calif., June 1970.

[28] R. McKenna, "An MOS quality and reliability program," *Microelectron. Rel.*, vol. 10, pp. 67-74, 1971.

[29] "Microelectronic failure rates," Rel. Anal. Cen., IIT Res. Inst., Chicago, Ill., Dec. 1970.

[30] Y.T. Yen, "Intermittent failure problems of four-phase MOS circuits," *IEEE J. Solid-State Circuits*, vol. SC-4, pp. 107-110, June 1969.

[31] G. Madland, "The most in MOS," *Electron. Products*, vol. 10, pp. 10-12, 1967.

[32] D.D. Forsythe, "Surface-charge induced failures observed on MOS integrated circuits," *Microelectron. Rel.*, vol. 8, pp. 339-347, 1969.

[33] D. Frohman-Bentchkowsky, "The metal-nitride-oxide-silicon (MNOS) transistor–Characteristics and applications," *Proc. IEEE*, vol. 58, pp. 1207-1219, Aug. 1970.

[34] J.J. Curry and H.E. Nigh, "Bias instability in two-layer MIS structures," in *Proc. 8th Annu. IEEE Rel. Phys.*, 1971, pp. 29-33.

[35] F. Faggin, D.D. Forsythe, and T. Klein, "Room temperature instabilities observed on silicon gate devices," in *Proc. 8th Annu. IEEE Rel. Phys.*, 1971, pp. 35-41.

[36] R.C. McCoy, Signetics Corp., Detailed Rel. Rep. PR930-001, Dec. 31, 1969.

[37] "User guidelines for quality and reliability assurance of LSI components," *Microelectron. Eng. Bull. 17,* Electron. Ind. Ass., Dep. Eng., MED-43 Committee on Microelectron. Rel. Characterization, Washington, D.C., Nov. 1970.

[38] G.L. Schnable and R.S. Keen, Jr., "Failure mechanisms in large-scale integrated circuits," *IEEE Trans. Electron Devices*, vol. ED-16, pp. 322-332, Apr. 1969.

[39] Staff article, "IC reliability increases despite rising complexity," *Comput. Design*, vol. 9, no. 4, p. 42, 1970.

[40] G. Flynn, "Forum on MOS," part I, *Electron. Prod. Mag.*, vol. 13, pp. 21-26, Dec. 21, 1970.

[41] J.L. Seely, "A survey of MOS process technologies," in *IEEE 71 Int. Conv. Dig.*, 1971, pp. 184-185.

[42] G.L. Schnable and R.S. Keen, "Aluminum metallization–Advantages and limitations for integrated circuit applications," *Proc. IEEE*, vol. 57, pp. 1570-1580, Sept. 1969.

[43] H.A. Lauffenberger, "LSI reliability assessment and prediction," in *Proc. 1970 IEEE Annu. Symp. Rel.*, vol. 3, 1970, pp. 364-380.

[44] P.L. Holden, "LSI reliability predictions," *Eval. Eng.*, vol. 9, pp. 16-18, 1970.

[45] *RADC Reliability Notebook*, vol. II, RADC-TR-67-108, p. 419, Sept. 1967.

[46] P.F. Manno, "Failure rate prediction for complex bipolar microcircuits," Rome Air Devel. Cen., Griffiss AFB, N.Y., Final Tech. Rep., RADC-TR-69-650, Oct. 1969.

[47] RADC Specification 2867, Rome Air Devel. Cen., Griffiss AFB, N.Y.

An IEEE SPECTRUM applications report

A look at automatic testing

Automatic testing of electronic devices has been a major factor not only in the overall improvement of product quality and reliability, but also in the dramatic lowering of product costs

Harold T. McAleer **General Radio Company**

This survey reviews the current state of automation in the testing of electronic components, networks, and circuits. Elements and characteristics of typical test systems, both hardware-controlled and computer-controlled, are described. Paths to be followed and pitfalls to be avoided in achieving automation are discussed in an effort to help the reader toward a better understanding of the subject and its broad applications. "How to automate successfully," a major theme in the report, places emphasis on economic justification.

Reprinted from *IEEE Spectrum,* vol. 8, pp. 64–78, May 1971.

The need for and desirability of automatic testing have been well established and require little justification. The testing function itself, which used to be considered somewhat of a poor relation, a necessary evil, has gradually emerged as a well-recognized and essential part of any manufacturing operation—particularly in electronic manufacturing. This testing function appears in many areas and in many different forms. It may be called incoming inspection, line inspection, final test, or quality control, and it may involve sampling tests or 100 percent inspection. From the standpoint of product quality, 100 percent inspection at every recognizable step in production would be ideal. This ideal was seldom possible because of the high costs involved, both in time and money. As the inspection process becomes automated, however, 100 percent inspection becomes economically feasible and an attainable goal, especially when one realizes how rapidly the cost of repair rises as a faulty item gets buried in the final product.

Experience with transistors offers an example. It has been calculated[1] that the cost of *not* subjecting transistors to 100 percent testing as received from the manufacturer, but rather sample testing them to an acceptable quality level (AQL) of 0.65 percent, can amount to 8½ cents per transistor purchased. For a user of a million transistors per year, this amounts to an annual *loss* of $87 500!

Integrated circuits provide another example. Back in 1969 when the price of integrated circuits varied from 85 cents to $3.50 each, depending on complexity, the typical failure rate, as received from the IC manufacturer, was around 1.5 percent. Now that the price is down to 20–85 cents each for the same units, the failure rate is up to 2–3 percent.* With an average IC count of 20 to 30 per board and an average time to diagnose and repair of 20 to 30 minutes, we can see the importance of culling out bad units as early in the manufacturing process as possible.

Since electronic equipment is used more and more in products involving safety considerations—spaceships, aircraft, automobiles, medical instrumentation—and the cost of failures may be measured not only in dollars but in human lives, the need for thorough testing (as well as sound design) becomes even more important.

Levels of testing

In general, five distinct levels of testing can be recognized, one in the engineering-design phase of a product, three in the manufacturing phase, and one in the postshipment phase. In the design phase, testing is required not only to prove out or modify the design, but also to evaluate and select the materials, components, or techniques used in the product. In the manufacturing phase, testing is required at the input to the production process to qualify the materials used in the product (incoming inspection). It is also required during the process to assure that all the steps have been successful (line inspection) and at the end of the process to qualify the final product (final inspection). In the postshipment phase, (service) testing is required, to establish that operation is proper or to diagnose failures and aid in repair.

Different test information is required at these different levels and different types of test equipment are often used. Engineering and incoming-inspection tests may require the acquisition and analysis of detailed measurement data, whereas line-inspection and final-inspection tests are more concerned with "go" or "no-go" decisions to maximize throughput.[2] Testing and test equipment used during the production process is sometimes separated into two types—equipment and techniques for the rapid sorting of items into "good" and "bad" categories and equipment and techniques for the slower diagnosis and repair of the bad items. Diagnosis and repair equipment is also used in the servicing of delivered products.

Distinctions between laboratory, production, and servicing equipment are disappearing, however, as design engineers are required to design and specify not only the product but also the test equipment and techniques.

Measuring systems

The block diagram of Fig. 1 shows the variety of items or functions to be found in any measuring system, manual or automatic. The major functions are within the blocks, and the words around the blocks show examples of hardware items that accomplish the functions. Although not perfect, the diagram provides a useful basis for discussion. The tinted blocks show the essential items: a device to be tested, a stimulus source to provide a signal for the device, a measurement instrument to quantify the response, and an operator to make the whole thing go.

The other blocks show the functions that can be added to ease the operator's task and automate the process. As equipment for these functions is added, the operator acts less as a mechanical part of the measurement process and can devote his attention to the results rather than the details of the measurement. If the results are used for automatic control of other equipment operating on the device under test, a process-control system can be achieved.

Device under test. It is useful to categorize the device under test according to the hierarchy shown in Fig. 1. Thus, a component is made out of materials, a network is made out of components, a circuit is made out of networks, etc. The categories can be further subdivided: components and networks can be linear or nonlinear, active or passive; circuits can be analog or digital, discrete or integrated. The hierarchy can be shown to be imperfect (for example, is a transformer a component, a network, or a circuit?) but in general any electronic device can be fitted into one of the categories.

Adaptor. The adaptor block is used to define items that interface the device under test into the measurement system. This category includes test boards, fixtures, or sockets that contact the leads or terminals of a component or network, or a multipoint prober that contacts the nodes of a microcircuit.

Input. Input equipment facilitates the insertion of the device under test into the adaptor. Examples are vibratory-bowl component feeders, rotary and linear transports, and other mechanical devices. Such items are usually associated with a sorting or binning device to deposit the tested unit into a useful location.

Condition. Conditioning equipment applies to the device under test a secondary stimulus, such as power,

* Typical user experience. However, for a premium of 5–10 cents each the customer can buy units with a guaranteed failure rate of less than 0.1 percent; i.e., the IC manufacturer will test them more thoroughly.

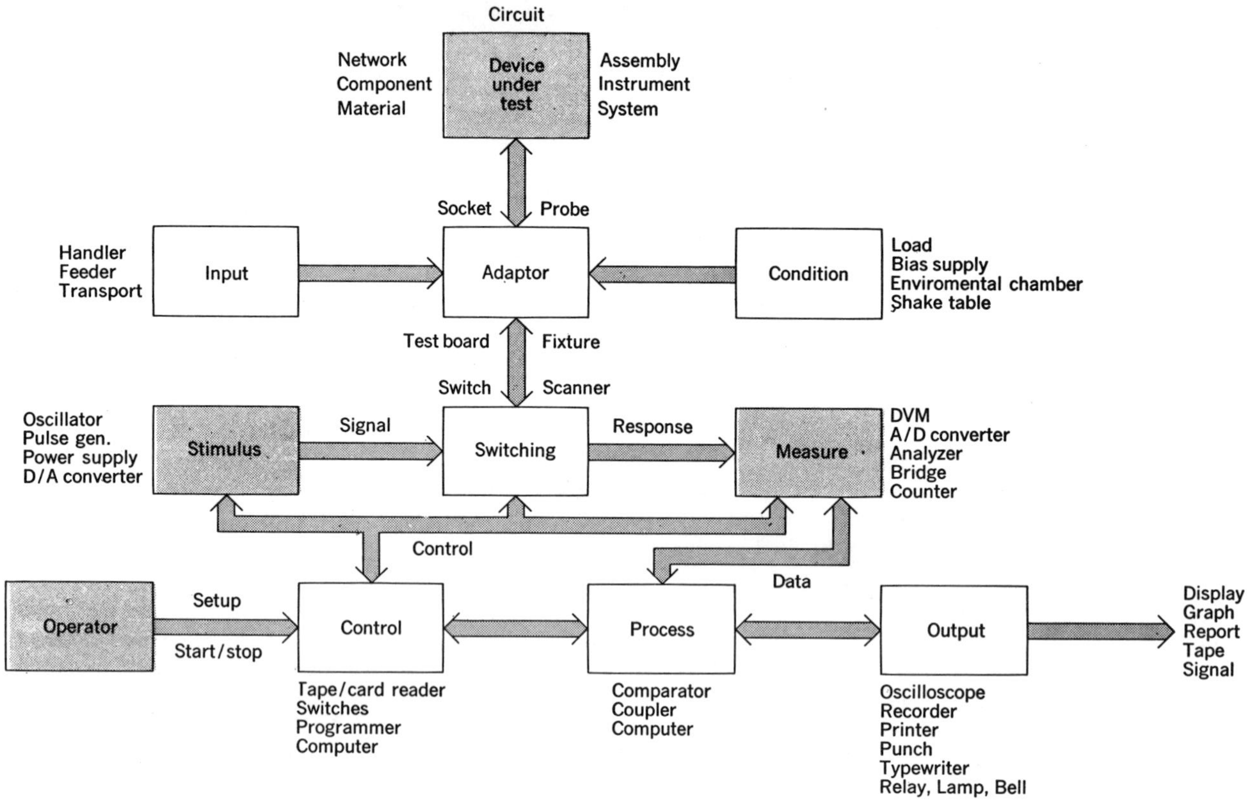

FIGURE 1. Functional block diagram of a manual or automatic measuring system.

bias, heat, cold, or shock, or sets up terminal load conditions.

Switching. Switching equipment is used to connect the device or device adaptor to the test system and to vary the connections of the device terminals. Examples are solid-state scanners (multiplexers), relay trees, crossbar scanners, and reed-relay scanners.

Stimulus. Stimulus sources for measurement systems are legion, and include dc supplies, oscillators, synthesizers, function generators, ramp, pulse, burst, and word generators, and D/A converters, among others. For use in automatic test systems such sources are often required to be programmable; that is, all their functions should be controllable by electric signals instead of (or in addition to) manual controls.

Measure. Measuring instruments are likewise legion; they include voltmeters, current meters, phase meters, impedance bridges, frequency counters, A/D converters, etc. These instruments should also be programmable. Data output in digital form is a useful, but not always necessary, characteristic of such instruments. The signal and measure functions are often combined in a single instrument, as for example in an automatic capacitance bridge or a digital ohmmeter.

Control. The control function can be accomplished by a variety of equipment in a variety of ways. Card readers and tape readers are often used to program and sequence automatic equipment with the actual program determined by the holes in the cards or tape. Patchboards, plug boards, and switches can be used as "memories" to set up test conditions and sequences. Cam-operated motor-driven switches are used as well as relay networks. Specialized digital-logic circuits are often applied to this function. The most versatile programmer/controller of all, the computer, will be discussed in a later section.

Process. Data from the measurement instruments must often be "massaged" or processed in some way. One example of processing is limit comparison. The comparator may be an analog or digital circuit. It compares the measured data with preset or programmed limits and determines whether the data are above, below, or within the limits and provides appropriate output signals (lamp illumination, contact closures, digital-logic signals, etc.). Another form of processing is coupling. The coupler may store and convert parallel measurement data (all digits at once) to serial data (one digit at a time) to operate card punches or tape punches or strip printers, or it may perform the opposite function (serial-to-parallel conversion) to operate gang punches or line printers or displays. Yet another form of processing is computation or data processing. The processor may perform arithmetic computations on the data to provide corrections or conversions or to somehow convert the data to more useful form.

Output. The output function of an automatic test system is also performed by a variety of equipment and appears in a variety of ways. One typical output function is that of display. Measurement results or test conditions may be displayed by means of oscilloscopes, panel meters, indicator lamps, digital indicators (visual display), or by ringing a bell (auditory display). Hard-copy output can be provided in the form of graphs, printed lists, or typed reports by means of recorders, printers, or typewriters. Output can consist of condensed data in the form of punched cards, punched tape, or magnetic tape for off-line processing by other equipment. Output can also consist of signals for on-line processing by local or remote equipment. Finally, output signals can be used to close the loop on the test process by activating sorters, automatic handlers, lasers, sandblasters, or other devices that affect the device under test.

Interface. One function not shown in the diagram is that of interface. This can be considered a miscellaneous or catchall category used to dignify any function that cannot be conveniently forced into one of the other categories. In general, it defines any item that is used to tie together some of the other items. This function becomes one of extreme importance in computer-controlled systems, as will be discussed later.

It should be noted that the diagram of Fig. 1 is a functional rather than a hardware diagram. The hardware diagram may be similar, but often it is not, since many hardware items combine several of the functions.

The computer

The largest single factor contributing to the revolution in the art of automatic testing has been the emergence of the computer, in particular the minicomputer, as an available and reliable item of hardware with immense, if not perfect, capability. It appears everywhere in automatic test systems, both for programming and controlling and for data processing. The computer however, is not always necessary, or even desirable, in some applications. How do you know when to use one? The following paragraphs discuss some considerations.

Operational flexibility. If the job of the test system is fixed, and well defined—for example, the sorting of resistors into prescribed categories—and the test system will be used for that one specific task forever, a computer may not be required. If simple and economical hardware—timers, programmers, bridges, comparators, etc.—exists to do the job, it should be used. It won't take much, however, to warrant a computer. When the number of possible controls and settings becomes large—nominal values, upper and lower limits, bias voltage, test voltage, test frequency, advance, dwell and soak times, etc.—the sheer number of knobs, dials, and switches to adjust can get out of hand and some form of sophisticated programmer will be required. For this task alone, the computer may represent an economical solution.

Changing requirements. When the test system has to be rapidly converted from one job to another (and enough hardware can be combined in one system to do all the jobs) the computer may be the most economical way to provide this rapid conversion.

Data manipulation and output. If the measuring instruments provide data in the right form and format and processing equipment and output devices (couplers, printers, etc.) are available, the computer is not necessary. If, however, the data must be massaged or corrected and a complicated output format provided, the computer may offer the only solution. In this regard the computer provides a side benefit: the ability to correct measurement data for fixed or systematic errors and thus improve the accuracy of the basic measurement system. It also offers the possibility of providing data analysis in the desired form and eliminating the need for further off-line processing.

Computer functions. The computer can perform two of the basic functions shown in Fig. 1, the control and process functions. It can perform the control function by means of a program stored in its memory and the process function by means of its central processor. To make use of these capabilities, however, two other elements are required: an interface system and software.[3]

FIGURE 2. Simplified block diagram of a typical computer-controlled component-test system emphasizing the role of the interface system.

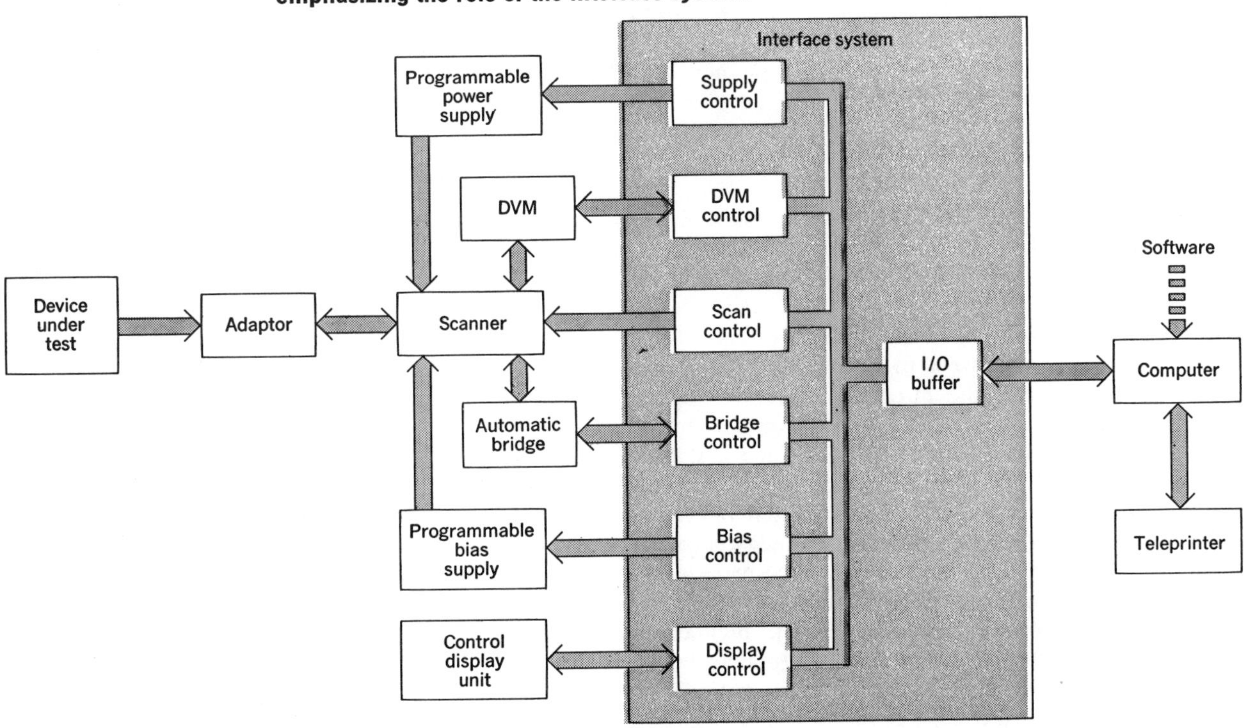

Interface system. Figure 2 is a simplified block diagram of a typical computer-controlled test system drawn purposely to emphasize the role of the interface system. The test system includes several programmable elements to perform the functions of Fig. 1. The interface system is required to wed these elements to the computer, since at this stage of the art neither computers nor test instruments are directly compatible.[4] Cost considerations in the design of general-purpose test instruments do not allow the inclusion of expensive interface circuits. As we grope our way toward standardization, however, test instruments are becoming available that make the interfacing task, if not unnecessary, at least easier. Similarly, the design of computers cannot anticipate all the possible interface requirements. In general, the interface circuits perform the functions of I/O (input/output) bus expansion, level conversion, decoding, storage, transfer, and control. By storing data signals, both from and to the computer, the interface system allows the (fast) computer to keep busy while the (slow) test instruments are doing their job.[5] When the test instruments are ready, either to receive instructions or give up their data, their interface control circuits can provide an "interrupt" signal and an identifying signal (flag) to the computer to request its attention.

Some minicomputers make provision for plug-in interface-circuit cards within their packages; others require an external unit. General Radio uses its 1761 Interface System in computer-controlled test systems. Hewlett-Packard provides plug-in interface cards for its computers and offers its 6936A Multiprogrammer System, a main-frame unit and a series of plug-in interface cards for its test instruments. Digital Equipment offers card-cage hardware and digital-circuit modules that can be used to construct interface systems, and other manufacturers offer similar items.

Expanding role of the computer. As the art progresses, it is becoming increasingly difficult to separate the elements of computer, interface, stimulus source, and response measurement, as well as to decide where one function leaves off and the other begins. Many automatic test sets of the future (and some of the present) will consist of card cages or main frames to house plug-in functional modules—memories, arithmetic elements, storage registers, stimulus sources, measurement modules, and peripheral-control circuits for input, control, and output devices.

Software. To make most efficient use of the limited core memory of the typical minicomputer, operating programs are often written in assembly language. (Such programming efforts are not trivial; it can take 4 to 6 man-months to produce 4000 words worth of debugged software and can cost $10 000 to $20 000, depending on how you value a man-month.) This approach works fine if the test set will be used with a few infrequently changed programs. If not, more memory, and perhaps the use of a high-level test-set language, may be needed.

Computer languages. Much has been written and said about computer languages,[6,7] the terminology of communication between the programmer and the computer, and the level of such languages. There are no strict definitions of language level. In general, high-level languages allow the programmer to write program statements in terms that are familiar to him, such as English words or mathematical equations, and with few rules of order or format, so that he can ignore the particular characteristics of the computer used to run the program. Low-level languages, on the other hand, force the programmer to adapt his problem to the characteristics of the machine to be used. They may use mnemonics or octal numbers and are related to the operation of the computer and its subsections. The use of higher-level languages may require several intermediate steps before a final operating program is prepared or run—that is, editing, debugging, punching tapes, loading tapes, etc. Given sufficient computer capability, however, the steps can be handled automatically (with little programmer action) by the use of a sophisticated operating-system program, and programs can be prepared and run at essentially the same time.

Table I shows some common types of programming languages. For further details, see Appendix A.

Test-set languages. One also can speak of programming and operating languages for automatic test systems. Test-set languages can be high level or low level, depending on their relationship to the application (problem) or to the test set (equipment). When you set the function knob of a digital counter to FREQUENCY, you are using a high-level language, related strongly to the application but of little flexibility. Similarly, setting the power switch to OFF uses a lower-level language. This concept has given rise to a host of test-set languages, Pol (Problem-Oriented Language), Tool (Test-Oriented Operator Language), Atlas (Abbreviated Test Language for Avionics Systems), and many others that have no names but do their jobs just as well. If the test set is dedicated to a specific application (capacitor testing, transformer testing, logic-circuit testing), the language can be made high level and use terms relating to the device under test (nominal value, transformer terminals, logic-drive pattern). If the test set is for more universal use, its language should be lower level and use terms relating to the instruments within the set (function, range, connection).

The program of a dedicated test set can provide two levels of language for the users of the set, a low-level language for the programmer, which will enable him to change certain operating constants or test sequences, and a higher-level language for the operator, which will enable him to control the set for a specific test. A partic-

I. Common programming languages

Type	Example	Typical Statement	Comment
Machine	PDP-8	100011101101	Machine code, object code, run code
Assembler	Pal III	HLT	Can make optimum use of memory and minimize run time. Error prone. Expensive. Great variation between programmers. Difficult to document
Compiler	Fortran Algol	D = A + B + C	Off-line compiling. Object program requires more memory and runs slower. Less variation between programmers. Easier to document. Compiler program may be expensive
Interpreter	Basic	IF E > 2 PRINT GO	Interactive for easy changes. On-line run. Requires even more memory and runs even slower

ularly useful mode of operation with such languages is an interactive or conversational mode, wherein programming questions are automatically displayed, printed, or typed by the set and replies are typed by the operator. The main resident program—the operating system—interprets the operator's instructions and runs the test.

Typical test systems

To illustrate the concepts described earlier, the following paragraphs discuss a few examples of commercially available automatic test systems. Excellent surveys are available in trade publications and commercial reports.[8-10] The examples will be discussed in order of the device under test in the hierarchy previously described.

Component tests. Although threatened by film technology, classical discrete components (resistors, capacitors, inductors, transistors, diodes, etc.) remain with us. Their form is changing, but their function remains the same. Many automatic systems are used in the manufacture of these components—some for measurement only and some for process control.

Diode test. Figure 3 shows a hardware-controlled system for the automatic sorting of semiconductor diodes. The system includes a handler/sorter, which includes a vibratory-bowl feeder, a rotary test table with three test positions, and a five-bin sorting mechanism. The instrumentation includes three test instruments; a capacitance comparator, which compares the junction capacitance of the diode to a preset limit; a stored-charge (switching-time) detector; and a diode classifier, which tests for short and open circuits, reversed polarity, peak inverse voltage, reverse current, and forward voltage drop. The results from the three instruments are combined in the diode classifier to provide bin-sort signals to the handler. These systems can test and sort up to 10 000 diodes per hour.

Capacitor tests. Figure 4 shows an automatic system for the testing and sorting of capacitors. The input, adaptor, and switching functions are performed by a manually loaded transport unit that grips the capacitor terminals and connects them to various soaking, measurement, and discharge busses. The transport is stepped by signals from a control unit, which also houses other interface circuits. Signals from the control unit also enable the transport to perform an output function: dropping the tested capacitors into sort bins. Programmable power supplies provide stimulus signals for the measurement of dielectric strength and leakage current, and also provide a conditioning bias voltage for capacitance measurement. An automatic bridge provides both stimulus and measure functions for the measurement of capacitance and loss, and a digital voltmeter measures leakage current. The digital output data of the automatic bridge and digital voltmeter are processed by digital limit comparators to provide sort decisions. The instruments are controlled by a program unit, which includes a card reader to interpret holes in a prepunched card to program voltages, soak time, instrument ranges, and sort limits, and an overriding series of thumbwheel switches to allow manual programming. A further output function is provided by a series of electromechanical counters, which display a running tally of the sort decisions. This system provides a good example of a hardware-controlled system for a specific inflexible purpose.

Another capacitor-testing system, shown in Fig. 5, is used for evaluation and life studies rather than physical sorting. Components are loaded on test boards, which are inserted into a test socket. A reed-relay scanner is used to connect the components to the stimulus and measurement instruments. A minicomputer and interface system control the stimulus and response instruments, process the measurement data, and provide a typewritten output report on a teletypewriter. The system software provides an initial test sequence that produces a printout of parameter values and indicates out-of-limit units, and a final test sequence that provides parameter values, deviations from initial values, and a statistical summary.

Network tests. One passive network of extreme importance is the connection network, which has the

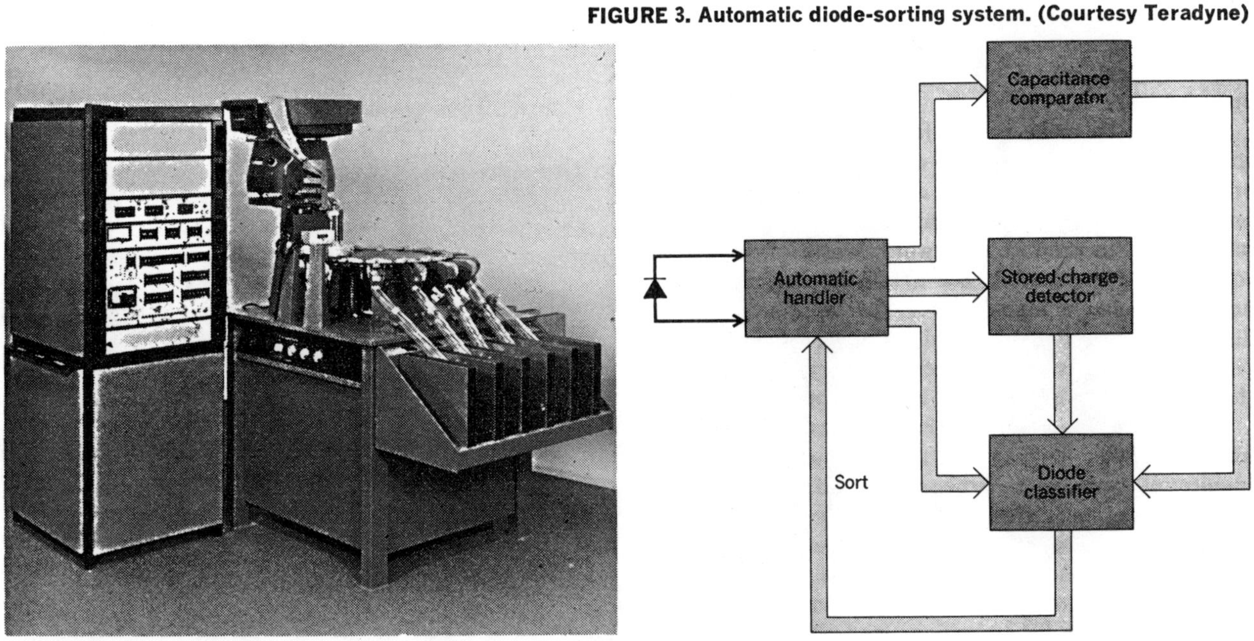

FIGURE 3. Automatic diode-sorting system. (Courtesy Teradyne)

mundane task of connecting some circuit nodes together and isolating others. This type of network appears in many forms—a cable, a wiring harness, a wired backplane or motherboard, a bare printed-circuit board, or a hybrid-circuit substrate before the components are attached. These networks are perhaps the most critical elements in electronic hardware and the ones most prone to error in manufacture, and therefore considerable effort has gone into the design of automatic test equipment to check them out. The stimulus and measurement units are relatively simple, since the measurements required are usually dc continuity, dc or ac dielectric strength (hipot), and dc leakage resistance; however, the control and switching units can get quite complicated. The less expensive systems ($20 000) use paper-tape readers and bar relays and can handle networks with up to 500 nodes. The more expensive systems ($100 000) use computers and random-access mercury-reed scanners and can handle networks with up to 100 000 nodes.

Another important passive network is the multipair communications cable. Figure 6 shows two versions of a computer-controlled test set for the measurement and evaluation of the capacitance parameters of such cables.[11] (Most of the important transmission characteristics are determined by these parameters.) A functional block diagram is shown in Fig. 7. The input function is provided by a large "fanning fixture" and the switching function is performed by a special-purpose reed-relay scanner. Stimulus and measurement functions are performed by an automatic capacitance bridge and the control and process functions are provided by an interface system and a minicomputer. A teletypewriter provides the output report. The set uses a high-level language to enable an interactive dialogue with the operator to determine test limits and desired output data. Although individual measurements can be printed, in the usual mode of operation all the data are condensed into a compact statistical summary. A typical output report is shown in Fig. 8.

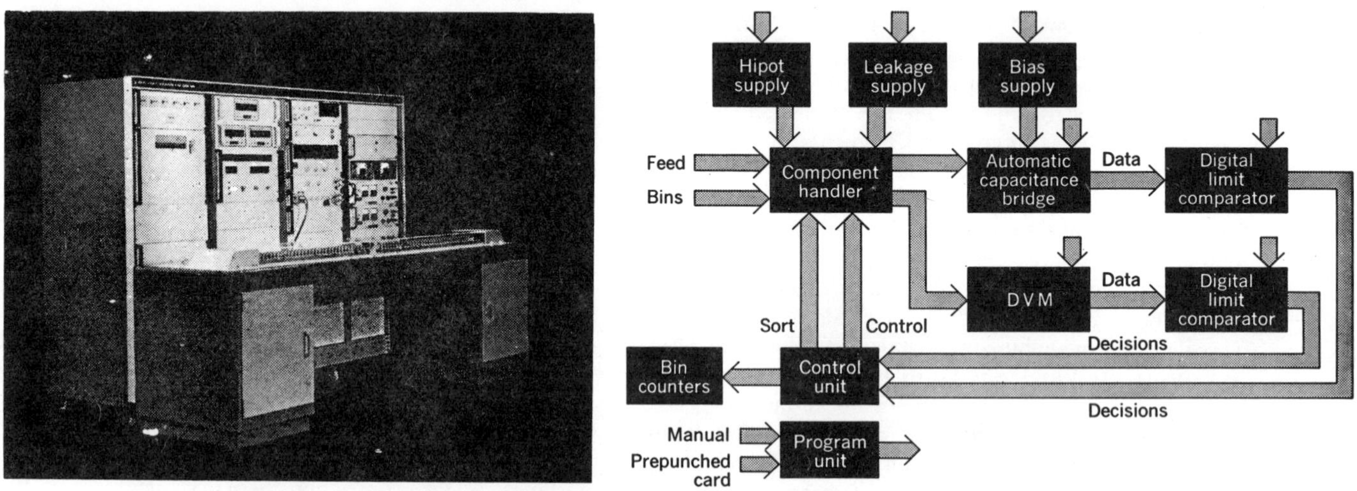

FIGURE 4. Automatic capacitor-sorting system, which measures and sorts capacitors to preprogrammed specifications. (Courtesy General Radio)

FIGURE 5. Computer-controlled system for life evaluation of capacitors. (Courtesy General Radio)

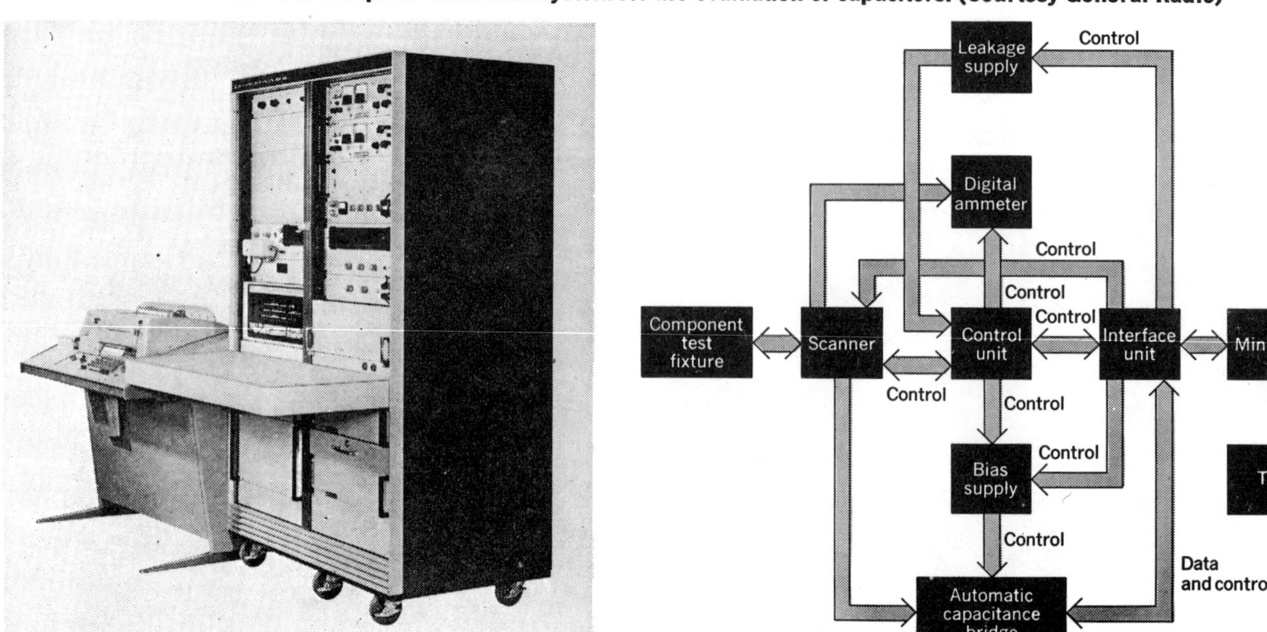

Circuit tests. Circuit testers usually fall into three categories, depending on the type of circuit to be tested—that is, digital, analog, or hybrid (combination).

Digital circuits. The explosion in the use of digital circuits, both discrete and integrated, has created an entire industry in test equipment with its own evolving terminology and jargon.[12] Digital-logic engineers speak of three basic types of testing[10]:

1. Functional testing, to check truth-table or "Boolean" operation of digital circuits by applying and sensing patterns of 1's and 0's on the pins of the circuit.
2. Parametric testing, to check dc or static parameters, such as forward and reverse currents, saturation and offset voltages, etc.
3. Dynamic testing, to check time-domain parameters, such as rise and fall times, propagation delay, etc.

Inexpensive ($10 000) functional testers now available apply a sequence of bit patterns to the input pins of both a known good circuit and the circuit under test and compare the output patterns. Only one input signal is altered between consecutive tests to avoid indeterminate states. As one author points out, however,[13] since it takes at least 2^n tests to provide a complete input sequence for a unit with n interdependent inputs, it might take 40 000 years to test a 60-input device, even at a 2.5-MHz test rate! The answer, of course, is to bring out internal logic nodes as test points to reduce the number of interdependent points to ten or so and the test time to milliseconds. Such testers are useful in the rapid sorting of bad from good units but are of little help in diagnostic repair.

The more expensive ($20 000 to $100 000) testers use tape readers or computers to exercise a predetermined test program for both sorting and diagnostic information.

Figure 9(A) is a photograph of the General Radio 1790 Logic-Circuit Analyzer, a high-speed functional tester for digital circuits.[14] This system, shown in block-diagram

FIGURE 6. Automatic systems for the measurement of capacitance parameters of multipair communication cable. (Courtesy General Radio)

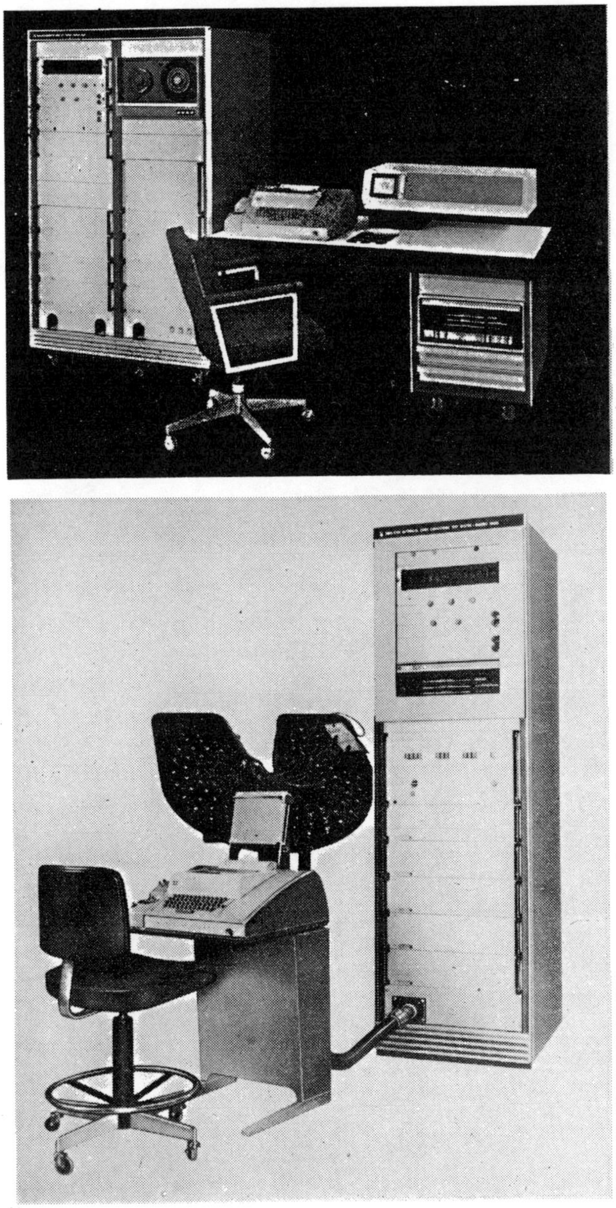

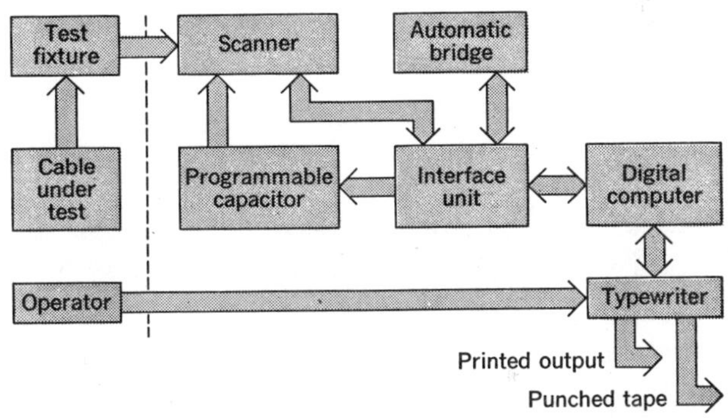

FIGURE 7. Block diagram of cable capacitance-measuring system shown in Fig. 6.

FIGURE 8. Statistical summary of results from cable capacitance-measuring system of Fig. 6.

```
IDENTIFICATION INFORMATION:
135382 ADMC-15ØØ L2 RD-WH UNIT #4  6/2Ø/69

PAIRS:   1ØØ
LENGTH (FEET):   75Ø

SELF TEST:  OK

MUTUAL
ERROR THRESHOLD:   3.Ø2Ø NF
MUTUAL DISTRIBUTION - CELL NOS ARE NF/MI +/-1

   L     71     73     75     77     79     81     83     85     87     89
   Ø      Ø      Ø      Ø      Ø      1     13     16     29     13     19

  91     93     95     97     99    1Ø1    1Ø3    1Ø5    1Ø7    1Ø9      H
   5      2      1      1      Ø      Ø      Ø      Ø      Ø      Ø      Ø

NUMBER OF PAIRS TESTED:   1ØØ       COEFF. OF DEVIATION:  3.987%
AVERAGE:   85.777  NF/MI
STD. DEVIATION:   3.32  NF/MI       LIMITS:  PULP XXX  PIC XXX

UNBALANCE TO GROUND DISTRIBUTION - CELLS ARE 1ØØ PF/15ØØ FT WIDE

   1    2    3    4    5    6    7    8    9    1Ø    11    12
  21   17   17   13    8    6    8    5    3     Ø     2     Ø

  13   14   15   16   17   18   19   2Ø
   Ø    Ø    Ø    Ø    Ø    Ø    Ø    Ø

STD. DEVIATION FROM BAL.:  43Ø PF/15ØØ FT
AVERAGE UNBAL.:   372 PF/15ØØ FT
PAIR 32:   +191Ø PF/15ØØ FT   +98Ø PF
PAIR 47:   +196Ø PF/15ØØ FT   +915 PF

PAIR-TO-PAIR UNBALANCE
CAP. UNBAL. LIMIT:  124.2Ø7 PF    SKIP LIMIT:   13.799 PF   (11.11%)
CAP. UNBAL. DISTRIBUTION--CELLNOS ARE CELL MAX. % OF CAP. UNBAL. LIMIT

   2.5    5    7.5   1Ø   12.5   15    17.5    2Ø    22.5    25
 4597   135   79     38   36     18    17       9     6       6

  27.5   3Ø   32.5   35   37.5   4Ø    42.5    45    47.5    5Ø    H
   2      3    2      Ø    2      Ø     Ø       Ø     Ø       Ø    Ø

STD. DEVIATION FROM BAL.:   3.644% of CAP. UNBAL. LIMIT
NUMBER OF TESTS:   495Ø    COMBINATIONS SKIPPED:   4358

SELF TEST:    OK
END OF TEST
```

form in Fig. 9(B), is an example of a test set developed by a manufacturer out of necessity for in-house use and then offered for sale after prove-in. The set includes a minicomputer for control and processing, a control panel, tape reader, display oscilloscope, and teletypewriter for operator communication and printout, a device adaptor for connecting the circuit under test, and an interface system that provides logic-level drive and sense circuitry for digital circuits with up to 240 pins.

The system communicates with the operator in four ways. The operator control panel contains control buttons and switches to control the system operation, and lamps to indicate the go/no-go results of a test. A display oscilloscope serves as alphanumeric output of test conditions and results, and is also used to display instructions to the system operator. A teletypewriter and paper-tape reader provide keyboard input, paper-tape input and output, and printed output where a permanent record of test results is desired.

A key feature of the system is the use of a high-level test-set language that allows test programs to be prepared by technicians familiar with the terms of digital-logic circuits. The language contains English-like commands, which perform functions such as the following:

1. Set specific inputs or outputs high or low.
2. Check specified outputs against output patterns specified in the program.
3. Generate program "loops" to allow selective repetition of part of a program.
4. Branch to another place in the program if specified output patterns exist.
5. Print instructions or error comments to the operator.
6. Insert specified delays in the program.

A typical program is given in Fig. 9(C). Two types of programs may be prepared: a simple truth-table program to allow quick sorting of good and bad circuits or a more complicated diagnostic program to facilitate trouble-shooting and repair. Programs can be prepared either on line or off line on paper tape. An interactive interpretive mode provided ("combined interactive system") allows generation, translation, and execution of a test program on line. An "autoprogramming translator" permits the use of a known good circuit to ease program preparation by automatically recording output states.

LSI circuits. Continuous evolution in the technology of integrated circuits, particularly in the manufacture of large-scale integrated arrays, has sparked much interest in automatic test systems, as well as some controversy.[15,16] The development of the devices to be tested is running somewhat ahead of the development of the test equipment.

The title illustration shows a typical test system used in this area, the Teradyne J283 Circuit Test System, called the "slot machine" because of its application in the testing of sequential logic. The system consists of three free-standing racks or "kiosks," a teletypewriter, and one or more test decks. One rack contains a minicomputer and magnetic-tape transport. Another contains a control

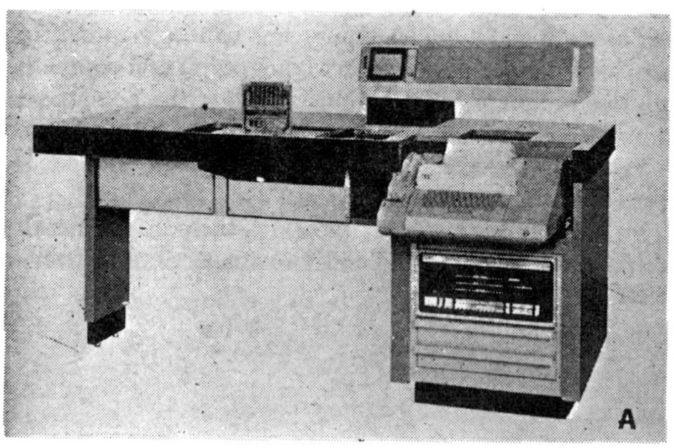

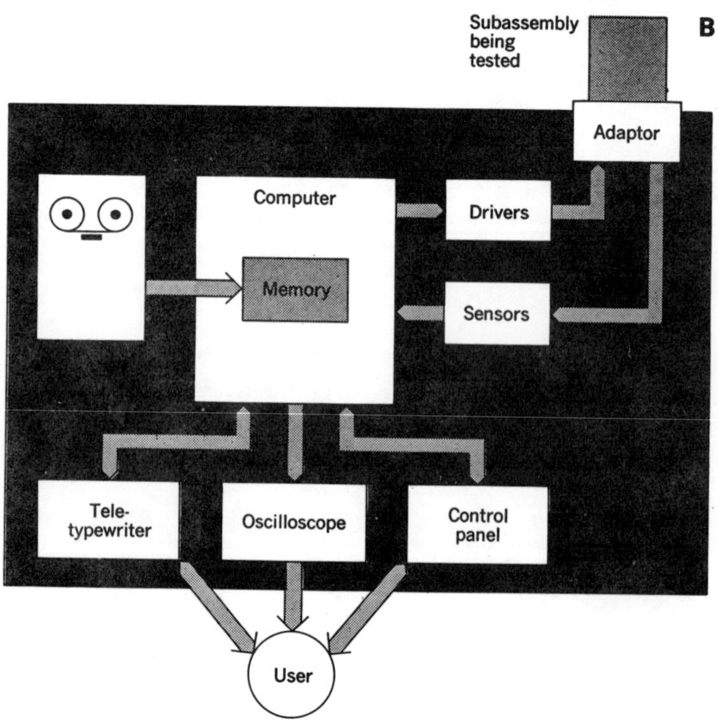

FIGURE 9. Automatic system for functional testing of digital logic circuits. A—System. B—Block diagram. C—Typical program. (Courtesy General Radio)

C

```
/BOARD NUMBER 1790-4140
/USE ADAPTOR 4
/PROGRAM REVISION 1
*I (1, 2, 14, 15, 17)
*O (20, 21, 22, 23, 30)
1; IH(#) OH (22, 23)
2; IL (14, 2)
PRINT MOVE S1 TO POSITION 5
AND PRESS CONTINUE
PAUSE 1
DO 5, 30
3; IH (2) IL (17) $
4; IL (1) OH (30)
5; IH (1) IL (2) OL (30)
DELAY 100
IGNORE (21, 22)
6; OL (20)
IF (23) 7
PRINT PROBE TEST POINT 1 AND IF HIGH,
THEN 1C8 IS BAD
OTHERWISE CHECK 1C6 !
PAUSE 2
.....................................
.....................................
END
```

unit, a CRO display unit, and a dc parametric measurement unit. The third rack contains multiplex relays, buffer amplifiers, and the drive and sense circuits for functional testing.

The system çan perform functional and parametric tests on digital circuits with up to 120 input/output terminals over a range of ±30 volts. The functional test rate is 50 kHz. A clock-rate tester is available in a separate kiosk to permit functional testing of dynamic MOS circuits with clock rates of 10 Hz to 8 MHz. An independent dynamic test system can also be incorporated to provide measurements of propagation delay and rise and fall times on 48-pin devices. Programs are prepared on magnetic tape, using the teletypewriter and display unit and a high-level test-set language.

Analog circuits. Figure 10 shows an example of the Hewlett-Packard 9500 Series Automatic Test System,[17] used to test analog circuits. A system is configured from a selection of building-block modules, including computers, stimulus sources, measurement units, processing units (called converting and conditioning units), switching units, and output units—most of which are commercial test instruments. Also shown is a block diagram of one of the simpler configurations. Figure 11 gives a program used to test the gain versus frequency of an audio amplifier. The program language is HP Basic, a modified version of the "Beginners All-Purpose Symbolic Instruction Code" developed at Dartmouth College in the mid-sixties. The software includes a 5500-word interpretive compiler and "software-driver" routines for the hardware elements.

Figure 12 shows the Instrumentation Engineering System 390, a computer-controlled system for the testing of printed-circuit boards—blank, digital, analog, or hybrid—or other forms of circuits. The system is another example of a modular building-block approach that allows a system to be configured from a selection of stimulus, measurement, switching, control, and output units. The system software uses Atlas language and can be used in a compile-before-run mode or in an interpretive direct-run mode.

Military systems. As is true of almost all our technology, many significant advances in automatic test systems have been spurred by the requirements (and the money) of the military services. Many military contractors (Northrop, LTV, General Dynamics, etc.) have furthered the art of automatic testing. Although this article is not directed specifically to such equipment or such applications, a brief description of one program may be useful.

The VAST (Versatile Avionics Shop Test) system[18,19] has been developed by PRD Electronics over the past ten years or so under several Navy contracts. About $100 million has been spent to date.

The system is intended for maintenance tests, both aboard ship and at supporting shore sites, on the avionic equipment in Navy aircraft, particularly the new Lockheed S-3A, the Grumman E-2C and F-14A, and the LTV A-7E. This project influences not only the testing of the avionic equipment but also the basic design and packaging of the equipment itself, since the suppliers of the aircraft and avionics equipment will be required to provide interface devices (device adaptors) and compatible test programs, and demonstrate that the equipment can indeed be tested and diagnosed on the VAST system.

Figure 13 shows a typical VAST test station. The hardware contains three basic sections:

1. The computer subsystem, which includes a general-purpose digital computer and two magnetic-tape transports.

FIGURE 10. Automatic test system for analog circuits. (Courtesy Hewlett-Packard)

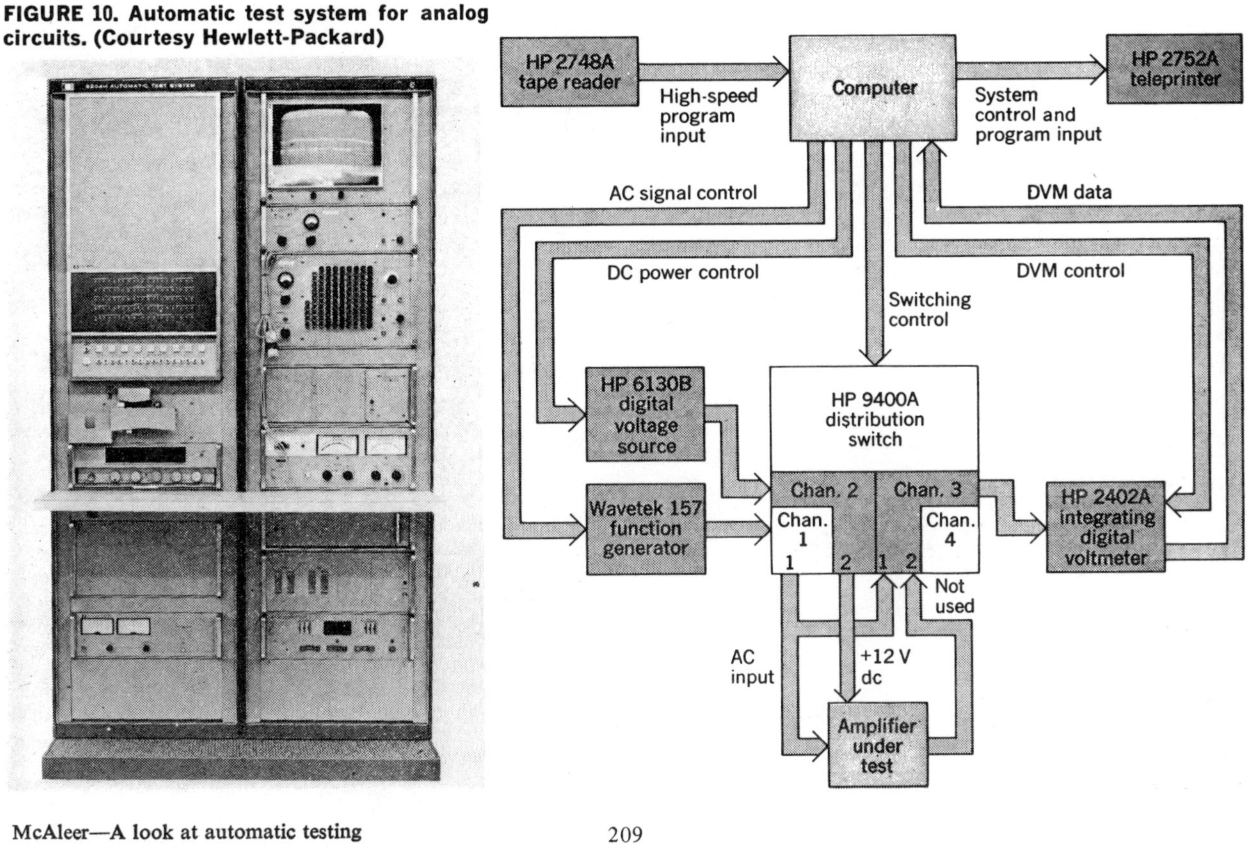

2. The data transfer unit, which represents the operator/machine interface. This unit includes a control panel with mode-select buttons, status indicators, and an ASCII keyboard; a display panel that provides a cathode-ray tube presentation of test data and instructions; and a maintenance panel to monitor autocheck results.

3. The stimulus and measurement section, which includes the basic building blocks used to configure a particular test station. These blocks are chosen from a list of about 50, which are divided into three categories: stimulus, measurement, and operational. The stimulus blocks include dc and ac power supplies, signal generators, function generators, pulse generators, delay generators, word generators, etc. The measurement blocks include digital voltmeters, frequency meters, power meters, impedance meters, spectrum analyzers, etc. The operational blocks include switching and interface modules.

Several sophisticated software techniques have been developed for the VAST system.[20] A high-level language, Vital (Vast Interface Test Application Language), is an extension of Atlas, which was developed in 1968 by Aeronautical Radio, Inc. The new language includes several sublanguages, or "dialects," to obtain the ease of programming of a high-level language but still retain the flexibility of harder-to-use low-level languages.

Programs are prepared off line. Several compiler programs are used to convert the dialect programs into VOSC (Vast Operating System Code) at a compiling station, which includes peripheral items and a PDP-8 minicomputer. The compiling station is in turn linked to a remote Univac 1108 computer. Source programs are entered on punched cards and are compiled into an object program (in VOSC) on magnetic tape. This object program is entered into the VAST test station along with an operating-system program that provides control of test-station operation and interprets the instructions of the object program during a test run.

Test complex. Most of the systems just described are "stand-alone" test systems. Some can include a few multiplexed test stations but in the main they are autonomous units. Another concept being considered, and in some cases implemented, is the test complex consisting of a central control station and remote test centers. Each test center can include a master test station and several "satellite" or "slave" test stations. The approach is intended to minimize cost by "putting all the computer in one place and developing the software once and for all." In practice, however, it has been found that the minicomputer itself is often the most economical item to use in a remote test center to communicate back to a central station. It also allows some degree of autonomy at the test center in case the central unit goes down.

Such a test complex is in operation at the Raytheon plant in Andover, Mass.[21] It is called IHATS (Improved Hawk Automatic Test System) and is used for factory test of missile components. Each test station includes a

FIGURE 11. Programs for system shown in Fig. 10. A—Test result. B—Program example.

A

```
AMPLIFIER FREQUENCY RESPONSE TEST
SPECIFIED GAIN IS 10, TOLERANCE 1.7%

CONNECT AMPLIFIER
SERIAL NUMBER IS?12


FREQUENCY        GAIN          % ERROR
 1000            9.8424        -1.576
 2000            9.85134       -1.48661
 3000            9.85929       -1.40711
 4000            9.83948       -1.60521
 5000            9.84841       -1.5159
 6000            9.81771       -1.82291       OUT OF SPEC
 7000            9.82665       -1.73349       OUT OF SPEC
 8000            9.82665       -1.73349       OUT OF SPEC
 9000            9.8247        -1.75301       OUT OF SPEC
 10000           9.83363       -1.6637
```

B

```
10 PRINT "AMPLIFIER FREQUENCY RESPONSE TEST"
20 PRINT "SPECIFIED GAIN IS 10, TOLERANCE 1.7%"
30 PRINT
40 G = 10
50 DVS (1, 0, 10)
60 PRINT "CONNECT AMPLIFIER"
70 PRINT "SERIAL NUMBER IS";
80 INPUT S
90 PRINT
100 PRINT
110 SWH (1, 2, 1, 0)
120 DVS (1, 12, 100)
130 PRINT "FREQUENCY", "GAIN", "% ERROR"
140 FOR F = 1000 to 10000 STEP 1000
150 SYN (1, 1, F, .283)
160 WAIT (50)
170 DVM (2, 1, V1)
180 SWH (1, 2, 2, 0)
190 DVM (2, 10, V2)
200 E = ((V2/V1-G)/G)*100
210 PRINT F, V2/V1, E
220 IF ABS(E)>1.7 PRINT "OUT OF SPEC"
230 PRINT
240 NEXT .F
250 END
```

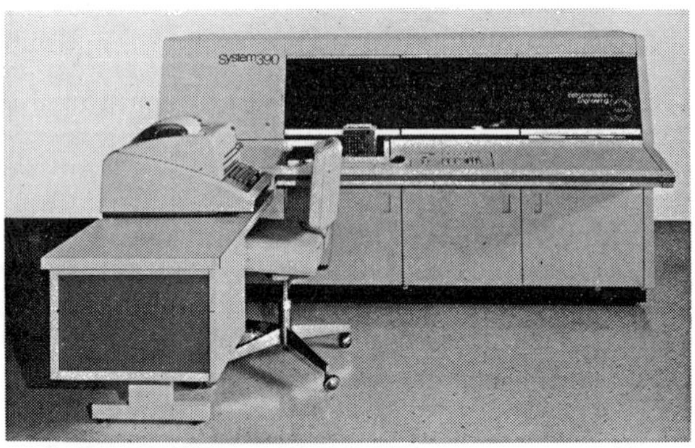

FIGURE 12. Computer-controlled system for testing hybrid circuits. (Courtesy Instrumentation Engineering)

FIGURE 13. VAST test station. (Courtesy PRD)

collection of stimulus, measurement, conditioning, switching, adaptor units, etc. (usually commercial items) plus specially designed interface circuits for control and communication with a "line computer" on a multiplexed basis. Several line computers in turn communicate with a central "collection computer" for data collection on magnetic tape. Programs are prepared and data are analyzed on an isolated "utility computer" and associated peripheral units. The system uses a special compiler language, TCOMP, based on Fortran IV, which allows program preparation by test engineers rather than programmers.

Cost. The prices of commercial automatic test systems seem to cluster about decade and half-decade values (1:3:10), with, of course, a continuum in between. Table II shows what to expect.

How to automate successfully

There are two basic approaches to achieving automation: the do-it-yourself approach and the buy-it-outside approach, with several gradations in between. In the do-it-yourself approach you attempt to do as much of the job in house as your time, talent, and budget allow. (Of course you have to buy something—from wire and solder to ICs or complete instruments or subsystems—but the basic responsibilities for design, procurement, fabrication, assembly, test, documentation, installation, training, and maintenance are yours.) In the buy-it-outside approach you attempt to contract for as many of these services as possible. Either approach can succeed or either approach can fail, depending on how you go about it. The following paragraphs discuss techniques that can make the second approach successful.

Getting started. First, study your task. Try the "man-from-Mars" technique and mentally step back as far as you can to visualize your operation as part of an overall system. Then gradually narrow your view to specific operations. Where does the device to be tested come from? Where does it go? What paper work goes with it? Why? What happens to it? What tests are to be performed? Why? What accuracies are required? Why? What reports are required? Why? This exercise can sometimes disclose unnecessary or redundant steps or operations no longer required because of changes in other parts of the organization. Or it may disclose new or impending requirements.

The chickens in the other yard. Find out what other people with similar problems have done. Read, study, telephone, visit. Present your problem informally to a few potential suppliers to get their ideas. Don't tell them how to do it; tell them the basic task. You can suggest ways but don't, at this stage, demand them. Others may have different ideas based on previous experience; you're not committed to follow them.

Put it in writing. Prepare a written specification for your automatic system. This exercise will clarify your thinking and provide an essential tool for you and the supplier. Again, specify what the system must do rather than how to do it. Separate your specifications into "must have" and "would like." In other words, don't demand 0.1 percent accuracy if 1 percent will do; don't demand four-week delivery if 12 weeks will do. You can get almost anything you want, but cost and time will increase exponentially as specifications get tougher.

Get competitive bids. Send your specification to as many vendors as you think are qualified to do the job. Give them time to absorb the request and invite them in to discuss it to make sure they understand the requirements. Keep prospective vendors separate unless you can handle a bidders' conference in a firm, unbiased manner.

In some ideal utopian world a prospective buyer would get a group of three or four prospective vendors together and say, "Look, fellows, I've got $37 500 and this is the job I'd like done. What can you give me?" The vendors would leave and come back later with their best proposals—all priced at about $35 000 (with optional extras to $45 000).

Or, better yet, he might say, "If I can get a machine to do these things in four seconds it's worth $37 500 to me. If you can get it down to two seconds it's worth $75 000!"

In the real world this does not happen often. A buyer is afraid that the vendors will combine their efforts to milk him of his money for a machine of lesser value. This, of course, is a small danger, since most vendors are engaged in a competitive life-or-death struggle and have no intention of cooperating with each other.

Get it in writing. Request a written proposal from your prospective suppliers in addition to a quotation. The proposal should list and describe the equipment to be provided, both electrically and physically. It should define the specifications of the system as a whole. It should describe the operation of the system from the operator's point of view (rather than the designer's). It should describe the other services provided, such as installation, training, warranty, and repair. It should include a list of recommended spare parts. Get as many competitive bids as possible.

Evaluate the bids. When you open the bids you'll be surprised. The prices will be higher than you expected, the accuracies will be poorer, the speed will be slower, you'll get less equipment, the equipment will be harder to use, and you'll get fewer services. Don't be dismayed; the approach can still succeed.

There are four basic considerations in evaluating the bids: the equipment, the supplier, the economics, and yourself.

The equipment. Lean toward standard off-the-shelf items whenever possible. Such items are usually well debugged and conservatively specified, and will have a good backup of instruction and maintenance manuals, available spare parts, and trained service technicians. You may require some modified or specially designed items, but it is important to appreciate the differences between these items and standard items. In this area you are highly dependent on the integrity of the supplier.

The supplier. You should thoroughly evaluate the

II. Commercial automatic test systems

Price Range, dollars	Description
10 000–30 000	Hardware systems of dedicated capability. Sometimes with tape readers, patchboards, or other forms of semiautomatic control. Printed or machine-readable output
30 000–100 000	Computer-controlled systems. One or more test stations and processed output
100 000–300 000	Computer-controlled with several test stations. Flexible capability and completely processed output
300 000–1 000 000	Multicomputer installations. Central control with remote test centers and satellite test stations

supplier. Where is he? What are his local sales and service facilities? What's his reputation? What else has he done like this? How did it work out? If you are contemplating a major expenditure, visit the suppliers' facilities. Talk to the people who will be doing your job.

The economics. Although you have probably made some form of economic analysis to establish your budget, do it again based on the actual bids and the actual throughput promised. Appendix B describes some methods used to compare investments. All methods require an estimate of the initial investment and the first-year savings.

The cost of the initial investment includes not only the billed price of the equipment but also the additional costs of installation, training, program preparation, spare parts, etc. As a rule of thumb, these costs can add 10 to 15 percent to the cost of the equipment.

The computation of operational savings requires an estimate of the cost of doing a job with the new equipment versus doing it the old way. The following factors are involved:

1. Number of units to be tested.
2. Number of types to be tested.
3. Average failure rate.
4. Old labor costs, including overhead.
5. New labor costs, including overhead.
6. Preparation time for test procedures.
7. Sorting good units from bad.
8. Troubleshooting and repair of bad units.
9. Tooling and maintenance.

Table III shows a savings calculation for the testing of logic-circuit boards. The comparison is between the use of manual test equipment and an automatic system. The results illustrate two things. First, the automatic equipment saves $20 742 in doing a year's work; second, it uses anly 398 hours in doing it. This reduction in time is, of caurse, the reason for the cost savings, but it also shows that the machine will be sitting idle most of the time. It will be underutilized and has the capacity to provide greater savings during the year if the work volume increases.

Armed with such an estimate, you can find an accounting expert who can calculate the payback period or discounted rate of return or net present value. You will usually find that you are losing money every day you are without the system—even a high-bid one. Amazingly enough, this is generally true (at least in the type of investment discussed here).

Yourself. You should evaluate yourself along with the bids. Can you assign a system "father" who will learn and master the system and nurture it as his own? (If it's foisted upon him by the "front office," it will never work.) Do you have, or can you develop, the capability to get the most out of the system and keep it running? Will you train new people as the present people get promoted or leave?

Follow-through. Once you've made up your mind and placed the order, get behind it solidly. Keep in touch with the supplier and work with him. Agree on an acceptance-test procedure—the simpler the better. If possible, visit his plant to look at and try the equipment before it is shipped; last-minute modifications are accomplished much easier at his plant than at yours.

Pilot operation. When your new system is installed, don't put it into full critical use at once; drive it slowly for the first "1000 miles." You'll find bugs, either in your operation of the system (cockpit errors), or in the set itself—early-life failures or some combination of switch positions that the designer never thought of. Work these problems out patiently and smoothly and get your operators used to the system. Then you can put it into use with full confidence.

Service and maintenance. If the system is really to do a job for you, keeping it running is of prime importance. Most vendors provide free service at your plant for 90 days. This usually works well since most of your problems will occur early and, if the system is in trial use, the few days (or weeks) it takes them won't hurt you much. Beyond the 90 days, service at your plant will take time and will cost you—unless you've signed up for a service contract (for about ½ percent of the system price per month) or unless you're willing to diagnose the difficulty and send the defective item back to the vendor. If the system is in critical use and down time is disastrous, the best approach is to fix it yourself. This requires people who know the system inside out and a good stock of spare parts. The spare parts can range all the way from a few relays to spare circuit boards to spare instruments.

Conclusion

This article has attempted to survey the state of automation in electronic testing and to discuss some of the characteristics of current approaches. The message is simple: automation is desirable and everyone should have it. Get it any way you can. Rent it, lease it, buy it, make it. It's not as easy to achieve as it may seem (nothing worthwhile is), but neither is it as hard. The rewards can be startling and will be in direct proportion to the effort put in.

III. Operating savings analysis, logic-circuit boards, annual cost

	Manual	Automatic
Preparation:		
Number of types	10	10
× hours per type	× 40	× 16
Total hours	400	160
× cost per hour	× 15	× 15
Cost	$6 000	$2 400
Sorting:		
Number of boards	5 000	5 000
× minutes per board	× 10	× 0.25
× hours per minute	× 0.017	× 0.017
Total hours	833	20.8
× cost per hour	× 15	× 10
Cost	$12 500	$208
Troubleshooting and repair:		
Number of failures	1 000	1 000
× minutes per failure	× 30	× 10
× hours per minute	× 0.017	× 0.017
Total hours	500	167
× cost per hour	× 15	× 15
Cost	$7 500	$2 500
Tooling and maintenance:		
Total hours	100	50
× cost per hour	× 10	× 15
Labor cost	$1 000	$650
+ material cost	500	1 000
Cost	$1 500	$1 650
Total cost	$27 500	$6 758
Total hours	1 833	398
Cost savings		$20 742

Appendix A. Software languages; some definitions*

Figure 14 is a generic block diagram applicable to all computer systems. Any system can be considered as a subsystem consisting of some of the elements in the diagram. Let us consider the function of each element, beginning with the lowest.

Base machine. The base machine consists of registers and decision-making circuits that perform the basic data-processing operations. This machine is controlled by signals for opening or closing gates to allow the data to flow between the machine registers. These signals are sometimes called the *microprogram* of the machine, and the language that describes this microprogram is called the *register-transfer language*. Programs for basic machines are very complex and are usually determined only by the computer manufacturer; recently, however, machines that can be microprogrammed by the user have been introduced.

Emulator. The emulator block usually consists of hardware decoding circuits or a read-only memory, which decodes certain bit patterns, called the *machine code*, to operate the base machine. These circuits determine what is usually called the *instruction set* of the machine. Instructions consist of a pattern of bits in an instruction word. Some special machines use a string of ASCII characters as the instruction set, so they may be programmed directly from an I/O device, such as a teletypewriter. A few machines have been built whose machine language is a high-level language. The emulator and base machine together make up what can be called the *real machine*. This is what we generally buy from a computer manufacturer.

Interpreter. The interpreter block defines a program, sometimes called an *operating system*, or *simulator*, which can be thought of as a means for converting the real machine into a new machine, called a *pseudo-machine*, whose effective machine code is different from that provided by the real machine. The machine code for the pseudo-machine, called *interpretive code*, is usually chosen to make the pseudo-machine perform more directly those functions desired by the application. As a consequence, the program in interpretive code requires fewer bits of memory than the same program would require in machine code directly.

This efficiency in memory space is obtained at the expense of both speed and flexibility. The pseudo-machine usually executes the program at a slower rate than would occur with the same program coded in machine code. Also, the pseudo-machine will tend to be less general than the real machine. The interpreter is designed for the effective performance of certain functions needed by the class of programs for which it was intended. Consequently, it performs certain other functions poorly and, in fact, it may not allow the programmer to perform still other functions at all.

Program store. The *program store* block consists of the medium in which the program is stored. It may be either core memory or paper tape.

Translator. The *translator* takes the source code generated by the programmer, often performs some rudimentary checking for errors, and translates the programmer's statements into interpretive code. Note that translation is not part of the real-time operating environment. It is performed once when the program is written, either on the machine in question or some other machine of a different type, or even manually by the programmer. If the translation is performed mostly by machine, the source language that the programmer uses can use terms very close to the terms in which he thinks about the problem. If this is the case, he is using *high-level language*.

Programmer. If the machine translation process is relatively simple (such as a basic assembler), then the *programmer* must first translate his programs into the language needed by the translator program. The closer the terms of the source language are to the terms in which the programmer considers the problem, the higher the level of the language. The further the programmer has to translate his thoughts into the terms of the machine, the lower the level of the language.

Degenerate cases. There are many practical systems where various parts of this hierarchy seem to be missing. In these cases the functions are actually absorbed by some other block. Some examples may illustrate this.

Missing interpreter. In many systems the interpreter or operating system block is missing. This occurs with most of the software provided by minicomputer manufacturers with their systems, including the following:

1. The assembler program requires the programmer to translate his problem into a language very close to machine code. It further translates his assembly-language code to machine code. When writing in assembly language the programmer has access to all the machine functions provided by the machine designers. For this reason, good assembly-language programmers can write efficient programs that use the machine resources efficiently. In small-computer systems most of the programs are usually written in assembly language, since efficient use of the limited machine resources is very important.

2. A Fortran compiler translates a higher-level language, Fortran, into machine code. This enables the programmer of algebraic problems to think more in terms of the problem and leave most of the translation to the Fortran translator or compiler. Good compilers generally use machine resources more efficiently than a poor assembly-language programmer and less efficiently than a good one.

Missing translator. There are several important software systems in which the translator appears to be missing. In these systems the program is stored directly in the source language of the programmer. Examples are most uses of Basic and Focal. In these systems the program will take much more memory space than a translated program, since the source language contains much redundant information. In addition, the system will usually run slower since each program statement must be processed each time it is executed. These systems have the advantage of being easy to change on line and are often more interactive than those in which translation is needed. There are several programming languages that are used in this way during program preparation and debugging. After prove-in, the program is translated into machine code for repetitive use.

Between these examples is a continuum of possibilities. The skill of the system designer determines which is the

* The thoughts in this Appendix were provided by R. G. Fulks of General Radio Company; the words were compiled by the author.

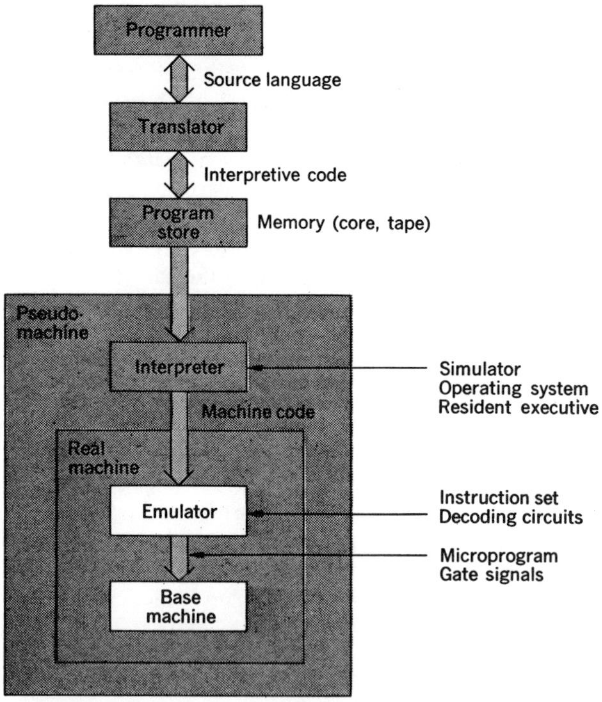

FIGURE 14. Block diagram of a computer system.

most appropriate for a given system problem. The right decision can result in the optimum use of program space and machine time.

Appendix B. Economic justification for automatic test equipment

Justifying an automatic test system from an economic standpoint is relatively easy. In fact, it can be said that automation, at almost any price, is worth it—or soon will be. Looking backward or forward ten years from any point of our lifetime we see a smooth increase (with time) of the cost per unit of production using skilled labor only, and a decrease of the cost per unit using a combination of semiskilled labor and automatic equipment. This seems a paradox, since the cost of automatic equipment itself might seem to be related to the increasing cost of the labor that goes into the equipment. That it is not is due to the insatiable demand for the products of technology, which causes, in turn, an increasing demand for automatic equipment and brings to bear the forces of "the economies of scale"; that is, the more items you make, the less their unit cost. Although the cost of the equipment goes up (after the initial decline from the first few developmental models), the efficiency of the equipment goes up even faster. This is further illustrated by the paradox of the equipment cost/product cost ratio. When diodes sold for $1.50 each you could buy a diode tester for $300. Now that diodes sell for a few pennies apiece, the test equipment costs $100 000! Morever, the $100 000 testers have helped achieve the price reduction of the diodes.

The real problem in automatic testing is not in justifying automatic equipment, but in selecting which automatic equipment to buy. Even so, we are still called upon to justify such expenditures to our supervisors. The following paragraphs may help.

The steps in justifying the purchase or construction of an automatic test system are identical to those in justifying any capital investment. Basically you have to compare the money saved by the new machine with the cost of the machine, and establish that the use of the money required to buy the machine is better than any other use you can think of. The preceding sentence makes it sound easy, but a rigorous analysis can get quite complicated, for here we leave the comfortable world of volts and ohms and enter the strange world of our professional colleagues in the accounting department, with its tax shields and discounted cash flows. We quickly find that what we thought was a dollar isn't a dollar at all. It may be a half dollar, or 75 cents, or even $1.25! This is the result of two aspects of our economic system: the corporate income tax and the time value of money. If you're a profit-making corporation, you can view any additional dollar of operating expense or saving in its incremental effect on the profit of the corporation. If you "save" a dollar you must give half of it to the Federal Government as tax and you wind up with only half a dollar. Similarly, if you spend a dollar for operating expense, you reduce your tax by half of it and you're only out half a dollar. Thus some of the operating savings and expenses in the use of capital equipment are viewed at half their value (tax shield).

The time value of money, on the other hand, shows us that a dollar to be saved during the next year is not worth a dollar today. In fact, if we can invest money and earn 10 percent on it, that dollar return spread over next year is worth only 95 cents today. Therefore, the time distribution of savings and expenses must be considered.

Three basic methods are currently used to estimate the relative profitability of capital investments: return on investment, payback period, and discounted cash flow.[22,23]

Return on investment. Several methods are used to estimate the percentage rate of return for a capital investment (in order to compare it with other possible investments). A simplified version is the MAPI formula,[24] wherein you use the estimated savings in the first year of operation to determine an "urgency rating," the percentage return in the first year. The following calculation shows the urgency rating for a $30 000 test system that will provide savings of $15 000 a year (based on ten-year straight-line depreciation).

Operating savings:	$15 000
Less depreciation:	3 000
Net operating advantage:	12 000
Less added tax:	6 000
After-tax return:	$ 6 000

$$\text{Urgency rating} = \frac{6000}{30\,000} = 20 \text{ percent}$$

Payback period. The payback period is the time necessary for the savings (after taxes) to pay back or make up for the initial investment. The shorter the period, the better the investment. Rather than savings, total cash flow is used; that is, depreciation charges are added to the actual savings, since these charges are an additional source of cash. (Depreciation dollars never leave the company; you've already paid for the machine.) Table IV compares two investments on this basis, a $30 000 system that will save the company $15 000 a year

IV. Payback-period investment approach

Invest-ment	Operating Savings	Depreci-ation	Net Return After Tax	Cash Flow	Payback Period, years
$30 000	$15 000	$3 000	$6 000	$ 9 000	3
50 000	20 000	5 000	7 500	12 500	4

V. Discounted cash flow technique

Year	Savings	Depreci-ation (sum of years)	Net Cash Flow After Taxes	Discounted Value at 20 Percent Return	
				Factor	Amount
1	$15 000	$10 000	$12 500	0.909	$11 363
2	15 000	8 000	11 500	0.751	8 637
3	15 000	6 000	10 500	0.621	6 520
4	15 000	4 000	9 500	0.513	4 873
5	15 000	2 000	8 500	0.424	3 604
Total:	$75 000	$30 000	$52 500		$34 997

and a $50 000 system that will save $20 000 a year.

Discounted cash flow. The discounted cash flow technique takes into account the smaller present value of future returns. The usual method is to tabulate all the future savings, select a desired interest rate (opportunity cost), and reduce the values of the future returns to present values by factors read from a discount table. You then add up all the discounted (reduced) returns. If the sum is greater than the investment, you'll do better than the selected interest rate; if it's less, you won't. Table V shows this calculation for a $30 000 system that will save $15 000 a year for five years and a desired intrest rate of 20 percent (returns beyond five years and salvage value are ignored).

Since the total discounted value (net present value) of $34 997 is greater than the investment of $30 000, the actual rate of return is greater than the 20 percent selected. This technique can be refined by choosing different interest rates and interpolating until the net present value equals the investment, thus determining the actual interest rate.

The preceding methods are fairly straightforward and relatively simple. The procedure can be made a good deal more complicated, however, if all the fine points are accounted for, such as the growth of operating savings from year to year as the equipment is utilized more (more volume, more shifts) or as the cost of labor *not* paid for increases, the salvage value of the equipment if and when you sell it, the interest-compounding period to be used (continuous, monthly, annually), nondepreciated start-up costs, varying tax rates, etc. If your company requires such refined analysis you'd best consult a friend in the accounting department. Some equipment manufacturers provide forms, tables, charts, and even computer programs to simplify the analysis, but, if you use them, make sure you apply the same techniques to all choices.

REFERENCES

1. Lucey, R., and Newth, D., "When to inspect all incoming semiconductors," *Electron. Equipment Eng.*, Mar. 1968.
2. Holt, A., and Stoughton, A., "Guidelines for the purchase of memory testing equipment," *Evaluation Eng.*, July/Aug. 1970.
3. Bartik, J. J., "Common computer interfaces," IEEE Computer Society, Mideastern Area, Cherry Hill, N.J., Mar. 1971.
4. Litzinger, J., "Hardware/software characteristics of a computer-controlled test system," IEEE Computer Society, Mideastern Area, Cherry Hill, N.J., Mar. 1971.
5. Bobroff, D. A., "Avoid pitfalls in computerized testing," *Electron. Design*, Aug. 16, 1969.
6. Wegner, P., *Programming Languages, Information Structures, and Machine Organization.* New York: McGraw-Hill, 1968.
7. Ross, D. T., "Fourth-generation software—a building-block science replaces hand-crafted art," *Computer Decisions*, Apr. 1970.
8. "IC test equipment buyer's guide," *Solid State Technol.*, Mar. 1970.
9. A.T.E. Reports, Box 746, Camden, N.J.
10. "Electronic industry manufacturing markets report," Theta Technology Corp., Wethersfield, Conn.
11. Fulks, R. G., and Lamont, J., "An automatic computer-controlled system for the measurement of cable capacitance," *IEEE Trans. Instrumentation and Measurement*, vol. IM-17, pp. 299–303, Dec. 1968.
12. *Glossary of Integrated Circuit and Related Terminology.* Motorola Semiconductor Products Inc., Phoenix, Ariz.
13. Johnson, W. R., Jr., "Proving out large PC boards," *Electronics*, Mar. 15, 1971.
14. Fichtenbaum, M. L., "A computer-controlled system for testing digital logic circuits," *NEREM Rec.*, pp. 38–39, 1969.
15. Robinton, M. A., "A critique of MOS/LSI testing," *Electronics*, pp. 62–64, Feb. 1, 1971.
16. Curran, L., "Readers reply on MOS/LSI testing," *Electronics*, Mar. 1, 1971.
17. *Hewlett-Packard J.*, Aug. 1969.
18. Loughlin, R. G., and McCoy, F., "The introduction of VAST," presented at WESCON, Los Angeles, Calif., Aug. 1970.
19. Fuller, E. J., and Proctor, J. A., "The evolution and development of a digital test capability for VAST," presented at the Automatic Support Systems Symposium for Advanced Maintainability, St. Louis, Mo., Oct. 1970.
20. Ellis, M. T., "Vital: A general-purpose multiple-dialect automatic test language," presented at the Automatic Support Systems Symposium for Advanced Maintainability, St. Louis, Mo., Oct. 1970.
21. Gundal, R. K., and King, J. F., "A central time-sharing mini-computer—provides data collection and automatic control over a variety of remote test systems," *NEREM 70 Tech. Appl. Papers.*
22. Ammer, D. S., *Manufacturing Management and Control.* New York: Appleton, 1968.
23. Helfert, E. A., *Techniques of Financial Analysis.* Homewood, Ill.: Richard D. Irwin, Inc., 1967.
24. Terborgh, G., *Business Investment Policy*, Machinery and Allied Products Institute, Washington, D.C., 1958.

Part VI
System Examples

Editor's Introductory Comments

Very few semiconductor memory system examples have been documented in the literature. The first paper (VI,1) is a very brief discussion of the main memory for the IBM System 370 Model 145. This paper mentions the use of an error-correcting code, a feature that very likely improves the system-level reliability of this memory by one to two orders of magnitude from the levels mentioned in the Part V Comments. Thus, system level MTBF for a 512 000-byte version of this memory is likely to be in the range 2–20 years, although component failures may occur as often as every few months on average.

The two remaining papers describe in more detail a bipolar memory system for the Illiac IV and an MOS memory system for general application.

TABLE 6-A

	Technology	Access/Cycle Time (ns)	System Pwr. (mW/bit)	Density (bits/cm^3)	Bits/Chip
IBM 360-85	Bipolar	40/80	5.5	5	64
IBM 370-145	Bipolar	125/150	0.6	54	128
COGAR 15C06	Bipolar	125/150	2.7	14	128
Fairchild-Illiac	Bipolar	188/200	3.8	5	256
Semi RAM 300	Bipolar	300/400	0.7	31	128
COGAR 30C06	MOS	250/400	0.4	70	1024
STD Logic-1103	MOS	450/550	0.5	32	1024
EM & M 3000	Core	300/650	0.5	50	—

Table 6-A. Comparative performance of alternative memory designs.

Table 6-A compares some existing memory system designs in performance, power consumption, and overall density. High density is desirable because it permits a large memory to be installed near the central processing unit. Signal propagation delays are thereby minimized. When forced air cooling is used, power density in a packaged system is limited to about 30 mW/cm^3. Most of these systems approach the limit. The superior combination of characteristics for the 370–145 memory is evident.

Main Monolithic Memory

JOHN K. AYLING AND RICHARD D. MOORE

Abstract—**The 128-bit NDRO random-access memory chip used in the IBM System 370 model 145 and System 7 is described. Consideration is also given to packaging power dissipation and systems environment.**

TWO-YEARS ago a 64-bit bipolar buffer chip was described as being a state-of-the-art memory chip in manufacturing[1] (see Fig. 1). This compares to the 128-bit main memory chip shown in Fig. 2, which is presently being made in quantity production for use in the IBM System 370 model 145 and System 7. This chip is fully functional in that it contains its own decoding, powering, and sensing circuitry. Chip dimensions are 123 mils × 113 mils × 15 mils. The chip requires 17 pads for powering, input, and output.

The carrier for the memory chip is a 1/2-in² 18-pin module. See Fig. 3. The module contains two 1/2-in ceramic substrates with two chips mounted on each substrate. The resultant 512-bit memory module is fully functional and is used as a basic building block for various storage-card configurations.

Fig. 4 illustrates some of the basic differences between the buffer memory module of two-years ago and the main memory functional module described in this paper. The buffer memory module has a capacity of 128 bits and dissipates an average of 2 mW/bit. The main memory module contains 512 bits of storage at a 1/2 mW/bit. The buffer module contains no decode while the main memory module is fully functional and contains decode circuitry. The organization of the main memory chip is shown in Fig. 5. The storage array is a 16 × 8 matrix, which has been defined for purposes of discussion as 16 words by 8 bits.

Power is supplied to the chip by the chip-select circuitry. Four address bits are used on chip to select one of 16 word drivers. Eight cells on the word line are driven into a selected condition. In a similar way, three address bits enable one of eight bit sense circuits on the chip. The intersection of the selected word and bit circuits determines which cell is logically addressed and ready for a READ or a WRITE operation.

During READ, the information stored in the selected cell is read out nondestructively through the final sense amplifier. During WRITE, data circuitry drives the selected cell to its desired state. Design innovations included in

Fig. 1. 64-bit buffer chip.

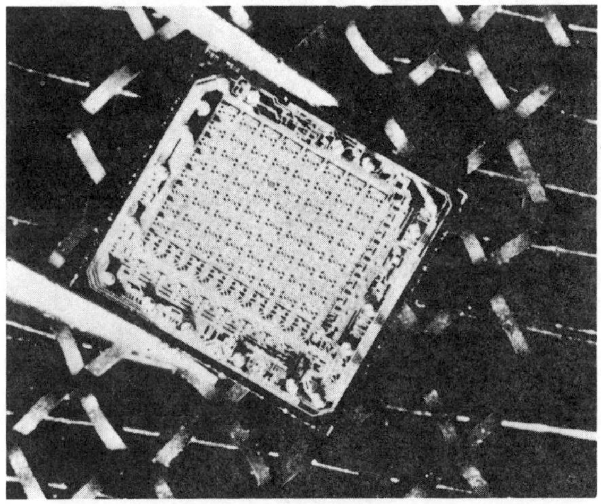

Fig. 2. 128-bit main memory chip.

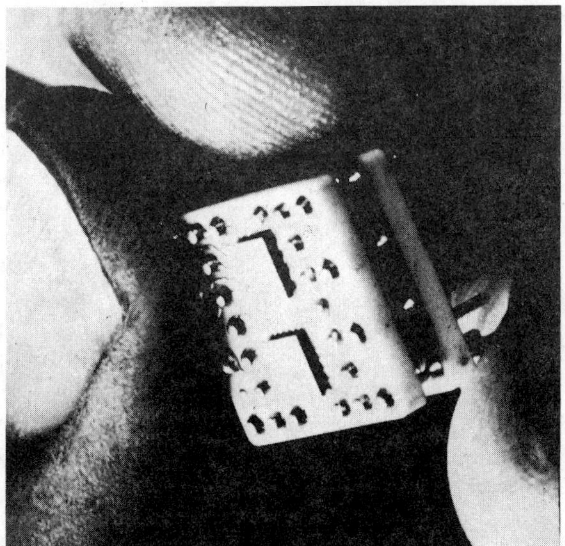

Fig. 3. 512-bit main memory module.

Manuscript received April 29.

The authors are with the IBM Corporation, East Fishkill Facility, Hopewell Junction, N. Y. 12533.

[1] R. D. Moore and J. K. Ayling, "High performance monolithic store," *ISSCC Digest Tech. Papers,* vol. 13, pp. 36–37.

Reprinted from *IEEE J. Solid-State Circuits,* vol. SC-6, pp. 276–279, Oct. 1971.

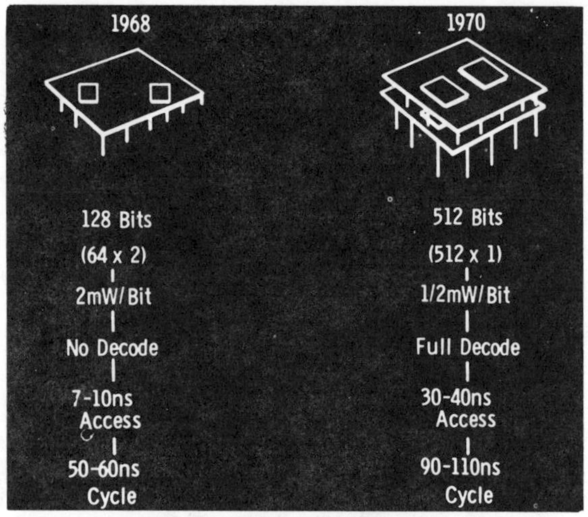

Fig. 4. Module features.

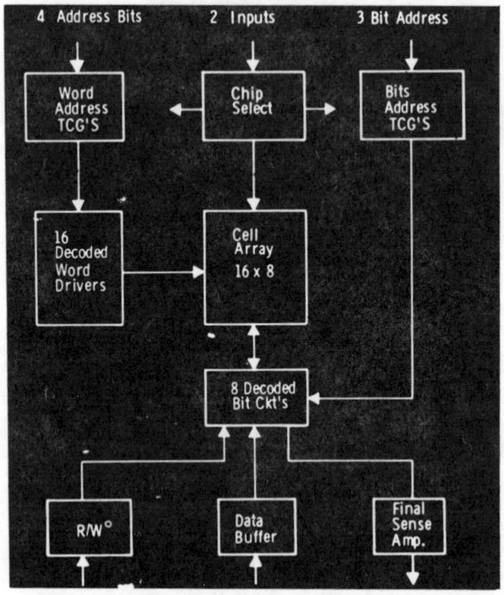

Fig. 5. Block diagram of array chip.

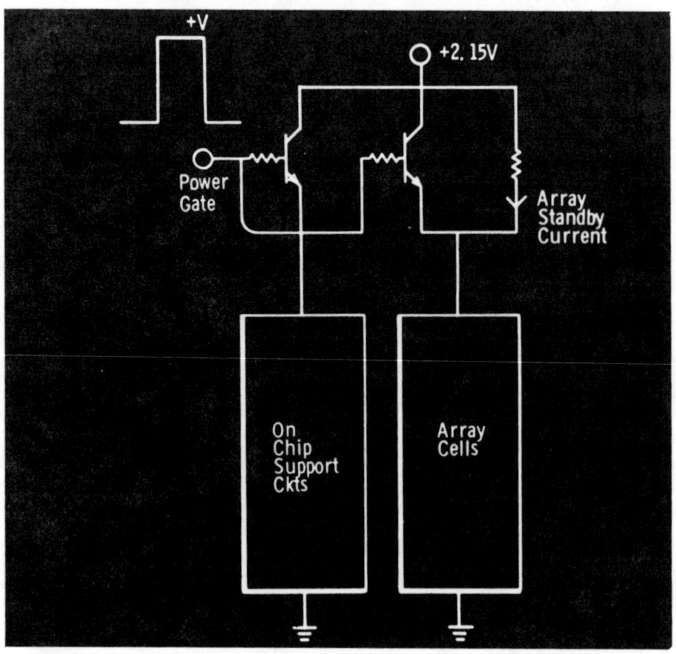

Fig. 6. Pulse/bilevel power.

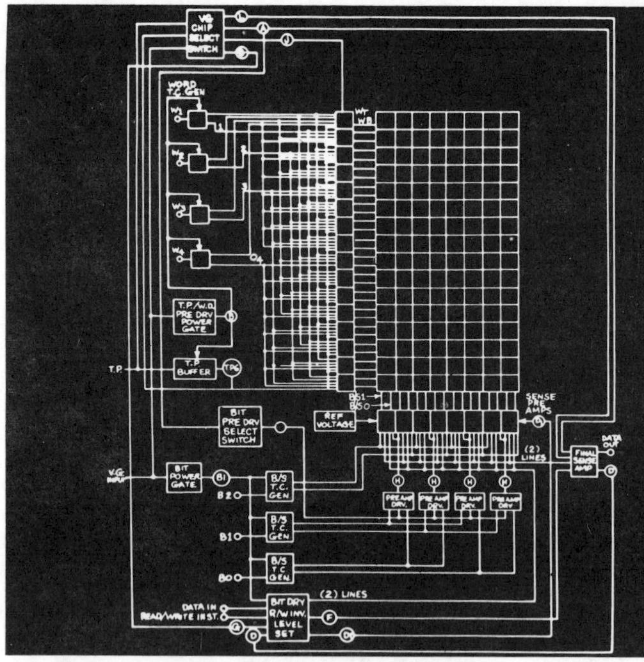

Fig. 7. Chip logic diagram.

this memory chip are the use of pulse powering and bilevel powering concepts as illustrated in Fig. 6. The periphery circuits, decode, driving, and sensing are pulse powered. That is, when not selected their power dissipation is zero. When a chip is being addressed, a select switch supplies voltage to the periphery circuitry and an operation can be performed. The storage array is bilevel powered. When not selected, a small amount of current is supplied to the storage cells just sufficient to maintain the stored information. When the chip is addressed additional current is supplied to the array to ensure a higher power level for reading or writing. These two concepts were instrumental in achieving the very low power-to-performance ratio.

Fig. 7 shows a more detailed description of the logical organization of the array chip. For a READ or a WRITE operation, the proper power gate (PG) and timing pulse (TP) signals enable the select switch, which supplies power to the periphery circuits. The word true-complement (T-C) generators develop the true and complement of the four-word address inputs. Each of the 16-word drivers contains a four-way OR, which, when driven by the T-C generators, ensures the addressed word driver is selected. One of eight bit SELECT circuits is chosen via a 2×4 matrix decode. As was mentioned, the intersection of the selected word line and selected bit line defines the addressed cell.

For a READ operation, the information stored in the selected cell is detected by the on-chip sense amplifier and fed out. During a WRITE operation, the WRITE circuitry supplies the appropriate signals to the selected cell and writes the supplied data.

Fig. 8 is a schematic of the chip select switch and word drive system. The select switch with the appropriate PG and TP signals applied supplies power to the on-chip

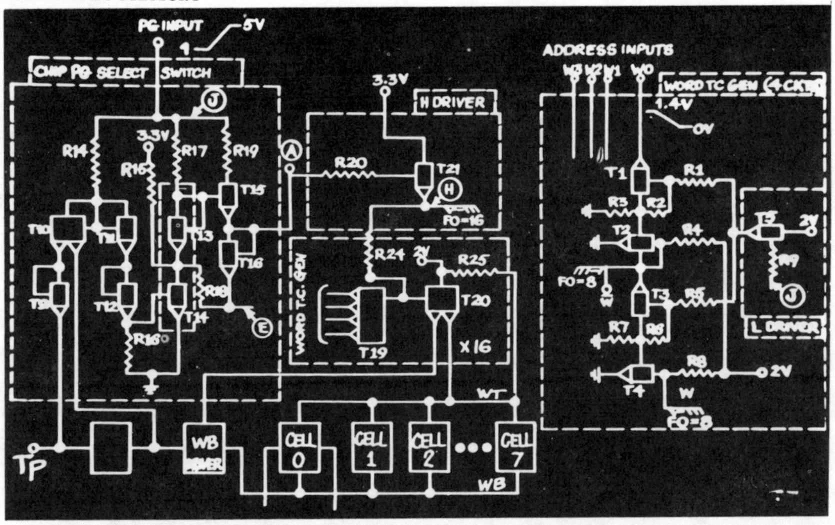

Fig. 8. Select switch and word drive schematic.

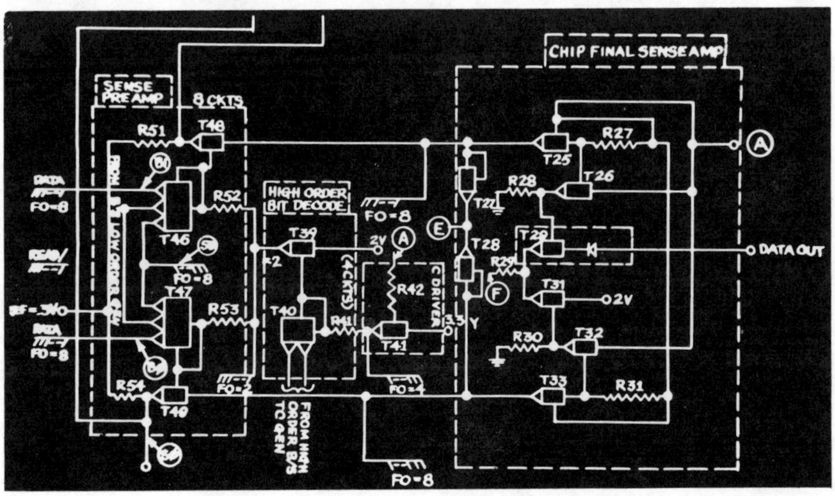

Fig. 9. Bit drive/sense schematic.

periphery circuits. The word driver supplies a small current to cells during standby via R_{25}. When a word driver is selected T_{20} turns on and the cell currents on the selected word line increase. This is the method used to bi-level power the storage array.

The word drive system is designed so that in standby all 16 word drivers are logically selected; that is, all T–C generator outputs are high. During a selection, the 15 unaddressed drivers are driven to the unselected state. This minimizes the word driver circuit delay in that the capacitances associated with T_{19} are precharged.

The bit drive and sensing system is shown in Fig. 9. The sense preamplifier performs a double function. It acts as a common-base differential amplifier during a READ operation and supplies the required signal to the selected cell during a WRITE operation. The eight sense preamplifiers are dot ORed into the final amplifier that amplifies the sense signal and generates the data-out pulse. The final sense amplifier is a low input impedance differential amplifier. As was mentioned, the storage module is a fully functional memory element.

Several different storage-card organizations have been developed using the module in conjunction with support modules. One such card is shown in Fig. 10. This is a 12 228 word × 1 bit card made up of 24 storage modules and 10 support modules. The 1 bit per card organization has two distinct advantages. First, it allows the select power to be dissipated over many cards in the system. Second, with the use of single bit error correction, it provides the maximum protection against multiple bit failures caused by malfunction of individual components. Fig. 11 illustrates the logical organization of this card. The 24 power gate drivers and the four TP drivers form a matrix so that only a single array chip per card is selected. A 4 × 1 matrix selects one chip on the selected module. Seven additional address lines are supplied to all modules and chips. All the chip data outlines are dot ORed into the final sense amplifier latch module.

A simplified 12 288 × 1 bit card timing diagram is shown in Fig. 12 for a READ operation followed by a WRITE operation. As can be seen, both the WRITE and the READ cycles are 105 ns. Card access time is 78 ns.

Fig. 13 illustrates the final card carrying package—a large board that has the storage capacity of 12 288 × 36

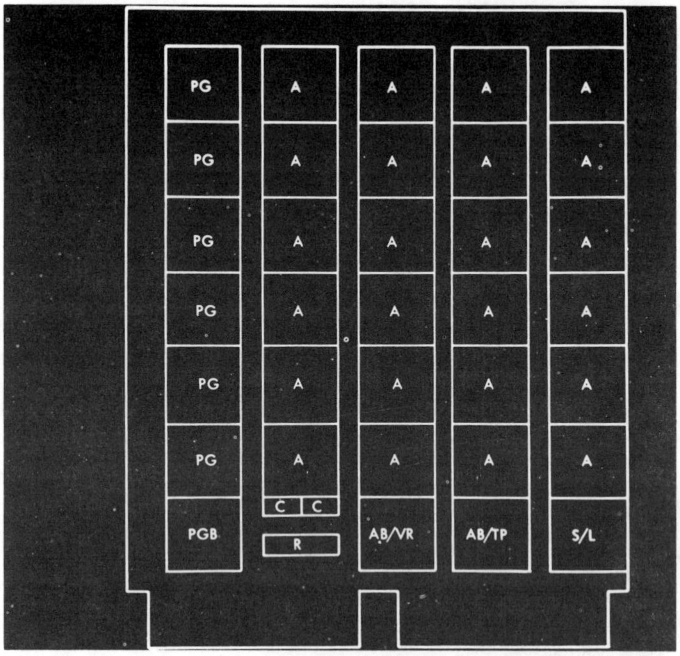

Fig. 10. 12 288 × 1 bit card outline.

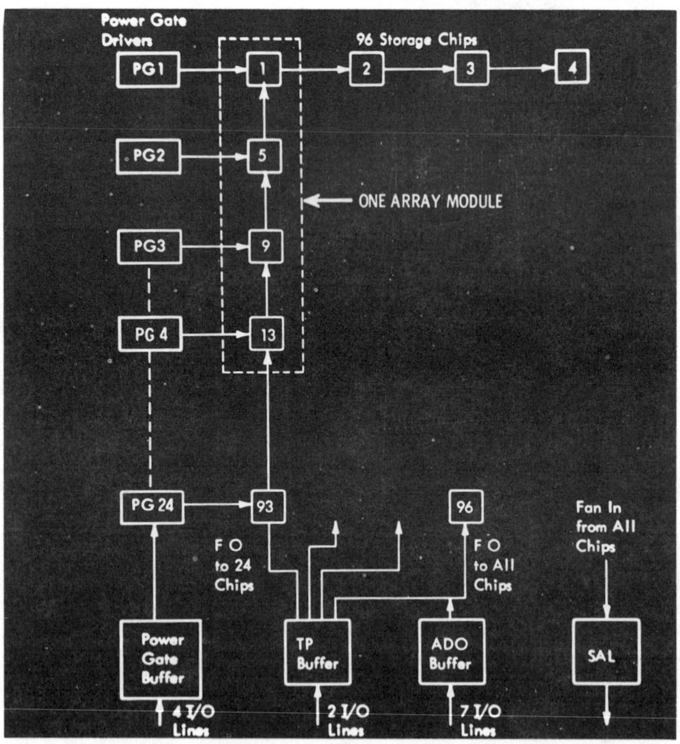

Fig. 11. 12 288 × 1 bit block diagram.

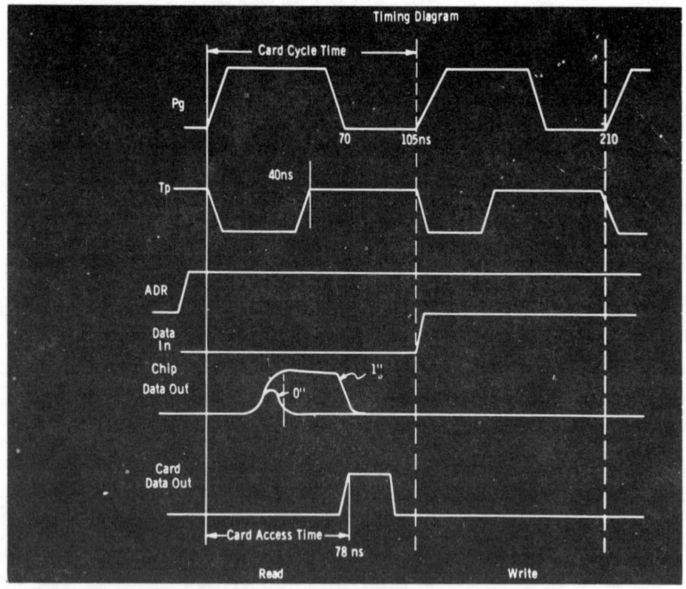

Fig. 12. Card timing diagram.

Fig. 13. 12 288 × 36 bit memory board.

Fig. 14. Monolithic memory cards.

bits. Fig. 14 shows five card types presently being manufactured for different applications. Shown here are cards with capacities of 8192 × 1, 12 × 1, 4096 × 2, 2048 × 2, and 512 × 9.

The main memory described here probably represents the largest commitment to the technology made to date. In many ways, it opens the door to future semiconductor memory development and a new phase in the contribution of semiconductor manufacturing for the computer industry.

Design Considerations Leading to the ILLIAC IV LSI Process Element Memories

REX RICE, FELLOW, IEEE, WENDELL B. SANDER, MEMBER, IEEE, AND FRANK S. GREENE, JR., MEMBER, IEEE

***Abstract*—This paper presents the design considerations for a 256-bit bipolar LSI memory component and the system-packaging technology by which the LSI components were assembled to provide seventy 131 072-bit 200-ns full-cycle main process element memories (PEM) for the ILLIAC IV computing system. The concepts utilized allow memory modularity from about 1024 bytes to over 3 000 000 bytes while maintaining an approximately linear price per bit over the whole range. The experience on LSI memory component and system testing is discussed. Also included are observations on experiences with component designs and prices during the PEM project.**

I. Introduction

FOR SEVERAL years the principal semiconductor manufacturers have been saying that large-scale integrated silicon memories are here. The word "here" has had to take more than one meaning. "Here" has definitely applied to components being available in semiconductor research laboratories and in feasibility demonstrations with small experimental memories. "Here" has applied in the case of IBM supplying LSI memories to itself for cache use in its large machines [8]. In this environment, the relatively high cost per bit to IBM was far less important than the performance speedup the LSI memory gave the system when used in conjunction with a large core memory. To the manufacturers of magnetic memories, and core memories in particular, "here" has a very different meaning. "Here," in the real sense in the memory industry, must imply commerical availability of large quantities of memory bits at a price per performance range that is competitive.

Many random access memory (RAM) classification schemes have been used. One useful, but nonexhaustive, presentation is found in Table I.

TABLE I

Speed range	Typical use	Technology
10–50 ns	registers stacks control memory etc.	bipolar silicon
50–100 ns	high-speed main memory registers stacks control memory etc.	bipolar silicon
100–200 ns	medium-speed main memory registers stacks control memory etc.	bipolar silicon
200–300 ns	main memory etc.	bipolar silicon MOS/bipolar silicon plated wire
300–500 ns	main memory etc.	bipolar silicon MOS silicon plated wire fast cores
500 ns and up	main memory etc.	cores MOS silicon

It follows that if a higher speed main memory system is available (with no special disadvantages) in high-volume production and at the same price per bit as a slower memory system then the faster speed system will dominate the market. Cores in the 1 μs and slower speed ranges do presently dominate the market with prices per bit in the 1-cent-and-under range. Cores in the 500 ns and faster ranges begin to approach 5 cents per bit and have severe operating problems. Some plated-wire magnetic memories perform in this speed range but production volume and price per bit has not settled out. Small bipolar silicon memory systems cycling at 15 ns are on the market. The price per bit of these memories is considerably above 5 cents at this time. Medium-size bipolar silicon memories in the 80-ns speed range are being used in large commercial computers. Large bipolar memory systems in the 100–200-ns cycle time range and priced near 5 cents per bit in large volume orders have come into existence during 1970. The technology was ready and needed only the push of a sufficiently large single contract to crystallize the details. The ILLIAC IV PEM contract performed this function [2].[1]

The very tight time scale on the ILLIAC IV and the desire to have as low risk a project as possible and still meet objectives caused a reexamination of component processes and packages. For its ILLIAC IV PEM, Fairchild backed away from more exotic silicon-processing techniques and tolerances and yet-to-be-developed (for production) component-packaging techniques. Multiple chips, flipped chips, beam leads, ceramic substrates with multiple interconnection layers, etc., were considered too risky on the time scale [6], [10]. MOS technology alone did not provide sufficient speed so bipolar technology was selected. The main gamble in this project was to predict yields of all-good 256-bit die so that production volume and costs were acceptable. Here again effort was

Manuscript received June 9, 1970. This work was supported by the Advanced Research Projects Agency, Rome Air Development Center, under Contract USAF 30(602)4144.

The authors are with Fairchild Research and Development Laboratory, Palo Alto, Calif. 94304.

[1] Fairchild is a supplier of LSI memory systems to Burroughs, the prime contractor on the ILLIAC IV computing system.

Reprinted from *IEEE J. Solid-State Circuits*, vol. SC-5, pp. 174–181, Oct. 1970.

concentrated on learning how to get good yields; exotic schemes for discretionary wiring, etc., were not considered [7], [9]. The considerations leading to the 256-bit bipolar component, the Fairchild 4100, are discussed in Section III. This component and the systems into which it will assemble fall into the 100–200 ns medium-speed memory category. The PEM memories represent a "first" against which newer technologies must compete on the basis of cost, speed, ease of repair, power, volume, etc.

II. ILLIAC IV PEM Specification Summary

The ILLIAC IV computer is composed of 256 individual process elements (PE) and PEM organized into a matrix. One quadrant (i.e., 64 PEs and PEMs) of ILLIAC IV is now being produced by Burroughs under an ARPA contract for the University of Illinois.[1] The memories for this quadrant have the following specifications.

Word Format: Each word is divided into two parts; the OUTER part (bits 0–7 and bits 40–63) and the INNER part (bits 8–39).

READ Operation: The memory accesses a specified location and returns the data on the output lines within 188 ns from the time of request.

WRITE Operation: The memory receives an address and a WRITE-INNER and/or a WRITE-OUTER signal and writes the appropriate part(s) into the addressed location; an unselected part in the addressed location remains unchanged.

Memory Cycle: The complete memory cycle is 200 ns.

Power: The total current drain for all supplies is under 200 amperes and +4.8 volts and −2 volts are supplied. The total power used in the PEM is under 500 watts. A maximum ripple of 30 mV peak to peak is allowed on either power supply.

Signals: All signals to and from the PEM are on 75-ohm transmission lines.

Physical Size: The memory size, exclusive of power and small cooling fans, is $15\frac{1}{2} \times 20\frac{1}{2} \times 5$ inches.

Timing: The timing specifications for the PEM are presented in Section V.

III. 256-Bit Bipolar LSI Memory Component

Computer-system clock speeds in the order of 100 ns and a main memory-system cycle time of no more than twice the basic clock speed or about 200 ns is highly desirable [3]. The complete memory cycle of 200 ns includes address decoding for binary address in the order of 17 binary bits (i.e., 131 072 words) and the latching of output data into a register. This requires a memory component with an internal decode and access time of approximately 100 ns. These considerations dictated the use of bipolar silicon technology.

The use of well-developed component processing and packaging technology was indicated to meet the required delivery schedule. The main new factor involved in the memory component development was complexity and size of the die (i.e., 0.139 × 0.111 inch) with the attendant problem of achieving reasonable yields. This die contains approximately 1400 devices. Single-layer metallization and TTμL manufacturing tolerances of 0.2-mil linewidths were chosen. The interconnection of the devices with single-layer metal was a major problem. To solve this layout problem, epitaxial resistors were chosen that allowed an internal device connection between the collector and the $+V$ power without requiring metallization [2]. Fig. 1 illustrates the cell design.

Another difficult problem was the choice of a component package. A 16-lead dual-in-line package had the largest number of leads available in a high-volume low-cost reliable component package. It was chosen because of availability. This decision necessitated the placing of sense, drive, decoding, and some controls on the chip. Fig. 2 presents a block diagram of the chip.

The resulting 256-bit memory component is presented in Fig. 3. It has the following characteristics.

Organization: 256 words by one bit, random access.

Address: Coincident address x and y, each accepting a six-variable address input (3 out of 6 code) [1].

Data: A single, bidirectional data INPUT/OUTPUT line is provided; control is by a READ/WRITE operation line (70-ns READ access, 40-ns WRITE).

Logic Levels: Compatible with Fairchild DTμL, TTμL, and CTμL circuits, data output has OR tie capability (open collector).

Power: 500 mW typical; single power supply = 5 volts nominal.

Package: Standard 16-lead dual-in-line; chip mounted face up and wire bonded.

IV. LSI Memory System Design and Packaging

It is common practice to equate the advantages of LSI with a circuit-to-circuit cost comparison where the LSI array circuit is replacing an equivalent integrated circuit. As shown in Fig. 4, this is a very misleading comparison. The total functional price-to-performance ratio of the complete computing system, not the cost per bit or the cost per circuit of the LSI circuit replacing the normal integrated circuit, is the thing upon which we should focus attention. As illustrated, the LSI array will replace the interconnections that normally occur on printed circuit boards, the connectors that connect the integrated-circuit boards, large numbers of wires that normally cost at least 6 cents each, installed, and finally the integrated circuits and components that are the only items normally considered. The cost comparison should not be just a circuit-for-circuit replacement comparison but a complete cost-replacement comparison for all the hardware including nuts and bolts and circuits replaced by LSI [4].

The Fairchild Digital Systems Research Department has, for several years, been exploring the use of large printed circuit boards with only two layers of copper (i.e., the two outside surfaces) and with plated holes for the complete system interconnection of very large computing systems [4], [5]. Two-layer boards are used both for component mounting and interconnections at the function level and also for the complete system interconnection where "mother boards" replace connections

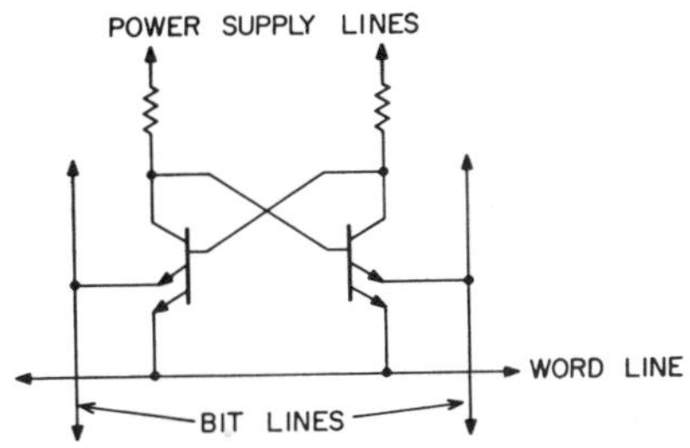

Fig. 1. Fairchild 4100 cell diagram.

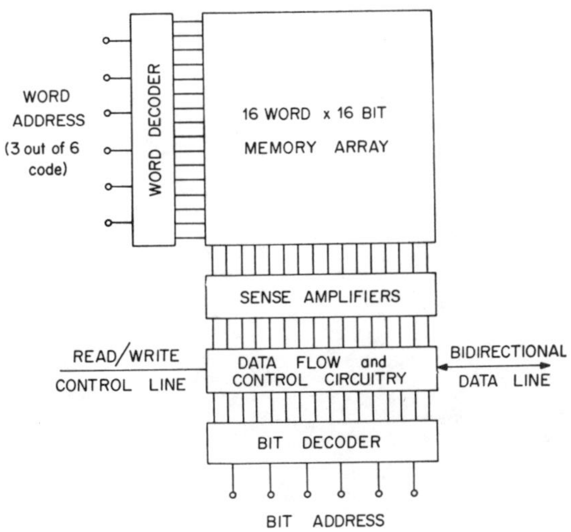

Fig. 2. Block diagram of 4100 256-bit memory component.

normally done with wire-wrap back panels. The object is to achieve minimum system interconnection cost coupled with maximum reliability. A printed circuit board of this type is illustrated in Fig. 5. It has the following characteristics.

Functional Circuit Board: 12 × 17 inch two-layer plated-through holes, glass epoxy.

Contacts: Each 17-inch side contains 100 contacts (total 200) for logic and power distribution use. One 12-inch side contains 50 contacts for test purposes only.

Number of Components: More than 200 dual-in-line packages may be mounted on a standard board. Interconnection density rather than component density dominates the layout (i.e., rarely are all package positions used).

The system interconnection of a multiplicity of the functional circuit boards is also done using a pair of two-layer printed circuit mother boards. A Fairchild proprietary cam-operated contact system is used to allow easy board insertion and to provide maximum reliability. Fig. 6 shows a system interconnection (i.e., mother) board for the ILLIAC IV PEM. This board contains the complete memory-system interconnections between memory boards and the control board. No wires, twisted pairs, or coaxial cables are used. A similar board with a different printed circuit pattern is used on the opposite side of the memory for power distribution within the memory. The interconnection base has the following characteristics.

Fig. 3. 4100 256-bit memory component.

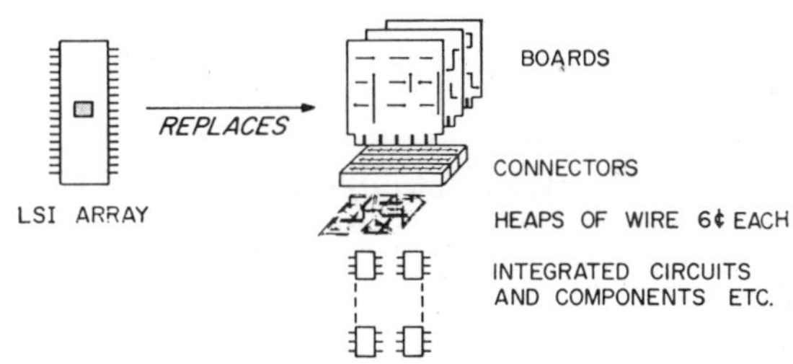

Fig. 4. Total system price, not price per circuit or cost per bit, is important.

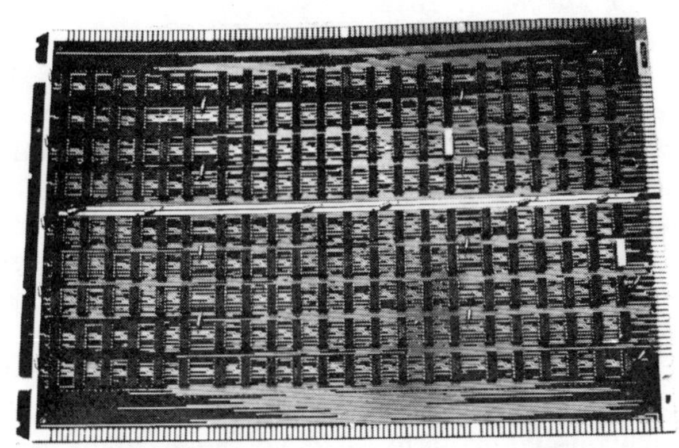

Fig. 5. 12 × 17-inch two-layer printed circuit board.

Mother Board: 17 inch × (length as required at $\frac{1}{2}$ inch per card) two-layer plated holes, glass epoxy.

Contact Mounting: Contacts spaced $\frac{1}{2}$ inch between columns, 100 per column are soldered directly to the interconnection mother board. Contact material is nickel and gold plated, beryllium copper. Spring pressure is 750 grams when engaged on a contact point that has a 30-mil radius. The springs clear the cards for zero insertion force when retracted by cam action.

Contact Operation and Protection: Contacts are operated

Fig. 6. End-view details of a two-layer mother board for interconnecting six-system function boards. Contacts are soldered to mother board and actuated by rotating cams. Three cams and five card guides are removed to illustrate detail.

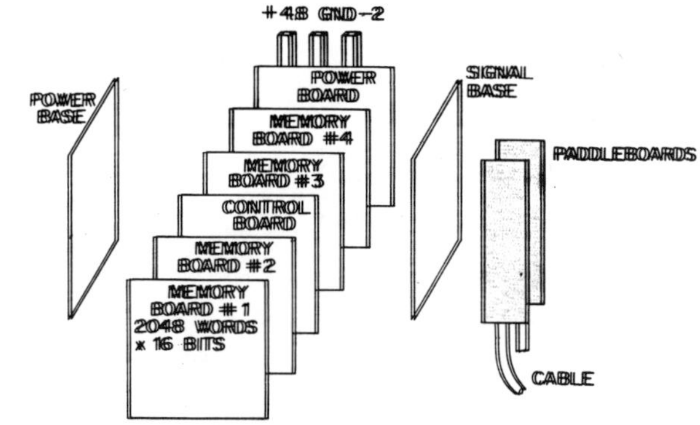

Fig. 7. Exploded diagram of ILLIAC IV PEM system parts.

Fig. 8. ILLIAC IV PEM memory card containing 2048 words of 16 bits.

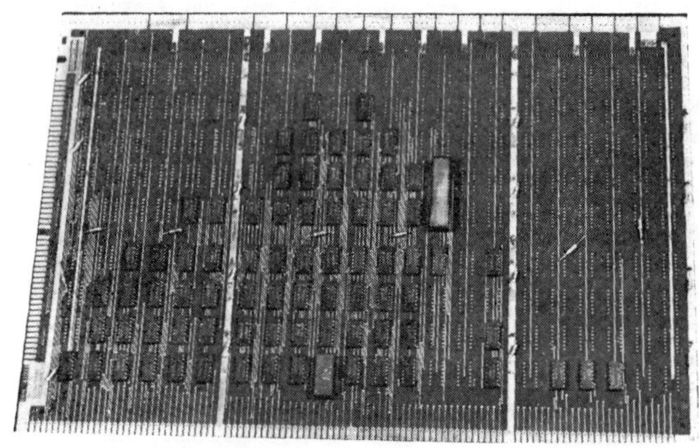

Fig. 9. ILLIAC IV PEM timing control board.

by a cam rod supported by a rod holder and a card-insertion guide. These parts are molded plastic and are assembled to the interconnection base.

An illustrative exploded view of the ILLIAC IV PEM is shown in Fig. 7. Since the packaging technology is modular, the two interconnection bases can be made longer to accommodate a larger number of memory cards. Designs have been made for the following.

Small Memories: 1024–4096 bytes.

Medium Memories: 2048–132 kbytes.

Large Memories: To 3 000 000 bytes.

The memory card used for the ILLIAC IV PEM is shown in Fig. 8. This board contains 2048 words of 16 bits each. Four such boards are used in the PEM to obtain 2048 words of 64 bits each. The arrays of 4100 memory packages are clearly distinguishable by their white surface. Note that integrated circuits and discrete components were used as required. Each memory board places a single unit of loading on input lines and puts full drive capability on output data lines out so that the 2048 × 16 bit array is completely self-contained and essentially isolated from external noise and loading problems. One board may be thought of as a 4096-byte memory.

Fig. 9 presents the timing control board for the PEM. This board does a portion of the binary address decoding and generates all necessary internal timing signals. One such board can drive 16 memory boards without fan-out problems. A duplication of the drive circuits on the control board allows memories of about 1 000 000 bits to be driven before a second control board is needed.

An actual ILLIAC IV PEM is illustrated in Fig. 10. A simulated one-half-million-bit memory is illustrated in Fig. 11 to provide a picture of memory modularity.

The electrical characteristics of the waveforms resulting from the combination of the system package and the components were given very special attention during design. To control noise on the power distribution system, the two-sided PC boards include an interlocking mesh for power and for ground with distributed capacitors for filtering. In the memory boards, particular attention was given to isolation of signal noise from one PC line to adjacent lines. In general, this was done by using a ground strip between adjacent address or data lines within the array. One of the most difficult problems was to route all of these interconnections using two layers. An extensive engineering design package using computer aids to augment the engineer had been developed previously and was used for this purpose. Fig. 12 presents sample waveforms obtained in the memory. The upper

Fig. 10. An ILLIAC IV PEM containing 131 072 bits organized into 2048 words of 64 bits.

trace is from a $+V$ power-distribution line within the memory array itself, where ± 100 mV of noise may be observed. The middle trace shows an address line in the matrix. The upper trace on the lower photo shows the data output line driving a 75-ohm load cabled to the computer. The lower trace is the initiate pulse running at a 5-MHz clock rate.

V. Testing

Specifying and executing tests for this first commerically available LSI memory presented some difficulties. The experience gained from magnetic core and other systems provided some data, but it was not clear that the interactions within the LSI chip and in the system interconnections would be the same as those observed in magnetic-type memories. A conservative approach has been taken on the ILLIAC project. 1) Each 256-bit memory component is thouroughly tested for dc and ac characteristics. A special tester was developed to exercise the component at full speed in an environment as closely duplicating the system as possible. This tester is known as WST for its inventor, W. Sander. In the WST a checkerboard pattern of alternating ZEROS and ONES is written into all 256 bits and then each bit is read and checked. An additional pattern of a diagonal of ONES with all other bits ZERO is written into all 256 bits and then checked on READ cycles. Additional patterns can be induced into the tester for component design purposes but are not utilized during 4100 acceptance testing. 2) After the memory boards are stuffed, flow soldered, and inspected they are exercised in a board tester. Here the memory is exercised at full speed as a 2048 $\times$ 16 bit array. A checkerboard, a checkerboard complement and an all ZERO or all ONE pattern are used. Corrections to any malfunctions are made and the board is placed in stock. Fig. 13 is a picture of the memory board tester with a board inserted. 3) Each complete memory system is exercised in a memory system tester (MST). Here all parts and the complete interchangeability of both control boards and memory boards are verified and tested.

Fig. 11. A simulated one-half-million-bit memory to illustrate modularity.

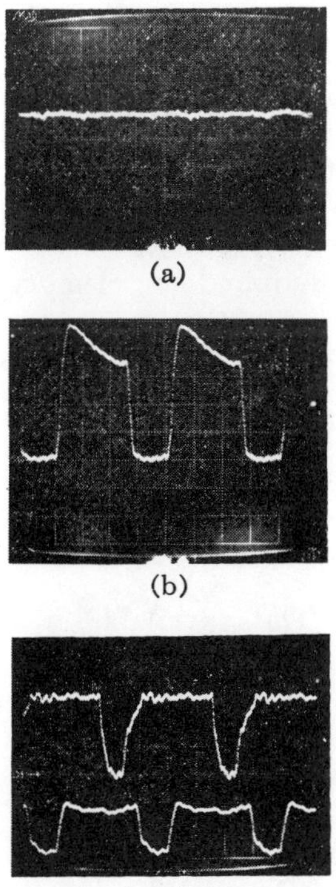

(a) (b) (c)

Fig. 12. Waveforms from an ILLIAC PEM. (a) $+V$ line in memory array 0.5 volt, 50 ns/div. (b) 4100 memory-package address line 1 volt, 50 ns/div. (c) Data out—1 volt, 50 ns/div. Initiate—2 volts, 50 ns/div. 5-MHz clock rate.

The final acceptance test for the ILLIAC IV PEM is accomplished in the MST. Fig. 14 shows the MST. The power supplies for the tester and for the memory under test can be seen below the shelf. The MST is in the foreground above the shelf with its control console on the right and test lights extending out on top. The top of the PEM can be seen at the right upper rear. Cables

Fig. 13. Memory-board tester used for functional debugging.

from the left side of the MST run between it and the front of the PEM and attach to the PEM on its right side. These cables simulate the actual cables in a system environment. Power is delivered to the PEM at the top and rear of the MST. As a safety precaution these connections are covered by plexiglass.

The conditions required for memory acceptance tests are as follows.

Input Signals:

initiate stable	0–50 ns
READ/WRITE stable	0–50 ns
address stable	0–50 ns
data IN stable	40–150 ns.

Output Signals: Data available at 188 ns and stable through 120 ns of next cycle.

Full Cycle: 200 ns nominal (may be any slower speed by delaying initiate).

Logic Voltages:

high level	+2.0 to 3.5 volts
low level	0.0 to −1.0 volt.

Temperature: Inlet +25°C, ±10°C (+77°F ± 18°F).

The memory-system tests are made by writing each test pattern into all 2048 words and then by reading all 2048 words and checking each and every bit for the correct output at the full 200-ns cycle speed. In the walking patterns the ZEROS or ONES are set, checked, and then reset before advancing to the next address. Each of the patterns listed below is run in this manner in sequence and then the whole group is repeated for two continuous hours without error before acceptance. The MST console allows selection and cycling on any one pattern, on any one address, or on any one memory board. In addition, errors can be ignored (for trouble diagnosis) or the MST can be set to run until an error is detected and then lock on that address and cycle continuously.

The test patterns run during memory acceptance are as follows:

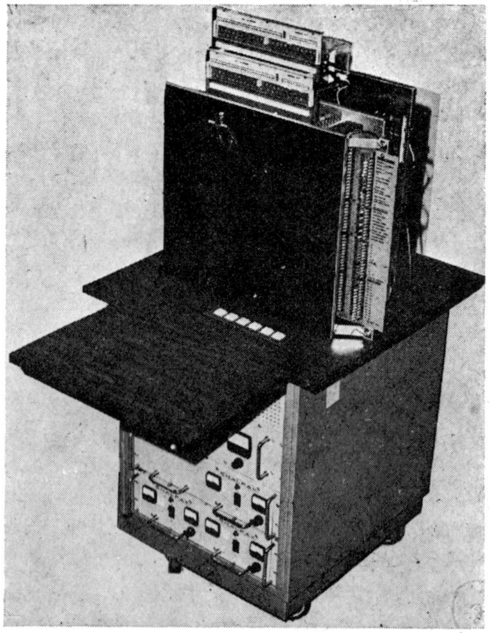

Fig. 14. Memory-system tester used for functional debugging and acceptance testing.

- all ZEROS
- all ONES
- walking ZEROS in a field of ONES
- walking ONES in a field of ZEROS
- checkerboard (alternate ZEROS and ONES)
- complement checkerboard (alternate ONES and ZEROS)
- double checkerboard (two ZEROS, two ONES)
- complement double checkerboard (two ONES, two ZEROS)
- column of ZEROS (in a field of ONES)
- column of ONES (in a field of ZEROS).

Preliminary experience with this LSI memory indicates that the all ZERO, all ONE, checkerboard, and checkerboard complement are the most useful debugging patterns. However, some system problems were turned up in each pattern. Much industry discussion is being conducted on bit interactions and testing patterns for LSI memories. Our experience to date shows this is not as bad a problem as anticipated. When the 2500-hour reliability tests are concluded on an ILLIAC IV PEM we will know much more about this subject.

VI. What We Have Learned

The first lesson to be learned from this LSI memory project is that it is not necessary to include all the exotic research items available in the semiconductor industry to get an early and viable product. For example, the use of a flipped chip with bumps or beams was not required, nor was the use of a ceramic substrate with multiple flipped chips and layers of metal interconnections with insulated cross-overs necessary [6], [10]. In fact, it is not clear at this time that more exotic techniques would be cheaper or more reliable, although they will provide more-dense memory systems. Since reliable performance at the lowest cost per bit is most important, it is not yet apparent what price increase one would be willing to pay to achieve greater density in ground-based general-purpose systems.

The second, and more surprising, lesson learned from this project may be stated as follows. Component manufacturers can supply individual components or wired LSI memory systems at nearly the same price. Fig. 15 illustrates why it is possible and practical to supply either components or wired and tested LSI memory systems at nearly the same price per bit. In any manufacturing process there are a series of parameters that may be controlled to produce a distribution of useful parts to a given specification. The probability curve in the figure illustrates this point. When the components are to be placed in unknown system environments, it is necessary to specify worst case tolerances and tests to ensure that all components will work in any system and in any location. This represents a "narrow" spectrum of useful device parameters and in terms of component/system specification is a dirty interface. On the other hand, if the same components can be placed in a system-packaging environment where all the electrical parameters are highly controlled and adjusted to take advantage of the specific component, a much broader distribution of production parts can be used. Here, the system-to-system specification must be controlled but this is reasonable to do since it is a much easier interface to define. LSI memories can certainly fall into this system category. In a controlled memory "system" environment the system reliability and performance will not be penalized by the use of wider tolerance components from the same production run, and the resulting higher yield of components will reduce costs. When the system packaging technolgoy is cheap enough, then the additional system-assembly cost can be less than or equal to the difference in cost between narrow and wide component tolerances. This happened, surprisingly, on the ILLIAC IV PEM project where we reliably used many 4100 components in the system that would not meet any normal and standard component specification. The implications of the above lesson are that the semiconductor user now has the choice of purchasing components or wired "black-box" memory systems at about the same price per bit. This will cause some adjustment in traditional design roles. In retrospect, it probably should not surprise us since a very similar development has evolved in the core memory business.

The third lesson learned was that the ability to disassemble the memory system by removing boards (i.e., portions of the stack) and the ability to replace individual components significantly reduced engineering design and debugging time. The memory stack for ILLIAC IV is composed of four planes (i.e., boards) that can be replaced or interchanged in about 30 seconds. Complete repair of a malfunction within a board can be made in 15 minutes. The two-layer printed circuit boards were easy to repair with wired changes and/or cuts in runs. Line termination

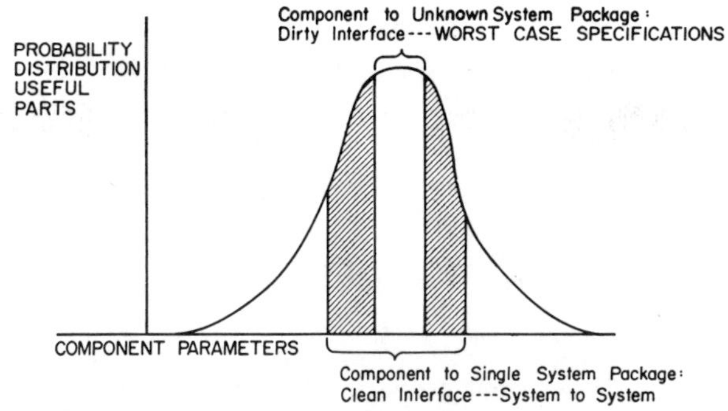

Fig. 15. Memory-component price per bit versus memory-system price per bit. Component manufacturers can supply memory components or wired memory systems at nearly the same price per bit by using a broader distribution of components in a tightly controlled system packaging environment.

and load resistors were easy to apply during final prototype test. These same features make field service of memory systems very easy and economical.

VII. Conclusions

The combination of a 256-bit bipolar memory device in a dual-in-line package together with a sophisticated use of two-layer printed circuit boards has provided the first commerically available random access LSI computer main memories. The ILLIAC IV PEM using this combination represents a first generation. It will be replaced by better, faster, and less expensive LSI memories in the near future. It has set a price per performance standard that future LSI memories must exceed. It is cost competitive with the high-speed core memories. The techniques utilized can supply main memories for systems ranging from minicomputers to ILLIAC IVs.

This development produced one surprising result. It was discovered that the price per bit can be approximately the same for either LSI memory components or for completely wired memory systems.

This first commerically available LSI memory system is not the most dense talked about, the fastest talked about, the slowest talked about, or the lowest powered. This first commercially available LSI memory system is a low-price 200-ns main random access memory, modular from 1024 bytes to over 3 000 000 bytes, supported with test equipment, easily field repairable, and in volume production in 1970.

Acknowledgment

The authors wish to express sincere thanks to Dr. R. Seeds and the Department of Digital Integrated Electronics for the design of the 4100, and to J. Gates, W. Baker, and R. Repass whose untiring efforts brought the production of this difficult LSI chip into a reality. We also extend our thanks to the many others at Fairchild whose extra efforts made the delivery of ILLIAC IV PEMs possible on a very tight time schedule. The cooperation and guidance of V. Smith and W. Wikiera of the Borroughs Corporation is greatly appreciated. Finally, we owe a debt of gratitude to Prof. D. Slotnick of the University of Illinois, Urbana, for his foresight and creative drive in bringing the ILLIAC IV into reality.

References

[1] F. S. Greene, Jr., and W. B. Sander, "Address selection by combinatorial decoding of semiconductor-memory arrays," *IEEE J. Solid-State Circuits* (Correspondence), vol. SC-4, pp. 295–297, October 1969.

[2] F. S. Greene, Jr., and N. V. Phan, "Megabit bipolar LSI memory systems," *1970 ISSCC Digest Tech. Papers*, p. 41.

[3] R. Rice, "Interaction between LSI and computer system architecture," in *Parallel Processor Systems, Technologies and Applications.*, Monterey, Calif., June 1969. (New York: Spartan, to be published).

[4] ——, "Impact of arrays on digital systems," *IEEE J. Solid-State Circuits*, vol. SC-2, pp. 148–156, December 1967.

[5] ——, "The packaging revolution—Pt. V: Simpler designs for complex systems," *Electronics*, 1966, pp. 109–116.

[6] T. Hart, Jr., D. Hillis, J. Marley, R. Lutz, and C. Hoffman, "A main frame semiconductor memory for fourth generation computers," *1969 Fall Joint Computer Conf., AFIPS Proc.*, vol. 35. Montvale, N. J.: AFIPS Press, 1969, pp. 497–488.

[7] R. L. Petritz, "Current status of large scale integration technology," *IEEE J. Solid-State Circuits*, vol. SC-2, pp. 130–148, December 1967.

[8] D. Gibson, "Considerations in block-oriented systems design," *1967 Spring Joint Computer Conf., AFIPS Proc.*, vol. 30. Washington, D. C.: Thompson, 1967, pp. 75–80.

[9] D. F. Calhoun, "The pad relocating technique for interconnection LSI arrays of imperfect yield," *1969 Fall Joint Computer Conf. AFIPS Proc.*, vol. 35. Montvale, N. J.: AFIPS Press, 1969, pp. 99–109.

[10] T. Asai, J. H. Freidrich, and J. D. Schmidt, "A hybrid LSI memory," *Digest, IEEE Computer Group Conf.*, 1968.

Design of a megabit semiconductor memory system

by D. LUND, C. A. ALLEN, S. R. ANDERSEN and G. K. TU

Cogar Corporation
Wappingers Falls, New York

INTRODUCTION

This paper describes a 32,768 word by 36 bit word Read/Write Memory System with an access time of 250ns, and a cycle time of 400ns.

The memory system is based on MOS technology for the storage array and bipolar technology for the interface electronics. A functionally designed storage array chip with internal decoding minimizes the number of external connections, thereby maximizing overall system reliability. The average power dissipation of the overall system is maintained at about 0.4mw per bit including all support circuitry dissipation. This is based on a card configuration of 102 modules with a maximum module dissipation of 600mw.

System status

At present test sites containing individual storage array chip circuits and single bit cross sections have been processed and are being evaluated. Although initial test results are favorable sufficient data has not been accumulated to verify all design criteria. Source-drain storage array chip production masks are in line with other levels nearing completion. Layouts of the bipolar support chips are complete and ready for generation of production masks.

System description

An isometric view of the complete 32,384 word by 36 bit memory system is shown in Figure 1. The total volume occupied by the system is 0.6 cu. ft., resulting in a packing density of approximately 2 million bits/cu. ft. A mechanical housing is provided for the eight multi-layer printed circuit cards that contain the memory storage elements and peripheral circuits. To facilitate insertion and extraction of cards a mechanical assembly is also included. The card connectors are mounted on a printed circuit interconnection board. All necessary system wiring is done on the outside surfaces of this

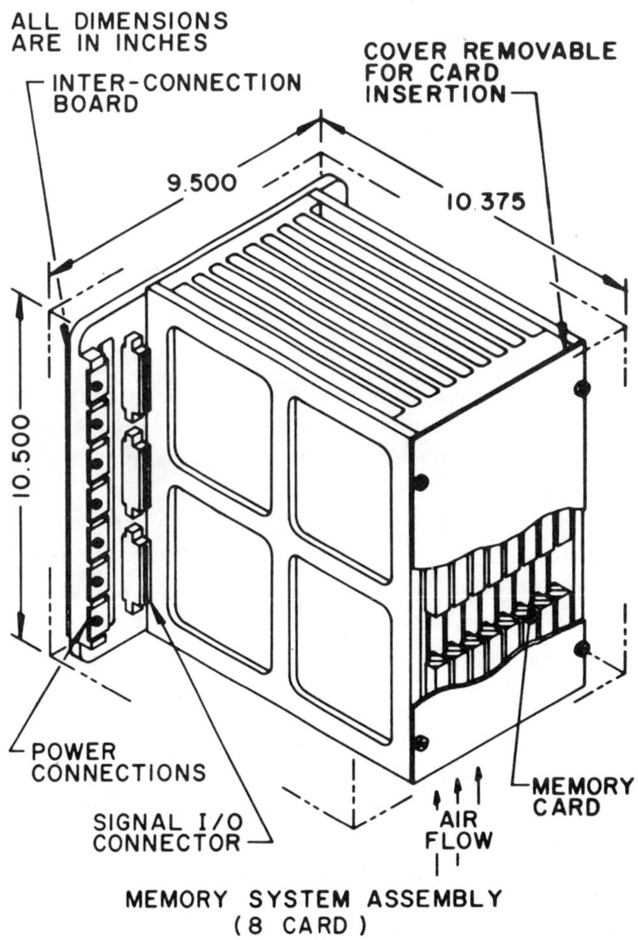

Figure 1—Memory system assembly

Reprinted with permission from *AFIPS Conf. Proc., Fall 1970 Joint Comput. Conf.*, pp. 53–62.

board with voltage distribution accomplished by the internal planes. Additional edge connectors are mounted in this board to accommodate I/O signal cabling via plug-in paddle cards. Power connections are provided at the outermost edge of the board.

Since the purpose of this design was to provide a large, fast, low-cost system for use as a computer main frame memory the following design constraints were observed:

Capacity

A one megabit capacity was chosen to be representative of the size of memory that is applicable to a fairly large, high-speed processor. It was decided that the system should be built from modular elements so that memory size and organization could be easily varied. An additional advantage of ease of servicing and stocking accrued from this approach.

Speed

A balance between manufacturability and system requirements was established in setting the performance objectives. This tradeoff resulted in a goal of 250ns. access time and 400ns cycle time.

Density

The density of memory cells should be maximized in order to create minimum cost per cell. An objective of 1024 bits of information was chosen as a reasonable goal using present LSI technology on a .125 in. × .125 in. chip. In order to keep the I/O signal count within reasonable bounds it was decided that address complementing and decoding should be included within the chip. The chip was structured 1024 words by one bit.

Memory card

A drawing of the basic modular unit, the memory card, is shown in Figure 2. The card is a multilayer printed circuit unit with two external planes for signal wiring and two internal planes for distribution of the three required voltages and ground. Ninety-eight double sided connecting tabs are situated along one edge of the card on a .150 in. pitch. These tabs provide for a mating connection with the edge connectors mounted on the interconnection board, and serve to electrically connect all supply voltages and signal wiring

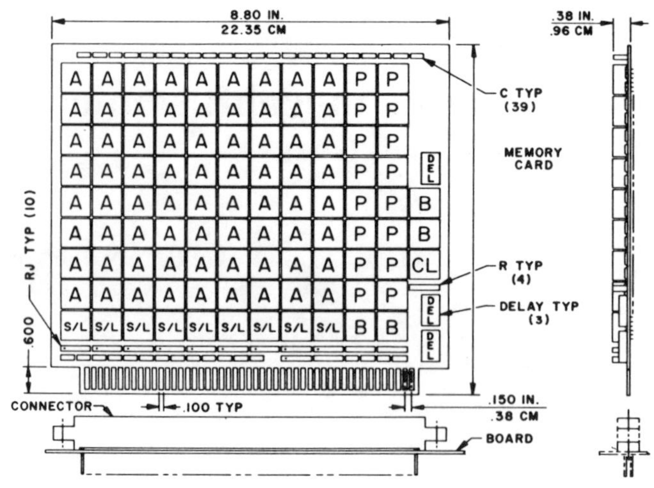

Figure 2—Memory card

to the card. The modules mounted on the card contain one or two chips each, solder reflow bonded to a wiring pattern on a ceramic substrate. Each module occupies a 0.7 in. square area. The 72 modules marked "A" contain the storage array with two chips of 1024 bits each included in each module. The "B" modules provide the primary stages of bipolar buffering while the "P" modules contain the secondary bipolar buffering and decoding. Modules "CL" and "DEL" provide for timing generation while the remaining "S/L" modules perform the sense amplification and latching functions.

Logic design

Memory system logic design was based on the modular card concept to provide easy upward and downward variation of total memory capacity. This card contains all necessary input buffering circuitry, timing circuits, storage elements, sensing circuits, and output registers. The card is structured so that smaller organizations can be obtained by depopulating modules. TTL compatible open collector outputs are provided to allow "wired-or" expansion in multiple card systems such as the 32K word by 36 bit system discussed here. Unit TTL compatible input loads help alleviate the problems of driving a multiple card system.

Card logic flow

A signal flow logic diagram for the 8192 word by 18 bit memory card is shown in Figure 3. Thirteen single rail address lines are required to uniquely determine one

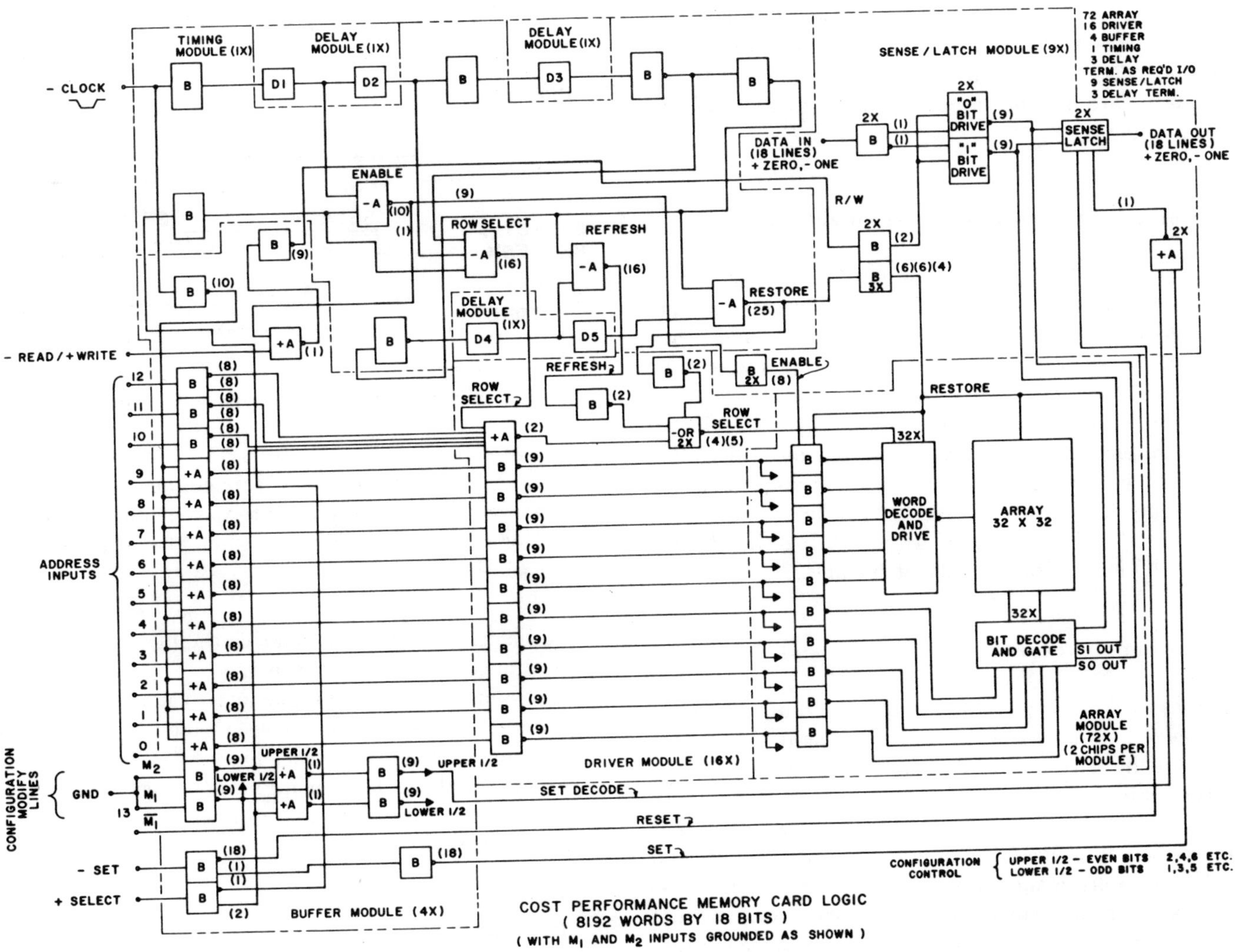

Figure 3—Cost performance memory card logic

of 8192 words. Four control lines are required as follows:

Select—causes selection of entire card.

Read/Write—determine the mode of operation to be performed.

Set—provides timing for the output data register.

Clock—generates timing for read and write operations as well as timing for cyclic data refreshing.

Thirty-six more lines are used for data-in and data-out.

Read operation signal flow

All input lines are buffered immediately upon entering the memory card. A second stage of address buffering is included on the card to allow fan out to all 144 storage array chips. Ten address lines (0–9) drive all storage array chips on the card in parallel, decoding to one of the 1024 bits stored on each chip. The remaining address lines (10–12) are decoded and combined with the timed Select pulse to create two Row Select signals which energize two of the sixteen rows of array chips on the card (two rows of chips per row of modules). Since there are nine array chips in each row, a total of eighteen bits are read out in each operation. The eighteen bits are transmitted to eighteen combination differential sense amplifier and latch circuits which are, in turn, wired to the card connector interface.

Write operation signal flow

Cell selection is performed in the same fashion during a write cycle as in a read cycle. However, instead of sensing the differential pairs associated with each bit position as in a read operation, the lines are pulsed by one of a pair of bit driver circuits. The magnitude of this excursion is sufficient to force the selected cell to the desired state as indicated by the condition of the data-in line.

Storage array chip logic organization

The storage array chip is organized in a 32 by 32 matrix of storage cells. Five input address lines are complemented upon entering the chip and then selectively wired to the word decoder drivers to provide a one-of-32 selection. These word drivers are also gated by Row Select so that only storage cells on a selected chip are energized. The remaining one-of-32 decoding function is performed on the cell outputs using the remaining five input address lines. The 32 outputs of this final gating stage are wire-ored together to the single differential pair of output bit lines.

Timing structure

Because the array chip is operated in a dynamic fashion, it is necessary to provide several timed lines for periodic refreshing of data and for restoration of the array chip selection circuits after a read or write operation. To minimize the number of lines required at the system interface, the timing generation circuits and delay lines are included on each memory card. These functions are implemented with nonsaturating current switch circuits for minimum skew between timed pulses. Tapped delay lines are used to chop and delay the input clock pulse. A total of four timing pulses are generated as described below:

Row Select: This line is used to turn on the array chip word and bit selection circuits during a read or write operation.

Refresh: This line is timed to follow the Row Select line and energizes all word selection circuits to refresh the array data.

Enable: The address inverters on the array chip are enabled by this line during a normal read or write operation. During the refresh portion of the cycle the absence of this pulse disables the address inverters so that all word selection circuits are simultaneously energized. This permits refreshing of data in all storage cells.

Restore: This line gates on load devices in all array chip selection circuits during the refresh portion of the cycle. These devices provide a recharging path for all the selection circuit node capacitances that were discharged during the immediately preceding operation, and for the node capacitances of the storage cells themselves.

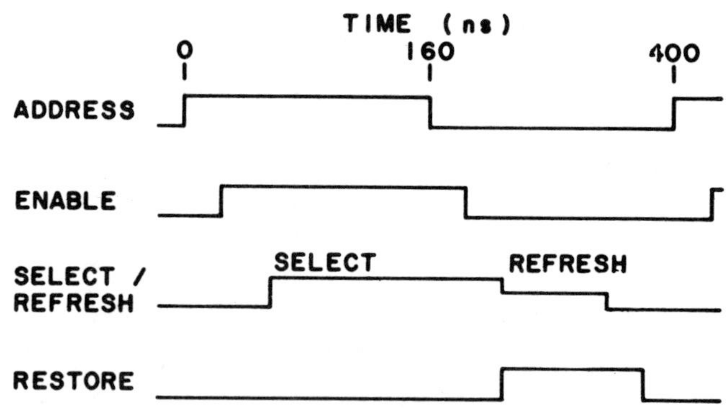

Figure 4—Storage array chip input timing

A diagram showing the relative timings of array chip input lines is shown in Figure 4.

A timing chart for the memory system interface is shown in Figure 5. It can be seen that two timed lines are required at this interface. The first is the Clock line from which all the aforementioned timings are derived. The second is the Set line which latches array data into the output register.

System operation

A block diagram for the complete 32K word by 36 bit memory system is shown in Figure 6. Eight memory

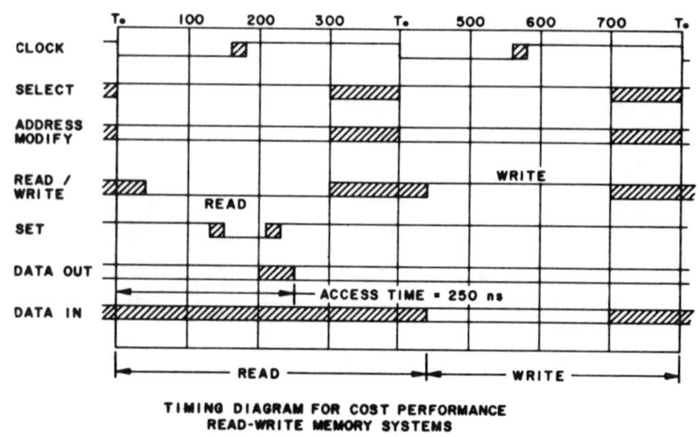

Figure 5—Timing diagram for cost performance read-write memory systems

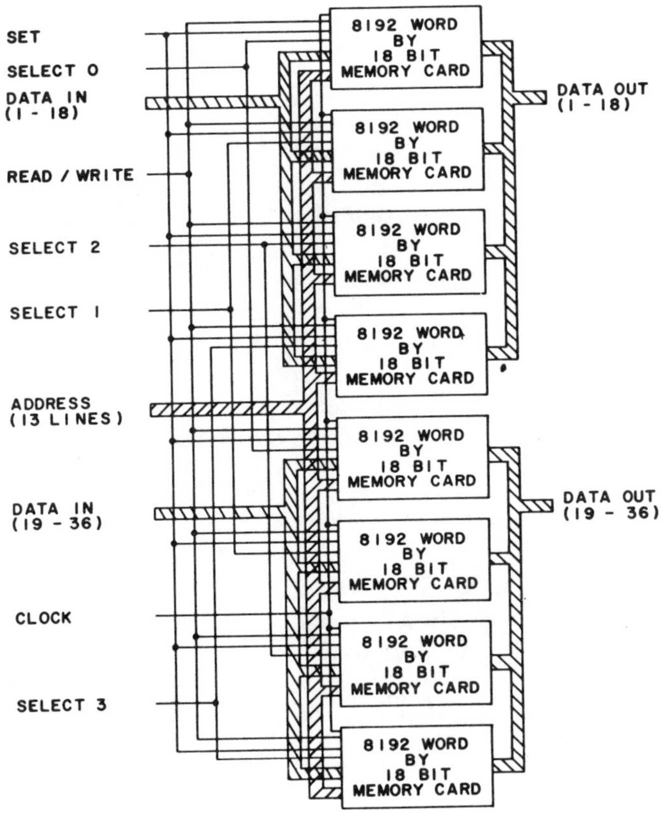

Figure 6—Memory system block diagram

cards, each containing 8192 words by eighteen bits are interconnected as shown to form the total system. All cards are addressed in parallel with four mutually exclusive Select lines energizing one pair of memory cards each cycle. Each card output is "wire-ored" with three other card outputs to expand word depth from 8192 words to 32,768 words.

Maximum access time is 250ns as measured from the +1.6 volt level of the input Clock leading edge transition. Minimum allowable cycle time is 400ns. and is measured in a similar manner from one leading edge Clock transition to the next. Since the Clock line provides refreshing of data, it is also necessary that a *maximum* Clock repetition time of 1.2μs be maintained, to avoid loss of information.

Circuit design

In the design of LSI memories the most important costs to be minimized are as follows:

Unmounted chip cost per bit
Chip carrier cost per bit
Printed circuit card cost per bit
Support costs per bit

The chip cost per bit is largely a function of the area of processed silicon required per bit of storage, the process complexity as measured by the number of masking or diffusion steps, and the chip yield. All of these factors strongly favor a MOS-FET chip process over bipolar process. For a given chip size the chip carrier costs, the printed circuit cost and the support costs are all inversely proportional to the number of bits per chip, thus the advantage of high-density MOS-FET array circuitry is overwhelming.

The chief drawback to MOS-FET circuits for semiconductor memories is their low gain-bandwidth compared with bipolar circuits using equivalent geometric tolerances. This shortcoming can be minimized by using bipolar circuits to provide the high-current drives to the MOS-FET array circuits, and by using bipolar amplifier circuits to detect the low MOS-FET sense currents. If the circuits are partitioned so that all the devices on a given chip are either bipolar or MOS-FET, no additional processing complexity is added by mixing the two device types within the same system. The use of bipolar support circuits also allows easy interfacing with standard bipolar logic signals, thus the interface circuits can match exactly standard interface driving and loading conditions.

Given an MOS-FET array chip, the two most important remaining choices involve the polarity of the MOS-FET device (n-channel or p-channel) and the gate oxide thickness. It is well known that the transconductance of n-channel devices is approximately three times that of equivalent p-channel device and thus the current available to charge and discharge capacitance is substantially greater. Since the substrate is backbiased by several volts in an n-channel device, the source-to-substrate and drain-to-substrate capacitances are also slightly lower, with the net result that n-channel circuits are a factor of two to three faster than equivalent p-channel circuits. This speed difference is critically important if address decoding and bit/sense line gating are to be included on the MOS-FET chip. Because the transconductance of a MOS-FET device, and consequently its ability to rapidly charge and discharge a capacitance, is inversely proportional to the gate oxide thickness, it is advisable to use the minimum thickness that the state of the art will allow; in this case 500 Angstroms was chosen as the minimum that would give a good yield of pinhole free oxide with adequate breakdown voltage. Other key device parameters are tabulated below:

V_t = 1.25V nominal with substrate bias
ρsub = 2Ωcm P type
γm = 33.5μa/v nominal
ρd = 7Ω/square N type

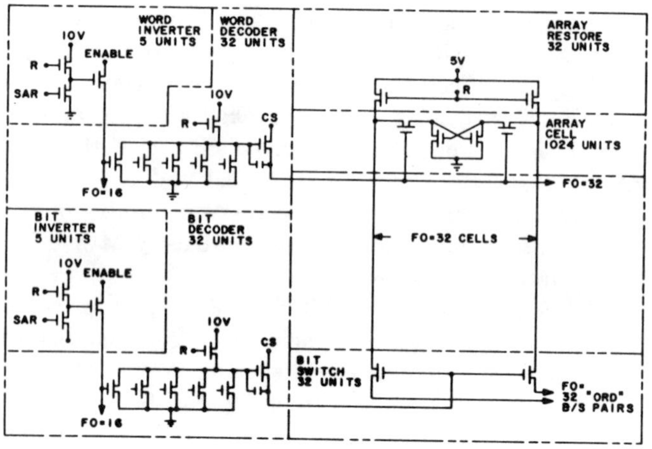

Figure 7—Array chip cross-section

Chip partitioning

Since it was desired that the same chip set be used to configure memory systems of different sizes, different word lengths, and different system speeds, many of the chip partitioning choices are obvious. The timing circuits, which are used only once per system, are contained on a separate chip. The sensing and bit/drive circuits are combined on one chip to allow easy expandability in the bit dimension. The array drivers are contained on a third chip type to allow easy expansion in the memory size, while general buffering and gating logic make up the fourth chip type. The most important chip-partitioning choice involves the dividing line between bipolar and MOS-FET circuits at the array chip interface. By including the array word-line decoding and the array bit/sense line gating on the array chip, the number of connections to the array chip can be greatly reduced, allowing the chip carrier wiring to be less dense and the chip pad size and spacing to be relaxed. The complexity of the bipolar support circuitry was reduced still further by including the address inverters on the array chip, with a small penalty in delay. If a MOS-FET sense amplifier/bit driver were included on the array chip, however, the increase in delay would be excessive, owing to the poor response time of MOS-FET high-gain amplifiers. In the design shown here, the cell sense current is gated to a bipolar sense amplifier for amplification and discrimination, and the cell nodes are driven through the same MOS-FET gating circuits to the desired state during the write operation. This arrangement requires that approximately 35 percent of the available array chip area be used for decoding and gating circuits, with the remaining 65 percent used for storage cells. Figure 7 shows a cross-section circuit schematic of the array chip. Included below are nominal chip parameters:

Address input capacitance . . . (including gate protective device) 4pf
Enable input capacitance . . . (depending on address) 2.75 pf or 20pf
Restore input capacitance . . . (including gate protective device) 57pf
Sense line input capacitance . . . 5.5pf
Select input capacitance . . . 8pf
Word line capacitance . . . 7.5 pf
Bit line capacitance . . . 2pf
Sense current . . . 150μa
Maximum gate protective device input 3400V

Storage cell

Typical MOS-FET storage cells are shown in Figure 8. In cell 8(a), T_1 and T_2 form the cross-coupled pair, while T_3 and T_4 gate the external circuitry to the cell nodes, either to sense the state of the cell by detecting the imbalance in the current through T_3 and

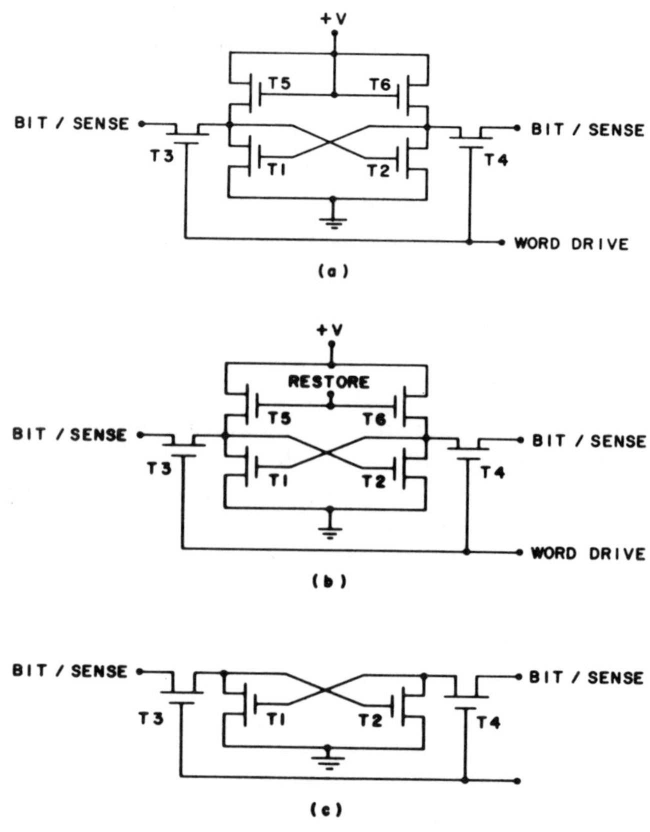

Figure 8—Storage cell configurations

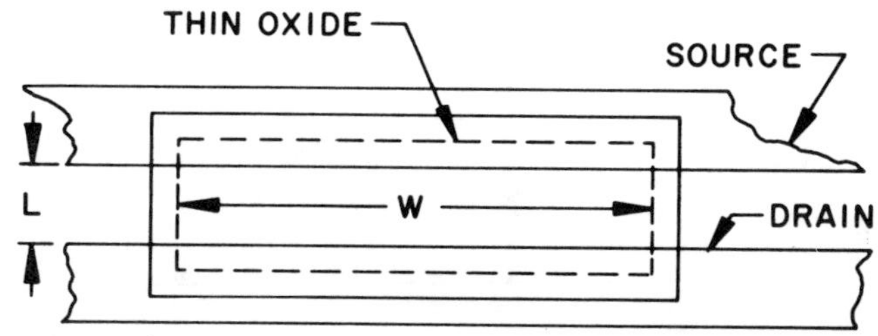

Figure 9—W/L ratio

T_4 or to write into the cell by pulling one node to ground while simultaneously driving the other cell node positive. The load devices, T_5 and T_6, replace the leakage current from the more positive node during stand-by. Since one of the load devices has full voltage across it at all times, the standby power dissipation of the cell will be quite high in comparison to the cell sense current unless the W/L ratio, Figure 9, of the load device (T_5, T_6) is made very small compared to the W/L ratio of the cross-coupled device (T_1, T_2). This, in turn, requires that either the load devices or the active devices or both occupy a large chip area. In addition, the standby load current flowing through the on-biased active device provides a voltage drop across that device, tending to unlatch the cell. This effect can be compensated for by increasing the value of all device thresholds, however, this will require a higher supply voltage to maintain the same standby current thereby increasing the power dissipation.

In cell 8(b), the standby power is reduced by pulsing the "restore" input at a clock rate sufficiently fast to replace leakage current from the cell node capacitance, while maintaining a low average power drain. The chief drawback to this cell is the five connections must be made to the cell, with a resulting increase in cell complexity over (a) above.

Cell 8(c) shows the configuration chosen for this memory. In this cell, both the word selection and the restore functions are performed through the same devices and array lines, by time sharing the word-select and restore line. During read-out, the cell operation is similar to 8(b) above. At the end of each memory cycle, however, all word lines are raised to the "restore" level for a period sufficient to recharge the cell node capacitances, then all word lines are dropped and the next memory cycle can begin. Selection of the "restore" level is dependent on the speed at which the cell node capacitance is to be charged and the sense line voltage support level required during restore. Too high a "restore" level creates a large current flow thru the restore devices lowering the sense line voltage used to charge the cell; too low a voltage prevents the cell node capacitance from reaching the required voltage for data retention. This cell employs fewer devices and less complex array wiring than either of the cells above, and thus requires substantially less silicon area. The disadvantage of this approach is that the restore function must be synchronized with the normal read/write function since they share the same circuitry. The average power cannot be made as low as in (b) above, since the restore current and the sense current are both determined by a common device, and the restore frequency is determined by the memory cycle time; however, the average power can be made significantly lower than with the static cell 8(a) above.

MOS-FET support circuits

The MOS-FET support circuits employed on the array chip are shown in Figure 10. A better understanding of the circuit operation will be gained by first considering the MOS-FET inverter circuit (Figure 10). At the end of a read/write cycle, the input address level is held down, the E level is down, and the R line is pulsed positive, charging node A to approximately +7 volts. When the R pulse has terminated, node A remains positive awaiting the next input. At the start of the read/write cycle, the address input is applied to T_1; if the address line is positive, node A quickly discharges through T_1, and when E is applied to T_3,

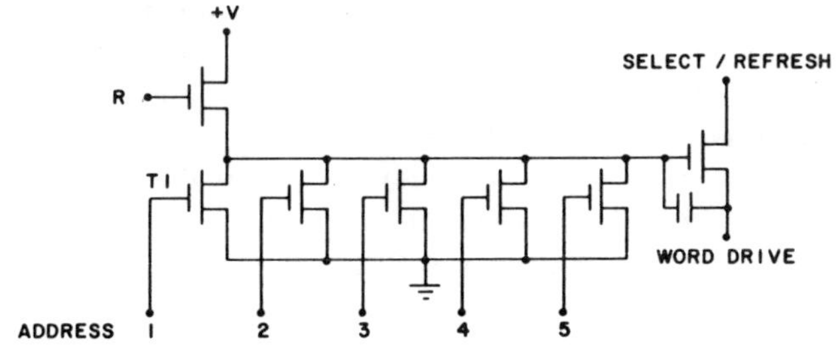

(b)

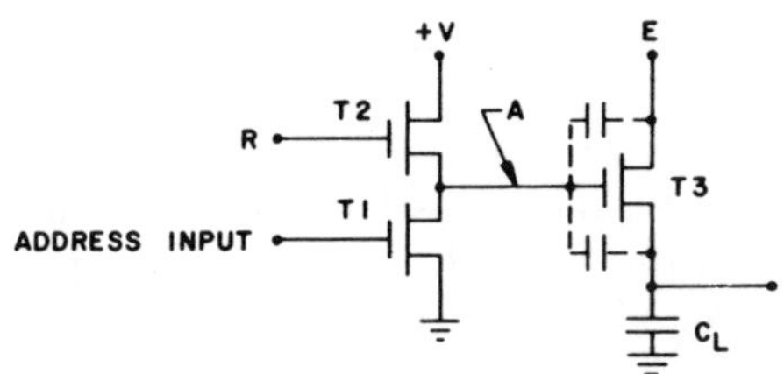

(a)

Figure 10—Array chip inverter-decoder circuits

T_3 remains non-conducting and the address inverter output remains at ground potential. If, however, the address input line is a down level, then node A remains charged to +7 volts, and both T_1 and T_2 are cut-off, while T_3 is biased on. When a positive E pulse is applied to T_3, current is capacitively coupled into node A from both the E node and from the output node, with the result that node A is driven more positive than either; thus T_3 remains strongly biased on, charging the output node capacitance to the level of E. When the positive E pulse is terminated, the same action quickly discharges the output to ground through the E line. At the end of the address pulse, a positive R pulse is again applied to T_2, restoring node A to +7 volts. This regenerative inverter has several advantages over a conventional source follower circuit; (a) the output up level is set by the level of the E input, and does not vary with the device threshold voltage; (b) the output rise time is nearly linear, since the gate-to-source bias on T_3 remains well above the threshold voltage throughout the transition, and (c) this same high conductance output device can be used to both charge and discharge the load capacitance. Since the leakage current from node A during a cycle is negligible, the final potential of node A, and thus the output drive current, is determined by the capacitor-divider action of the gate-to-source, gate-to-drain, and gate-to-substrate capacitances associated with device T_3. Any of these capacitances can be artificially increased to optimize the circuit operation. The operation of the decoder circuit (Figure 10b) is similar to the inverter just described, with the bi-level chip select/refresh line replacing the E input discussed previously. Thus, a single word line is selected to the higher (Select) level during the Read/Write portion of the cycle, while *all* word lines are selected to the lower (Refresh) level during the Restore portion of the cycle. Thus the cell input/output devices are biased to a low impedance to provide maximum sense current during readout, and to a higher impedance to reduce the power dissipation and maintain the necessary sense line voltage during the restore operation.

Protective devices are used on all gate inputs to the array chip to reduce the yield loss from static electricity during processing, testing, and assembly. Because the array chip uses a P-epitaxy grown on a $P+$ substrate it was possible in this system to replace the usual RC protective device with a more favorable zener type. This device is an $N+$ diffusion diffused at the same time as the source-drain diffusions and exhibits a low internal impedance when its depletion region intersects the $P+$ substrate. The required reverse breakdown voltage is obtained by controlling the depth of the $N+$ diffusion. When driven with an impedance equivalent to a human body, approximately 1000 ohms, gate protection is

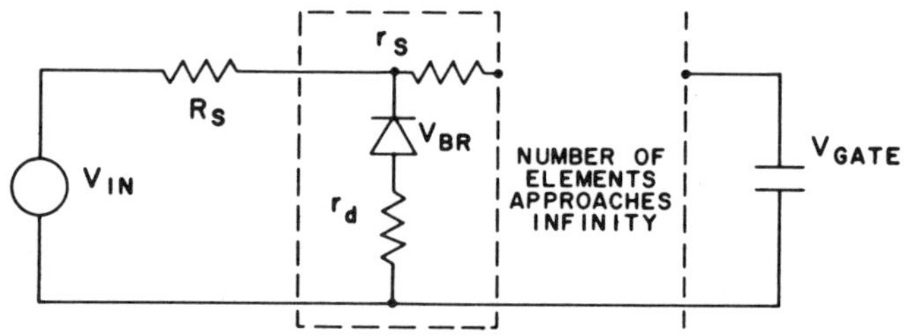

Figure 11—Gate protective device

provided for input levels up to 3400 volts. Figure 11 and equation 1 represent the characteristics and operation of this type protective device as presented during the IEEE, International Electron Devices Meeting, October 29 thru 31, 1969.[1] For analysis the device is arranged as a series of distributed elements; each element containing sub-elements r_s, r_d, and V_{BR}.

$$V_{gate} = V_{BR} + \{(V_{in} - V_{BR})(r_s r_d)^{1/2} [\cosh(r_s\gamma/r_d)^{1/2}]^{-1}\} / R_s + (r_s r_d)^{1/2} \quad (1)$$

In this design γ, the number of elements, was set at nine with the following sub-element values:

r_s = 4.27 ohms

r = 61.2 ohms

V_{BR} = 30 volts

The maximum capacitance before breakdown is 1.25pf.

Bipolar support circuits

Because of the critical timing relationships required among the Select/Refresh, Enable, Address, and Restore pulses to the array chip, all timing pulses are generated *on each card* by a special timing chip and three tapped delay lines. This arrangement allows each card to be fully tested with the timing circuits that will drive it, and minimizes any interaction between cards in a multi-card system.

The TTL compatible buffer chip allows interfacing conveniently with the TTL compatible logic of the using system, and minimizes the loading which the memory presents to the system.

A schematic cross-section of the Drive and Sense circuits is shown in Figure 12. The Driver module, when addressed, selects a row of nine Array Chips from a low impedance TTL output. The ten address inputs to the

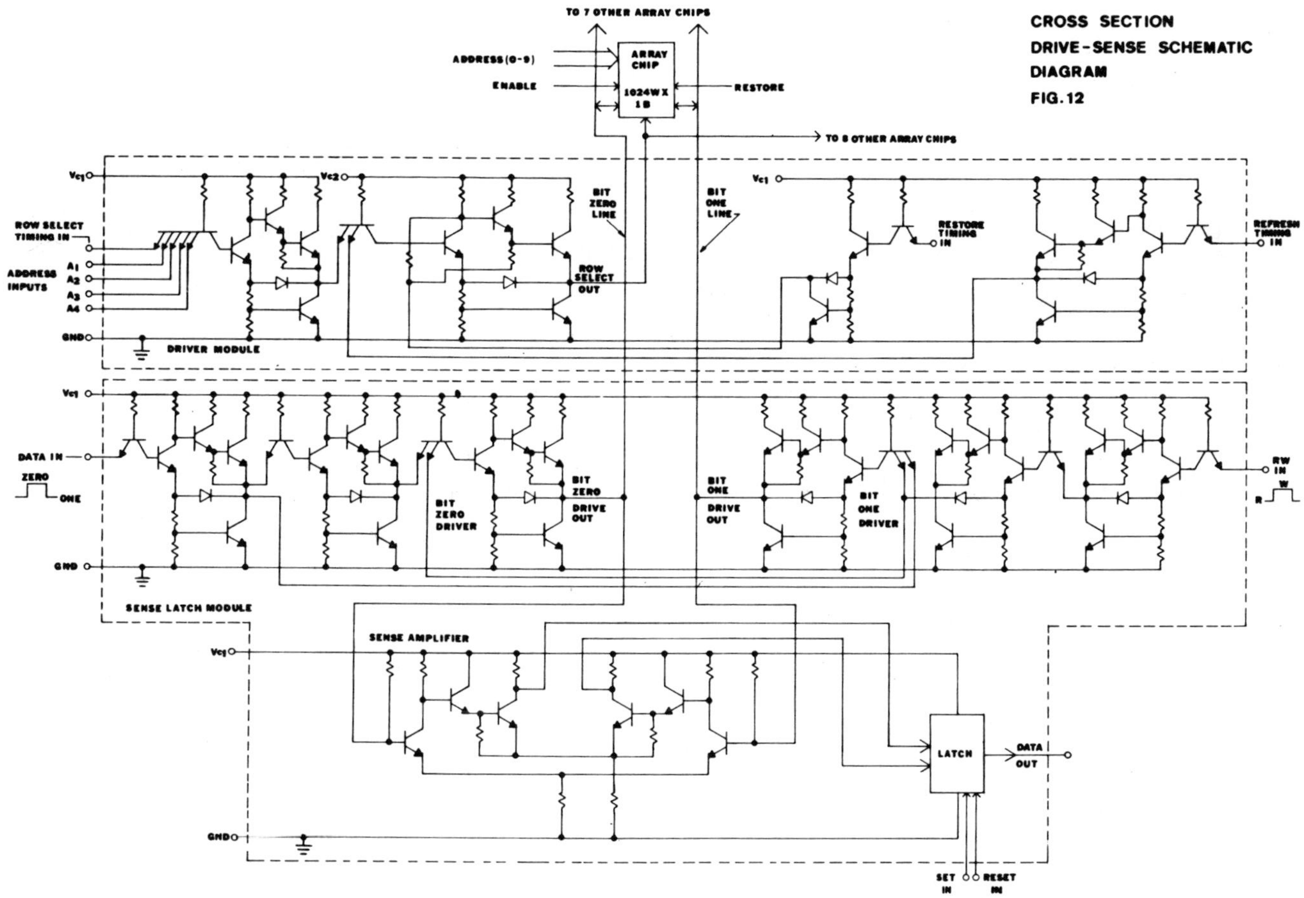

Figure 12—Cross-section drive-sense schematic diagram

Array Chip serve to select one bit of the 1024 bits/chip. The write pulse permits the data-in to be loaded differentially into the single bit which has been addressed. The removal of the write pulse turns off both the "one" and "zero" bit drives with the low impedance active pull ups rapidly charging the capacitance of the bit lines to a voltage level required for the read mode of operation.

The sense amplifier requires a minimum differential signal of 50 microamps to detect a "one" or "zero" stored in the addressed bit. This information is transferred to a set-reset latch which is included to increase the "data-good" time to the using system.

During a portion of every cycle not used for read/write operation the timing chip provides refresh and restore timing pulses which turn on all the driver modules on the memory card to a lower voltage level, and perform the refresh operation previously discussed.

All four of the bipolar support chips are packaged one-chip per chip carrier, to allow flexibility in configuring various size memory systems. In all cases, the power density is limited to 600 mw per chip carrier,

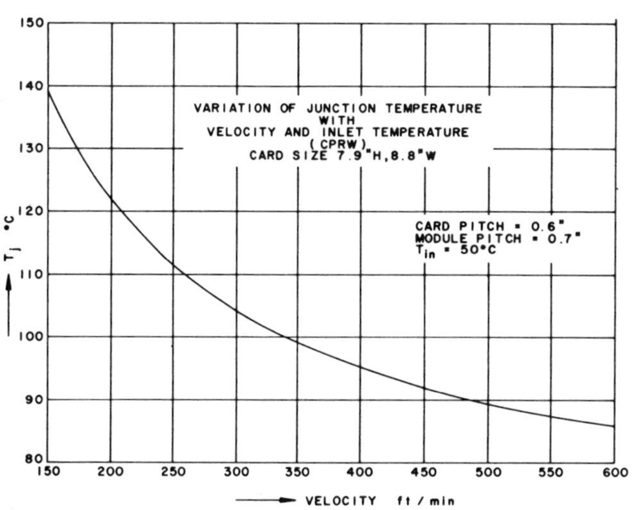

Figure 13—Variation of junction temperature with velocity and inlet temperature

a level which allows for convenient forced-air cooling. Because the limiting heat factor is the junction temperature of the bipolar support circuits all cooling considerations are in respect to this parameter. Figure 13 illustrates the junction temperature as a function of air flow thru the system.

CONCLUSION

The memory system described here is but one of many possible sizes and organizations that can be created using the same modular approach. If desired, several smaller organizations can be used within the same system without significant cost penalties. The system approach to memory design has created an optimum condition wherein each individual component is matched to the other components with which it must interact. This approach also yields a memory with a simple, effective, easily usable set of interface requirements. It is anticipated that increasing yields will allow prices competitive with magnetic storage for high-performance main memories. This low cost, coupled with high performance and density, makes a powerful combination for use in future system designs.

REFERENCE

1 M LENZLINGER
Gate protection of MIS devices
Presented at International Electron Devices Meeting
Washington D C 1969

Part VII
Serial Memories and Content-Addressable Memories

Editor's Introductory Comments

Probably the largest selling form of semiconductor memory to date is the dynamic MOS shift register (VII,1). Compared to random-access memories, shift registers are easy to design, easy to manufacture, and easy to test. Consequently, competition is heavy, prices are low, and unit sales are large. In the long run, conventional MOS shift registers may cost more per bit than random-access memories, since RAM's can be more densely packed. If the rate of innovation in RAM's slows, the heavy burden of engineering and marketing costs will be reduced, competition will increase, and RAM's should sell for less per bit than the conventional shift register.

New forms of serial memory are described in VII,2 and VII,3. These offer the prospect of matching or surpassing RAM's in density, along with potential application to analog storage and imaging problems. Commercial products will begin to appear in the next year or two.

Content-addressable memories (CAM), often called associative memories, have been the subject of study for many years. If CAM were available at a price per bit not much higher than RAM, it might be widely used. As shown in VII,4 and VII,5, CAM is certainly feasible from a technical standpoint. However, the low density of the CAM shown in VII,5 (16 bits without decoders on a chip which could carry 256 bits of fully decoded RAM) suggests that CAM is still far more expensive per bit than RAM. A 512-bit MOS CAM chip without decoding is described elsewhere.[1]

[1] J. L. Mundy, J. F. Burgess, R. E. Joynson, and C. A. Neugebauer, "Low-cost associative memory," in *IEEE Int. Solid-State Circuits Conf., Dig. Tech. Papers*, 1972, pp. 58–59.

MOS LSI Shift Registers **1402A, 1403A, 1404A**

INTEL CORP. 3065 Bowers Avenue, Santa Clara, California 95051 • (408) 246-7501

SEPTEMBER 1971

1024 BIT DYNAMIC SHIFT REGISTER

- **Guaranteed 5 MHz Operation over Temperature Range**
- **Low Power Dissipation --.1 mW/bit at 1 MHz**
- **DTL, TTL Compatible**
- **Low Clock Capacitance -- 140 pF**
- **Low Clock Leakage -- ≤ 1 μA**
- **Inputs Protected Against Static Charge**
- **Standard Packaging -- 8 Lead Metal Can, 16 Pin Ceramic Dual In-Line**
- **Three Standard Configurations -- Quad 256 Bit -- 1402A, Dual 512 Bit -- 1403A, Single 1024 Bit -- 1404A**

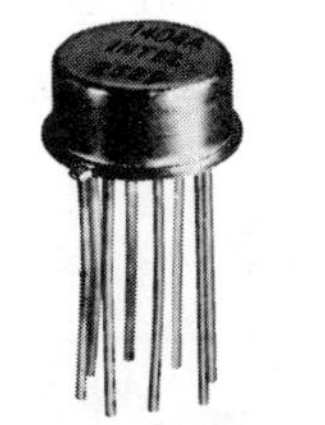

1403A, 1404A

1402A

PIN CONFIGURATION

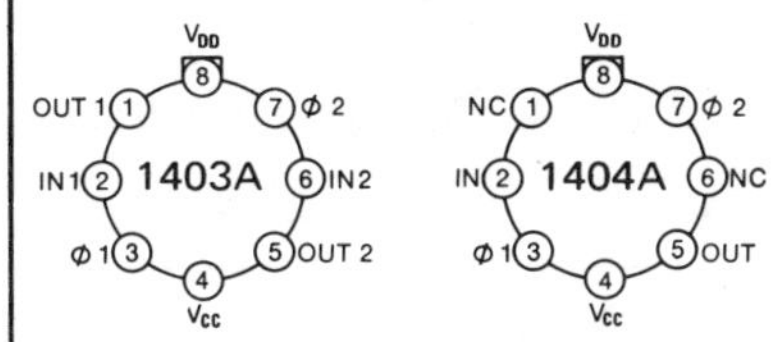

Top View

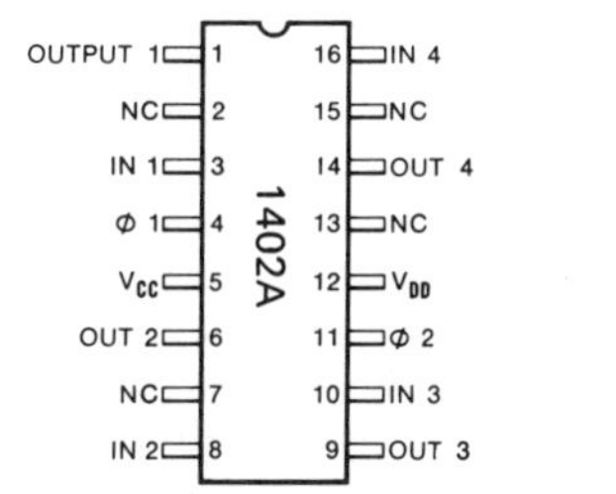

Top View

PACKAGE OUTLINES

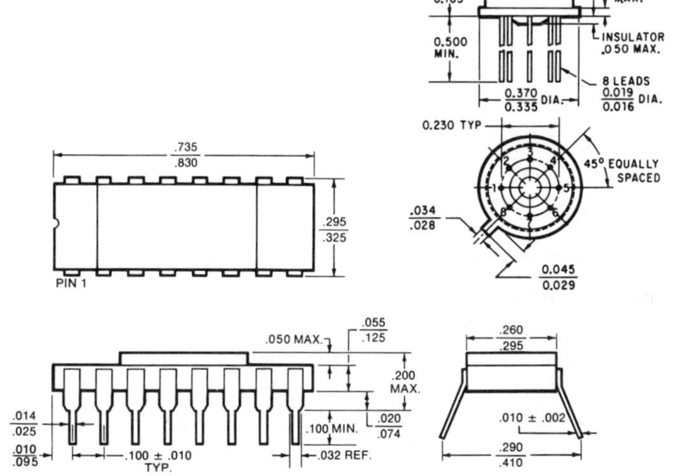

The 1402A/1403A/1404A are direct pin for pin replacements for the 1402/1403/1404. Due to "on chip" multiplexing the data rate is twice the clock rate. Data is shifted one bit on each clock pulse (both ϕ_1 and ϕ_2).

The 1402A/1403A/1404A family is ideally suited for usage in low cost serial memories or delay line applications. A high speed data rate of 5 MHz is easily obtained at the power supplies of +5V and −5V. The 1402A/3A/4A are capable of operating at the power supply voltages of +5V, −9V as well as +5V, −5V.

The input to the shift register can be driven directly by standard bipolar integrated circuits (TTL, DTL, etc.). The design of the output stage provides driving capability for both MOS and bipolar IC's.

Use of low threshold **silicon gate technology** allows high speed (5 MHz guaranteed) while reducing power dissipation by a factor of **two** and clock input capacitance by a factor of **three** compared to equivalent products manufactured by conventional technologies.

Maximum Guaranteed Ratings[1]

Case Temperature Under Bias	0°C to +85°C	Data and Clock Input Voltages and Supply Voltages with respect to V_{CC}	+.3V to −20V
Storage Temperature	−65°C to +160°C		
Power Dissipation [2]	600 mW @T_A=25°C		

D.C. Characteristics T_{CASE} = 0°C to +85°C, V_{CC} = 5V ±5%, unless otherwise specified

V_{DD} = −5V ±5% or −9V ±5%

SYMBOL	TEST	MIN.	TYP[3]	MAX.	UNIT	CONDITIONS
I_{LI}	Input Load Current		< 10	500	nA	T_A =25°C
I_{LO}	Output Leakage Current		<10	1000	nA	V_{OUT}=0.0V, T_A =25°C
I_{LC}	Clock Leakage Current		10	1000	nA	Max. V_{ILC}, T_A =25°C
V_{IL}	Input "Low" Voltage	V_{CC}−4.2		V_{CC}−10	V	
V_{IH}	Input "High" Voltage	V_{CC}−2		V_{CC}+.3	V	

V_{DD} = −5V ±5%

SYMBOL	TEST	MIN.	TYP	MAX.	UNIT	CONDITIONS
I_{DD1}	Power Supply Current		40	50	mA	T_A=25°C — Output at Logic "0", 5 MHz Data Rate, 33% Duty Cycle, Continuous Operation, V_{ILC}=V_{CC} −17V
I_{DD2}	Power Supply Current			56	mA	T_C=0°C — (same as above)
V_{ILC}	Clock Input Low Voltage	V_{CC}−15		V_{CC}−17	V	
V_{IHC}	Clock Input High Voltage	V_{CC}−1		V_{CC}+.3	V	
V_{OL}	Output Low Voltage		−.3	0.5	V	R_{L1}=3K to V_{DD}, I_{OL}= 1.6 mA
V_{OH1}	Output High Voltage Driving TTL	2.4	3.5		V	R_{L1}=3K to V_{DD}, I_{OH}= −100 μA
V_{OH2}	Output High Voltage Driving MOS	V_{CC}−1.9	V_{CC}−1		V	R_{L2}=4.7K to V_{DD} (See p. 6 for connection)

V_{DD} = −9V ±5%

SYMBOL	TEST	MIN.	TYP	MAX.	UNIT	CONDITIONS
I_{DD3}	Power Supply Current		30	40	mA	T_A =25°C — Output at Logic "0", 3 MHz Data Rate, 26% Duty Cycle, Continuous Operation, V_{ILC}=V_{CC}−14.7V
I_{DD4}	Power Supply Current			45	mA	T_C =0°C — (same as above)
V_{ILC}	Clock Input Low Voltage	V_{CC}−12.6		V_{CC}−14.7	V	
V_{IHC}	Clock Input High Voltage	V_{CC}−1		V_{CC}+.3	V	
V_{OL}	Output Low Voltage		−.3	0.5	V	R_{L1} =4.7K to V_{DD}, I_{OL} = 1.6 mA
V_{OH1}	Output High Voltage Driving TTL	2.4	3.5		V	R_{L1}=4.7K to V_{DD}, I_{OH}=−100μA
V_{OH2}	Output High Voltage Driving MOS	V_{CC}−1.9	V_{CC}−1		V	R_{L2} =6.2K to V_{DD}, R_{L3} =3.9K to V_{CC} (See p. 6 for connection)

Note 1: Stresses above those listed under "Maximum Guaranteed Rating" may cause permanent damage to the device. This is a stress rating only and functional operation of the device at these or at any other condition above those indicated in the operational sections of this specification is not implied. Exposure to absolute maximum rating conditions for extended periods may affect device reliability.

Note 2: For operating at elevated temperatures the device must be derated based as shown on page 3. In typical applications a case temperature of 85°C corresponds to ambient temperatures of 75°C for the ceramic dual-in-line package and 70°C for the low profile TO-99. When operating at V_{DD}=−5V±5% the maximum duty cycle is 33% and at V_{DD}=−9V±5% the maximum duty cycle is 26%. In applications the duty cycle should be a minimum to reduce power dissipation. Duty cycle= $[t_{\phi PW} + ½(t_R + t_F)]$ x clock rate.

Note 3: Typical values are at T_A =25°C and at nominal voltages.

Typical Characteristics

I_{DD} CURRENT VS DATA RATE

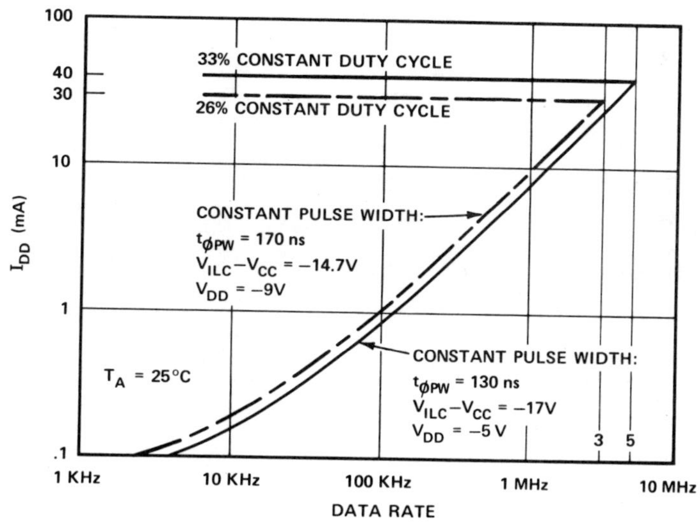

POWER DISSIPATION / BIT VS SUPPLY VOLTAGE

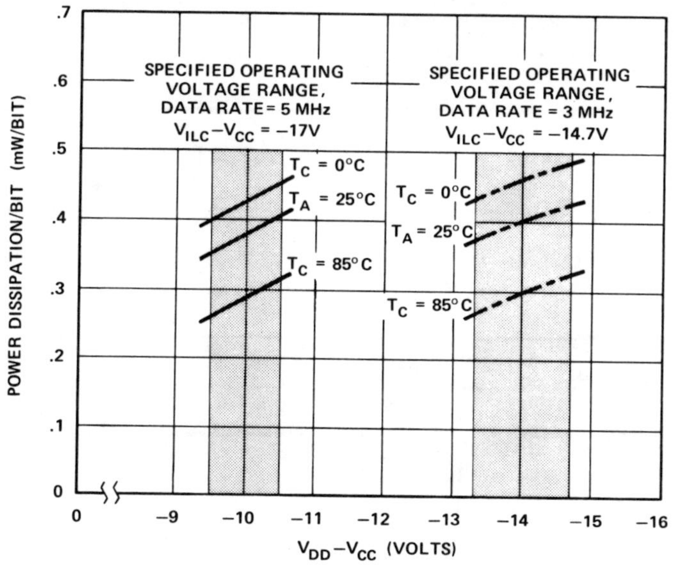

POWER DISSIPATION / BIT VS CLOCK AMPLITUDE

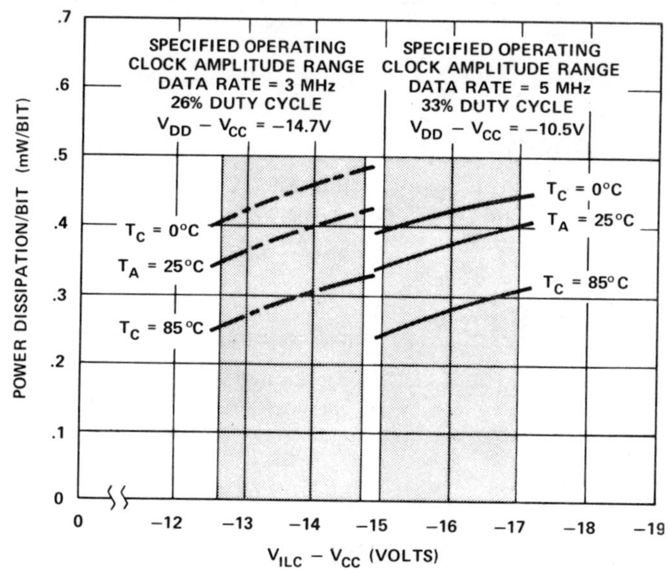

MAXIMUM PACKAGE POWER DISSIPATION

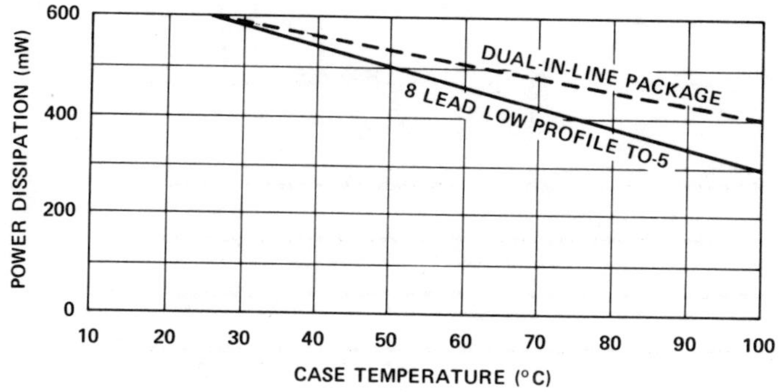

POWER DISSIPATION / BIT VS TEMPERATURE

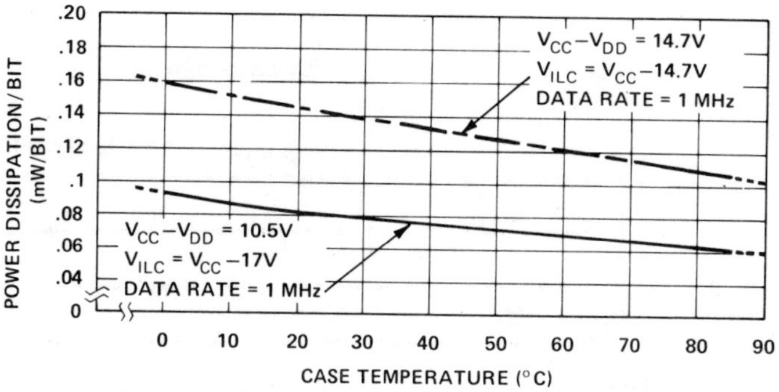

A.C. Characteristics $T_A = 0°C$ to $70°C$; $V_{CC} = +5V \pm 5\%$

SYMBOL	TEST	$V_{DD} = -5V \pm 5\%$ (Test Load 1)		$V_{DD} = -9V \pm 5\%$ (Test Load 2)		UNIT
		MIN.	MAX.	MIN.	MAX.	
Frequency	Clock Rep Rate		2.5		1.5	MHz
Frequency	Data Rep Rate	Note 1	5.0	Note 1	3.0	MHz
$t_{\phi PW}$	Clock Pulse Width	.130	10	.170	10	μsec
$t_{\phi D}$	Clock Pulse Delay	10	Note 1	10	Note 1	nsec
t_R, t_F	Clock Pulse Transition		1000		1000	nsec
t_{DW}	Data Write Time (Set Up)	30		60		nsec
t_{DH}	Data To Clock Hold Time	20		20		nsec
t_{A+}, t_{A-}	Clock To Data Out Delay		90		110	nsec

CAPACITANCE[2] $V_{CC} = +5V \pm 5\%$, $V_{DD} = -5V \pm 5\%$ or $-9V \pm 5\%$, $T_A = 25°C$

SYMBOL	TEST	TYP.	MAX.	CONDITIONS	
C_{IN}	Input Capacitance	5 pF	10 pF	$V_{IN} = V_{CC}$	f = 1 MHz
C_{OUT}	Output Capacitance	5 pF	10 pF	$V_{OUT} = V_{CC}$	
C_{ϕ}	Clock Capacitance	110 pF	140 pF	$V_{\phi} = V_{CC}$	
$C_{\phi 1 \phi 2}$	Clock to Clock Capacitance	11 pF	16 pF	$V_{\phi} = V_{CC}$	

Note 1: See page 5 for guaranteed curve. Note 2: This parameter is periodically sampled and is not 100% tested.

Switching Characteristics

Conditions of Test

Input rise and fall times: 10 nsec
Output load is 1 TTL gate

TEST LOAD 1

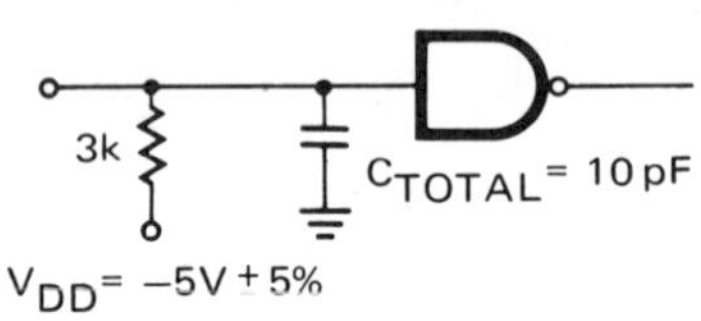

TEST LOAD 2

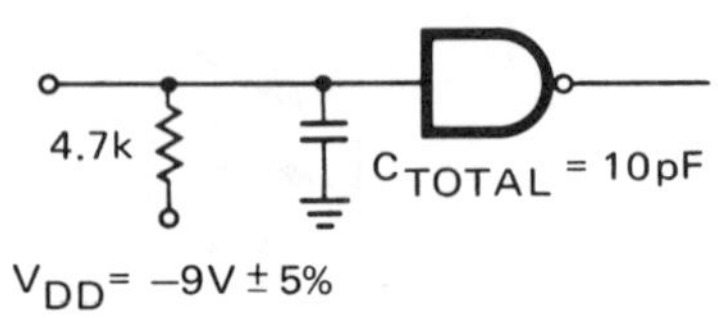

Timing Diagram

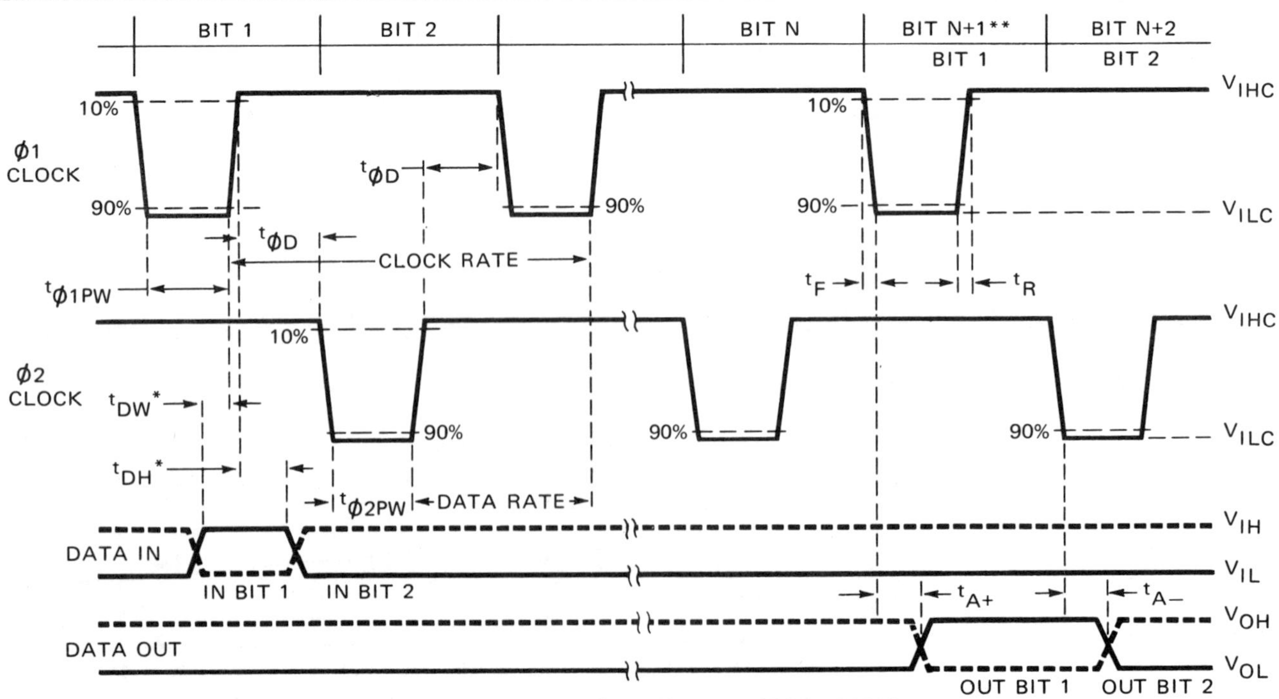

*t_{DW} and t_{DH} same for $t_{\phi 2}$ **N = 256 for 1402A, N = 512 for 1403A, N = 1024 for 1404A

Typical Characteristics

MINIMUM OPERATING DATA RATE OR MAXIMUM CLOCK PULSE DELAY VS TEMPERATURE

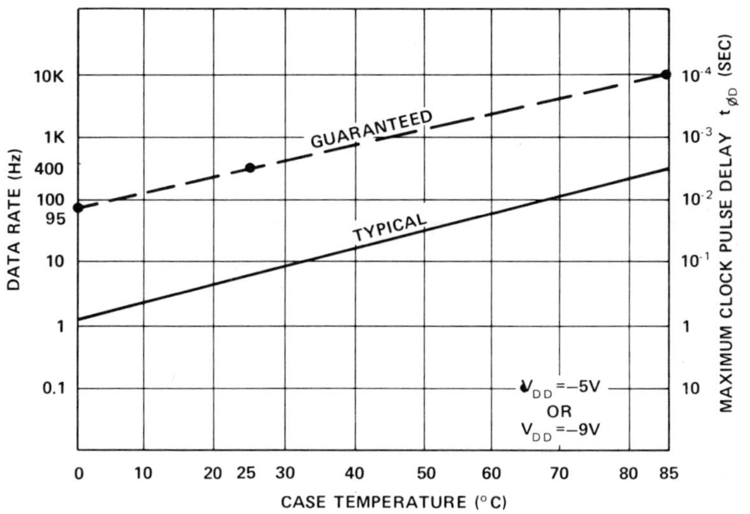

MAXIMUM DATA RATE VS CLOCK AMPLITUDE

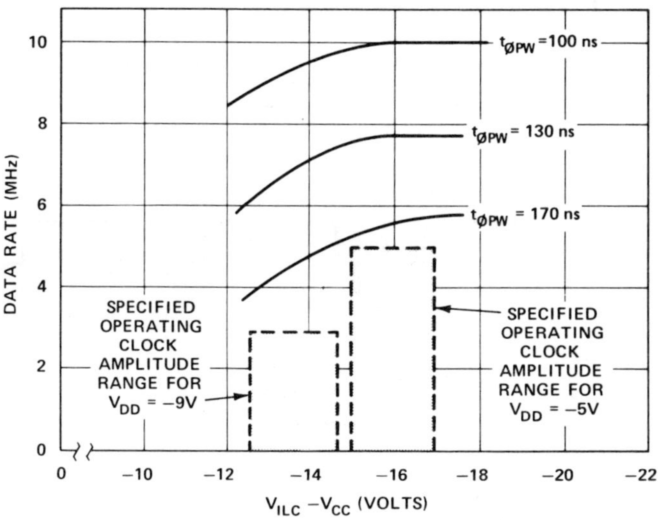

Circuit Schematic

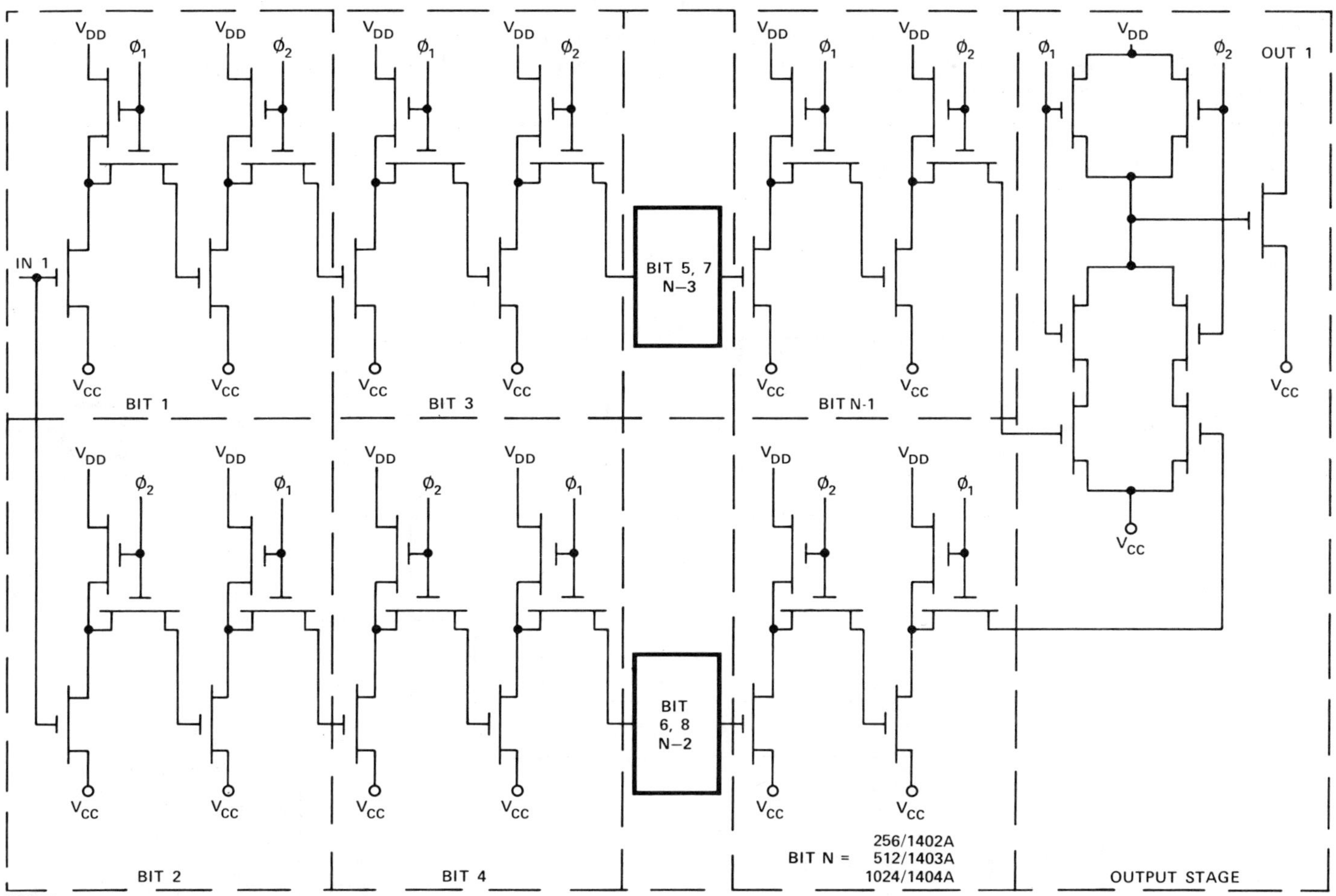

NOTES:

1. N = 1024 ON 1404A
2. N = 512 ON 1403A Schematic for second register same as above.
3. N = 256 ON 1402A Schematic for second, third and fourth registers same as above.

Typical Applications

Multiplexed 4-Bit MOS Shift Register

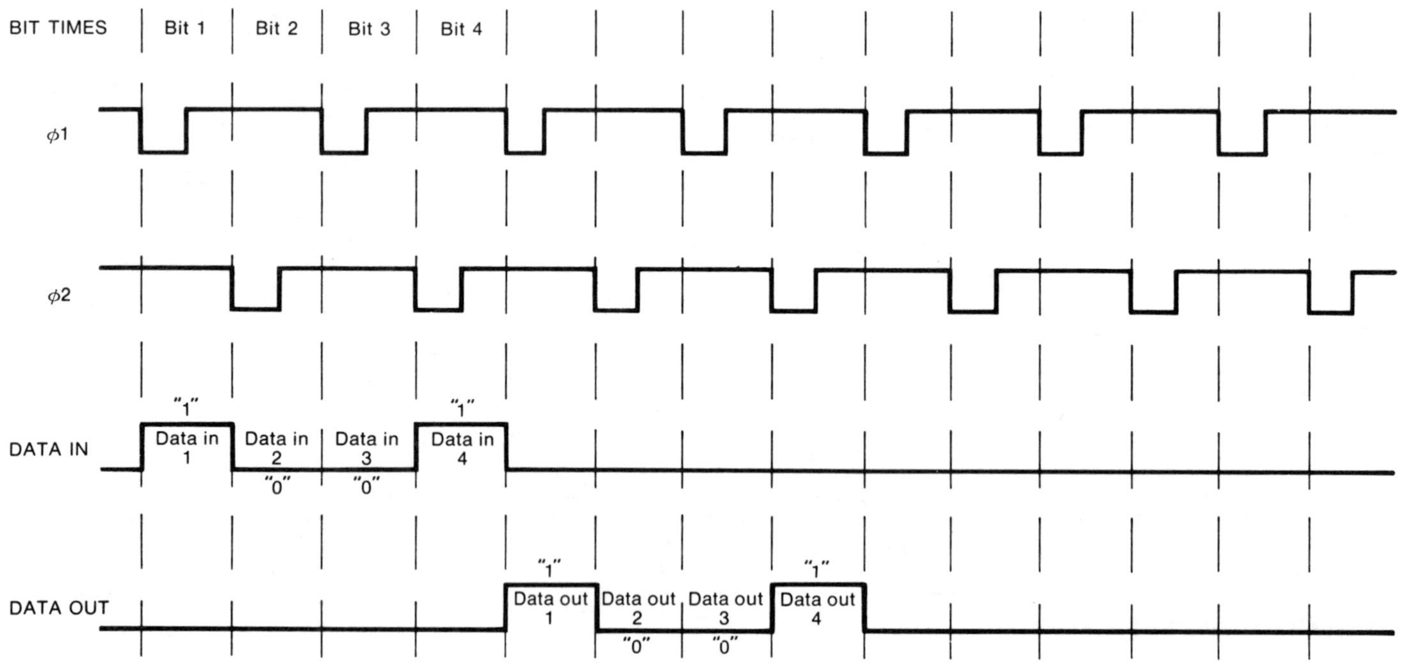

Figure 1.

Figure 1 is a simplified illustration of the timimg of a 4-bit multiplexed register showing input output relationships with respect to the clock. If data enters the register at Ø1 time, it exits at Ø1 time. (Beginning on Ø1's negative going edge and ending on the succeeding Ø2's negative going edge).

DTL/TTL to MOS to DTL/TTL Interface

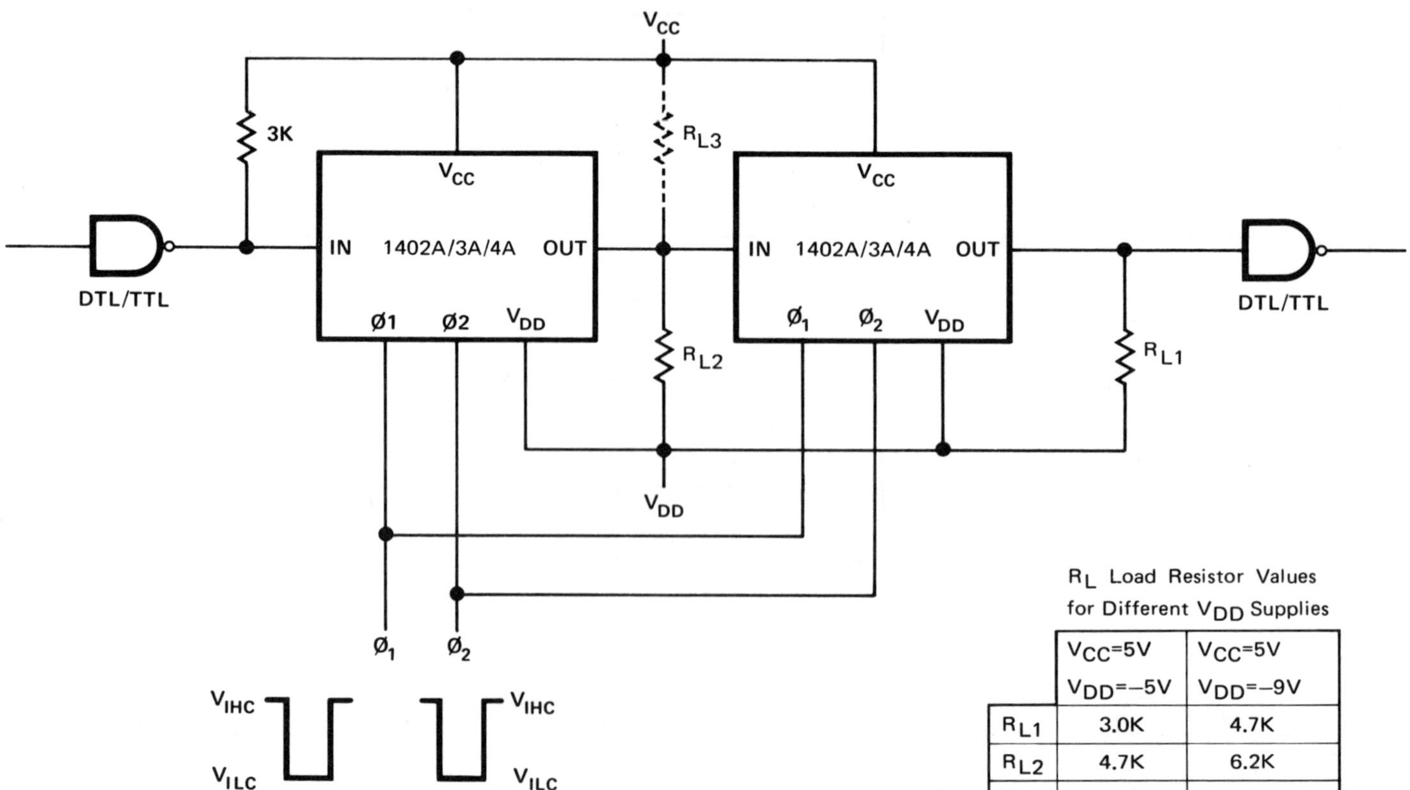

R_L Load Resistor Values for Different V_{DD} Supplies

	V_{CC}=5V V_{DD}=–5V	V_{CC}=5V V_{DD}=–9V
R_{L1}	3.0K	4.7K
R_{L2}	4.7K	6.2K
R_{L3}	Not required	3.9K

Two Phase Clock Generator

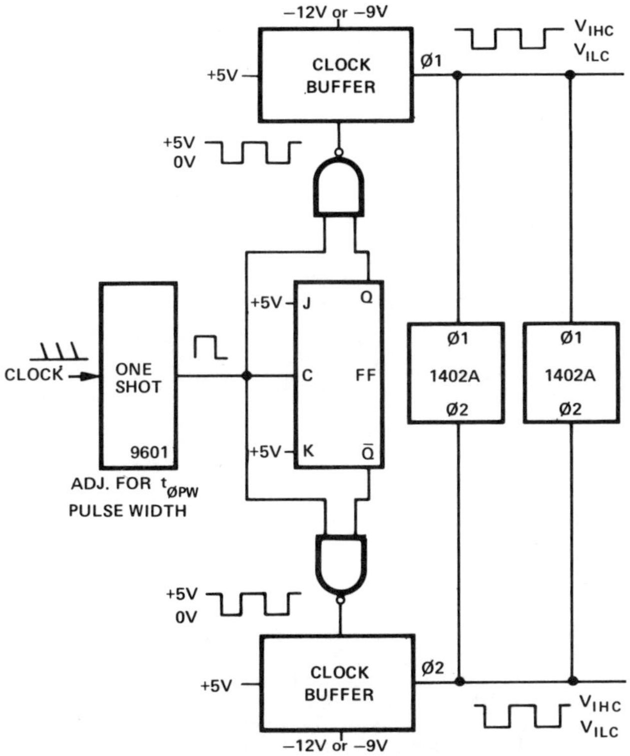

High Speed Clock Buffer

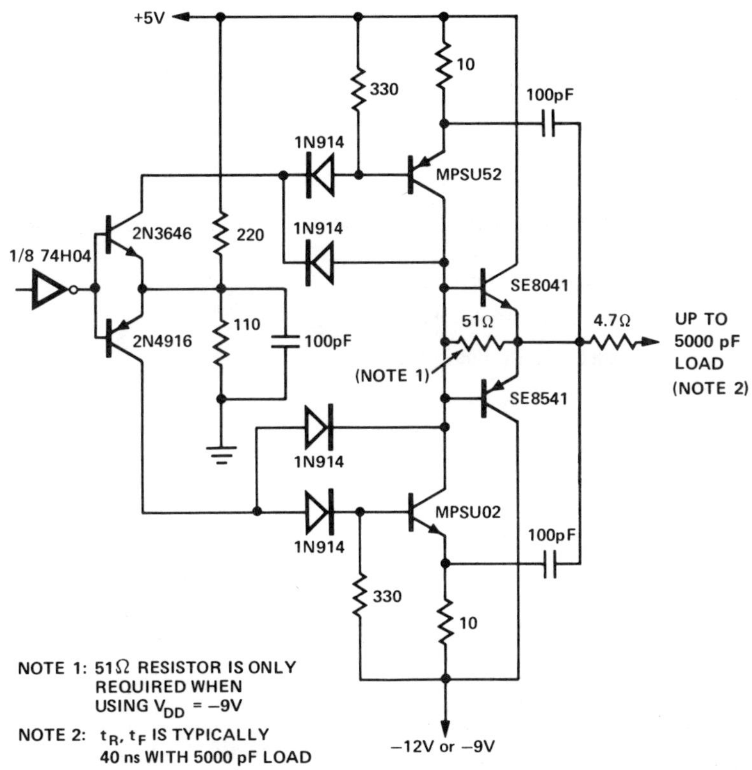

Write/Recirculate Logic

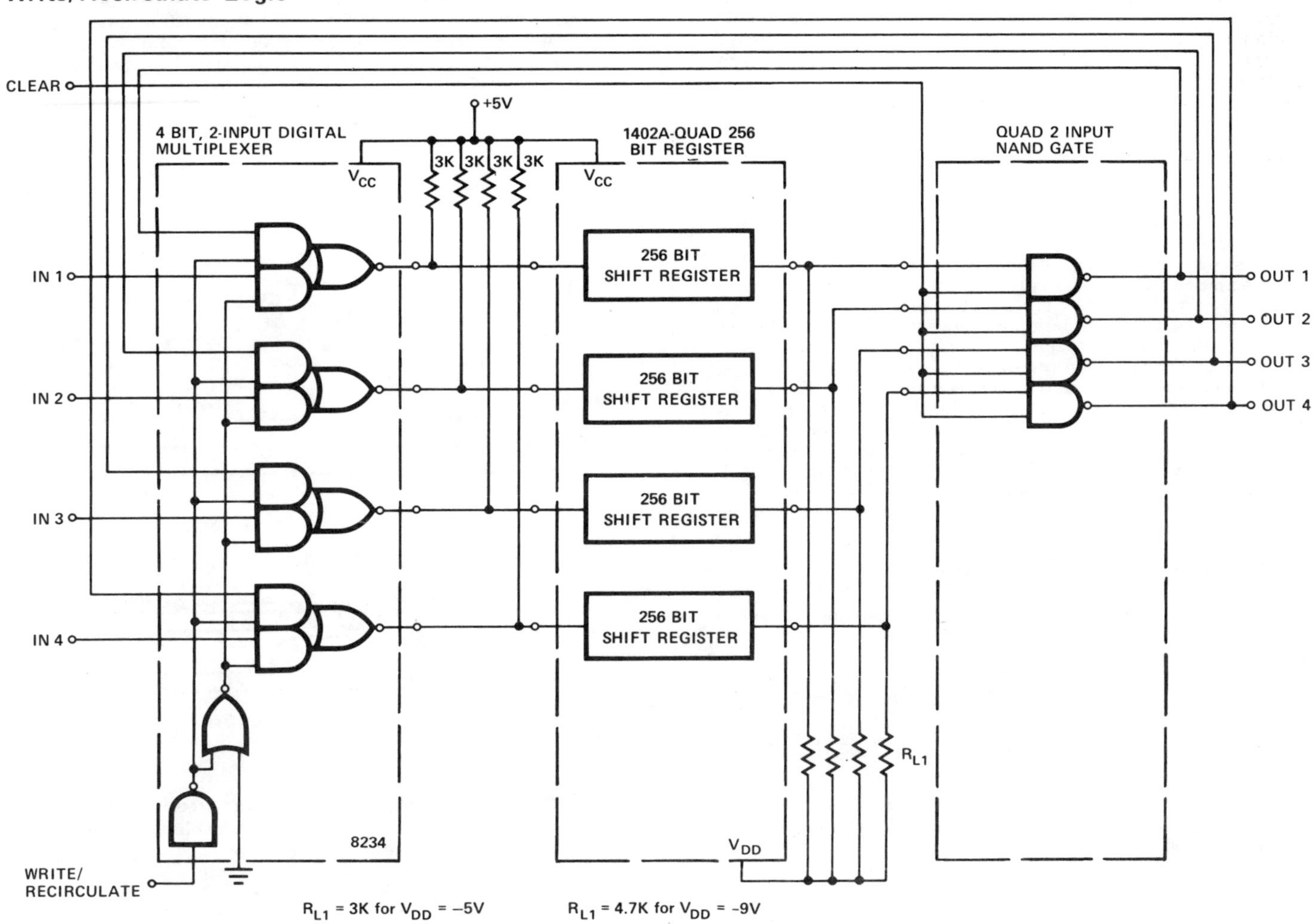

Integrated MOS and Bipolar Analog Delay Lines using Bucket-Brigade Capacitor Storage

F. L. J. Sangster

Philips Research Laboratories

Eindhoven-The Netherlands

A NEW ELECTRONIC variable delay line for analog data processing has been realized in integrated-circuit form.

Its basis is a chain of storage capacitors and charge-transfer circuits, acting as an analog shift register with externally variable shift rate. This configuration is known as a *bucket-brigade* circuit[1,2] because of the resemblance to a fire-brigade of old.

A feature of the new device is that information is stored in an array of capacitors not directly as charge level, but rather as a charge deficit. This concept leads to a simple circuit (Figure 1*a*) with only one transistor per storage capacitor[3], which is less complex than analog delay lines described elsewhere[4,5].

The delay line uses two complementary clock signals with a frequency equal to the sampling frequency applied to the input signal. Functionally, the device provides a delay line in which bandwidth and delay can be interchanged between wide limits. Signal delay can be accurately controlled or if needed, can be changed electronically.

As the storage capacitor is located between collector and base of the switching transistor, the delay line is simply a series connection of transistor with enlarged parasitic *Miller* capacitance, which is easily realized in integrated circuit form. Figure 1*b* shows a cross section of a bipolar integrated *bucket*. The enlarged junction between base and collector acts as the storage capacitor.

The performance of the circuit is, among other things, dependent on the interaction between successive signal samples proceeding along the capacitor chain. Although this interaction was only a small effect per stage in our first IC, it accumulated to an unacceptable level after hundreds of stages. This phenomenon, caused by the parasitic capacitance between collector and emitter, results in imperfect signal transients and reduces the bandwidth to less than the maximum value of half the clock frequency. Figure 3 shows a new integrated circuit in which parasitic coupling between the storage stages has been reduced to a sufficiently low level by locally narrowing the aluminum interconnection layer. This circuit contains 72 *buckets* with a capacitor value of 25 pF, a sampling circuit, an amplifier and an output buffer stage integrated together on a 2x2 mm chip. It was designed to obtain an experimental delay element for various applications. The amplifier included on the chip is necessary to compensate for signal attenuation in the *bucket* brigade. Charge deficits appearing at the emitter of a bipolar *bucket* are not fully transferred to the storage capacitor at the collector side, as a small amount of charge is supplied via the base contact. This causes some loss of information during each transfer, resulting in a considerable attenu-

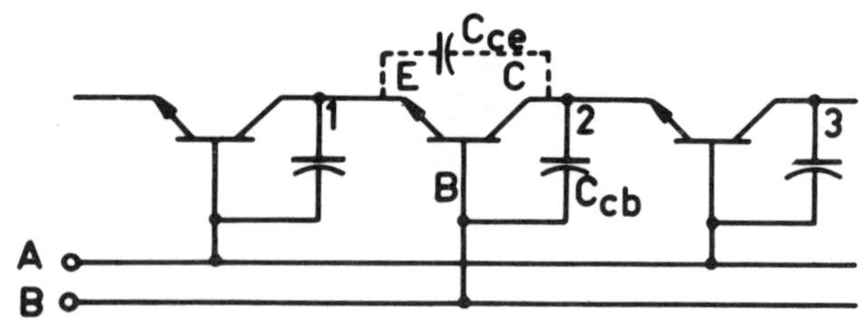

A

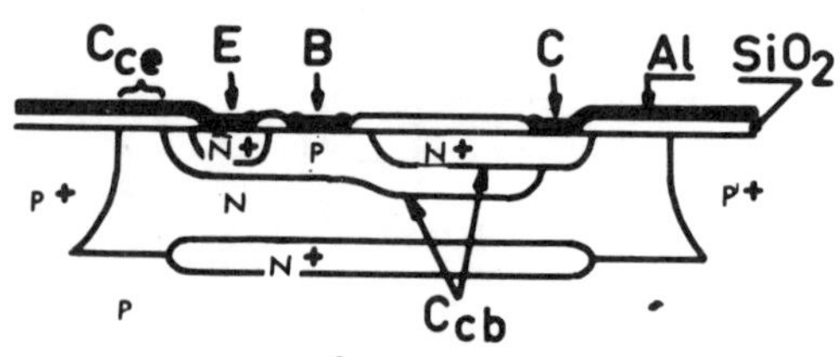

B

***FIGURE 1**—A bipolar **bucket-brigade** circuit is illustrated in (a). Sectional view of one monolithic integrated **bucket** is shown in (b).*

[1] Janssen, J. M. L., "Discontinuous Low-Frequency Delay Line with Continuously Variable Delay," *Nature*, p. 148-149; Jan. 26, 1952.

[2] Hannan, W. J., et al, "Automatic Correction of Timing Errors in Magnetic Tape Recorders," *IEEE Transact. MIL*, p.246-254; July-Oct., 1965.

[3] Sangster, F. L. J., and Teer, K., "Bucket-Brigade Electronics-New Possibilities for Delay, Time Axis Conversion and Scanning," *IEEE J. Solid State Circuits*, p. 131-136; June, 1969.

[4] Krause, G., "Analogue Memory Chain: A New Circuit for Storing and Delaying Signals," *Electronics Letters*, p. 544-546; Dec., 1967.

[5] Mao, R., et al, "Integrated MOS Analog Delay Line," *ISSCC DIGEST OF TECHNICAL PAPERS*, p. 164-165, Feb., 1969.

[6] To be shown at the Physics Exhibition, Alexandra Palace, London, March 2-5, 1970.

Reprinted from *1970 ISSCC Digest*, pp. 74, 75, 185.

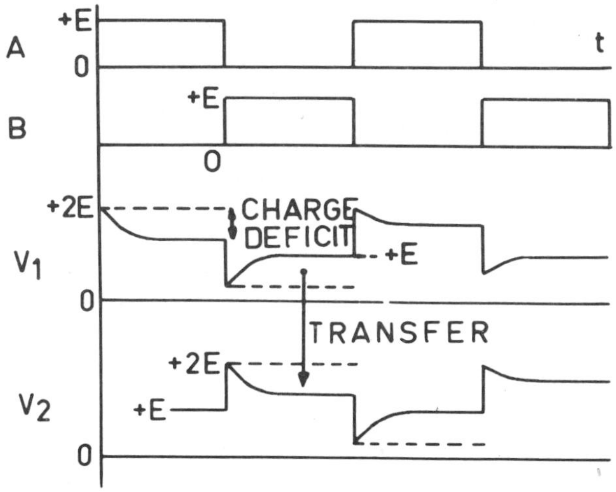

FIGURE 2—Principle of charge-deficit transfer.

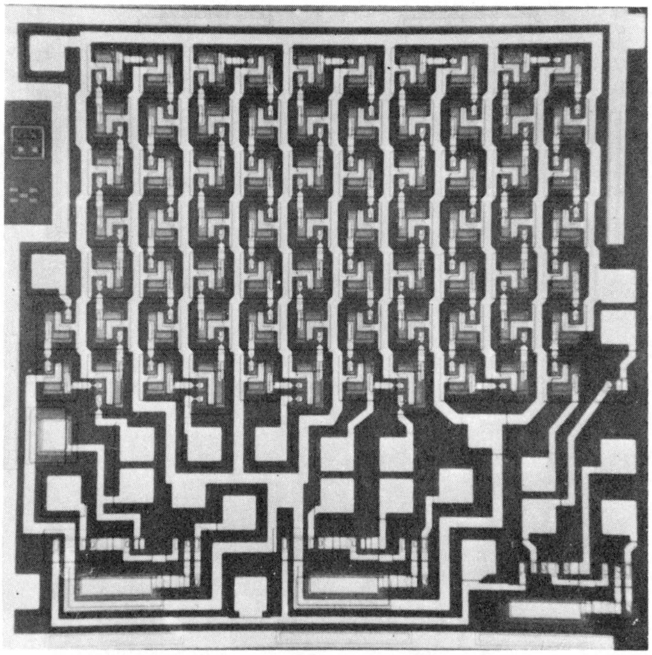

FIGURE 3—Bipolar integrated *bucket-brigade* delay line with 72 stages; chip size, 2 x 2 mm.

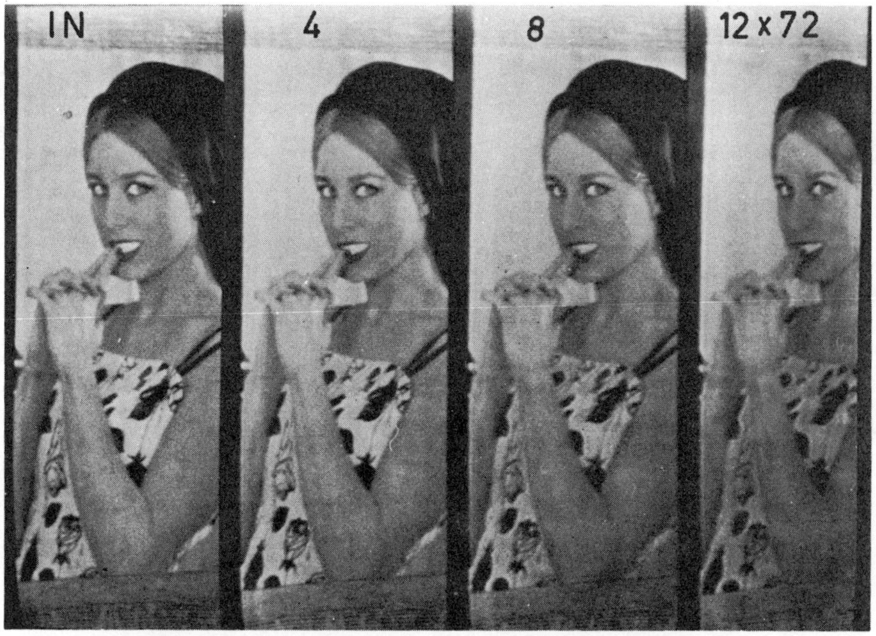

FIGURE 4—Delayed video pictures. left to right, original signal and output signals after 4, 8 and 12 *ICs* (each with 72 stages), respectively.

ation of the signal in large arrays; e.g., by a factor of 3 after 100 stages.

Some characteristics of this IC are: clock pulses 2 V, minimum and maximum clock rate 10 kHz and 30 MHz, signal-to noise ratio after 144 stages better than 60 dB; linearity after 144 stages within 1%. This circuit has been used successfully in time-fault correction systems for video recorders and in several other applications. To illustrate its performance a TV picture obtained with this bipolar circuit is shown in Figure 4. The four parts of the picture, from left to right represent the input signal and output signals after 288, 576 and 864 *buckets,* respectively. Signal degradation is small, even after several hundreds of stages[6].

MOS Integrated Circuit

In MOS technology, the *bucket-brigade* concept leads to an even simpler configuration than possible in bipolar technology. Input and output circuits with fewer components are possible. Dut to the absence of dc gate currents, attenuation is negligible even after hundreds of stages and no amplifiers are necessary. Figure 5*a* represents a complete MOS delay line. A cross section of an MOS *bucket* is shown in Figure 5*b*. The storage capacitor is formed by the enlarged aluminum gate layer and the P area underneath. A striking feature of the MOS *bucket* is that no interconnection pattern between adjacent stages is necessary as the drain of one stage also forms the source of the following one. Figure 6 shows a P-channel MOS integrated circuit comprising 72 stages with 8-pF storage capacitors on a 1.5 x 2.4 mm chip. This circuit uses two clock signals of 5-V amplitude. It has been tested between clock frequencies of 100 Hz and 3 MHz and it has been used experimentally for speech processing and audio delay[6].

The MOS circuit uses larger clock voltages and has more dissipation at a given frequency (roughly a factor of 10) than the bipolar circuit; this however is not a severe problem in audio applications. Its speed is not high enough for most video applications, except for color signal delay—part of signal handling in European PAL color-TV receivers; at present glass delay lines are in use for that purpose.

Acknowledgments

The author is very much indebted to the microcircuits group led by J. G. van Santen for fabricating the experimental samples, and gratefully acknowledges the valuable contributions of H. Heyns and C. Mulder in the layout design of the integrated circuits and of W. Ruiterkamp in the electronic experiments.

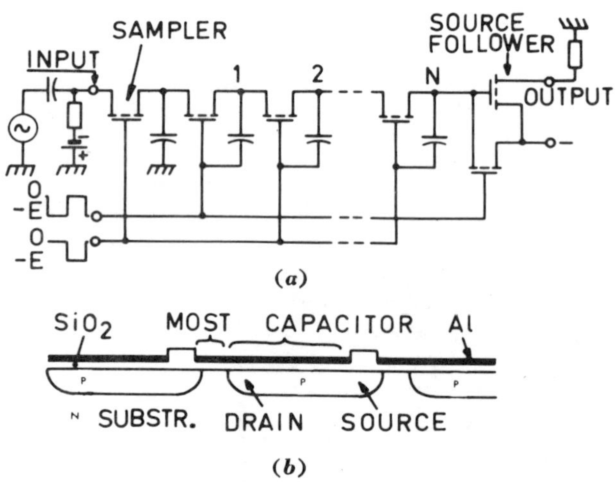

***FIGURE 5*—Circuit configuration of a MOS delay line is shown in (*a*); sectional view of the integrated circuit in (*b*).**

***FIGURE 6*—MOS integrated *bucket brigade* delay line with 2 x 36 stages: chip size, 1.5 x 2.4 mm.**

Charge-coupled devices— A new approach to MIS device structures

Charge-coupled devices require no diffusion for their active parts and can be conceptually fabricated with two masking steps. These new devices are finding application as image sensors and memory elements

W. S. Boyle, G. E. Smith **Bell Telephone Laboratories, Inc.**

Recent advances in materials and processing have resulted in a new class of information-handling structure—the charge-coupled device. This three-layer structure creates and stores minority carriers, or their absence, in potential wells near the surface of the semiconductor. The minority carriers move from under one electrode to a closely adjacent electrode on the same substrate when a more negative voltage is applied to the adjacent electrode. Because of their high transfer efficiency, these devices have already found application as image sensors. In addition, there is every expectation that memories made by use of the stored-charge concept will be less expensive and faster, and will require less power than a magnetic counterpart now in use.

Rarely do important technological applications follow major advances in our understanding of the fundamental physics and chemistry of a phenomenon or a materials system. Rather, it is more likely that the improved understanding from fundamental studies will give rise to some rudimentary device structures that may not be used for several years before the original idea finds widespread application. In the past year, the IMPATT and transit-time oscillators, for example, have become important as sources of power for millimeter waves. These devices had their genesis in work performed by W. Shockley and W. T. Read a decade earlier. The injection laser, first demonstrated in 1962, is only now becoming a device in which the understanding of a heterojunction materials system has proceeded to a point at which continuous room temperature operation is possible.

Similar delays are apparent in the most important of all semiconductor technologies: low-level signal processing. Throughout the past decade, integrated-circuit technology has been dominated by bipolar transistor structures that comprise metal and silicon dioxide films on silicon and use a process technology of diffusion and oxide-masking. There has been little tendency to vary this system even though the many changes in materials and processing have been explored experimentally.

The reason for the delay between structural concept and application is easy to find. Although it may be possible to demonstrate the feasibility of an idea in a novel device structure, its wide application depends on a complete understanding of both the physical phenomenons involved and the materials and processes required to fabricate the structure. For these reasons, progress in major device technologies tends to await the development of a large amount of background information. Improve-

Reprinted from *IEEE Spectrum*, vol. 8, pp. 18–27, July 1971.

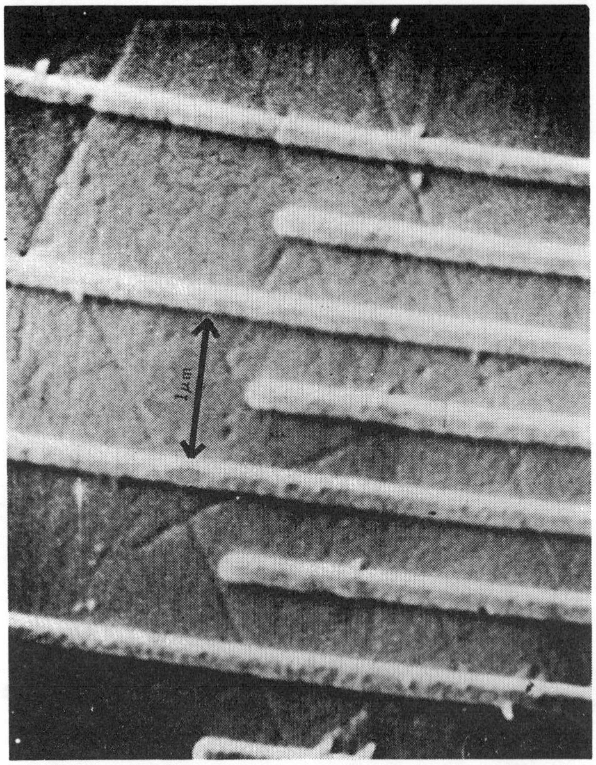

FIGURE 1. A piezoelectric surface wave transducer with metal stripes formed by electron-beam resist techniques.

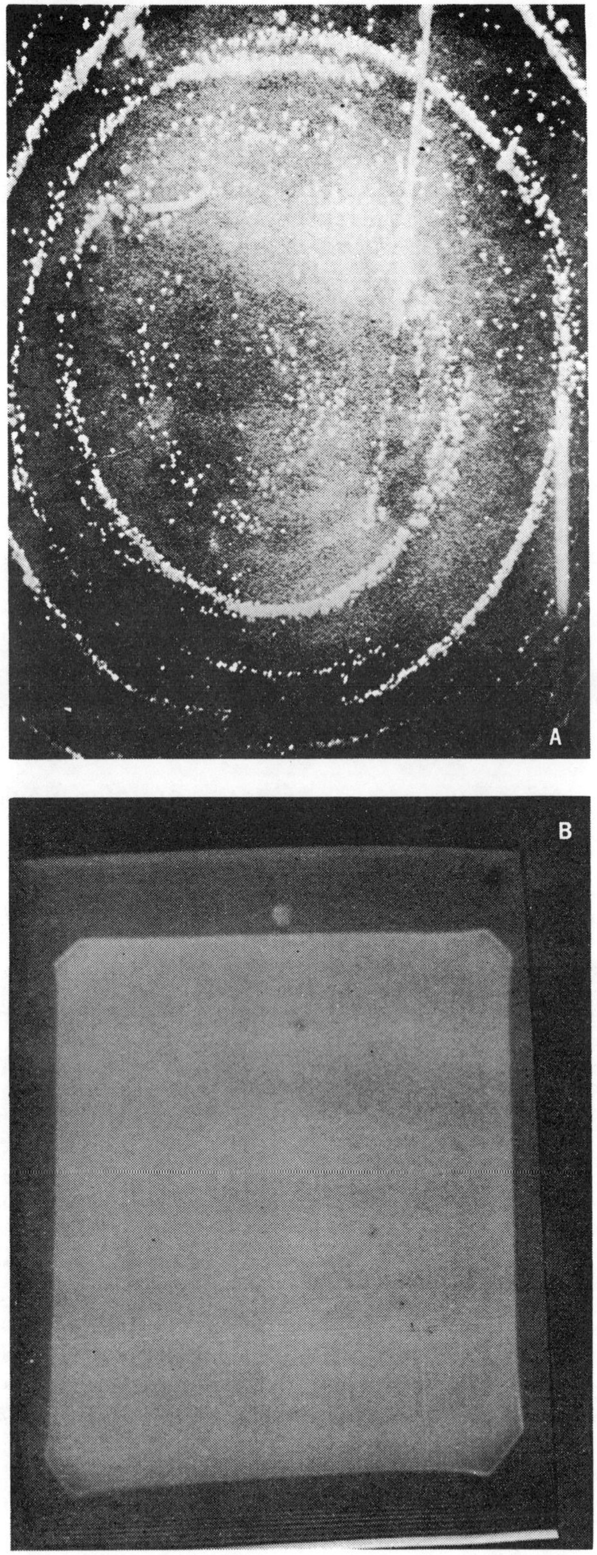

FIGURE 2. (Above right) A—Video display of excess currents from defects in a silicon-diode array camera tube. B—The same display from a defect-free array.

ments then occur in quantumlike jumps. Integrated-circuit technology is at the stage where in the next few years we are going to see just such a major change. The following areas, in which significant progress has been made, lead to this conclusion.

Semiconductor progress

First, after more than a decade of intensive study, we are beginning to understand interface states in the simplest systems—such as silicon dioxide on silicon. This understanding not only encompasses the physical description, but also includes the phenomenological knowledge of how to preserve desirable properties under severe stress. Of particular importance are dual dielectrics, which give the benefits of the low surface-state density available with silicon dioxide together with the high stability from less permeable materials, such as silicon nitride and aluminum oxide. Semiconductor devices can now be made that do not require the protection of a hermetically sealed enclosure. This single development changes the whole nature of integrated electronics by allowing a new freedom in interconnecting chips; no longer do the connections have to pass through vacuum seals. In this way a much higher order of integrated circuit is possible, with complete subsystems such as digital processors and memories taking the form of one large hybrid circuit.

The ability to fabricate structures with a thickness of less than one micrometer is a second factor. For some time it has been possible to produce insulating films of high quality in this thickness range, but only recently has ion implantation produced well-controlled doping profiles with such dimensions. An improvement in precision of at least one order of magnitude over any other process for both the number of impurity ions and their position has been clearly demonstrated.

Furthermore, new techniques in pattern generation have decreased lateral dimensions for masking operations. Figure 1, for example, shows a device fabricated by A. Broers *et al.*[1] that consists of a metal pattern laid down on a piezoelectric material to form a transducer. The individual fingers are 0.17 μm wide and are spaced on 0.5-μm centers. He uses a fine electron beam to expose a specially developed photoresist material, poly-methyl methacrylate, and then proceeds in a conventional way to employ this material as a mask to etch the patterns. This kind of fine geometric control makes it possible to fabricate devices with improved high-frequency performance and to make integrated circuits with higher packing density. It also presents the opportunity of fabricating structures with lateral dimensions that are comparable to depletion-layer widths and oxide thickness, a factor of importance in at least one device family.

A third area of progress is in the preservation of the homogeneity of the semiconductor material throughout the fabrication of device structures. All high-temperature process steps—such as diffusion and oxide growth—degrade the crystalline perfection of starting material. Only recently, however, did the full extent of this damage become evident or even important. The problem was brought very forcefully to our attention recently when we attempted to fabricate silicon-diode-array camera tubes—arrays of 500 000 individual diodes on a single slide of silicon. In the first devices we made, defects were a serious problem. Figure 2A shows the image of the leakage current obtained from a target made three years ago with the best technology we had at that time. Each one of the white spots is a single diode with excess leakage current. A normal leakage current is about 10^{-14} ampere. The main source of the difficulty has been traced to clusters of imperfections introduced at the time that high-

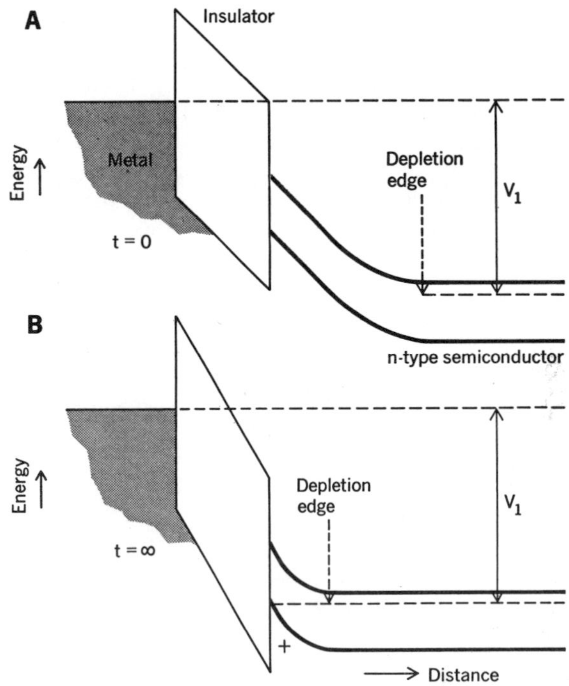

FIGURE 5. Plots of electron energy as a function of distance through an MIS structure. In (A), t = 0, no charge is stored at the surface. Charge is stored at the surface in (B), t = ∞.

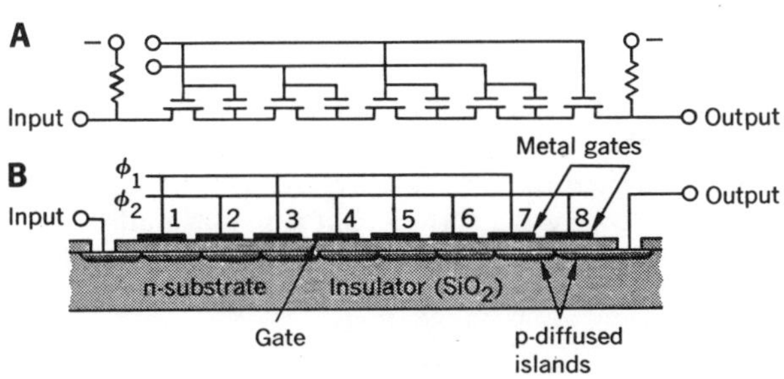

FIGURE 3. A—Circuit diagram for an IGFET bucket brigade. B—Cross section of the integrated circuit form.

FIGURE 4. Cutaway of a charge-coupled device. The dashed line represents both the edge of the depletion region and the potential distribution. Voltages are typical.

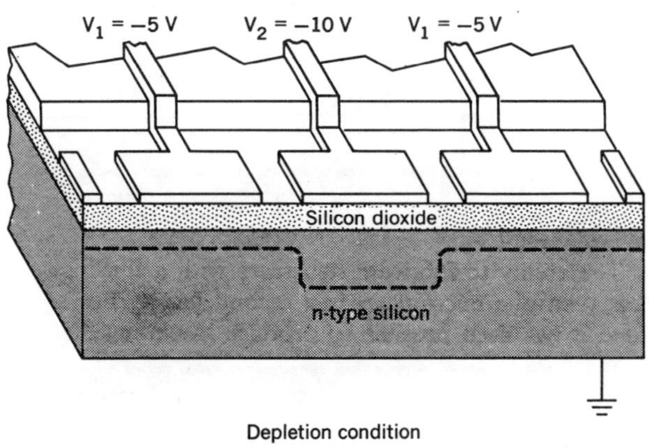

FIGURE 6. A theoretical plot of depletion width (x_d), surface potential (ϕ_s), and capacitance (C) of an MIS structure as a function of charge at the interface (Q′). Q is the maximum amount of charge that can be stored under conditions described in the text.

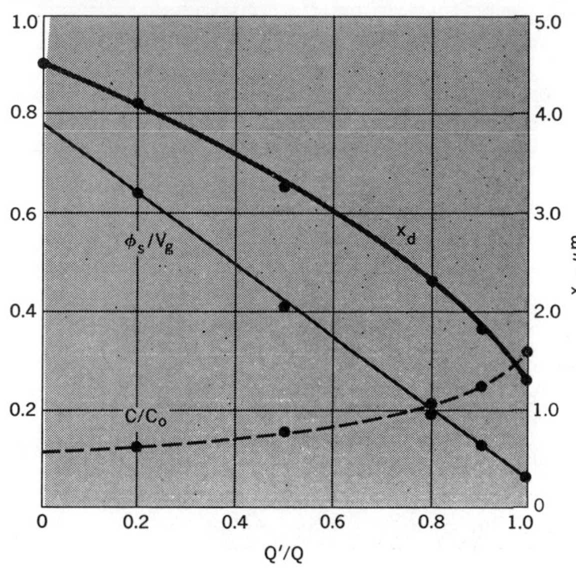

temperature oxidation takes place. Subsequent development efforts have led to nearly complete elimination of such defects (Fig. 2B).

These advances have come about from trying to improve an existing technology. Where these possibilities will lead is uncertain; however, to be more specific, we shall discuss a class of information-handling devices that make use of most of these advances. The common factor in this device family is that information is represented by stored charge. Several such structures have been reported in the past year. One class uses stored charge in an x-y array of elements for a random-access memory.[2] There are many commercially available insulated-gate field-effect transistor (IGFET) dynamic memories in which advantage is taken of the large off-resistance of an IGFET to store charge on a gate capacitance.

Another element, reported by P. T. Panousis,[3] consists of a metal oxide semiconductor (MOS) capacitor that is charged and interrogated through a bipolar switch. A second class of devices includes shift registers, in which charge is passed from element to element in a linear array.[4,5] An integrated-circuit version has been reported by F. L. J. Sangster.[6,7] It consists of either bipolar or IGFET transistors that are connected in series, with capacitors connecting base to collector or source to drain. Figure 3 illustrates this circuit for IGFETS. All of these devices are conventional circuit elements connected by wires, even though they can be put in integrated-circuit form.

Charge-coupled devices

A third class, charge-coupled devices (CCD), has been reported by the writers,[8] and the principle has been confirmed experimentally.[9] These devices represent a conceptually simple form of functional charge-storage device that cannot be constructed from discrete components. The device stores a minority-carrier charge in potential wells created at the surface of a semiconductor and transports the charge along the surface by moving the potential wells. In its simplest form the structure that accomplishes this consists of an array of closely spaced metal electrodes that overlay an insulator deposited on a uniformly doped semiconductor substrate. Figure 4 shows an n-type bulk (minority carriers are holes) device in a situation wherein a sufficiently large negative bias potential has been applied to all the electrodes to produce inversion, and the center electrode has a slightly larger applied potential. There is no inversion layer, only depletion, because a negligible number of minority carriers are present. This is not a steady-state situation; nevertheless, the surface will remain depleted for times of the order of seconds before thermally generated minority carriers accumulate because the present silicon art produces material with low densities of surface and bulk-generation centers. If minority carriers are introduced through any one of several means to be described shortly, they will collect at the surface in the potential minimum defined by the excess potential on the central electrode. The potential-energy diagram for the metal-insulator-semiconductor (MIS) structure is shown in Fig. 5. The Fermi energy of the metal appears on the left; on the right the band edges of the insulator and semiconductor are shown as functions of distance perpendicular to the surface. On the far right the Fermi level in the n-type semiconductor is shown. The bias voltage V_1 is the applied potential difference between the semiconductor and metal. Figure 5A shows the potential distribution in the absence of collected minority carriers; Fig. 5B illustrates the potential distribution when a saturated number of minority carriers has been collected. In the saturation condition shown in Fig. 5B, distribution of minority carriers is such that their diffusion current away from the surface is exactly balanced by their drift toward the surface. If less than the saturation value is accumulated, the net flow is toward the surface, but if the saturation value is exceeded, there is a net flow into the undepleted bulk,

FIGURE 7. Cutaway of a charge-coupled device in (A) the storage condition and (B) the transfer condition.

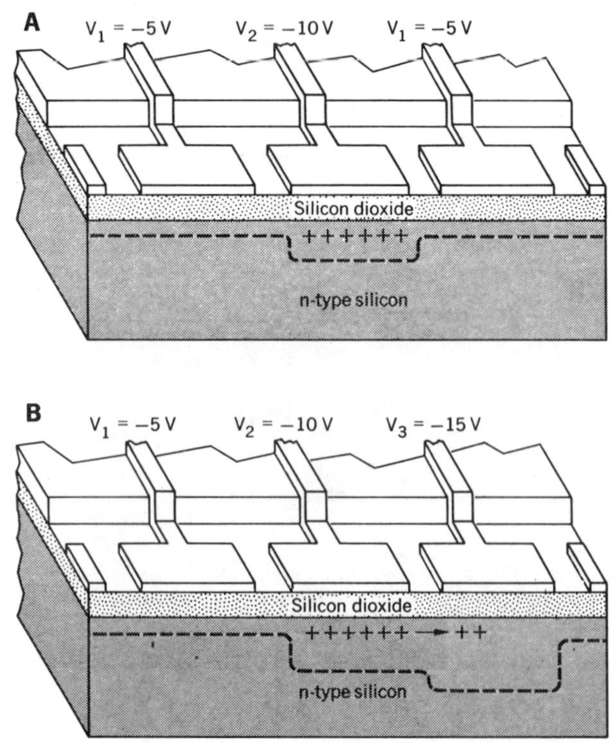

FIGURE 8. A three-phase charge-coupled device.

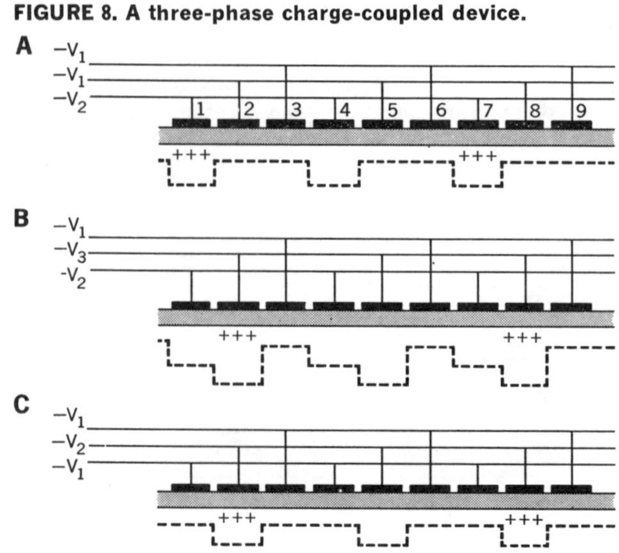

where the minority carriers recombine, just as in a forward-biased p-n diode. With the collection of minority carriers in the depletion well, the depletion width decreases as shown in the potential-level diagram; therefore, the differential capacity increases and, as shown, the surface potential decreases. Note that the minority-carrier accumulation held at the surface cannot change with rapid changes in electrode-to-bulk voltage. Therefore, the capacitance phenomenon consists only of mobile charge removal or addition at the depletion edge as the electrode bias is increased or decreased.

In Fig. 6, these three quantities—depletion width x_d, differential capacity C, and surface potential ϕ_s—are plotted as a function of the ratio of accumulated minority carriers to the saturation value. The plot here is for the particular values of doping density $N_d = 5 \times 10^{14}\ \text{cm}^{-3}$, gate bias $V_g = 10$ volts, oxide thickness $x_o = 2000$ Å. This results in an oxide capacitance of $C_o = 1.7 \times 10^{-8}$ F/cm². The Debye length, which is a measure of the sharpness of the depletion edge, is 5×10^{-6} cm, which is much less than the depletion width. For convenience, ϕ_s and C are plotted as normalized values to the gate bias and oxide capacity, respectively. As minority carriers

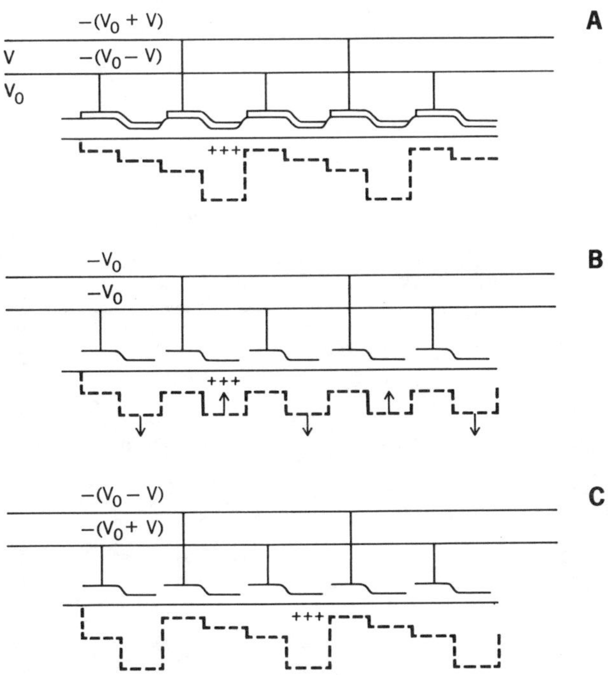

FIGURE 9. A two-phase charge-coupled device.

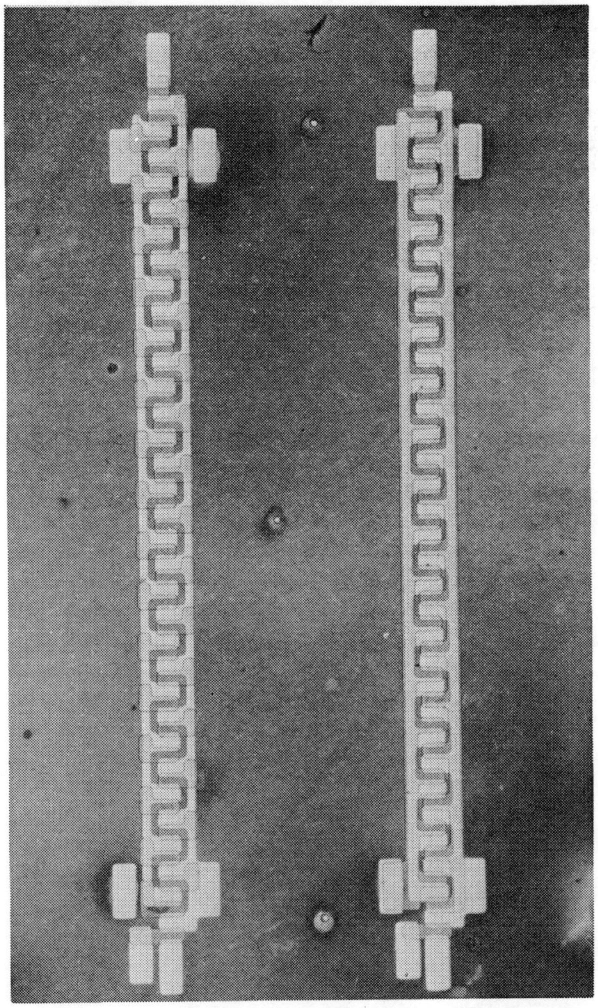

FIGURE 11. A four-phase charge-coupled device that uses silicon-gate technology.

FIGURE 12. Input schemes for charge-coupled devices.

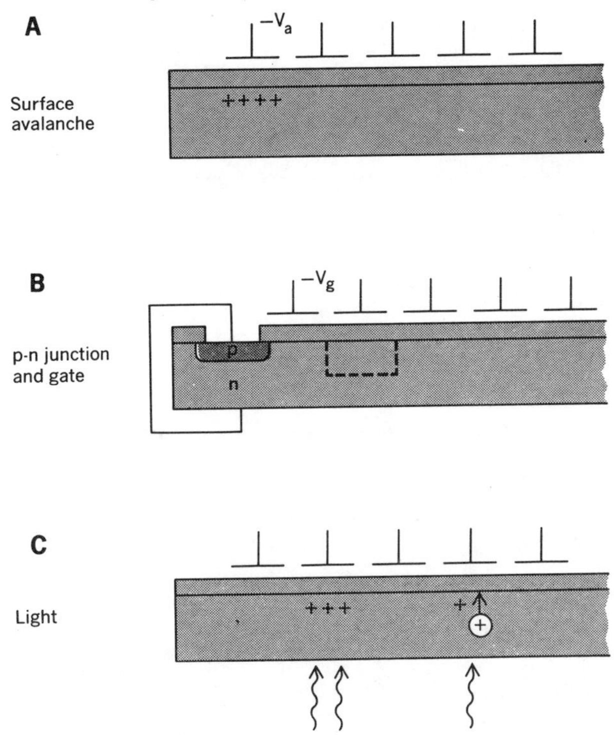

FIGURE 10. Basic overlapping metallization structure. The oxide under plates 1 and 3 would be made thicker and plates 1 and 2, 3, and 4 would be connected to make the device shown in Fig. 9.

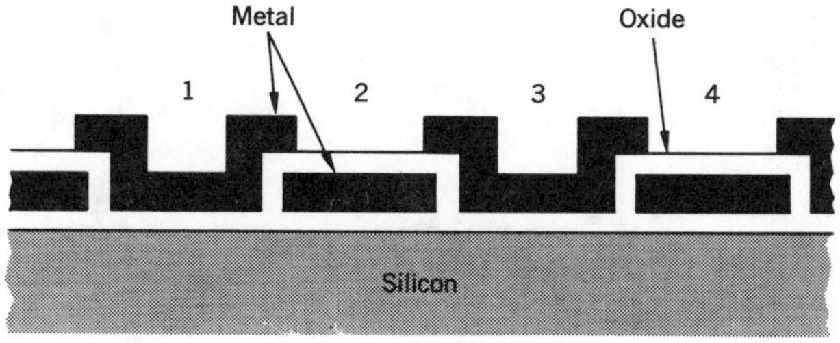

accumulate, the differential capacity increases by a factor of 3, and the surface potential by a factor of 8. These increases provide a large signal for determining the presence or absence of charge. The physical position of this charge in the storage condition is indicated by the plus signs in Fig. 7A.

The interesting feature of this structure is that it is possible to place the metal electrodes close enough together and to apply voltage differences of sufficient magnitude to obtain the potential distribution shown in Fig. 7B. In this case, the bias potential V_3 exceeds V_2, which causes carriers to be transferred from one electrode to the next. Subsequently, the potentials on the electrodes can be readjusted so that the quiescent storage site is located at the third electrode. Figure 8 shows just such a sequence, which effects transfer along a linear array. The electrodes are connected in groups of three and operated with a three-phase voltage supply to give direction to the transfer operation. The left column shows the potentials applied to each electrode. The response V_1 is sufficient to provide a depletion region in the semiconductor. Although V_2 is larger than V_1 and produces the storage site previously described, V_3 is still larger and effects the transfer. In Fig. 8A, charges are stored under electrodes 1 and 7; no charge is stored under electrode 4. In Fig. 8B, the potentials on 2, 5, and 8 are increased so that the charges move over one position. In Fig. 8C, electrodes 2, 5, and 8 have become the storage sites. In this way, coded information can be shifted along the linear array. A variety of geometries other than this three-phase structure utilize the same basic concept. One (Fig. 9) is a two-phase system in which the potential on adjacent electrodes alternates between $V_0 - V$ and $V_0 + V$. The electrodes have steps along their length; that is, the oxide thickness is stepped so that a different potential appears beneath each individual electrode.[19,21] The shifting operation is analogous to that obtained with the triplet geometry previously described.

Figure 10 shows a basic structure that implements the two-phase device by connecting the electrodes either directly, capacitively, or with an external dc bias. The use of overlapping electrodes also reduces the separation between plates from a photolithography tolerance to an insulator thickness. Alternatively, the device can be used with a four-phase drive. Figure 11 shows such a device in which silicon-gate technology is employed.

At the start of an array of shifting electrodes the charge can be generated by any one of a number of means. Some of these are shown in Fig. 12. In Fig. 12A, charge is generated through the application of a large enough pulse to produce avalanche breakdown in the semiconductor. Figure 12B illustrates a means of injection that uses the left electrode to produce an inverted region adjacent to a p-n diode formed in the bulk. The p region is wired to the

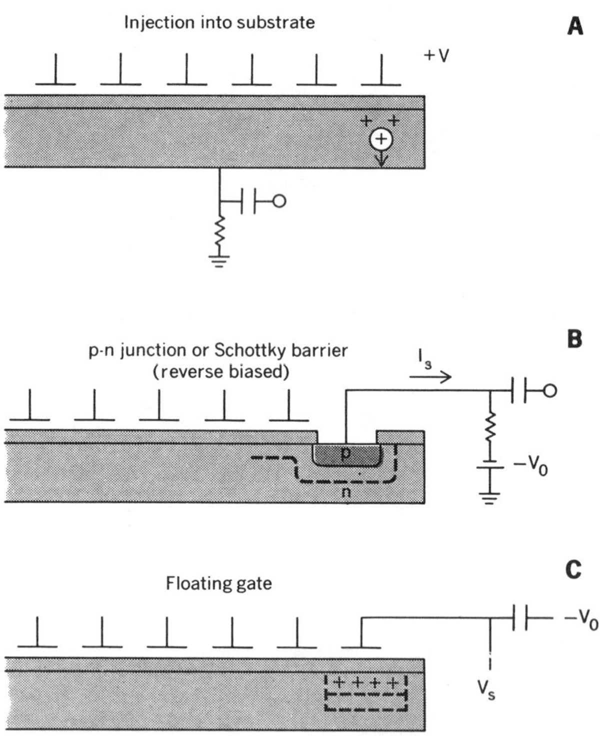

FIGURE 13. Output schemes for charge-coupled devices.

FIGURE 14. Longitudinal cross section of a charge-coupled 8-bit shift register.

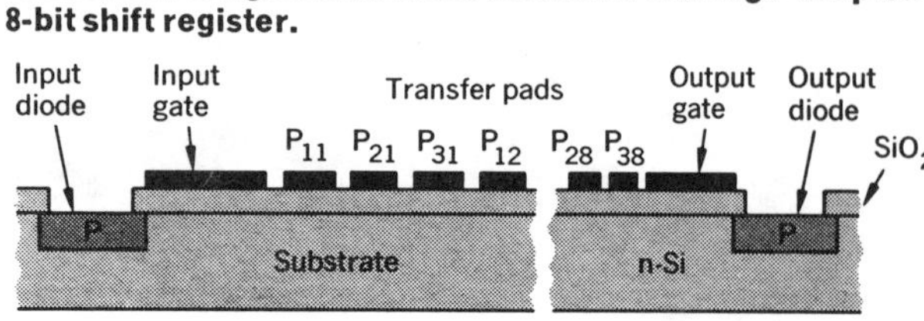

FIGURE 15. Photograph of the fabricated device of Fig. 14.

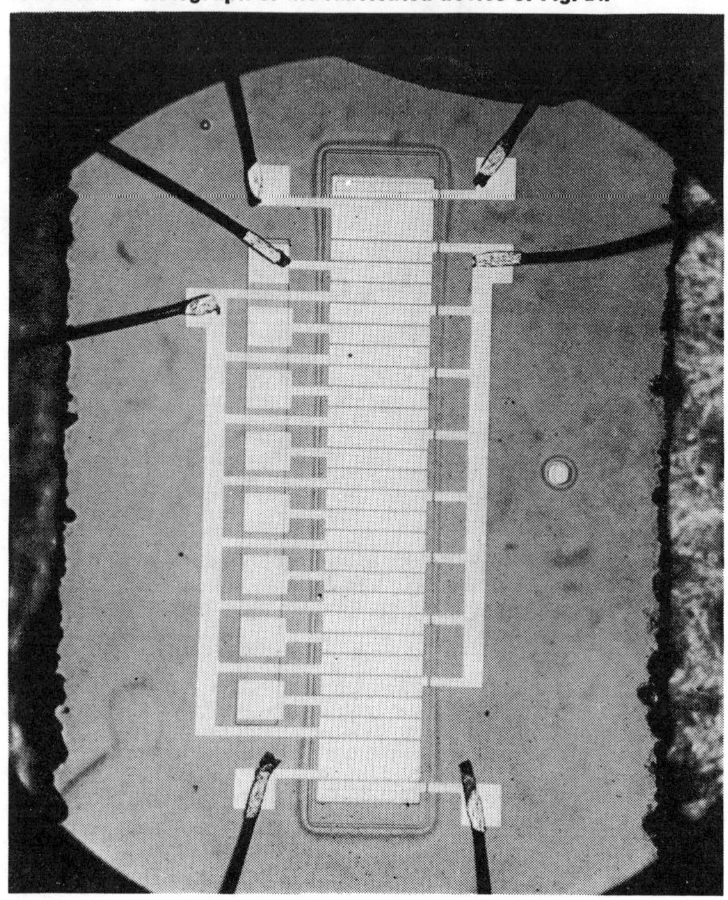

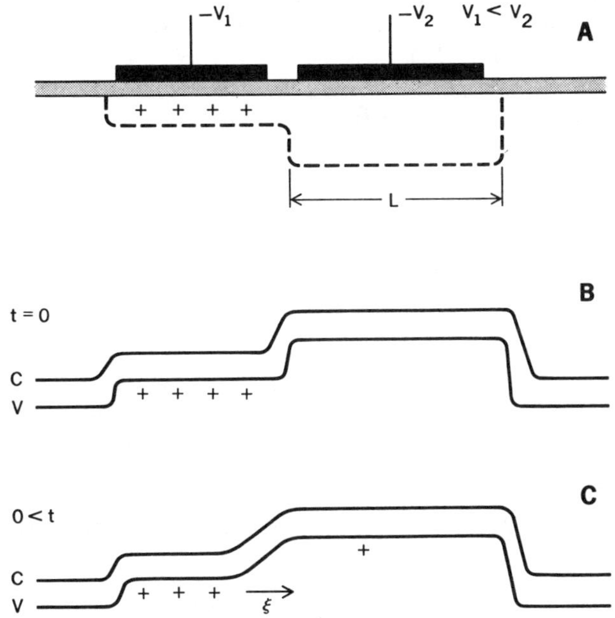

FIGURE 16. The transfer of charge from one plate to another. C and V refer to the conduction and valence bands.

FIGURE 17. The fraction of charge remaining under a plate as a function of normalized time. The abscissa is plated as $\sqrt{\tau}$, which spreads the region where $\tau < 1$, and results in straight parallel lines over the diffusion portion.

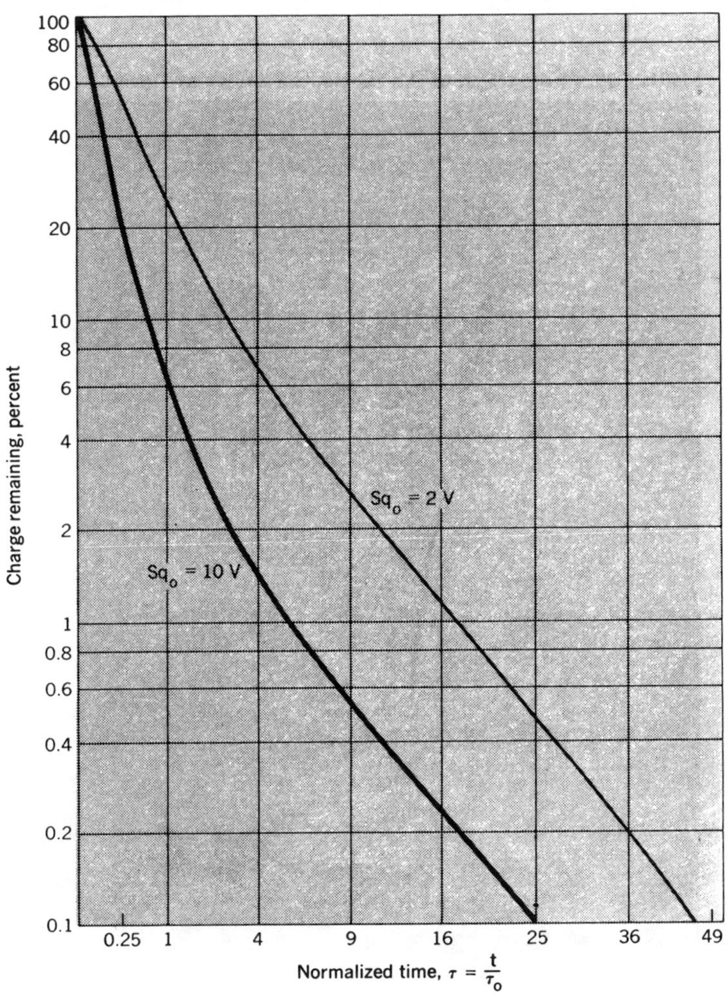

bulk and becomes a source of minority carriers to support the inversion. A minority-carrier current flows along the surface to the right and delivers a minority-carrier accumulation to the region under the next electrode, which is biased to produce a potential well. This is similar to gating charge from source to drain in an IGFET. Figure 12C shows that the minority carriers can be generated optically and collected under the electrodes. In this latter mode of operation the device operates as an imaging device and has the built-in features of image storage and scanning.

Figure 13 illustrates methods of detection. In Fig. 13A, the substrate connects to ground through a resistor. When charge is transferred to the last electrode, a positive voltage causes holes to be injected into the substrate, and a current passes through the ouput resistor. In Fig. 13B, a diode at the end of the line is reverse-biased to a voltage $-V_0$, which is more negative than any of the surface potentials used for transfer. When charge is transferred to the diode position, it produces an output current current I_s in the external circuit. In Fig. 13C, the fact that the capacitance of the MOS structure changes with charge is used in a capacitive division circuit, where the voltage V_s could be connected, for example, to the gate of an IGFET.

A cross section of one of the device structures studied[10] is shown in Fig. 14. The transfer pads are arranged for a three-phase operation, P_{11}, P_{21}, P_{31}, $\cdots$, P_{38}, with every third pad connected together as shown in Fig. 10, where the three-phase drive was illustrated. Injection is controlled by the potential on the input gate, which can induce an inversion layer that connects the first plate to a source of minority carriers at the input diode. A symmetrical arrangement is used for detection of the output signal with a reverse-biased diode.

Figure 15 illustrates this structure. In the central section, where the transfer pads are located, there is a thin (1200-Å) oxide and beyond that region a thick (5000-Å) oxide, which prevents depletion in regions other than beneath the transfer electrodes. All of the second electrodes in transfer triplets are connected together by means of a diffused crossunder. In this early device, the plates are 250 μm by 50 μm and are separated by 3-μm gaps.

Of particular interest are the mechanisms that inhibit the complete transfer of charge from one plate to the next. Our early work indicates that this is not a simple problem and depends on at least two distinct factors. The first involves the dynamics of the free carriers that undergo motion through both the mechanisms of diffusion and drift. The second involves trapping effects at interface states. Our preliminary data indicate that, with the material system and geometries we have used, both these effects are present but are not sufficiently large to preclude useful applications. We have, for example, made observations on structures where the total amount of charge left behind at each transfer is less than 0.1 percent at shifting times of 2 μs.

A qualitative feeling for the magnitude of these two effects may be obtained by first considering the dynamics of minority carriers. Figure 16A illustrates the distribution of carriers the instant after bias has been increased on the transfer electrode of length L. Figure 16B shows the corresponding potential distribution. In Fig. 16C, the potential distribution has changed because a few of the minority carriers have moved out through the action of

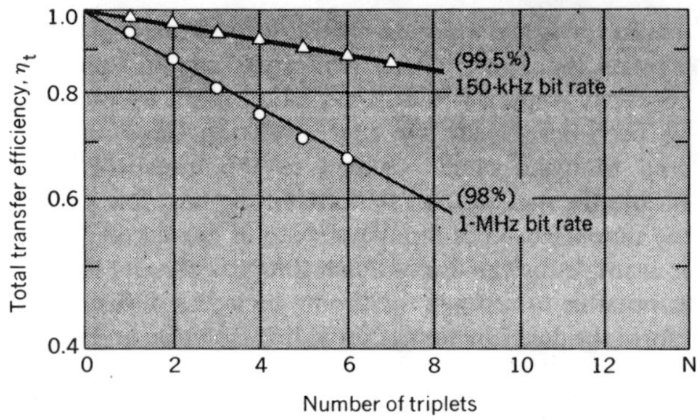

FIGURE 18. Transfer efficiency as a function of the number of bits for 150 kHz and 1 MHz.

FIGURE 19. Transfer efficiency as a function of the number of bits for various operating voltages.

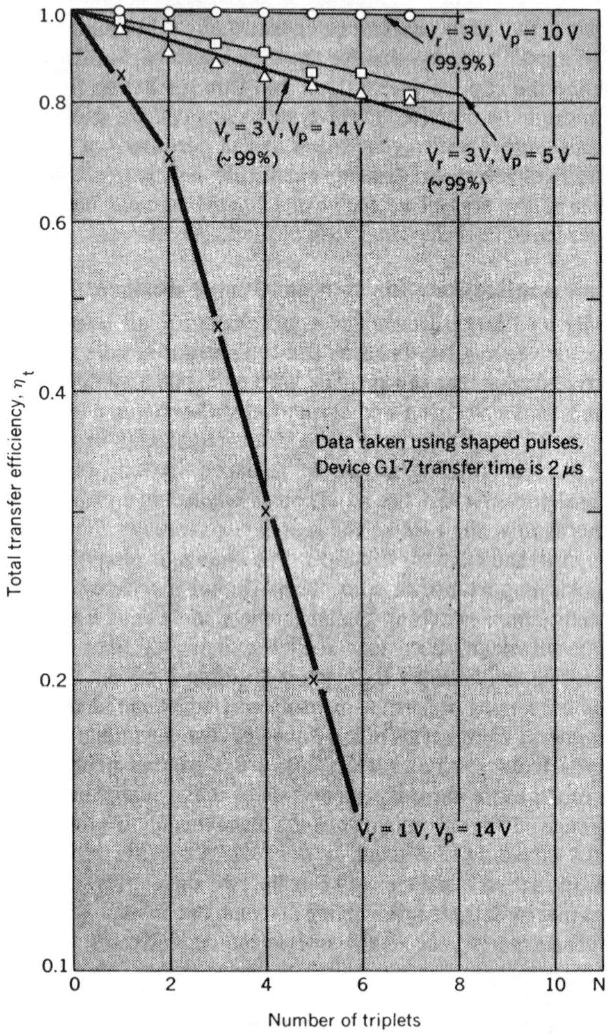

diffusion alone. At this point in time, however, the rearrangement of potential is such as to provide a drifting field for some of the remaining carriers over part of their path. This effect, as well as the enhanced transfer efficiency resulting from fringing fields that exist in finite geometries, are discussed qualitatively in Ref. 8. Mathematically, this means solving the transport equation in the presence of a potential distribution that is a function of the free carrier density. Approximate solutions to these coupled equations show that the drift component is much larger than the diffusion component except for the last fraction of charge to be transferred.[11,12,20] A plot[12] of the amount of charge remaining at a site as a function of time after a square-wave pulse is applied to an adjacent electrode is shown in Fig. 17. The normalization time is $\tau_o = L^2/\mu u_o$, where L is the length of a plate, μ is the mobility, and u_o = 1 volt. The quantity $V_s = Sq_o$ is a measure of the initial quantity of charge stored. There is a rapid drop initially, which is faster for a large amount of charge (storage voltage = V_s = 10 volts) than for the smaller amount (V_s = 2 volts). When only a small amount of charge remains, the slopes change as both become diffusion-limited. These calculations essentially agree with the experimentally observed charge-transfer efficiencies.

The upper limit of the amount of charge lost to surface states can be estimated as follows: The maximum number of charges stored N_m is the oxide capacity times the gate voltage divided by the electron charge. For typical values this is 2×10^{12} charges/cm^2. If the surface is kept under inversion at all times, only those states that have a time constant that corresponds to the time allowed for shifting will be effective in reducing transfer efficiency. With a density of states of 10^{11} cm$^{-2}\cdot$ eV^{-1} and if a band kT wide has the right time constant, the maximum fraction lost will be 2×10^{-3}.

The transfer mechanism has been studied[13] in an optical injection experiment whereby a light source, which moves along the array, provides a reproducible amount of injection at any one of the electrodes in a sequence. Figures 18 and 19 show some data that have been obtained in this way. In Fig. 18, the total efficiency plotted as an ordinate is the ratio of the amount of charge recovered at the collecting diode when the light beam is positioned at one of the electrodes divided by the charge obtained when the beam is positioned at the last electrode. The abscissa is the number of triplets along which the charge has passed. The voltages used are the same at

FIGURE 20. A 96-bit, three-phase CCD. The enlarged portion shows the output end of the device.

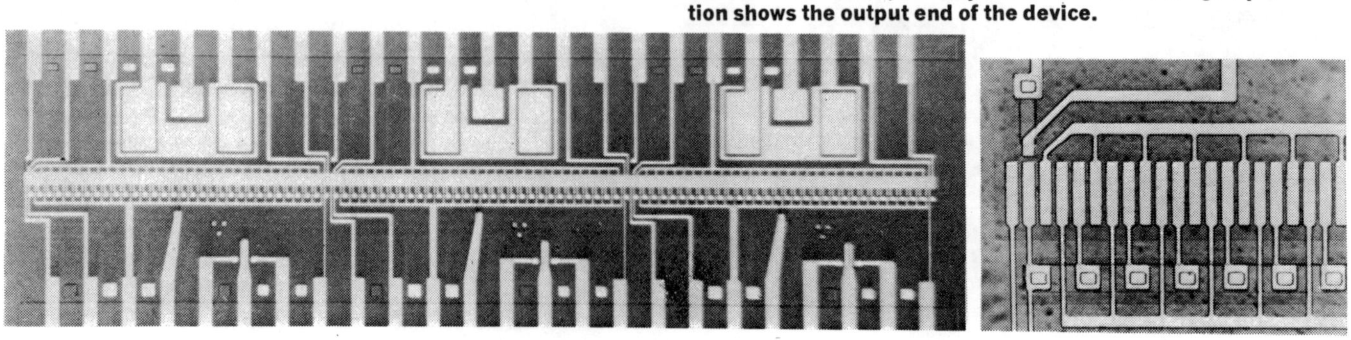

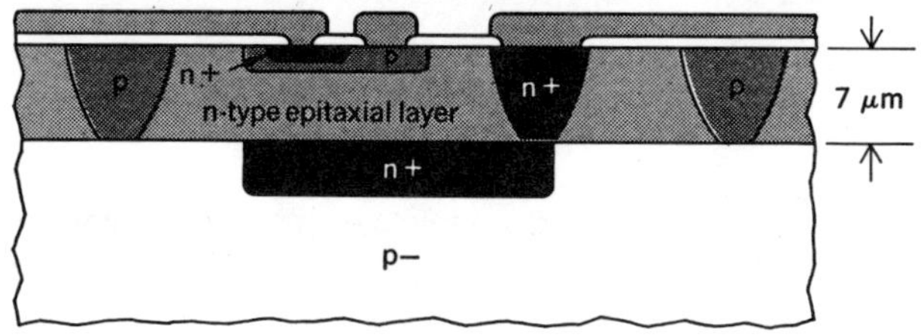

FIGURE 21. A cross section of a junction-isolated planar transistor.

both frequencies and the percentage, in parentheses, is the transfer efficiency per electrode. We do not have enough data as yet to map out in detail the frequency dependence of the transfer process; however, the early data and calculations indicate that the transfer efficiency falls off rapidly with frequency changes. The transfer phenomenon is even more complicated for the transfer efficiency as a function of resting and transfer potential. Figure 19 shows that the transfer falls off very rapidly as the resting potential is reduced from 3 volts to 1 volt. There is a further complication in the variation of the transfer with the transfer potential V_p (with a constant bias potential). An unexplained maximum occurs for a transfer potential of 10 volts. Some of the complexity may arise from the relatively large separation between electrodes, which results in poorly defined potentials in the regions of free oxide surface lying between the electrodes. Experiments under way use structures similar to Fig. 11 and will give much better control of the potential distribution over all regions through which the carriers move. In this way, some of the anomalous results may be resolved. These better-defined conditions will provide a new tool to explore experimentally and directly new facets of interface states. Not only should we be able to determine surface-state densities very close to the band edge, but we will also be able to examine the capture and emission cross sections as a function of the electric field parallel to the surface.

Advantage has been taken of these high transfer efficiencies in applying the charge-coupled-device concept to image sensors[14] (Fig. 12). Figure 20 illustrates a three-phase 96-bit linear array. Charge is transferred from right to left in this photograph, with input–output diodes located at intervals of 32 bits. This device has 10-μm plates with 3-μm spacings between them and has been run at 2 MHz with better than 98 percent efficiency.

The device structure that uses the charge-coupled process is significant because it makes use of some of the basic mechanisms in semiconductors, and it may also suggest some new experimental tools for basic measurement. However, it also typifies the direction that the technology is taking at this time. It does this in a number of ways.

Trend to simpler structures

For the past ten years the majority of integrated circuits have used a basic structure that has evolved directly from planar discrete-device technology. Figure 21 shows a cross section of a junction-isolated planar transistor. A deep p-type diffusion is formed at the periphery of the n-p-n transistor at the center of the diagram. A buried collector, shown by the n^+ region at the center, is contacted by another deep diffusion through the epitaxial layer. In total, five separate diffusion steps are required, each with its accompanying oxide masking and photolithography steps. After all the desired impurity profiles have been developed, two more masking steps are required to make ohmic contact to the transistor and delineate the metalization pattern. In the past few years, some simplification of bipolar structures has taken place. For example, by the use of much thinner epitaxial layers, it is possible to arrange for the n^+ collector diffusion to perform the dual function of providing isolation and making contact to the collector. Even further simplification has been reported recently by B. T. Murphy *et al.*[15] The field-effect devices have from their earliest days been conceptually easier to fabricate—a standard insulated-gate field-effect transistor requires only one diffusion for the establishment of the source and drain regions.

Charge-coupled devices require no such diffusion for their active parts and can be conceptually fabricated with two masking steps—one for the metal pattern, another to define the regions where thick and thin insulating film is required. In practice, IGFET-type structures are used for input–output and regeneration at the periphery of large arrays. Since these elements constitute only a small fraction of the area of a chip, overall yields should be high because of the basic simplicity of the CCD elements.

New applications for semiconductor devices

By and large, the major application for all semiconductor devices has been in the time-honored role of an active device that can provide gain or act as a switch. Indeed, it is surprising and somewhat disheartening to find that all integrated circuits are still recognizable in terms of discrete building blocks of resistors, capacitors, and transistors. At last the situation is beginning to change. This is now the case in the search for elements that will perform the memory function. The change is also of particular importance in some large digital machines, such as electronic switching systems, where there is at least an order-of-magnitude more memory elements than logic gates or switches. In the past, a bistable flip-flop circuit has been used to form a memory cell without the use of magnetic elements. Today, most of the fast memories made from semiconductors make use of this principle. A much more natural approach is to make use of charge storage. Some promising results have been obtained in MIS structures by using a dual dielectric and storing charge at the interface between the two dielectrics or in a floating metal gate sandwiched between two insulators.[16,17] Unfortunately, successful operation depends on being able to pass current in a controlled manner through one of the insulators. Three more direct approaches were mentioned earlier. There is every expectation that memories made by the use of the stored-charge principle will be less expensive and faster, and will require less power for operation than any of the magnetic memories that are now in existence. It should also be possible, of course, to make use of the shifting principles that are inherent in the charge-coupled devices to devise new ways of interrogating memories other than the coordinate address approach that is used now.

New approaches to signal processing

As measured by the fundamental physical limits, all present digital machines are extremely wasteful of power.

A figure of merit for a logic gate is equal to power times switching delay. This quantity has the units of energy required to process one bit of information. In practice, the lowest delay product of the best semiconductor devices lies in the range of 10^{-12} joule. R. Landauer[18] has shown by phenomenological reasoning that no more than a few times kT should be required for the switching function. Because kT is approximately 10^{-20} joule, there are eight orders of magnitude between today's performance and what is theoretically possible. Nowhere else, at least in the field of electronics, is there such apparent inefficiency! One does not have to search far for the reasons for such poor practical results. Because we are using junctions to obtain nonlinearities—in other words, switching action—the potential variations must be at least several kT/q volts or a fraction of a volt. The smallest structures that can be fabricated have dimensions that produce a capacitance of 10^{-12} farad. At each switching operation, CV^2 of energy must be dissipated and this is approximately equal to 10^{-12} joule.

One system for performing logic that circumvents these practical limits for junction devices is the one that uses magnetic bubbles for logic.[22]

The energy dissipated in the magnetic material per switched bit may be as small as 4×10^{-14} joule. Similarly, a highly idealized charge-coupled device in which charge is transported by a moving sine-wave potential well gives the same loss per bit (at 10 MHz) on the semiconductor slice. Both calculations neglect power-supply loss.

These charge-coupled devices had their origin in the search for semiconductors with device mechanisms similar to those obtainable from magnetic domains. Whether they will be able to remove the eight orders of magnitude between practice and theoretical limits is not yet clear. One thing, however, is evident: Much is still left to be done in getting a better understanding of the fundamental physics and chemistry of semiconductor structures. This is an extremely fruitful field for the development of devices of practical importance.

REFERENCES

1. Broers, A. N., Lean, E. G., and Hatzakis, M., "1.75 GHz acoustic-surface-wave transducer fabricated by an electron beam," *Appl. Phys. Lett.*, vol. 15, p. 98, 1969.

2. Wood, J., and Ball, R. G., "The use of insulated-gate field-effect transistors in digital storage systems," *ISSCC Dig. Tech. Papers*, pp. 82–83, 1965.

3. Panousis, P. T., "A TRIM bipolar charge storage memory," presented at International Electron Devices Meeting, Washington, D.C., October 28–39, 1970.

4. Janssen, J. M. L., "Discontinuous low-frequency delay line with continuously variable delay," *Nature*, vol. 149, pp. 148–149, Jan. 26, 1952.

5. Hannan, W. J., Schanne, J. F., and Woywood, D. J., "Automatic correction of timing errors in magnetic tape recorders," *IEEE Trans. Military Electronics*, vol. MIL-9, pp. 246–254, July/Oct. 1965.

6. Sangster, F. L. J., "Integrated MOS and bipolar analog delay lines using bucket-brigade capacitor storage," *ISSCC Dig. Tech. Papers*, pp. 74–75, 1970.

7. Sangster, F. L. J., and Teer, K., "Bucket-brigade electronics—New possibilities for delay, time-axis conversion, and scanning," *IEEE J. Solid-State Circuits*, vol. SC-4, pp. 131–136, June 1969.

8. Boyle, W. S., and Smith, G. E., "Charge-coupled semiconductor devices," *Bell Syst. Tech. J.*, vol. 49, pp. 487–493, 1970.

9. Amelio, G. F., Tompsett, M. F., and Smith, G. E., "Experimental verification of the charge-coupled device concept," *Bell. Syst. Tech. J.* vol. 49, p. 593, 1970.

10. Tompsett, M. F., Amelio, G. F., and Smith, G. E., "Charge-coupled 8-bit shift register," *Appl. Phys. Lett.*, vol. 17, p. 111, 1970.

11. Engeler, W. E., Tiemann, J. J., and Baertsch, R. D., "Surface charge transport in silicon," *Appl. Phys. Lett.*, vol. 17, p. 469, 1970.

12. Strain, R. J., and Schryer, N. L., "A nonlinear diffusion analysis of charge-coupled device transfer," *Bell. Syst. Tech. J.*, July/Aug. 1971.

13. Amelio, G. F., private communication.

14. Bertram, W. J., "Application of the charge-coupled device concept to solid-state image sensors," *1971 IEEE Internat'l Conv. Dig.*, pp. 250–251.

15. Senhouse, L. S., Kushler, D. L., and Murphy, B. T., "Base diffusion isolated transistors for low power integrated circuits," *IEEE Trans. Electron Devices*, vol. ED-18, pp. 355–358, June 1971.

16. Kahng, D., and Sze, S. M., "A floating gate and its application to memory devices," *Bell Syst. Tech. J.*, vol. 46, p. 1283, 1967.

17. Frohmann-Bentchkowsky, D., "A fully-decoded 2048-bit electrically-programmable MOS-ROM," *ISSCC Dig. Tech. Papers*, pp. 80–81, 1971.

18. Landauer, R. W., *IBM J. Res. Develop.*, vol. 5, p. 183, 1961.

19. Kosonocky, W. F., and Carnes, J. G., "Charge-coupled digital circuits," *ISSCC Dig. Tech. Papers*, pp. 162–163, 1971.

20. Kim, C. K., "Carrier transport in charge-coupled devices," *Dig. Tech. Papers*, pp. 158–159, 1971.

21. Tompsett, M. F., "A simple charge regenerator for use with charge-coupled and bucket-brigade shift registers and the design of functional logic arrays," *ISSCC Dig. Tech. Papers*, pp. 160–161, 1971.

22. Bobeck, A. H., and Scovil, H. E. D., "Magnetic bubbles," *Sci. Am.*, vol. 224, pp. 78–90, June 1971.

Integrated-Circuit Content-Addressable Memories

JAMES T. KOO

Abstract—**This paper describes integrated-circuit content-addressable memory designs employing simple memory cells that are compatible with peripheral circuits designed for existing IC random-access memories. Also described is an expandable resolver circuit that is capable of automatically resolving multiple responses and performing other special logic operations. Operating algorithms of the proposed content-addressable memory are also discussed.**

Introduction

WORD-organized random-access memory (RAM) is the most common type of present-day random-access memory. Information is either written into or read out from the word defined by the address. This location-addressable mode of memory operation requires extensive bookkeeping and imposes a stringent limitation on the capabilities of data-processing machines. Content-addressable mode of memory operation differs from the location-addressable mode of operation in the following ways.

1) The address is not needed for storing and retrieving information.

2) During the retrieval process, information is simultaneously sensed from all specified memory locations and processed to determine if it is associated with the desired information.

3) All associated information in the specified section can be retrieved, i.e., it is a multivalue output process.

A computer memory capable of the content-addressable mode of operation is called content-addressable memory (CAM) or associative memory. Content-addressable read is also called match or search operation.

A CAM is an efficient hardware substitute for a RAM plus a searching algorithm. Associated information can be retrieved from a RAM by means of a serial-by-word scan of the entire memory and a determination of the relevance of the information. This process would require N memory and logic cycles for a memory of N words.

In a CAM, data are retrieved by specifying the information content of an arbitrary portion of its structure. This portion of the data is called the selected pattern or key. While search "1" or search "0" is carried out in the key, search DON'T CARE is carried out in the remaining portion of the word. Search DON'T CARE is also expressed as search "ϕ" and matching signals are always produced at these bits. The portion of the memory word possessing associative properties may be limited only to a portion of the word defined by memory construction. Depending on the extent of the bounded region, CAM can be classified as in Table I [1].

Suggested applications for which CAM appear to be ideally suited are numerous [1]–[3]. It has been shown [3] that data-processing machines employing CAM systems can be two orders of magnitude faster than the standard machines in solving Dirichlet boundary value problems.

Though accessing of memory by an address is not necessary for a CAM system, CAMs have an address system and can be operated as RAMs. With this type of system, the content-addressable mode operations are subdivided into two steps. An address is obtained as a result of search, and then the memory is operated as a RAM using the address obtained.

Priority must be assigned to handle the multiple match situation. Certain pieces of information must be assigned with the highest priority while the rest of the matching pieces of information are temporarily suppressed. Later, this highest priority word must be suppressed to allow retrieval of lower priority words. Therefore, management of priority must be included in every memory cycle.

Since no bookkeeping is kept for a CAM, methods must be provided to determine if the memory is full, or location(s) of empty words are available for write. This is also part of CAM management.

From the above discussion, the simplest algorithm for an ideal CAM operating in the content-addressable mode will have the following necessary steps:

1) search for match
2) retrieve or store the desired information
3) alter the priority structure and manage the memory.

Analysis

Problems encountered in realizing a CAM can be closely related to each step of the algorithm.

During the SEARCH operation, information is sensed from every memory word and is compared with the SEARCH key. A nondestructive readout (NDRO) cell is necessary for a practical CAM system. The SEARCH operation also requires CAM cells to perform the compare logic operation within the cell. In memory organization, the cells should be accessed a word at a time as a word organized RAM and the whole memory at once as a CAM. The CAM cell must communicate with the bit line(s) and with other CAM cells within the same word. Therefore, four logic signal lines per cell is the minimum for a cell in a binary memory system. These are word line, intercell signal line (match line), and two bit lines to provide "1," "0," and DON'T CARE states. Since the signal from the SEARCH key is transmitted to every cell of the

Manuscript received April 4, 1970; revised June 2, 1970. This work is part of the author's Ph.D. dissertation submitted to Lehigh University, Bethleham, Pa., in 1969.
The author is with Bell Telephone Laboratories, Inc., Allentown, Pa. 18103.

Reprinted from *IEEE J. Solid-State Circuits*, vol. SC-5, pp. 208–215, Oct. 1970.

memory during the SEARCH operation, each cell must require a minimal amount of input energy, otherwise the problem of large number fan-out becomes insurmountable. Noise, power, component variation, and leads interconnection are some of the problems that also should be considered.

In order to successfully retrieve or store the desired information after a SEARCH operation, the problem of multiple response and priority assignment must be solved. It would be objectionable to serially detect all the matching signals as this would resemble the serial operation of the RAM-plus-algorithm system. Therefore, a parallel-processing technique and a simple algorithm must be used to allow the CAM system to compete successfully. The resolving system must be sufficiently flexible to allow change of priority assignment and perform various logic functions. If IC technology is to be used, the hardware used must be of repetitive and expandable nature.

To successfully operate the CAM system, methods or algorithms must be found for locating empty positions to store new information, retrieving data other than the highest priority word in a list of matching words, restoring the CAM to quiescent state after each operation, etc. From the speed, complexity, and economics considerations, the CAM must surpass the RAM-plus-algorithm system to be a desirable system.

Solutions to the above-mentioned problems are the purpose of this study. An IC–CAM system will be proposed here together with all the operating algorithms.

Previous Work

Cryogenic [4]–[8] and magnetic [9], [10] technologies have been employed to achieve CAM systems with varying degrees of success. Other techniques [11]–[14] have been devised in attempts to realize CAM cells. Recently, semiconductor [15]–[17] CAM systems have been proposed. The major difficulties have been circuit complexity and cost.

Various algorithms have been proposed for resolving multiple responses and for orderly retrieval [18]–[22]. Due to a decrease in the cost of semiconductor devices, hardware approaches to the multiple responses have been proposed [23]–[25]. This paper will describe complete CAM systems including the resolver–encoder and operating algorithms.

CAM Cell

Fig. 1 shows a proposed content-addressable memory cell employing complementary insulated-gate field-effect transistor (CIGFET) technology. The portion of the circuit in the dotted-line box is the matching logic circuit of the cell. Memory cells employing single-channel (p-channel) IGFETs or bipolar transistor technologies are shown in Figs. 2 and 3. Fig. 4 shows symbolically the manner in which these cells would be organized in the memory.

All three circuits operate on essentially the same

TABLE I

Associative Region	Percent of Information Contents Specified	Operating Modes
partial	total	tag mode
partial	partial	partial tag mode
total	total	identical mode
total	partial	fully associative

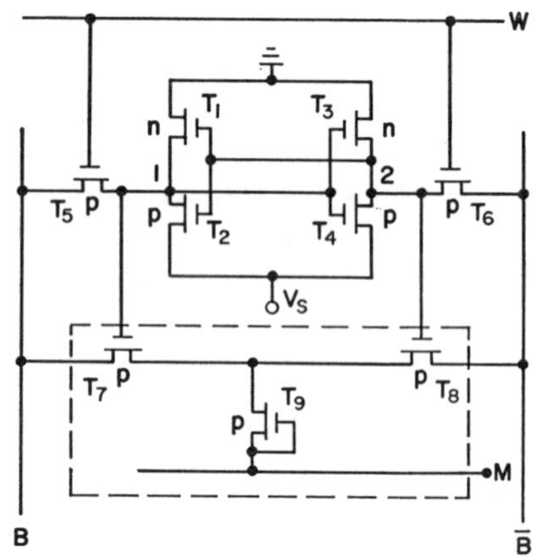

Fig. 1. Proposed CIGFET CAM cell.

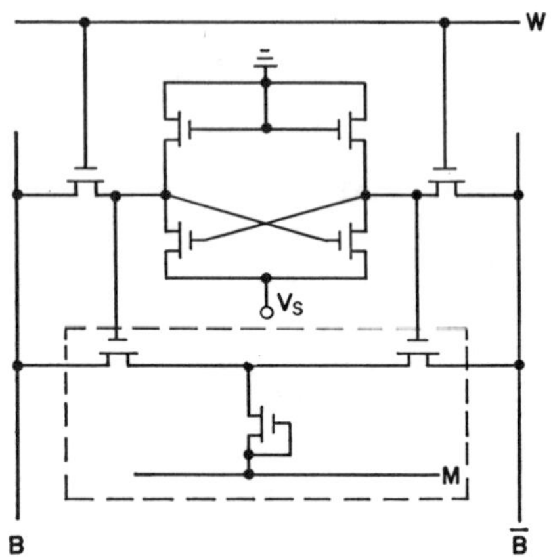

Fig. 2. Proposed IGFET CAM cell.

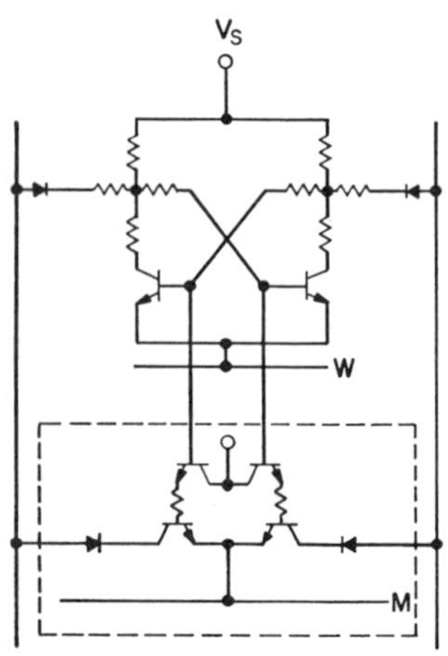

Fig. 3. Proposed bipolar transistor CAM cell.

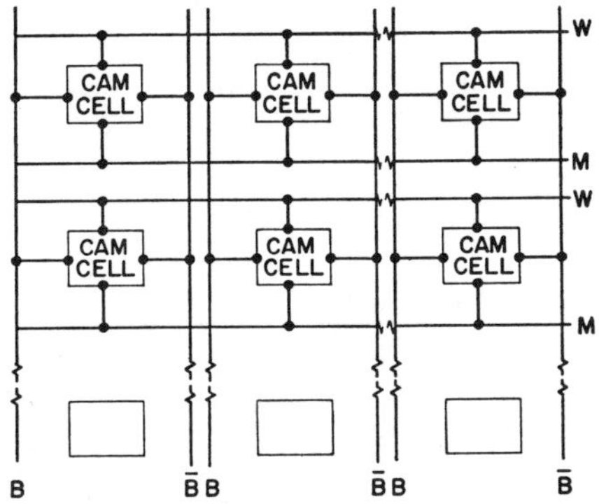

Fig. 4. Memory organization of proposed CAM.

principles. The cells consist of a flip-flop as the storage medium. The two bit lines are connected to the flip-flop via two gates controlled by the word line. A differential sensing amplifier is used to detect the signal output to reduce the common mode noise induced in high-speed operation. The same two bit lines also serve the double binary input of the write and the match logic circuit of the cell. The match line of all the bits in a word is wired together in OR logic. The operation of the complementary cell will be described in detail.

Transistors T_1 and T_2 in Fig. 1 constitute the basic inverter stage. Transistors T_1, T_2, T_3, T_4 form a flip-flop having two stable states with the potentials of nodes 1 and 2 (or Fig. 1) either at V_s or at ground. A binary "1" is stored in the cell when node 1 is at V_s and node 2 at ground potential Binary "0" is stored when node 2 is at V_s and node 1 at ground potential. During the quiescent state, the potential of the word line is kept at V_s. Since the two access transistors T_5 and T_6 are p-channel devices, they are in the OFF state and the memory cell is isolated from the bit lines B and $\bar{B}$. Both bit lines are kept at ground potential during standby and READ cycles. Transistors T_7, T_8, and T_9 constitute the match logic circuit. The gate terminals of transistors T_7 and T_8 are connected to that of transistors T_4 and T_2, respectively. Transistor T_7 will be on if "0" is stored in the cell, and transistor T_8 will be on if "1" is stored in the cell. Transistor T_9 is connected in a diode configuration in a diode-coupled logic circuit. Current flows in the direction from either transistors T_7 and T_8 through transistor T_9 to the match line. The match line M normally remains at ground potential unless there is a mismatch between the information being searched and that stored in the cell.

READ

The information of the cell is retrieved when the cell is accessed by lowering the potential of the word line to ground.

WRITE

The WRITE operation is accomplished by accessing the word line and simultaneously raising the potential of either the B or $\bar{B}$ line to V_s depending on whether a "1" or a "0" is to be written.

Search "1"

During a search "1" operation, the potential of the B line is raised to V_s while that of the $\bar{B}$ line remains at ground. If a "1" is stored in the cell, then transistor T_7 is off and the match line M remains at ground potential indicating a match. Notice that transistor T_8 is also off as all three terminals are at ground potential. Without transistor T_9, transistor T_8 will prevent the potential of the match line from rising higher than V_{tp} if there is a mismatch somewhere in the same word, where V_{tp} is the threshold voltage of the p-channel IGFET. If a "0" is stored in the cell, then transistor T_7 is on and current will flow from the bit line B through transistors T_7 and T_9 to the M line indicating a mismatch.

Search "0"

During a search "0" operation, the potential of the $\bar{B}$ line is raised to V_s while that of the B line remains at ground.

Search "ϕ"

During a search "ϕ" operation, both bit lines B and $\bar{B}$ remain at ground potential. Regardless of the information stored in the memory cell, the match line will remain at ground potential indicating a match.

The delay incurred during SEARCH operation is only the transit time for carriers to cross the channel of the IGFET and the transmission-line delay of the match line.

It is observed that the potentials of the B and $\bar{B}$ lines during a match "1," match "0," and match "ϕ" are exactly that of write "1," write "0," and standby or READ cycle, respectively. Therefore, no additional bit logic circuits are needed to perform the match operation. This feature means 1) low cost, 2) simplicity in memory control circuits and memory operation, and 3) reduction in number of inputs to the memory. Essentially, the SEARCH operation of the memory is carried out by the WRITE circuits.

The match logic section of the circuit consumes no dc power in any phase of the operation, requires no extra control signals, and does not alter the operating algorithms of the cell when used in a RAM.

The values of the transconductances of transistors T_7, T_8, and T_9 are not critical and only affect the current delivered to the match line. This will determine the speed of the SEARCH cycle. By adding another logic section of the memory cell (Fig. 5), simultaneous searches can be carried out [26].

Information can be written into the proposed cells even if the cell is not accessed by raising one of the bit lines to sufficiently high potential. This property means that it is possible to write the same information into the same bit of all the words in the memory. This function is useful in setting or resetting a particular bit of all the words in the memory. Since the potential necessary to write all the words is much higher than the potentials needed in accessed READ and WRITE operations, large operating margin is obtained for the memory cells.

Employing the same circuit for the WRITE and match

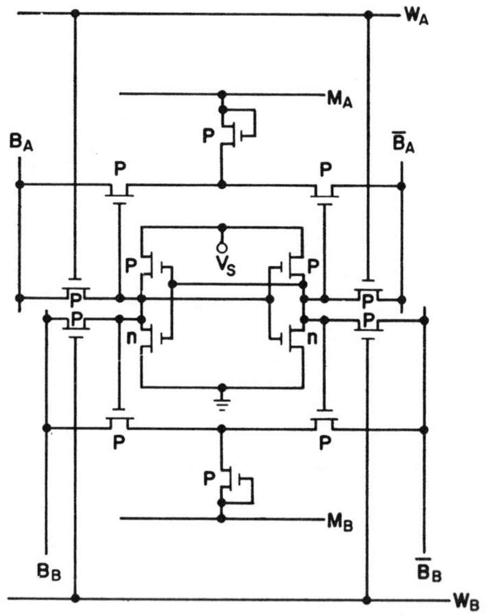

Fig. 5. Simultaneous search CAM cell.

operation further reduces the cost and simplifies the memory construction and operation.

Resolver–Encoder

The function of the resolver is to allow one element of the matching set of words to become accessible during the SEARCH operation in a multiple-match situation. Since the address of the chosen element may contain the desired information, the resolver–encoder proposed here not only resolves the multiple matches, but also provides the address of the chosen element.

Fig. 6 shows a simplified version of the proposed resolver–encoder circuit employing complementary IGFET technology. Circuits performing the same function employing bipolar transistor technology are shown in Fig. 7. Both circuits are shown for the purpose of demonstrating a principle only.

The principle of operation of the CIGFET resolver–encoder will be described here in some detail. The four inputs of the circuit labeled 00, 01, 10, and 11 are normally at V_s potential. Ground potential when applied to any of the inputs will cause the circuit to indicate a match. These terminals are so labeled that when a match signal appears at any one of the four inputs, the output at terminals 1 and 2 is the address label of that input terminal. (Ground potential is "0.") The output at terminal 3 indicates if there is a match signal at any of the four inputs.

If two or more match signals appear at the inputs, then the address of the higher priority input signal will appear at terminals 1 and 2, and a match signal will also appear at terminal 3. The priority of the four input terminals ranging from the highest to the lowest is 11, 10, 01, and 00.

For example, if match signals appear at both the 10 and 01 inputs, then both signal lines, *b* and *c*, tend to rise towards the V_s potential. However, once the diode D_1 conducts and raises the potential of terminal 1, transistor

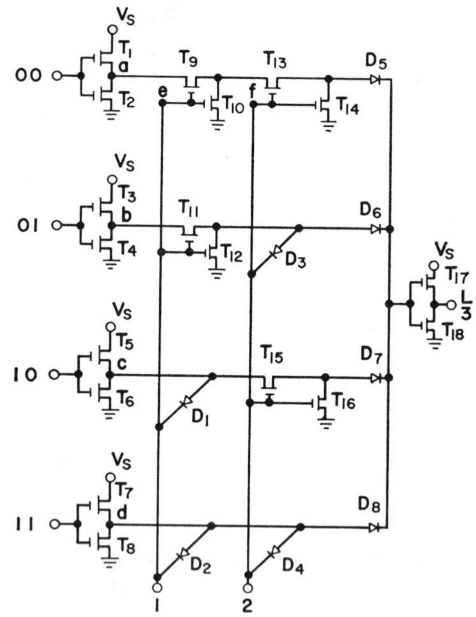

Fig. 6. Proposed CIGFET resolver–encoder.

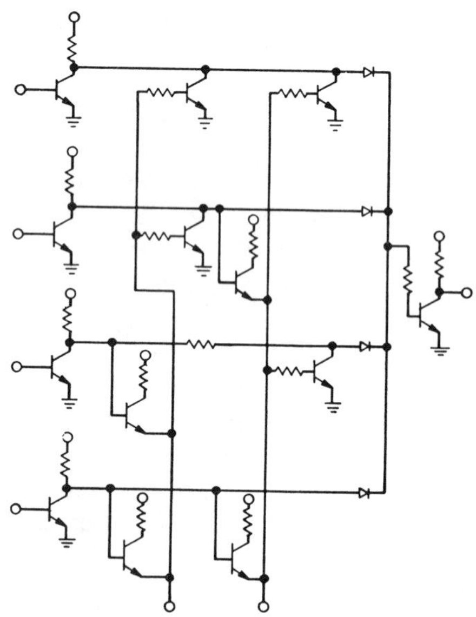

Fig. 7. Proposed bipolar transistor resolver–encoder.

T_{11} will turn off and the address that will appear at terminals 1 and 2 is that of the higher priority input, in this case 10. Since transistor T_{15} is on, there is also a match signal output at terminal 3.

For more than two match signals at the inputs, the address of the highest priority input signal will appear at terminals 1 and 2 with a match signal output at terminal 3. The operation is identical to that of the two-input signal case.

The output at terminal 3 is used either to indicate the presence of at least one match, or to drive the next stage of the resolver–encoder in a cascade for resolving more than four match lines.

To resolve more than four match lines, it is necessary only to use more identical circuits. Fig. 8 shows the manner

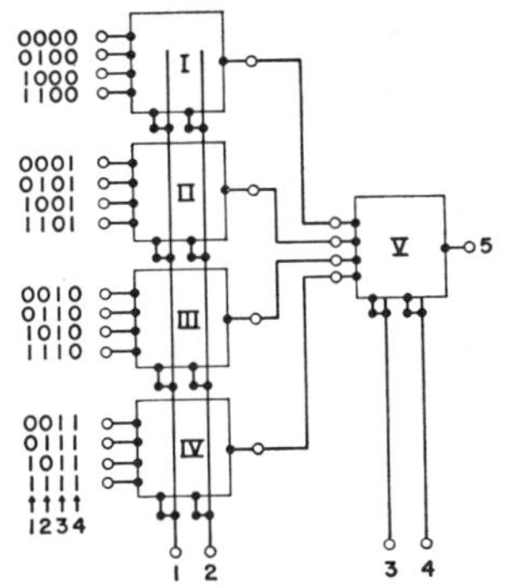

Fig. 8. Method of resolver–encoder interconnection.

in which these circuits would be connected together for a 16 match line case.

The expandability of the proposed resolver–encoder depends on the fact that terminals 1 and 2 can also serve as the inputs to the circuit. If the input to terminals 1 and 2 are 01, then there will be an output at terminal 3 if and only if there is a match signal present at the 01 input. This particular property also enables arbitrary changing of priority and partitioning of the memory as described in the next section.

The hierarchy of priority of the circuit for the 16-input case is the same as that for the original four-input case. The input that has the largest binary label has the highest priority.

Due to the fact that the data lines are also inputs to the resolver, the priority of the resolver can be arbitrarily changed at will by applying the proper signals at the data lines. For example, normally the 1010 input has higher priority than the 0101 input. However, if the 01 is applied to terminals 1 and 2, then the opposite is true. As a matter of fact, only signals from inputs having 01 as the first two binary bits can propagate through the resolver. This feature will allow search of any particular section of the memory at will as described in the next section.

Since the resolver–encoder circuit is modular in nature, it is suitable for integrated-circuit technology.

The operating principles for the single channel IGFET resolver–encoder are identical in nature to that of the CIGFET. A bipolar resolver–encoder employing the same operating principles has been built to be operated compatibly with the bipolar CAM cells and will be described in the experiment.

Operation of Proposed Memory

The various operations of the proposed memory will be described here. Fig. 9 shows the block diagram of a CAM. Circuitries such as the word-decoding circuitry, bit-sensing circuitry, and bit-write circuitry are similar to those use in existing RAM.

In CAM, there is a SEARCH logic circuit. The SEARCH logic circuitry receives the data to be searched (matched), then converts it to the proper form, level, and sequence and generates any control signals required by the memory

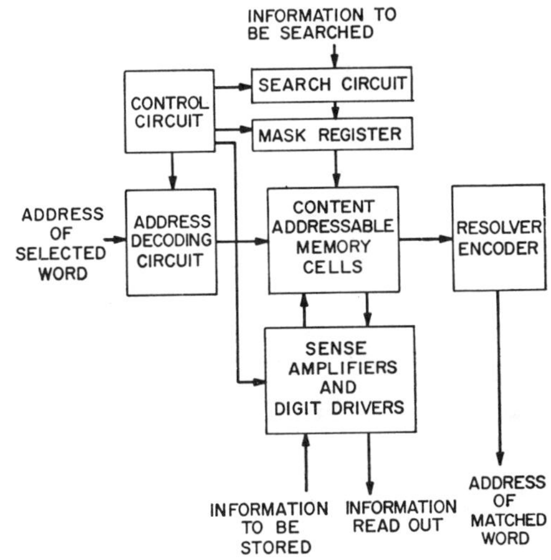

Fig. 9. Block diagram of a general CAM system.

cell to perform the SEARCH or match operation. There is no need for this circuitry in the proposed memory. The same bit lines are used for READ–WRITE and for match, and the bit-write circuit is used for the search circuit.

For those CAMs that are capable of match on partial specification in the associative region a mask register is required. For the CAMs proposed, the DON'T CARE condition required by the memory cell is provided by not activating the bit-write circuit at the desired bit. Therefore, there is no need for special circuitry for the mask register because the bit-write circuit serves this dual function.

The management bits keep track of various operations of the memory. They usually range from one to several bits per word. They can be either of standard or special memory cell designs. For the IC–CAM, standard cells are preferred to reduce the number of IC circuit types needed. One or two standard bits are used for the management purpose in the proposed CAM. Fig. 10 shows the proposed CAM with SEARCH and mask logic functions accomplished by RAM digit circuits.

There are three writing operations for a content-addressable memory. They are 1) location-addressed WRITE, 2) WRITE into any empty location, 3) content-addressed WRITE. The WRITE algorithms for the proposed content-addressable memory are as follows.

A. Location-Addressed WRITE

In this operation information is stored into the memory at a definite location (address).

Command:

1) Apply the address and activate the word decoding circuitry.
2) Apply the information and activate the bit-write circuitry.
3) Reset the word and bit circuitries.

Erasing can be accomplished by writing "0" into every bit of the word.

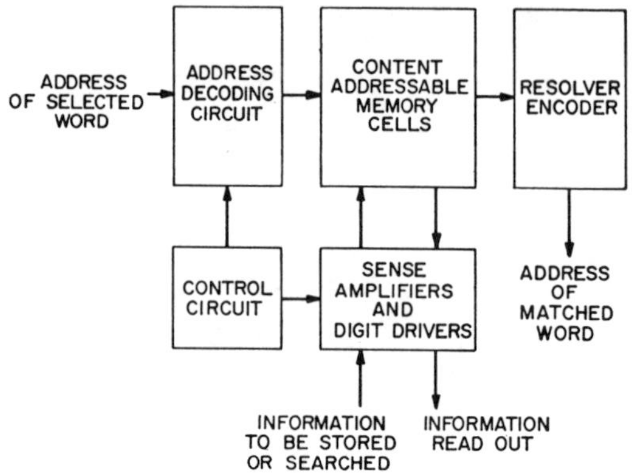

Fig. 10. Block diagram of proposed IC–CAM system.

B. WRITE Empty Locations

Command:

1) Search for an empty location by activating the bit-write circuit for "0" in all bit locations.
2) Follow the algorithm of location-addressed WRITE using address obtained in step 1).

In this algorithm, an empty location is assumed to be a location where all bits of the word are "0". However, if a word containing all "0" bits is a possible information word (i.e., it is desirable to store an all "0" bit word in the memory), then another management bit per word would be required. "1" will be written into this management bit whenever information is written into a word. Then step one of this algorithm would be changed to search "0" for this management bit and "ϕ" for rest of the word.

C. Content-Addressed WRITE

In this operation, desired information is to be stored or changed in words matching the SEARCH key.

Command:

1) Apply the SEARCH key and activate the bit-write circuit. Search for "0" for the management bit, and "ϕ" for that portion of the word that is not part of the SEARCH key.
2) Apply the address obtained and activate the word decoding circuit.
3) Apply the information and activate the bit-write circuit. Write "1" into the management bit.
4) Repeat steps 1–3 until no match is found.
5) Reset all the management bits to "0" by raising the line $\bar{B}$ of that bit to sufficiently high potential.

Information can be retrieved from a content-addressable memory either by location-addressed READ as in a random accessed memory or by content-addressed READ. The operating algorithms for the proposed CAM is as follows.

D. Location-Addressed READ

In this operation, information is retrieved from a definite location of the memory.

Command:

1) Apply the address and activate the word decoding circuitry.
2) Activate the bit sensing amplifiers.
3) Reset the word and bit circuitries.

E. Content-Addressed READ

In this operation all information associated with a searching key is to be retrieved.

Command:

1) Apply the SEARCH key and activate the bit-write circuit. Search "0" for the management bit, "ϕ" for rest of the word, which is not part of the search key.
2) Apply the address obtained and activate the word decoding circuitry.
3) Activate the bit sensing amplifiers and bit-write circuit of the management bit for writing "1" into the management bit.
4) Repeat steps 1–3 until no match is found.
5) Reset all the management bits to "0" by raising the potential of the $\bar{B}$ of the management bit to sufficiently high potential.

All above algorithms are standard operating sequences for a content-addressable memory. The CAM described here differs from previously proposed CAMs in that no external algorithms are required for resolving multiple matches. A matching word can be accessed for every operating cycle of the memory.

Due to the characteristics of the proposed resolver–encoder, the proposed CAM can be operated as 1) a partitioned content-addressable memory, 2) a queue content-addressable memory, and 3) a stack content-addressable memory.

A partitioned content-addressable memory is a memory that allows the SEARCH operation in a section of the memory only. The queue content-addressable memory is a first in, first out (FIFO) memory. The stack memory is a first in, last out (FILO) memory.

Partitioned operation of the proposed CAM system is the result of the property that the address output of the resolver–encoder is also an input. It is shown in the description of the resolver–encoder that the priority of the resolver–encoder can be arbitrarily changed. If a binary number is applied to a subfield of the output, then the resolver–encoder will resolve only those match signals occurring within the set of words specified by the field. For example, if a "1" is applied to the first bit of a 7-bit address output of the resolver–encoder, then the resolver–encoder will respond to match signals coming from only those words ranging from 65 to 128, i.e., those words having "1" in the first bit of their address. Similarly, an application of two bits to the address output of the resolver–encoder will divide the entire memory into four equal sections. For a subfield of N bits applied to the resolver–encoder, the memory is divided into 2^N equal sections. These N bits can be any subfield of the address (e.g., the first, third, and fourth bit of the address). However, the input must have the first bit specified in each group of two bits between the most significant bit positions and the least significant bit of interest. Within the partitioned memory, the original priority is preserved, i.e., the larger the address, the higher the priority.

Partitioned operation of the proposed memory utilizes the flexibility of the resolver–encoder in which the priority can be changed. Queueing and stacking operation of the memory are based on the fact that the priority of the resolver–encoder for a definite set of words has a definite order.

When a sequence of words is written into empty locations employing Algorithm B, the first word is written into the highest priority empty location among all empty locations, and the second word is written into the second highest priority empty location among all the empty locations, and so on. When a search is carried out to retrieve the information, the same group will respond again, and the first word having the highest priority will be retrieved first. When the first word is removed from the matching list by means of the management bit, the second word in the sequence becomes the highest priority word and will be retrieved next, and so on until the original set of words is retrieved in the original sequence. This inherent first in, first out (FIFO) nature of the proposed resolver–encoder is the basis of the queueing operation of the memory.

If two resolvers are connected in the opposite sense with their priority reversed, then the largest address has the highest priority in one resolver–encoder and the lowest priority in the other. Now, if one is used in searching for empty locations for writing and the other for retrieving, the sequence of the original set of words will be reversed. This FILO nature of operation is the basis for a stack memory.

Since the match output is available also from the resolver–encoder, existing algorithms [27] employing this signal to perform sorting, search maximum, minimum, greater than, less than, nearest to, between limits can be used. In some of the algorithms proposed elsewhere, the information of the existence of multiple matches is necessary. Since a double binary system if employed in the proposed CAM, and the bit lines of the same bits of all the words are wired together in OR logic fashion, multiple matches can be detected by measuring the common mode signal of those DON'T CARE bit locations. While a 10 or 01 signal at the B and $\bar{B}$ lines indicates "1" and "0", respectively, a 11 signal would indicate multiple matches. It is further noticed that the DON'T CARE condition is exactly the condition required by the sensing amplifiers to detect the signals. Therefore, multiple matches exist if and only if the 11 signal appears at one of the DON'T CARE bit locations. (Notice that multiple matches caused by two or more identical words would not be detected as multiple matches. But this is a trivial case since identical words in a CAM contain no more information than a single word.)

TABLE II
RANDOM ACCESS AND CONTENT-ADDRESSABLE MEMORY COMPARISONS

256 watts × 16 bit/W, Word Organized	Random Access	Content-Addressable
Cell complexity	6 transistors/bit	9 transistors/bit
Number of chips per plane	9	105
READ or WRITE cycle time	105 ns	105 ns
SEARCH cycle time	20 μs	0.4 μs
Power per bit	0.75 mW	0.75–0.88 mW/bit
Relative cost per bit	1	1.25

EXPERIMENT AND CONCLUSION

An 80-bit bipolar-transistor content-addressable memory system has been constructed to show the feasibility and to verify the proposed algorithms. The memory is organized into 16 words with 5 bits per word (four information bits and one management bit). Due to cost, availability, and other practical reasons, discrete devices and printed circuit boards are used. The modularity of the design is demonstrated by the fact that only two types of printed circuit boards are used to construct the experimental CAM system. One circuit board is a bipolar memory cell board, and the other is a bipolar resolver–encoder circuit board. Both circuits employ components of values that can be easily fabricated by integrated-circuit technology.

The model CAM system operates as designed and verifies all the algorithms described in the previous chapter. The successful operation of the model CAM system also shows the feasibility of an IC–CAM system.

Table II compares the relative cost performance of a IGFET CAM system proposed here with a IGFET RAM system. The SEARCH time described here is the time required to retrieve one matching word from the memory. For the RAM, the time required is, on the average, equal to 128 READ cycle time.

REFERENCES

[1] W. L. McDermid and H. E. Peterson, "A magnetic associative memory," *IBM J. Res. Develop.*, vol. 5, pp. 59–63, January 1961.
[2] A. G. Hanlon, "Content-addressable and associative memory systems," *IEEE Trans. Electron. Comput.*, vol. EC-15, pp. 509–521, August 1966.
[3] G. Estrin and R. H. Fuller, "Some applications for content-addressable memories," *Fall Joint Computer Conf., AFIPS Proc.*, vol. 24. Baltimore, Md.: Spartan, 1963, pp. 495–508.
[4] A. E. Slade and H. O. McMahon, "A cryotron catalog memory system," *1956 Proc. Eastern Joint Computer Conf.*, vol. 10, pp. 115–120.
[5] H. G. Kadereit and K. Goser, "A continuous film memory cell for superconductive associative memories," *Proc. IEEE*, vol. 56, pp. 121–122, January 1968.
[6] A. R. Sass and E. M. Nagle, "Flux shuttling—New method for the nondestructive readout of superconductive memory cells," *IEEE Trans. Magn.*, vol. MAG-3, pp. 268–271, September 1967.
[7] C. C. Yang and J. T. Tou, "Cryogenic associative memory," *J. Franklin Inst.*, vol. 284, pp. 109–121, August 1967.

[8] R. W. Ahrons, "Superconductive associative memories," *RCA Rev.*, vol. 24, pp. 325–354, September 1963.

[9] J. Kiseda, H. Peterson, W. Seelback, and M. Teig, "A magnetic associative memory," *IBM J. Res. Develop.*, vol. 5, pp. 106–121, April 1961.

[10] D. O. Smith and K. J. Harte, "Content-addressed memory using magneto- or electro-optical interrogation," *IEEE Trans. Electron. Comput.* (Correspondence), vol. EC-15, p. 123, February 1966.

[11] M. H. Lewin, H. R. Beelitz, and J. A. Rajchman, "Fixed associative memory using evaporated organic diode arrays," *Fall Joint Computer Conf., AFIPS Proc.*, vol. 24. Baltimore, Md.: Spartan, 1963, pp. 101–106.

[12] G. G. Pick, "A semipermanent memory utilizing correlation addressing," *Fall Joint Computer Conf., AFIPS Proc.* Washington, D. C.: Spartan, 1964, pp. 107–121.

[13] M. H. Lewin, H. R. Beelitz, and J. Guarracini, "Fixed resistor-card memory," *IEEE Trans. Electron. Comput.*, vol. EC-14, pp. 428–434, June 1965.

[14] D. Gabor, "Associative holographic memories," *IBM J. Res. Develop.*, vol. 13, pp. 156–159, March 1969.

[15] E. S. Lee, "Associative techniques with complementing flip-flops," *Spring Joint Computer Conf., AFIPS Proc.* Baltimore, Md.: Spartan, 1963, pp. 381–394.

[16] R. F. Herlein and A. V. Thompson, "An integrated associative memory element," *ISSCC Digest Tech. Papers*, pp. 42–43, February 1969.

[17] R. Igarashi and T. E. Yaita, "An integrated MOS transistor associative memory system with 11 ns 100 ns cycle time," *Spring Joint Computer Conf., AFIPS Proc.*, vol. 30. Washington, D. C.: Thompson, 1967, pp. 499–506.

[18] E. H. Frei and J. Goldberg, "A method for resolving multiple responses in a parallel search file," *IRE Trans. Electron. Comput.*, vol. EC-10, pp. 718–722, December 1961.

[19] R. R. Seeber and A. B. Lindquist, "Associative memory with ordered retrieval," *IBM J. Res. Develop.*, vol. 6, pp. 126–136, January 1962.

[20] L. Johnson and M. McAndrew, "On ordered retrieve from an associative memory," *IBM J. Res. Develop.*, vol. 8, pp. 189–193, April 1964.

[21] M. H. Lewin, "Retrieval of ordered lists from a content-addressed memory," *RCA Rev.*, vol. 23, pp. 215–229, June 1962.

[22] H. S. Miller, "Resolving multiple responses in an associative memory," *IEEE Trans. Electron. Comput.* (Short Notes), vol. EC-13, pp. 614–616, October 1964.

[23] H. Weinstein, "Proposals for ordered sequential detection of simultaneous multiple responses," *IEEE Trans. Electron. Comput.* (Correspondence), vol. EC-12, pp. 564–567, October 1963.

[24] C. C. Foster, "Determination of priority in associative memories," *IEEE Trans. Comput.* (Short Notes), vol. EC-17, pp. 788–789, August 1968.

[25] W. Hilberg, "Simultaneous multiple response in associative memories and readout of the detector matrix," *IEEE Trans. Electron. Comput.* (Correspondence), vol. EC-15, pp. 117–118, February 1966.

[26] N. K. Natarajan and P. A. V. Thomas, "A multiaccess associative memory," *IEEE Trans. Comput.*, vol. C-18, pp. 424–428, May 1969.

[27] A. D. Falkoff, "Algorithms for parallel-search memories," *J. Ass. Comput. Mech.*, vol. 9, pp. 488–511, October 1962.

Bipolar LSI Memory 3104

INTEL CORP. 3065 Bowers Avenue, Santa Clara, California 95051 • (408) 246-7501

OCTOBER 1971

HIGH SPEED 16 BIT CONTENT ADDRESSABLE MEMORY

- **Organization – 4 Words x 4 Bits**
- **Max. Delay of 30 nsec Over 0° C to 75° C Temperature**
- **Open Collector Outputs – OR Tie Capability**
- **High Sinking Current Capability – 15 mA max.**
- **Low Input Load Current – 0.25 mA max.**
- **DTL & TTL Compatible**
- **Bit Enable Input – Bit Masking**
- **Standard 24 Pin Dual In-Line**

The Intel 3104 is a high speed 16 bit Content Addressable Memory (CAM). It is a linear select 4 word by 4 bit array which is designed to compare data on its inputs with data already stored in its memory and to indicate a match when these data are identical. This equality search is performed on all bits in parallel. The 3104 can also be used as a read/write RAM with linear selection addressing.

LOGIC SYMBOL

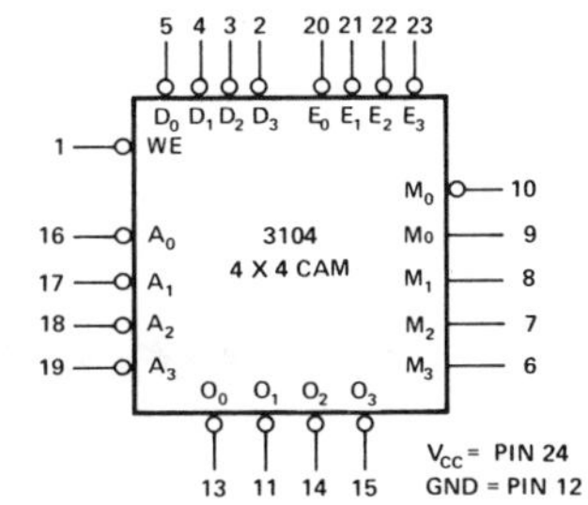

PACKAGE OUTLINE

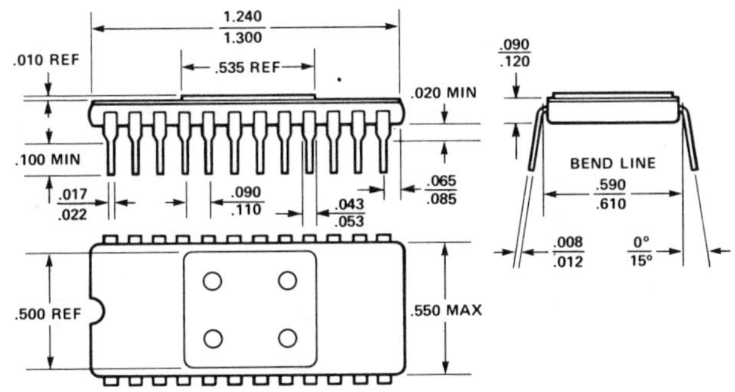

PIN CONFIGURATION

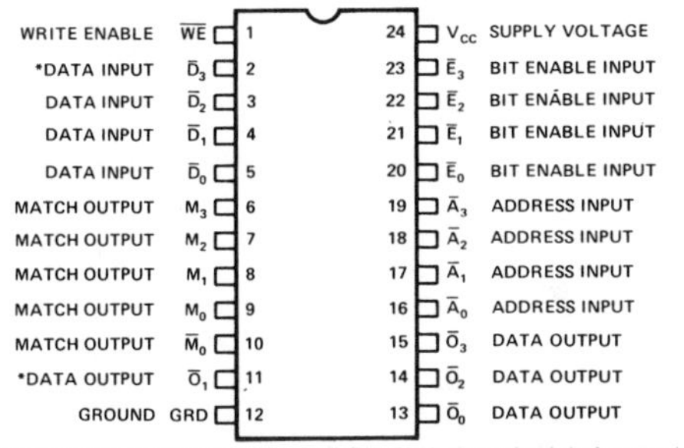

*DATA IN and DATA OUT are of the same logic levels. For a chip that is not selected, the data output is at a high level.

LOGIC DIAGRAM

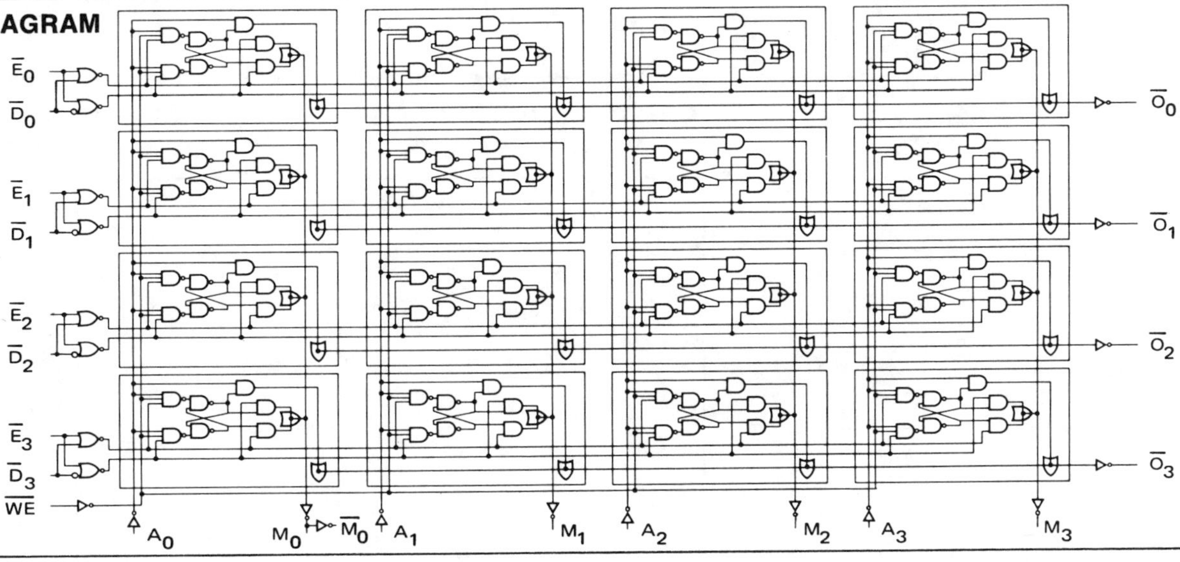

Absolute Maximum Ratings*

Temperature Under Bias	−65°C to +125°C
Storage Temperature	−65°C to +160°C
All Output or Supply Voltages	−0.5 to +7 Volts
All Input Voltages	−0.5 to +5.5 Volts
Output Currents	100 mA

*COMMENT:

Stresses above those listed under "Absolute Maximum Rating" may cause permanent damage to the device. This is a stress rating only and functional operation of the device at these or at any other condition above those indicated in the operational sections of this specification is not implied. Exposure to absolute maximum rating conditions for extended periods may affect device reliability.

D.C. Characteristics T_A = 0°C to +75°C, V_{CC} = 5.0V ±5%; unless otherwise specified.

SYMBOL	PARAMETER	LIMIT MIN.	LIMIT TYP.	LIMIT MAX.	UNIT	TEST CONDITIONS
I_{FA}	ADDRESS INPUT LOAD CURRENT			−0.25	mA	V_{CC} = 5.25V V_A = .45V
I_{FE}	BIT ENABLE INPUT LOAD CURRENT			−0.25	mA	V_{CC} = 5.25V V_E = .45V
I_{FW}	WRITE ENABLE INPUT LOAD CURRENT			−0.25	mA	V_{CC} = 5.25V V_W = .45V
I_{FD}	DATA INPUT LOAD CURRENT			−0.25	mA	V_{CC} = 5.25V V_D = .45V
I_{RA}	ADDRESS INPUT LEAKAGE CURRENT			10	μA	V_{CC} = 5.25V V_A = 5.25V
I_{RE}	BIT ENABLE INPUT LEAKAGE CURRENT			10	μA	V_{CC} = 5.25V V_E = 5.25V
I_{RW}	WRITE ENABLE INPUT LEAKAGE CURRENT			10	μA	V_{CC} = 5.25V V_W = 5.25V
I_{RD}	DATA INPUT LEAKAGE CURRENT			10	μA	V_{CC} = 5.25V V_D = 5.25V
I_{CEX}	OUTPUT LEAKAGE CURRENT (ALL OUTPUTS)			50	μA	V_{CC} = 5.25V V_{CEX} = 5.25V
V_{OL}	OUTPUT "LOW" VOLTAGE (ALL OUTPUTS)			0.45	V	V_{CC} = 4.75V I_{OL} = 15mA (Three other Outputs Sink 15mA each)
V_{IL}	INPUT "LOW" VOLTAGE (ALL INPUTS)			0.85	V	V_{CC} = 5V
V_{IH}	INPUT "HIGH" VOLTAGE (ALL INPUTS)	2.0			V	V_{CC} = 5V
I_{CC}	POWER SUPPLY CURRENT			125	mA	V_{CC} = 5.25V
C_{IN}**	INPUT CAPACITANCE		5		pF	V_{IN} = +2.0V, V_{CC} = 0.0V f = 1 MHz
C_{OUT}**	OUTPUT CAPACITANCE		8		pF	V_{OUT} = +2.0V, V_{CC} = 0.0V f = 1 MHz

**This parameter is periodically sampled and is not 100% tested.

Typical D.C. Characteristics

INPUT CURRENT VS. INPUT VOLTAGE

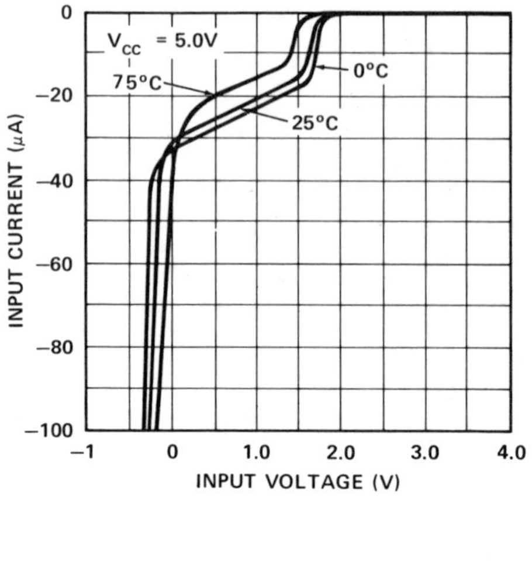

INPUT THRESHOLD VOLTAGE VS. TEMPERATURE

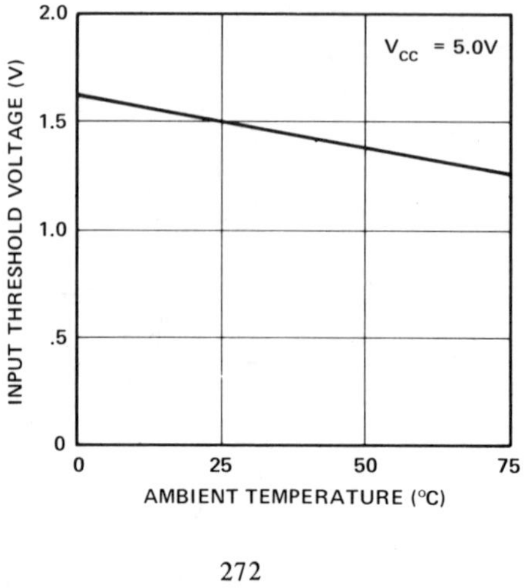

OUTPUT CURRENT VS. OUTPUT "LOW" VOLTAGE

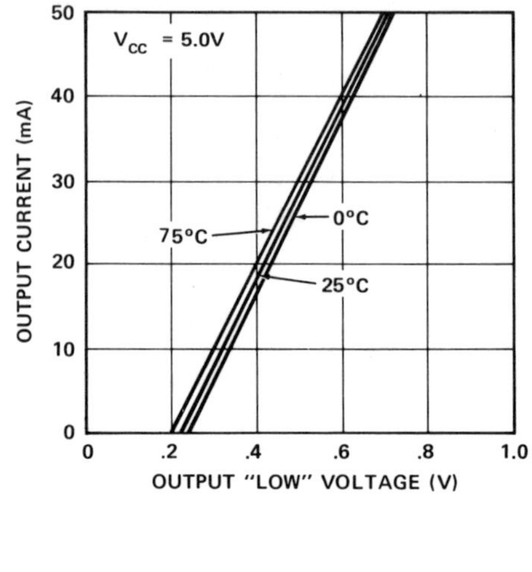

Switching Characteristics

Conditions of Test:

Input Pulse amplitudes · · 2.5V

Input pulse rise and fall times of 5 nanoseconds between 1 volt and 2 volts

Speed measurements are made at 1.5 volt levels

Output loading is 15 mA and 30 pF

15mA Test Load

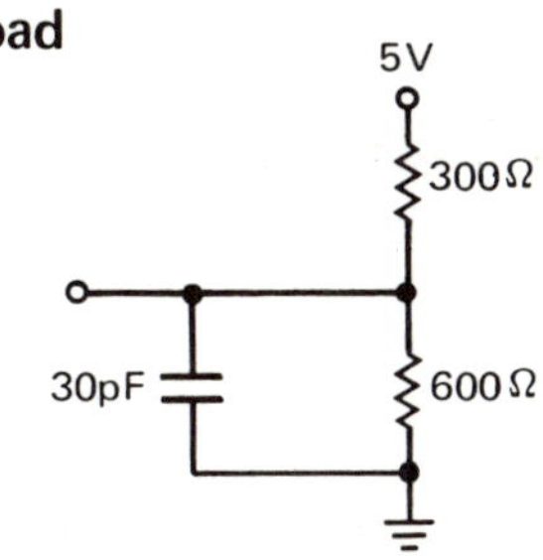

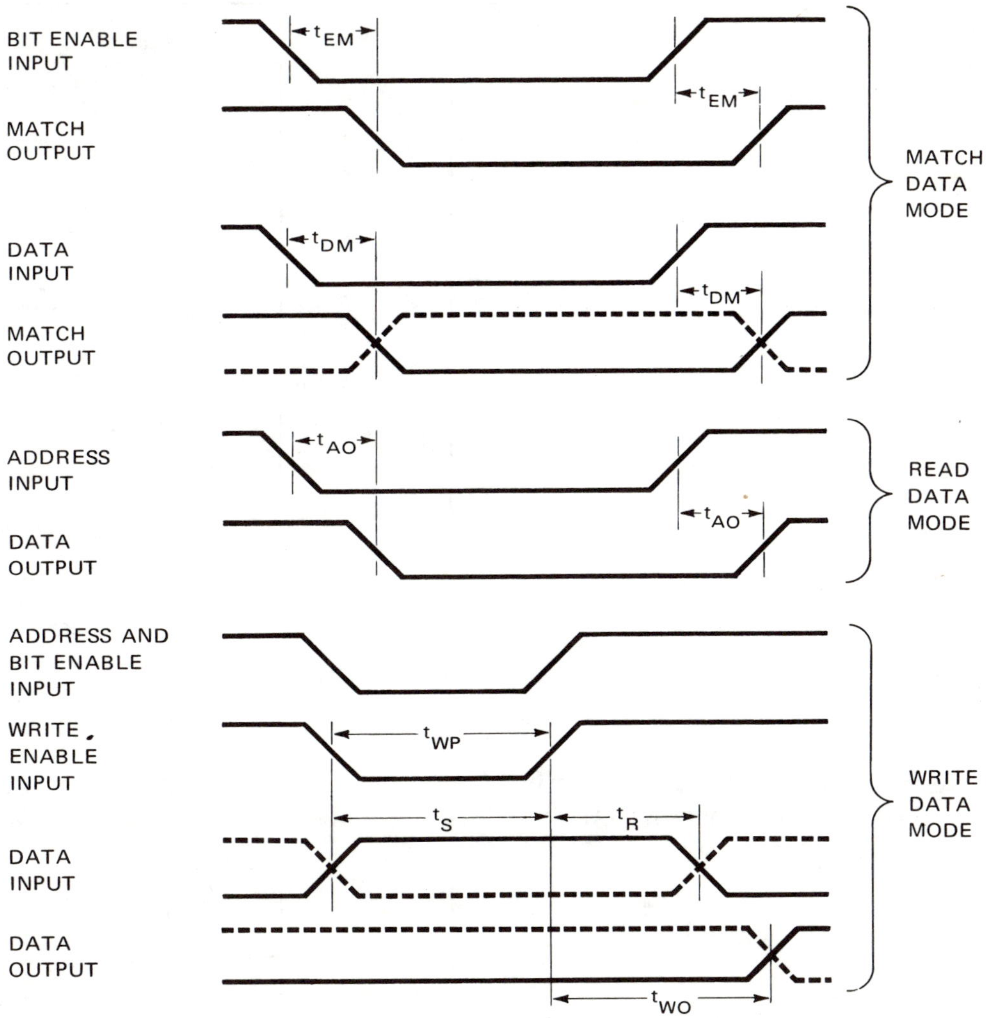

A.C. Characteristics T_A = 0°C to +75°C, V_{CC} = 5.0V ±5%; unless otherwise specified.

SYMBOL	PARAMETER	LIMITS			UNIT
		MIN.	TYP.[1]	MAX.	
t_{EM}	BIT ENABLE INPUT TO MATCH OUTPUT DELAY		15	30	ns
t_{DM}	DATA INPUT TO MATCH OUTPUT DELAY		16	30	ns
t_{AO}	ADDRESS INPUT TO OUTPUT DELAY		14	30	ns
t_{WP}	WRITE ENABLE PULSE WIDTH	40	25		ns
t_{WO}	WRITE ENABLE TO OUTPUT DELAY		–	40	ns
t_S	SET-UP TIME ON DATA INPUT		–	40	ns
t_R	RELEASE TIME ON DATA INPUT	0	–		ns

Note 1. Typical values are at nominal voltages and T_A = 25°C.

Typical A.C. Characteristics

BIT ENABLE INPUT TO MATCH OUTPUT DELAY VS. TEMPERATURE

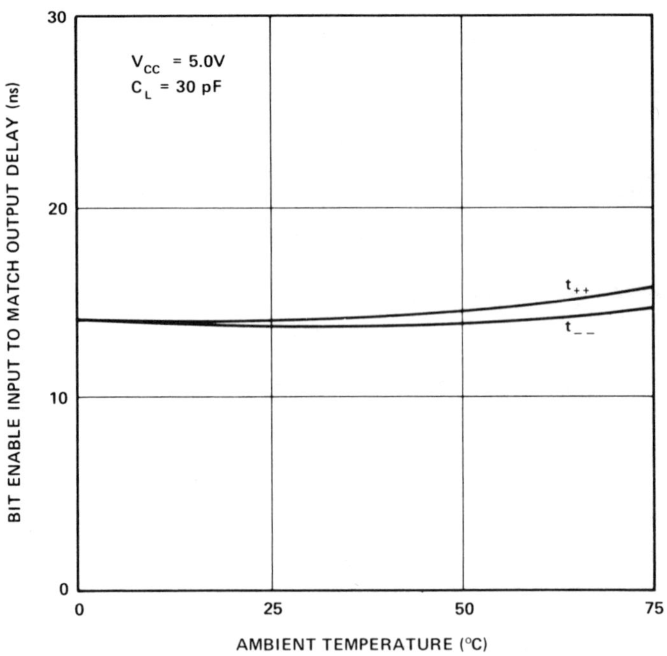

DATA INPUT TO MATCH OUTPUT DELAY VS. TEMPERATURE

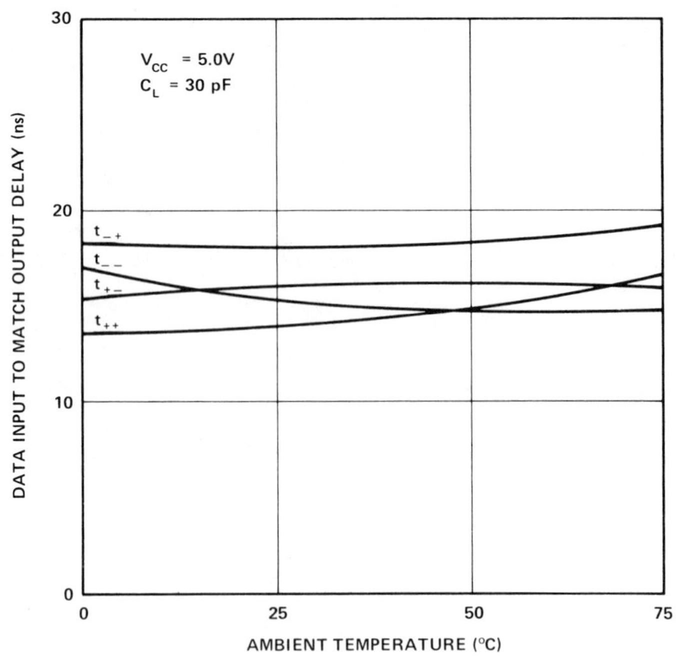

ADDRESS INPUT TO DATA OUTPUT DELAY VS. TEMPERATURE

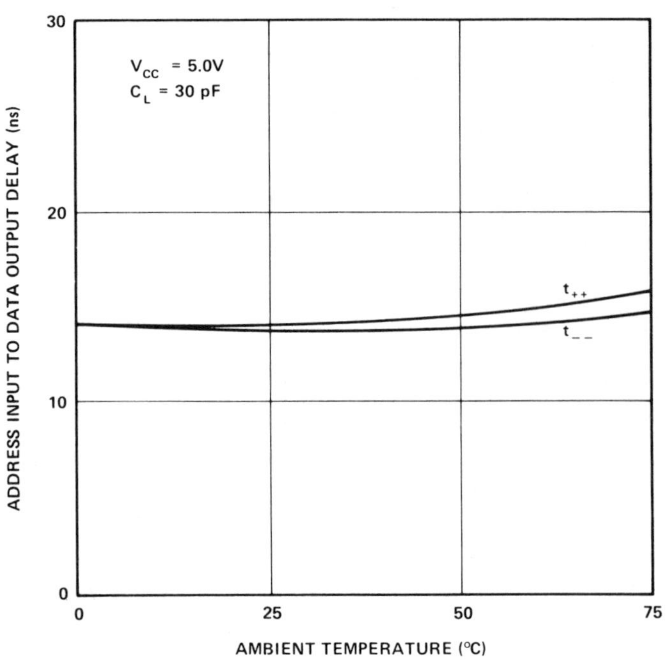

WRITE ENABLE PULSE WIDTH VS. TEMPERATURE

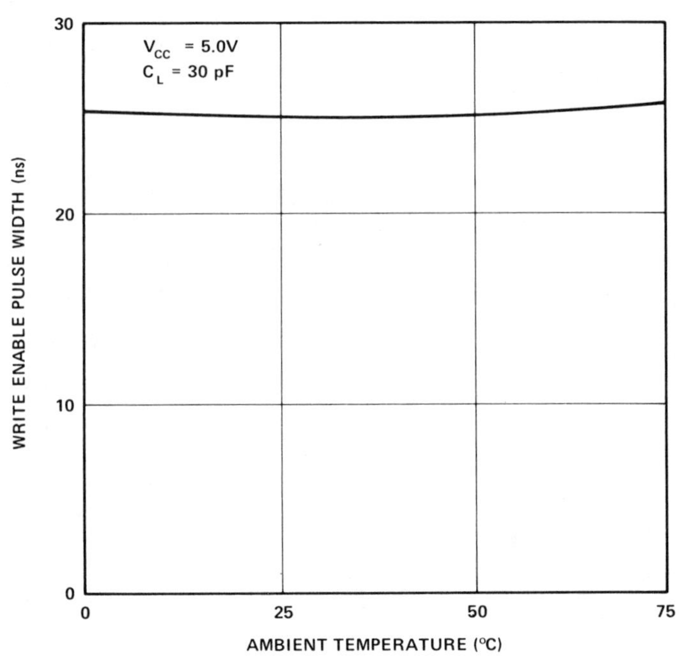

Circuit Schematic

3104 CAM

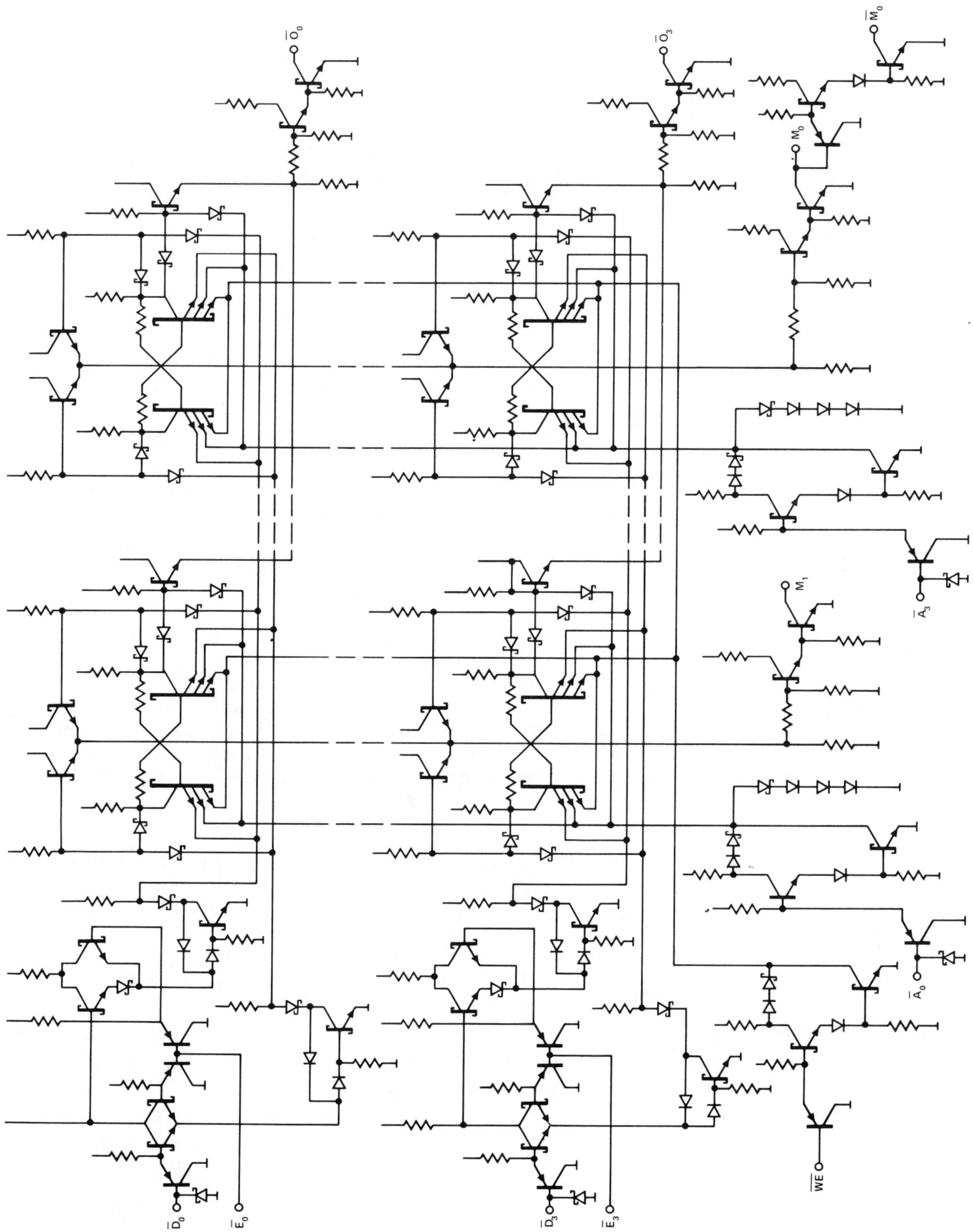

Application Information

BASIC OPERATION

Matching

The key feature of the 3104 is its ability to indicate the presence of a match at the M_0-M_3 outputs if the inputs on $\overline{D}_0-\overline{D}_3$ match the data already stored. When an equality search is being performed on all four bits in the word, all the bit enable inputs ($\overline{E}_0-\overline{E}_3$) are held low. Under this condition the match outputs (M_0-M_3) associated with all the matched words go high. For example:

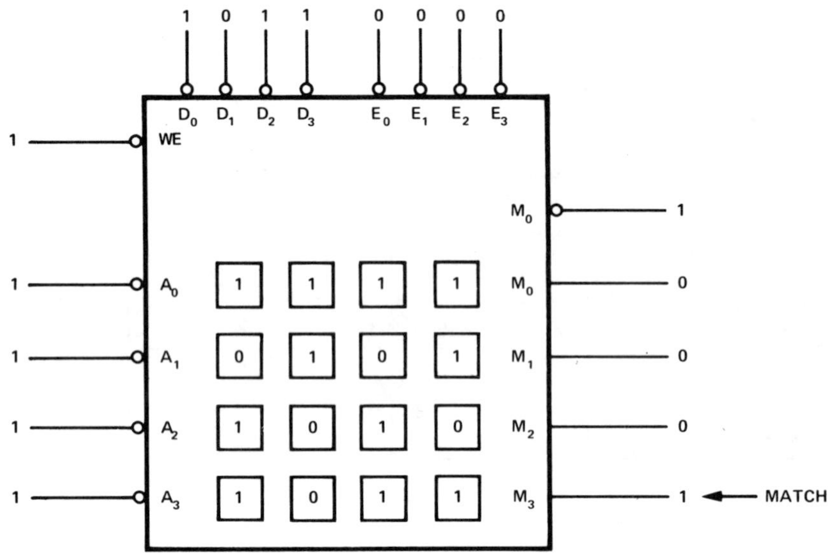

The match is indicated at M_3. This is because the content of the 4th word is identical to that of the four input lines.

Bit Masking

There are cases where searching is to be performed on only a part of the stored word. In this case, the bit masking feature should be used. Since the comparison only works for those bits whose corresponding bit enable line (E) is held low, a bit can be masked if its enable line is held high. This means that all the bits whose enable lines are held high are ignored during the matching process. In the example above, if $\overline{E}_1$ had been held high, M_0 would have indicated a match also, because the "1" in bit position two would have been ignored and the first word would effectively have matched the data input. The bit masking also works when new data is being written into the memory. Only the bits with a low on the bit enable line will be altered in a write process; the others remain unchanged.

Writing Information Into Memory

The 3104 has a linear select address and with the use of the bit enable lines ($\overline{E}_0-\overline{E}_3$) allows the user to write data in any one of the sixteen bit positions. The conditions for writing information are:

(a) Write Enable ($\overline{WE}$) low
(b) The appropriate address line low ($\overline{A}_0 \overline{A}_1 \overline{A}_2$ or $\overline{A}_3$)
(c) The appropriate bit enable line low ($\overline{E}_0 \overline{E}_1 \overline{E}_2$ or $\overline{E}_3$)

Actually data can be written in any column, row, or specific location by lowering the write enable ($\overline{WE}$) and the appropriate address and bit enable lines.

Nondestructive Readout of Information

For nondestructive readout, $\overline{WE}$ should be held high, and the required address should be low. The data will appear at the data output ($\overline{O}_0-\overline{O}_3$) and can be read during an equality search or a read operation.

If a single word is addressed that word will appear at the data outputs. If multiple addressing is used, then the word appearing at the data output is the AND (positive logic) or the OR (negative logic) of the addressed words.

Open Collector Outputs

All outputs are uncommitted collectors which allows for OR-Tie capability on all the outputs when expanding the memory. Whether the outputs of the 3104 are tied together or not, an external resistor must be used to attain a high on the output.

Application Information

Bit Expansion

Expansion of the number of bits in a word is accomplished by OR-tying the corresponding match outputs of several 3104 memories. In this case a non-match in any single 3104 (low condition on match output) will force the common match line to ground, indicating a non-match for the complete word. The common match line will remain high only if all the 3104's connected to the line contain a match.

It is possible to provide a single match output when expanding the number of bits in a word, by making use of the $\overline{M}_0$ output as shown in the configuration below. Since the $\overline{M}_0$ output is just an inverter of the M_0 output, it is also affected when you wire-OR the M_0 output. If the words are stored in staggered locations so that part of each word appears in address A_0, and all the $\overline{M}_0$ outputs are connected together, then the $\overline{M}_0$ output becomes a "no match" signal. This means that anytime the "no match" pin is low, a match is present on at least one of the four match outputs (M_0,M_1,M_2,M_3). If the "no match" pin is high, this indicates that there is no match present at any of the four match outputs.

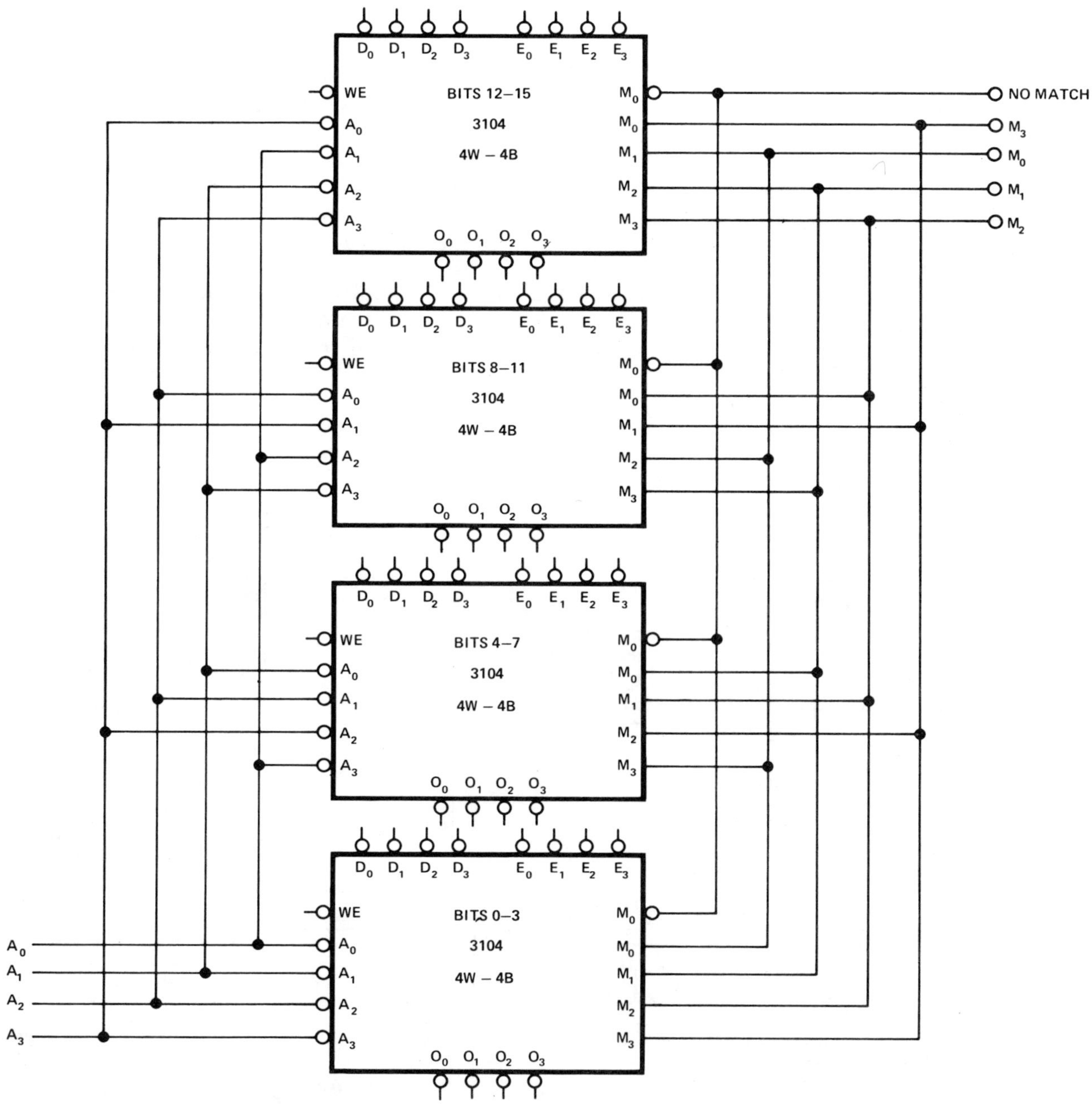

Word Expansion

The 3104 can be expanded to make a system of many words and many bits. This is done by connecting all the data input and data output lines in parallel for corresponding bits in all memories, and connecting the address and match lines for corresponding words in all 3104 devices.

Application Information

SYSTEM APPLICATIONS

The 3104 CAM is particularly useful in applications where continuous searching is required at high speeds. In implementing cache-type buffer memories, page tables, and other types of memories requiring a table lookup for address relocation, the 3104 can be used for both the search (3104 used as a CAM) and data (3104 used as a linear select, read/write RAM) portions of the memory. CAMS can also be used in pattern recognition systems where masked searching is important as well as in systems where a code conversion of sparse codes is useful.

Application of CAM in Table Lookup

In the configuration shown in Fig. 1 the 3104 is used in a table lookup application where the CAMS on the left side (8-4 bit words) are used for searching for a matching word, and the CAMS on the right side are used to read or write data in a corresponding word.

Loading the Memory

Open collector gates (address select inputs) are used to select a single search word and its corresponding data word simultaneously. The match line is inhibited so that the match outputs of the CAM do not affect the address line. The "write search enable" and the "write data enable" lines are strobed while data is present on the "search DATA IN" and "DATA IN" lines. The information in these lines are then written into the word that has been addressed in the memory. This process is repeated, selecting a different word, until all the search data and data have been loaded into 8 word positions in each memory.

Table Lookup Matching

With the write enable lines high, the data to be searched for is applied to the "search DATA IN" lines. This input is simultaneously compared with all 8 words stored in the search CAM. If a match is detected, and the Match Enable line is high, the corresponding address line is selected in the data CAM which reads out the data word on the "DATA OUT" lines.

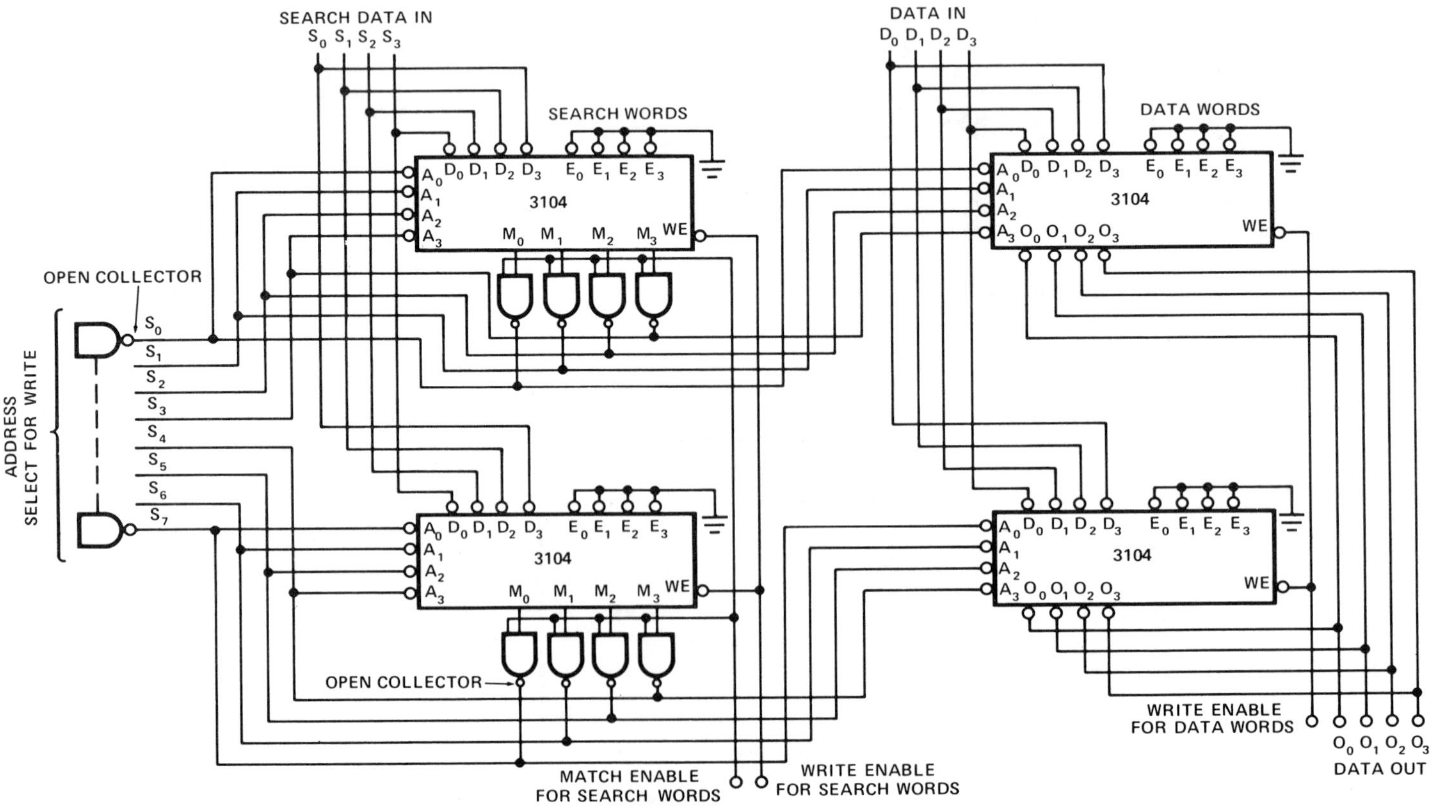

Fig. 1

Part VIII
Editor's Conclusions and Projections

Semiconductor memories are going to dominate the market for computer main memory. But it may take longer to reach this situation than some projections indicate.

The more optimistic projections are based on the forecast performance and cost advantages of semiconductor memory. These are illustrated in Fig. 8-1. The 1975 data for core memory systems are taken from a paper by an expert proponent of this technology [1]. The forecasts for semiconductor memory systems are this author's. There is unlikely to be any argument concerning ranges of memory system cycle time shown, save that those for semiconductor memory may be pessimistic. Concerning factory cost, the cost model of IV,3 applied to the memory designs of III,8 and VI,1 shows that very little technical progress is needed for these or closely similar designs to reach the cost levels of Fig. 8-1 for dynamic memory and flip-flop memory, respectively.

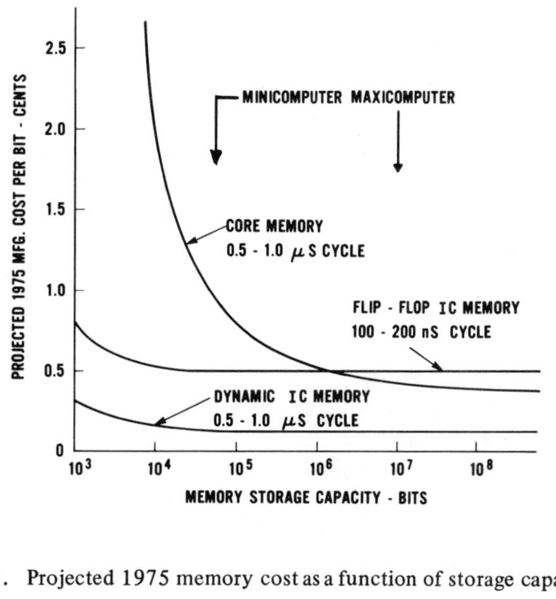

Fig. 8-1. Projected 1975 memory cost as a function of storage capacity.

TABLE 8–A

POSITIVE FACTORS	NEGATIVE FACTORS
1. MANY SUPPLIERS	1. LARGE DEMAND
2. SECOND SOURCES FOR POPULAR PRODUCTS	2. SHORTAGE OF SKILLED DESIGNERS
3. ONE TECHNOLOGY FOR WHOLE MEMORY	3. NOT IN-HOUSE AT MANY COMPUTER MANUFACTURERS
4. PROVEN IC MANUFACTURING CAPABILITY	

Table 8-A. Factors affecting availability of semiconductor memory.

TABLE 8–B

PERFORMANCE:	BETTER THAN MAGNETICS - TRANSLATES DIRECTLY TO BETTER SYSTEM PERFORMANCE
COST:	LESS THAN MAGNETICS AT 64 KBITS/MODULE
RELIABILITY:	1-10 YEARS MTBF; MODULAR FOR IMPROVED MAINTAINABILITY
AVAILABILITY:	MANY OUTSIDE SUPPLIERS
OTHER:	SHORT SYSTEM DESIGN CYCLE NO "MAKE OR CUY" PROBLEMS

Table 8-B. Semiconductor memory for minicomputers? (Typically 64 kilobits per memory module.)

TABLE 8–C

PERFORMANCE:	MAGNETICS CAN GIVE COMPETITIVE SYSTEM PERFORMANCE USING BUFFERS AND/OR INTERLEAVED ACCESSES.
COST:	IN-HOUSE MAGNETICS VS. PURCHASED SEMICONDUCTORS.
RELIABILITY:	UNPROVEN FOR 10^7 BIT MEMORIES; ERROR CORRECTION HIGHLY EFFECTIVE.
AVAILABILITY:	NOT IN-HOUSE (EXCEPT IBM); REQUIREMENTS CHALLENGE TOTAL U.S. IC MANUFACTURING CAPACITY.
OTHER:	LONG SYSTEM DESIGN CYCLE.

Table 8-C. Semiconductor memory for large computers? (Typically 10^6–10^7 bits per memory module.)

The cost and availability of semiconductor memory is affected by the fact that most computer manufacturers do not have in-house semiconductor memory manufacturing capability. On the other hand, many do have in-house magnetic memory manufacturing. Such a computer system manufacturer, in comparing memory costs, must compare magnetic memory from his own factory with purchased semiconductor memory. If the semiconductor manufacturer is to survive, his selling price must be two to three times his factory costs. Thus, the cost advantage of semiconductor memory may be lost to the system manufacturer due to these considerations.

Factors affecting the availability of semiconductor memory are shown in Table 8-A. The traditional patterns in the semiconductor industry, with many independent suppliers and multiple sources for popular products, are a distinctly favorable factor. Another positive factor is the fact that only one basic technology (silicon integrated circuit) of proven manufacturability is required. Most other approaches to electronic memory require mingling of two sophisticated technologies, e.g., magnetics and semiconductors. With a single technology, there are perhaps fewer opportunities for delays in production.

On the negative side, availability of semiconductor memory is inhibited by the large demand for successful designs, rapid evolution of new designs, and a shortage of highly skilled semiconductor memory designers. Finally, there is the "not in-house" factor mentioned earlier.

Tables 8-B and 8-C provide a brief summary of the factors bearing on system manufacturers' evaluation of semiconductor memory for use in their products. An arbitrary binary distinction is made between minicomputers and maxicomputers; a more precise assessment would take into account the continuous spectrum of computer memory applications.

First consider small computer systems; these usually incorporate a single main memory directly accessed by the central processing unit (CPU). Thus, any improvement in access and cycle time provided by semiconductor memory can lead directly to improved system performance. Referring to Fig. 8-1, we see that semiconductor memory is likely to be distinctly cheaper per bit than core memory at a module size of 64 kbit. (Most small computers offer memory units in approximately 64-kbit increments.)

Concerning reliability, Fig. 5-1 and the discussion in the Editor's Comments of Part V indicate that a memory system mean time between failures (MTBF) of 1–10 years could be expected now for a typical 10^5-bit semiconductor memory. This is adequate for many applications. The modular nature of semiconductor memory is an advantage as far as maintainability is concerned.

Factors relating to availability of semiconductor memory have been discussed earlier. A major consideration for minicomputer manufacturers is that most of these firms do not have in-house memory manufacturing capability in any technology. Thus, the "make or buy" problem raised earlier does not apply to most minicomputer manufacturers. Finally, since the computer system design cycle is about two years for minicomputers, semiconductor memories might be incorporated into production computers relatively quickly.

The above observations suggest that semiconductor memory is likely to be incorporated quite quickly into minicomputers. Projecting a 1975 market for 2×10^4 minicomputers, each with 10^5 bits of semiconductor memory, we find a market for 2×10^9 bits of semiconductor memory. At 0.5¢ per bit from semiconductor manufacturer to minicomputer manufacturer, total sales come to $10 million in 1975. This is barely sufficient to support one semiconductor memory manufacturer, and is negligible compared to 1971 semiconductor industry sales of $1.3 billion. Clearly, if semiconductor memory is to be a viable technology, other large markets must be conquered.

The obvious major market is maxicomputer main memory. The total 1972 worldwide sales of random-access memory are in the range of 30×10^9 bits, mostly for medium and large computer systems. Table 8-C summarizes the considerations relevant to possible adoption of semiconductor memory for medium and large computers. Concerning performance, magnetic memories can give system-level data transfer rates equivalent to those available from semiconductor memory by using a relatively small high-speed buffer (cache) memory and/or interleaved access to multiple magnetic memory banks. Cost considerations are weighted by the comparison of in-house magnetics versus purchased semiconductors. Major computer manufacturers such as Univac, Honeywell, Control Data, Burroughs, and National Cash Register all have magnetic memory manufacturing capability, but none for semiconductors. Only IBM among major U.S. computer manufacturers has in-house semiconductor memory manufacturing capability.

The reliability of present commercial semiconductor memories in dual inline packages probably is unacceptable for 10^7 bit memories. Error correction is highly effective if the memory system is properly partitioned. Automated manufacturing techniques and feedback of failure analysis results can be helpful; the reliability of semiconductor memory components is likely to improve dramatically as experience accumulates. Although the reliability of semiconductor memory may be questionable, maintainability of these memories in large systems should be easy. Therefore, overall system availability (determined by mean time between failures—MTBF, and mean time to repair—MTTR) may be superior for semiconductor memories due to their simple modular structure. Improved system availability appears to be a major motivation for IBM in its swing to semiconductor memories.

A sudden major shift to semiconductor memories would challenge total U.S. integrated circuit manufacturing capacity. Such a shift is unlikely, however, because the system design cycle for medium and large computers is 3–5 years. The above considerations all suggest that semiconductor main memories will be adopted only slowly into original equipment designs of medium and large computers, excepting IBM computers.

Important potential markets for semiconductor memories are not considered in the above discussion. One of these is the market for add-on memory units, directly compatible with existing computer systems, that are sold or leased directly to the computer system user. Semiconductor memories are invading this market by offering lower cost and better performance than magnetic memories. The potential market is very large.

A second major potential market for semiconductor memories is in computer terminals, peripherals, and small stand-alone digital systems. Past systems in these categories seldom used much memory because the cost per bit of small digital memories (e.g., 10^2–10^5 bits) was prohibitive. With the advent of low-cost semiconductor memories and other digital components, new types of systems not economically feasible previously are being developed and marketed. Examples may be found in areas such as electronic calculators, equipment controllers, communication systems, and prosthetic devices. The potential market here depends upon innovation by subsystem and system designers; the impact of the new technology in this area may be far greater than it is for computer systems.

Reference

[1] L. F. Slattery, "Magnetics vs. semiconductors as the memory elements of the future," in *Proc. IEEE Comput. Group Conf.*, 1970, pp. 94–95.

Author Index

Subject Index

EDITOR'S BIOGRAPHY

David A. Hodges (S'59–M'65) was born in New Jersey in 1937. He received the B.E.E. degree from Cornell University, Ithaca, N.Y., in 1960, and the M.S. and Ph.D. degrees in electrical engineering from the University of California, Berkeley, in 1961 and 1966, respectively.

From 1965 to 1966 he was a Teaching and Research Associate at the Technical University of Denmark. From 1966 to 1970 he was at Bell Telephone Laboratories, first in the components area at Murray Hill, N.J., then as Head of the System Elements Research Department at Holmdel, N.J. Now he is Associate Professor of Electrical Engineering and Computer Sciences at the University of California, Berkeley. He also currently is Editor of the IEEE JOURNAL OF SOLID-STATE CIRCUITS and is a consultant to Signetics Corporation and Vidar Corporation. His research interests are in the area of digital components and systems.

Dr. Hodges was the recipient of Outstanding Paper Awards from the IEEE International Solid-State Circuits Conference in 1963 and 1969.